The Companion to Development Studies

T0138840

The Companion to Development Studies is essential reading in the field of development studies. This indispensable resource offers succinct, up-to-date, and insightful chapters that reflect the diverse voices and perspectives informing the field and the dynamic interplay of theory, policy, and practice that characterises it.

This fourth edition brings together contributions from an impressive range of renowned international experts and emerging voices at the forefront of development studies to deliver engaging, interdisciplinary, and provocative insights into this challenging field. The 98 chapters spanning both theory and practice offer readers accessible discussions of the core issues, emerging trends, and key debates of the discipline. Divided into nine sections of: theories and their contentions; histories and discourses of development; actors and institutions; identities and practices; people and the planet; the economics of development; conflict, violence, and peace; the changing landscape of development; and approaches to policy and practice; this timely new text provides easy to use summaries of some of the major issues encountered in this rapidly growing and changing field.

The *Companion* serves students and scholars across various disciplines, including development studies, geography, politics, international relations, sociology, anthropology, and economics. It offers incisive analysis and critical insights, equipping those working in development policy and practice with the knowledge and understanding they need to navigate and address contemporary global challenges.

This textbook is supported by flexible, online resources for teaching and learning such as tutorial guides, key concept videos, and a filmography.

Emil Dauncey is Lecturer in Geography and Environmental Studies at The Open University, UK.

Vandana Desai is Professor of Human Geography at Royal Holloway, University of London, UK.

Robert B. Potter (1950–2014) was a leading researcher in urban and development geographies – with particular reference to Caribbean development studies – and an academician of the Academy of Social Sciences. He was Professor Emeritus at Reading University, Head of School (2008–2012), and the recipient of the higher doctorate degree of DSc from the University of Reading for his outstanding contributions to the field.

The Companion to Development Studies

Fourth edition

Edited by Emil Dauncey,
Vandana Desai and
Robert B. Potter

Routledge
Taylor & Francis Group

LONDON AND NEW YORK

Designed cover image: gettyimages.ca/Holly Wilmeth

Fourth edition published 2024
by Routledge
4 Park Square, Milton Park, Abingdon, Oxon, OX14 4RN

and by Routledge
605 Third Avenue, New York, NY 10158

Routledge is an imprint of the Taylor & Francis Group, an informa business

First edition published by Hodder Arnold 2002

Second edition published by Hodder Education 2008

Third edition published by Routledge 2014

British Library Cataloguing-in-Publication Data
A catalogue record for this book is available from the British Library

Library of Congress Cataloging-in-Publication Data
Names: Dauncey, Emil, editor. | Desai, Vandana, 1965– editor. |
    Potter, Robert B., editor.
Title: The companion to development studies/edited by Emil Dauncey,
    Vandana Desai, and Robert B. Potter.
Description: Abingdon, Oxon ; New York, NY : Routledge, 2024. | Earlier edition
    published: 2014. | Includes bibliographical references and index.
Identifiers: LCCN 2023047546 (print) | LCCN 2023047547 (ebook) |
    ISBN 9780367244231 (hardback) | ISBN 9780367244248 (paperback) |
    ISBN 9780429282348 (ebook)
Subjects: LCSH: Economic development. | Development economics. |
    Sustainable development. | Political development. | Globalization.
Classification: LCC HD75. C655 2024 (print) | LCC HD75 (ebook) |
    DDC 338.9—dc23/eng/20240105
LC record available at https://lccn.loc.gov/2023047546
LC ebook record available at https://lccn.loc.gov/2023047547

ISBN: 978-0-367-24423-1 (hbk)
ISBN: 978-0-367-24424-8 (pbk)
ISBN: 978-0-429-28234-8 (ebk)

DOI: 10.4324/9780429282348

Typeset in Sabon
by Apex CoVantage, LLC

Access the Support Material at: www.routledge.com/9780367244248

Rob worked on the third edition (2014) of the Companion right
up to the last few weeks of his life. He played a fundamental
role in its growth and success right from the inception of the
Companion on a back of an envelope in 2000. He brought his
extensive experience of publishing more than 30 books and
monographs and over 250 journal articles and book chapters
to the Companion. His single-mindedness, enthusiasm passion
for geographical perspective in understanding development,
and ethos of service to the needs of the wider development
community were underlined with his superb organisation skills,
professionalism, and meticulous attention to detail. He was an
excellent collaborator, who supported the work of younger
scholars and was an excellent mentor to early career academic
colleagues. He was strongly committed to the production of
books that would make a difference and be read widely. His own
vision of the discipline was one that emphasised integration and
inclusivity, stressing the interdependencies between people and
their environments. Rob argued that academics from the Global
North need to engage more with colleagues from universities
in the Global South, to treat their institutions and publications
with due respect and seriousness, and to publish enthusiastically
in their journals. He tried to encourage this diversity and
inclusivity in the Companion too. He is greatly missed by family,
friends and colleagues and is a sad loss to the wider discipline.

Contents

Preface

It is with great pleasure that we present the revised and updated fourth edition of *The Companion to Development Studies*. The first edition of the Companion was published in 2002, followed by subsequent editions in 2008 and 2014. Each edition has been well received by students, academics, researchers, policy makers, and practitioners. We hope this edition continues this legacy but also builds on it, to reach yet more diverse audiences and to reflect more diverse experiences and perspectives.

Almost ten years have passed between this edition and its predecessor. This is in part due to the natural spacing of editions but also due to our commitment to ensuring a thorough and comprehensive update in a context of significant, sometimes dramatic change. Approximately two-thirds of the chapters in this edition are entirely new, specially commissioned for this edition. Many of the other chapters have been revised for this edition, and there are a few where the original text has been preserved because they remain of contemporary salience. This balance reflects the nature of the field as one of both continuity and change. As is reflected in this new edition, while some ideas endure and find resonance with new audiences, others emerge over time, challenging or reinterpreting past understandings. However, the changes in this edition are not just a result of time but also the significant global events that have transpired since its inception.

In the years since the third edition, the world has witnessed unprecedented events and undergone some profound social transformations. The COVID-19 pandemic touched us all. It both exacerbated and drew attention to enduring inequalities. These disparities not only underscored the differential impacts of the pandemic across different sections of society but also differences in the capacity and willingness to respond. This highlighted the critical need for global cooperation, equitable resource distribution, social justice, and shared responsibility. The effects of the pandemic were further complicated by the pre-existing economic context.

The pandemic did not arrive in a vacuum but on the back of the 2008 financial crisis, an era of austerity, to be followed by a cost-of-living crisis gripping the world at the time of writing. These crises laid bare the fragility of our global economic systems, the privilege of some and disadvantage of many. They reinvigorated questions about economic inequalities, the resilience of global communities, and our ability and willingness to protect and support the most vulnerable. Beyond these broad global challenges, specific movements and environmental concerns have also risen to greater prominence.

Black Lives Matter, spurred by events such as the tragic murder of George Floyd, have brought systemic racism to the forefront. In development studies, this has contributed to an intensified focus on decolonisation, reinvigorating discussions around development's deep entanglements with colonialism and their present-day implications. Similarly, the #MeToo movement has amplified conversations around gender-based violence and inequality, helping to shift gender from peripheral interest to mainstream concern in development studies. Extreme weather events around the world have given a new urgency to the climate crisis, in public discourse at least. There is an emerging awareness that although the effects of climate change are felt by some more than others, everybody is affected, and everybody must take responsibility for action if the worst consequences are to be avoided.

All this takes place within the context of what feels like a new world order. The rise of populism has underlined social divides and sparked debates about the costs and benefits of competing forms of nationalism and globalism. Wars in Syria, Yemen, Ukraine, and elsewhere have highlighted shifting global power dynamics and the humanitarian challenges arising from such conflicts, including displacement and poverty.

These global shifts have not only changed political landscapes but have left an indelible mark on the field of development studies They have introduced new challenges, whilst reinvigorating longstanding concerns. This context perhaps more than ever underscores the relevance of the theories, debates, and practices that constitute the field of development studies. Consequently, this edition of the Companion is both reflective and forward-looking.

This edition of the Companion engages with the drive towards decolonisation by better reflecting the diverse voices, perspectives, and knowledges that shape the discipline and those whose lives it touches. And development's complex and contentious relationship with colonialism is a recurrent theme across all sections of the book. Like its predecessors, this edition has a section dedicated to exploring relationships between development and the environment. Environmental concerns, once peripheral in development studies, are now firmly centre stage such that the climate crisis and environmental degradation remain recurrent themes throughout the book. Moreover, this edition addresses the context of multipolarity and fragmentation, by re-examining the role of the state, recognising the influence of a wider range of actors and institutions, and challenging assumptions about where power may lie. This exploration reveals both concerns but also reasons for optimism. Our hope is that this edition of the Companion can play a small part in equipping students of development with the knowledge and skills they need to bring about meaningful change.

With this goal in mind, this edition of the Companion also responds to the changing context of learning and teaching, particularly in higher education. This edition incorporates new resources aimed at supporting and enriching the learning experience. These include key concept videos, a filmography, and tutorial guides to ensure that the content remains accessible, engaging, and relevant in the digital age.

While acknowledging the changing contexts and events that shape the contemporary concerns of development studies, this edition of the Companion also reflects continuities, in engaging with enduring challenges, questions, and debates. Furthermore, the overarching objective of the Companion has remained the same: to produce a book that truly serves as a companion. It helps readers navigate the field by offering a broad yet critical overview of key debates and themes in the field. Yet we would not want to claim that this edition is exhaustive or entirely objective; the contributions contained herein may at times be partial or contested, but this is by design. Above all, the Companion aims to foster critical thinking and debate, to challenge, provoke, and inspire. With this objective in mind, this edition should not be treated as a standalone reference collection but as a Companion to accompany the reader through a wider literature. Here it is important to recognise the many other good books that enrich this vast and ever-growing field, books that have not only inspired us but have also invigorated the entire discipline.

We began this preface by discussing how the context has shaped the content of this edition. But the context has also touched the lives of those who have contributed to it. At this point, we would like to thank our contributors, many of whom managed to find the time, energy, and will to contribute in contexts of personal struggle and change. We must also extend thanks to Claire Maloney, Prachi Priyanka, and Andrew Mould at Routledge and Dominic Dauncey for his administrative support. We are grateful for their enduring patience and support in bringing this book into being. We are also grateful to our dear colleague Alasdair Pinkerton (Royal Holloway), for his contributions in the early stages of this book's development. Last and by no means least, we extend thanks to Professor Rob Potter. As academics, we are all standing on the shoulders of giants. Sometimes we are lucky enough to get to work with them. This fourth edition is dedicated to Rob Potter, a true giant in the field of development studies and beyond.

Emil Dauncey and Vandana Desai 2023

Figures

Tables

Contributors

Yaw Adjei-Amoako, Department of Geography, University of Reading, UK.

Azadeh Akbari, Department of Public administration, University of Twente, the Netherlands.

Magali N. Alloatti, Observatory for Migration, State University of Santa Catarina, Brazil.

Maha Rafi Atal, Department of Management, Society and Communication, The Copenhagen Business School, Denmark.

Celia Bartlett, Development Policy and Practice, School of Social Sciences and Global studies, The Open University, Milton Keynes, UK.

Anthony Bebbington, The Graduate School of Geography, Clark University, Massachusetts, USA.

Stephen Brown, Faculty of Social Sciences Political Studies, University of Ottawa, Canada.

Francesco Burchi, German Institute of Development and Sustainability, Bonn, Germany.

John Cameron, Department of International Development studies, Faculty of Arts and Social Sciences, Dalhousie University, Halifax, Canada.

Pádraig Carmody, Department of Geography, Trinity College Dublin, Ireland/School of Tourism and Hospitality, The University of Johannesburg, South Africa.

Lakshmi Charli-Joseph, Laboratorio Nacional de Ciencias de la Sostenibilidad, Instituto de Ecología, Universidad Nacional Autónoma de México, Ciudad de Mexico, Mexico.

Kristen E. Cheney, International Institute of Social Studies, Erasmus University, Rotterdam, The Hague, The Netherlands.

Frangton Chiyemura, Development Policy and Practice, School of Social Sciences and Global Studies, the Open University, Milton Keynes, UK.

Julia Ngozi Chukwuma, Economics, School of Social Sciences and Global Studies, the Open University, Milton Keynes, UK.

Jennifer Cole, Department of Health Studies, Royal Holloway, University of London.

Dennis Conway, Department of Geography, Indiana University, Bloomington, USA.

John Cook, Climate Change Communication Research Hub, Monash University, Australia.

Sarah Cook, Institute for Global Development, University of New South Wales, Sydney.

Stuart Corbridge, Vice-Chancellor and Warden, Durham University, Palatine Centre, Durham, United Kingdom.

Andrea Cornwall, School of Global Affairs, King's College London, UK

Ruth Craggs, Department of Geography, Kings College, London, UK.

Charlotte Cross, Development Policy and Practice, School of Social Sciences and Global Studies, the Open University, Milton Keynes, UK.

Emil Dauncey, Geography and Environmental Studies, School of Social Sciences and Global Studies, the Open University, Milton Keynes, UK.

Kevin Deane, Economics. School of Social Sciences and Global Studies, the Open University, Milton Keynes, UK.

Vandana Desai, The Geography Department, Royal Holloway, University of London, UK.

Stephen Devereux, Centre for Social Protection, Institute of Development Studies, University of Sussex, UK.

Jakob Dirksen, Oxford Poverty and Human Development Initiative, University of Oxford, UK.

Martin Dittus, Oxford Internet Institute, University of Oxford, UK.

Radha D'Souza, Westminster Law School, University of Westminster UK.

Ruth Evans, Department of Geography, University of Reading, UK.

Adam Fejerskov, Danish Institute for International Studies, Copenhagen, Denmark.

Ines A. Ferreira, Department of Economics, University of Copenhagen, Denmark

Ben Fine, Department of Economics, School of Oriental and African Studies, UK/Wits School of Governance, University of Witwatersrand, South Africa.

Katherine Foo, Social Science and Policy Studies, Worcester Polytechnic Institute, Massachusetts, USA.

Erika Forsberg, Department of Peace and Conflict Research, Uppsala University, Sweden.

Augustin Kwasi Fosu, ISSER, University of Ghana, Accra, Ghana.

Elizaveta Fouksman, Centre for Public Policy Research, School of Education, Communication and Society, King's College London.

Levi Gahman, Department of Geography and Planning, University of Liverpool, United Kingdom/Institute for Gender and Development Studies, University of West Indies, Trinidad and Tobago.

Des Gasper; International Institute of Social Studies, Erasmus University Rotterdam, Netherlands.

John D. Giblin, Department of Global Arts, Cultures and Design, National Museums Scotland, UK.

Emma Gilberthorpe, School of Global Development, University of East Anglia, Norwich, UK.

Tom Goodfellow, Department of Urban Studies and Planning, University of Sheffield, UK.

Mark Graham, Oxford Internet Institute, University of Oxford, UK.

Catherine Fallon Grasham, School of Geography and the Environment, University of Oxford, UK.

Ali Burak Güven, Department of Politics, Birkbeck, University of London, UK.

Syed Haider, School of Art, Media and American Studies, University of East Anglia, UK.

Daniel Hammett, Department of Geography, University of Sheffield, UK.

Vanja Hamzić, School of Law, Gender and Media, SOAS, University of London, UK.

Nik Heynen, Department of Geography, University of Georgia, USA.

Stephanie Ketterer Hobbis, Sociology of Development and Change, Wageningen University

Lummina Horlings, Faculty of spatial sciences, University of Groningen, The Netherlands.

Rory Horner, Global Development Institute, University of Manchester, UK.

Katja Hujo, United Nations Research Institute for Social Development (UNRISD), Geneva, Switzerland.

Elizabeth Humphrys, Social and Political Sciences, School of Communication, University of Technology Sydney.

Cecile Jackson, School of Global Development, University of East Anglia, Norwich, UK.

Ben Jones, School of Global Development, University of East Anglia, Norwich, UK.

Gareth A. Jones, Department of Geography and Environment, London School of Economics and Political Science, UK.

Sara de Jong, Department of Politics and International Relations, University of York, UK.

Pallavi Joshi, Development Policy and Practice, School of Social Sciences and Global Studies, the Open University, Milton Keynes, UK.

Abigail Kabandula, Josef Korbel School of International Studies, University of Denver, Colorado, USA.

Dinar Kale, Development Policy and Practice, School of Social Sciences and Global Studies, the Open University, Milton Keynes, UK.

Bereket Kebede, School of International Development, University of East Anglia, Norwich, UK.

Madhu Purnima Kishwar, Manushi/Indian Council of Social Science Research, New Delhi, India.

Thomas Klak, Department of Environmental Studies, University of New England, Maine, USA.

Andrew Lawrence, University of Witwatersrand School of Governance, Johannesburg, South Africa.

Philipp Lepenies, Otto Suhr Institute of Political Science, Department of Political and Social Sciences, Freie Universität, Berlin, Germany.

Susan Levy, School of Humanities, Social Sciences and Law, University of Dundee, Scotland, UK.

David Lewis, Department of International Development, London School of Economics and Political Science, UK.

Ernesto López-Morales, Department of Urban Planning, University of Chile, Santiago de Chile, Chile.

Don D Marshall, Sir Arthur Lewis Institute of Social and Economic Studies, the University of the West Indies, Kingston, Jamaica.

Adrian Martin, School of Global Development, University of East Anglia, Norwich, UK.

Ana Luíza Matos de Oliveira, Faculdade. Latino-Americana de. Ciências Sociais. Brazil.

Emma Mawdsley, Department of Geography, University of Cambridge, UK.

Rónán McDermott, Department of Global and Local Governance, University of Groningen, The Netherlands.

Cheryl McEwan, Department of Geography, Durham University, UK.

Colin McFarlane, Department of Geography, Durham University, UK.

Simon McGrath, School of Education, University of Nottingham, UK.

Cathy McIlwaine, Department of Geography, Kings College London. UK.

Paula Meth, Department of Urban Studies and Planning, University of Sheffield, UK.

Giles Mohan, Development Policy and Practice, School of Social Sciences and Global Studies, the Open University, Milton Kenyes, UK.

William Monteith, School of Geography, Queen Mary University of London, UK.

David Neilson, School of Social Sciences, University of Waikato, New Zealand.

Patricia Northover, Sir Arthur Lewis Institute of Social and Economic Studies The University of the West Indies, Kingston, Jamaica.

Adwoa Ofori, Discipline of geography, Trinity College Dublin, Ireland.

Sanna Ojanperä, Oxford Internet Institute, University of Oxford, UK.

Uzoma Okoye, Department of Social Work, University of Nigeria Nsukka, Nigeria.

Kei Otsuki, Department of Human Geography and Planning, Faculty of Geosciences, Utrecht University, The Netherlands.

Theo Papaioannou, Development Policy and Practice, School of Social Sciences and Global Studies, the Open University, Milton Keynes, UK.

Elissaios Papyrakis, International Institute of Social Studies (ISS), Erasmus University Rotterdam, The Hague, The Netherlands.

Jane Parpart, Department of Conflict Resolution, Human Security and Global Governance, McCormack Graduate School, University of Massachusetts, Boston, USA.

Kamna Patel, Development Planning Unit, Faculty of the Built Environment, University College London, UK.

Sarah Peck, Geography and Environmental Sciences, Northumbria University, Newcastle, UK.

Filiberto Penados, Development Studies Programme, Galen University, Belize.

Robert B. Potter, Department of Geography and Environmental Science, University of Reading, UK.

Marcus Power, Department of Geography, Durham University, UK.

Raktim Ray, The Bartlett Development Planning Unit, University College of London, UK.

Michael Redclift, Department of Geography, King's College London, UK.

Jonathan Rigg, School of Geographical Sciences, University of Bristol, UK.

Susanne Schech, College of Humanities, Arts and Social Sciences, Flinders University, Australia.

Karsten Schulz, Governance and Innovation, University of Groningen, The Netherlands.

Frans J. Schuurman, Cultural Anthropology and Development Studies Department, Radboud University, Nijmegen, Netherlands.

Sarah E. Sharma, Department of Political Studies and Department of Global Development Studies, Queen's University, Kingston, Ontario, Canada.

Timothy M. Shaw, Institute on Comparative Regional Integration Studies, United Nations University, Belgium.

Alan Shipman, Economics, School of Social Sciences and Global Studies, the Open University, Milton Keynes, UK.

James D. Sidaway, Department of Geography, National University of Singapore.

Kalim Siddiqui, Department of Accounting, Finance, Logistics and Economics, School of Business, Education and Law, University of Huddersfield, UK.

David Simon, Geography Department, Royal Holloway, University of London, UK.

Jesús Mario Siqueiros-García, Instituto de Investigaciones en Matemáticas Aplicadas y en Sistemas, Universidad Nacional Autónoma de México, Mexico.

Shelda-Jane Smith, University of Liverpool, Power, Space, and Cultural Change Unit, UK.

Thomas Aneurin Smith, School of Geography and Planning, Cardiff University, Wales, UK.

Susanne Soederberg, Department of Political Studies and Department of Global Development Studies, Queen's University, Kingston, Ontario, Canada.

Lorenzo Squintani, Wubbo Ockels School for Energy and Climate, University of Groningen, The Netherlands.

Kate Symons, Edinburgh Climate Change Institute, University of Edinburgh, Scotland UK.

Jonathan R. W. Temple, is a freelance economist based in Bristol, UK.

Louise Waite, School of Geography, University of Leeds, UK.

Carole Sandrine White, Marine Management Organisation, London/School of Global Development, University of East Anglia, Norwich, UK.

Howard White, The Global Development Network, New Delhi, India.

Brendan S. Whitty, University of St Andrews School of Management, University of St Andrews, Scotland, UK.

Katie Willis, Geography Department, Royal Holloway, University of London, UK.

Annelies Zoomers, Department of Human Geography and Planning, Faculty of Geosciences, Utrecht University, The Netherlands.

Part 1

Theories and their contentions. Editorial introduction

Development studies is a large and expanding field, drawing on ideas, concepts, and debates from across various philosophical and disciplinary traditions. At the heart of these discussions lie foundational questions such as: what is development? What constitutes 'good change', and how can societies progress towards it? This first section of the Companion, dedicated to theories of development, examines these questions. It brings together a collection of chapters that describe, explain, critically interrogate, and evaluate some of the many theoretical frameworks that shape our understanding of development.

At its core, a theory is a systematic framework for understanding and interpreting phenomena. It offers explanations for observed phenomena, predicts future events, and provides a foundation upon which further knowledge can be built. In the context of development studies, theories offer lenses through which the complexities of development processes can be defined, analysed, interpreted, and understood. Development studies, by its very nature, involves an encounter with diverse disciplines – from economics and sociology to anthropology, geography, and political science. This interdisciplinary nature, while enriching the field with diverse perspectives, also brings the challenge of reconciling sometimes conflicting viewpoints.

Theories are not mere academic exercises; they often serve as the bedrock upon which practical strategies, policies, and interventions are formulated. They suggest objectives and pathways for development, guiding policymakers in crafting interventions that resonate with socio-cultural, political, and economic realities. Without theoretical insights, development initiatives risk being myopic and misdirected.

However, theories may also come to influence realities, by defining, naming, and shaping the phenomenon they seek to explain. The effects of this may not always be benign. Theories can foster understanding and guide positive change, but they can also perpetuate biases, reinforce stereotypes, or narrow our perspectives. Thus, while theories provide frameworks for understanding, it is important to recognise their potential to both illuminate *and* obscure.

The relationship between theory and development studies has seldom remained static, reflecting both emerging insights and advancements in the field but also broader political shifts that frame the relationships foregrounding knowledge production. From the early focus on economic growth as the primary goal and indicator of development to more recent approaches that consider social, cultural, and environmental factors, the field has continuously adapted, refined, and expanded its theoretical foundations.

DOI: 10.4324/9780429282348-1

Power dynamics, global interdependencies, institutional roles, and historical legacies emerge as recurring themes across the chapters contained in this section of the Companion. These shared concerns underscore the complexities and contestations inherent in development studies but also highlight the interconnectedness of global systems, processes, and relationships that give rise to and shape theories. This is perhaps most evident in the increasing prominence of the decolonisation agenda. Emerging in response to the historical legacies of colonialism, this agenda challenges the traditional Eurocentric biases embedded in many development theories. It advocates for a revaluation from diverse cultural and regional standpoints, emphasising indigenous and local knowledges, plurality of perspectives, and experiences. This shift underscores the importance of inclusivity and diversity in theoretical explorations but also draws attention to the legacies of colonialism that shape human and environmental conditions in the present.

It might be argued that development studies is unique as an evolving discipline that draws on multidisciplinary perspectives, its normative orientations, and its particular tethering to a world or policy and practice. As the world continues to change, so too will the theories that seek to explain it. It might be hoped that future theories in development studies will be characterised by an increased inclusivity, interdisciplinarity, and a greater focus on equity and justice in knowledge production. Embracing diverse voices and perspectives will be crucial in ensuring that development theories remain relevant, as well as effective in addressing the challenges of the future. Whilst there is growing attention to these important concerns, the extent to which they will become central to theories – and potentially transcend them to influence policy and practice – remains less certain.

While the chapters in this section stand alone in their exploration of specific theories, together they paint a broad picture of the field of development studies, inviting readers to engage critically with the ideas presented over time and to contribute to the ongoing debates that shape our understanding of development in the past, present, and future.

Chapter 1

Theories, strategies, and ideologies of development

An overview

Robert B. Potter

Introduction

A major characteristic of the multi-, inter-, and cross-disciplinary field of development studies since its establishment in the 1940s has been a series of sea-level changes in thinking about the process of development itself. This search for new theoretical conceptualizations of development has been mirrored by changes in the practice of development in the field. Thus, there has been much debate and controversy about development, with many changing views as to its definition and the strategies by means of which, however development is defined, it may be pursued. In short, the period since the 1950s has seen the promotion and application of many varied views of development. And the literature on development theory and practice has burgeoned (see, for instance, Hettne, 1995; Preston, 1996; Cowen and Shenton, 1996; Potter et al., 2008; Peet and Hartwick, 2009; Chant and McIlwaine, 2009; Nederveen Pieterse, 2010; Thirlwall, 2011; Potter et al., 2012). A major theme is that ideas about development have long been controversial and highly contested.

It is also necessary to stress that development covers both theory and practice, that is both ideas about how development should or might occur and real world efforts to put various aspects of development into practice. This is conveniently mapped into the nomenclature suggested by Hettne in his overview of *Development Theory and the Three Worlds* (1995). In reviewing the history of development thinking, he suggested that 'development' involves three things: *development theories*, *development strategies*, and *development ideologies*.

Development theories

If a theory is defined as a set of logical propositions about how some aspect of the real world is structured or the way in which it operates, *development theories* may be regarded as sets of ostensibly logical propositions, which aim to explain how development has occurred in the past and/or how it should occur in the future (Potter et al., 2008). Development theories can either be *normative* – that is they can generalize about what should happen or what should be the case in an ideal world – or *positive,* in the sense of dealing with what has generally been the case in the past. This important distinction is broadly exemplified in the figure that accompanies this account (see figure). Hettne (1995) remarks that 'development studies is explicitly normative' and that teachers, researchers and practitioners in the field 'want to change the world, not only analyse it'

DOI: 10.4324/9780429282348-2

(Hettne, 1995: 12). The arena of development theory is primarily, although by no means exclusively, to be encountered in the academic literature, that is in writing about development. It is, therefore, inherently controversial and contested.

Development strategies

On the other hand, *development strategies* can be defined as the practical paths to development which may be pursued by international agencies, states, non-government organisations and community-based organisations, or indeed individuals, in an effort to stimulate change within particular areas, regions, nations, and continents. Thus, Hettne (1995) provides a definition of development strategies as efforts to change existing economic and social structures and institutions in order to find enduring solutions to the problems facing decision-makers. As such, Hettne argues that the term 'development strategy' implies an actor, normally the state. In order to sound less top-down, it is necessary to think in terms of a wider set of development-oriented actors, including all those listed earlier.

Development ideologies

Different development agendas will reflect different goals and objectives. These goals will reflect social, economic, political, cultural, ethical, moral, and even religious influences. Thus, what may be referred to as different *development ideologies* may be recognised. For example, both in theory and in practice, early perspectives on development were almost exclusively concerned with promoting economic growth. Subsequently, however, the predominant ideology within the academic literature changed to emphasise wider sets of political, social, ethnic, cultural, ecological, and other dimensions of the wider human processes of development and change. Theories in development are distinctive by virtue of the fact that they involve the intention to change society in some defined manner. One of the classic examples is the age-old battle between economic policies that increase growth but widen income disparities, and those wider policy imperatives that seek primarily to reduce inequalities within society. All such efforts to effect change reflect some form of ideological base.

Development thinking

Perhaps the sensible approach is to follow Hettne (1995) and to employ the overarching concept of *development thinking* in our general deliberations. The expression 'development thinking' may be used as a catch-all phrase indicating the sum total of ideas about development; that is, including pertinent aspects of development theory, strategy, and ideology. Such an all-encompassing definition is necessary due to the nature of thinking about development itself. As noted at the outset, development thinking has shown many sharp twists and turns during the 20th century. The various theories that have been produced have not commanded attention in a strictly sequential-temporal manner. In other words, as a new set of ideas about development has come into favor, earlier theories and strategies have not been totally abandoned and replaced. Rather, theories and strategies have tended to stack up, one upon another, co-existing, sometimes in what can only be described as very convoluted and contradictory manners. Thus, in discussing development theory, Hettne (1995: 64) has drawn attention to the 'tendency of social science paradigms to accumulate rather than fade away'.

Development studies and disciplinary revolution/evolution

The characteristics of development studies as a distinct field of enquiry can be considered in a somewhat more sophisticated manner by referring to Thomas Kuhn's ideas on the *structure of scientific revolutions*. Kuhn (1962) argued that academic disciplines are dominated at particular points in time by communities of researchers and their associated methods, and they thereby define the subjects and the issues deemed to be of importance within them. He referred to these as 'invisible colleges', and he noted that these serve to define and perpetuate research which confirms the validity of the existing paradigm or 'supra-model', as he referred to it. Kuhn called this 'normal science'. Kuhn noted that only when the number of observations and questions confronting the status quo of normal science becomes too large to be dealt with by means of small changes to it will there be a fundamental shift. However, if the proposed changes are major and a new paradigm is adopted, a scientific revolution can be said to have occurred, linked to a period of what Kuhn referred to as 'extraordinary research'.

In this model, therefore, scientific disciplines basically advance by means of revolutions in which the prevailing normal science is replaced by extraordinary science and, ultimately, a new form of normal science develops. In dealing with social scientific discourses, it is inevitable that the field of development theory is characterised by evolutionary, rather than revolutionary, change. Evidence of the persistence of ideas in some quarters, years after they have been discarded elsewhere, will be encountered throughout the development literature. Given that development thinking is not just about the theoretical interpretation of facts but rather about values, aspirations, social goals, and ultimately that which is moral, ethical, and just, it is understandable that change in development studies leads to the parallel *evolution* of ideas, rather than *revolution*. Hence, conflict, debate, contention, positionality, and even moral outrage are all inherent in the discussion of development strategies and associated plural and diverse theories of development.

Approaches to development thinking

There are many ways to categorise development thinking through time. Broadly speaking, it is suggested here that four major approaches to the examination of development thinking can be recognised, and these are shown in the accompanying figure (see Potter et al., 2008). The framework first maps in the distinction previously made between *normative development theories* (those focusing on what *should be* the case) and *positive theories* (which ponder what has *actually been* so). Another axis of difference between theories is seen as relating to whether they are *holistic* or *partial*, and most partial theories emphasise the economic dimension. This is also intimated in the figure.

These two axes can be superimposed on one another to yield a simple matrix or framework for the consideration of development thinking/strategies/theories, as shown. Following Potter et al. (2008: Chapter 3), as noted, four distinct groupings of development theory can be recognised by virtue of their characteristics with regard to the dimensions of holistic–economic and normative–positive. The approaches are referred to here as: i) the classical–traditional approach; ii) the historical–empirical approach; iii) the radical political economy–dependency approach; and, finally, iv) bottom-up and alternative approaches. Following the argument presented in the last section, each of these approaches may be regarded as expressing a particular ideological standpoint, and can also be identified by virtue of having occupied the center stage of the development debate at particular points in time. Classical–traditional theory, embracing dualism, modernization theory, top-down conceptualizations, the new right and neo-liberal imperatives, is seen as stressing the economic and, collectively, existing mid-way between the normative and positive poles. In direct contrast, according to this framework, radical–dependency approaches – embracing neo-Marxism – and the articulation of the modes of production are seen as being more holistic. At the positive end of the spectrum exist those theories that are basically historical in their formulation and that purport to build upon what has happened in the past. These include core–periphery frameworks, cumulative causation and models of transport evolution, especially the mercantile model. In contrast, once again, are theories that stress the ideal or what should be the case. These are referred to as 'alternative approaches', and basic needs, neo-populism, 'another development', ecodevelopment, and sustainable development may be included in this category.

Conclusion

Many diverse and varied approaches to development remain in currency today and in many different quarters. Hence, in development theory and academic writing, left-of-center socialist views may well be more popular than classical and neo-classical formulations, but in the area of practical development strategies and policies, the 1980s and 1990s saw the implementation of neo-liberal interpretations of classical theory, stressing the liberalisation of trade, along with public-sector cut-backs. Such plurality and contestation are an everyday part of the field of development studies. In the words of Hettne (1995: 15), 'theorizing about development is therefore a never-ending task'.

Guide to further reading

Hettne, B. (1995) *Development Theory and the Three Worlds: Towards an International Political Economy of Development*, Second Edition, Harlow: Longman. Briefly introduces the concepts of development theories, strategies and ideologies (pp. 15–16), before presenting an overview of Eurocentric development thinking, the voice of the 'other', globalization and development theory, and 'another development'.

Peet, R. and Hartwick, E. (2009) *Theories of Development: Contentions, Arguments, Alternatives*, Second Edition, New York and London: Guilford Press. The two substantive sections, making-up chapters 2–7, review what are referred to as Conventional (Keynesian economics to neo-liberalism, modernisation) and Nonconventional/Critical (Marxism, socialism, poststructural, postcolonialism, feminist) theories of development.

Potter, R.B., Binns, T., Elliott, J. and Smith, D. (2008) *Geographies of Development, Third Edition*, Harlow: Prentice Hall. Seeks to stress the plural and contested nature of development theory and practice. Chapter 3 overviews theories and strategies of development, stressing their diversity and value-laden character. The structure of the account is based on the figure employed in the present chapter.

Preston, P.W. (1996) *Development Theory: An Introduction*, Oxford: Blackwell. The first part of the book treats social theory in general terms. Thereafter, contemporary theories of development are summarized followed by what are referred to as new analyses of complex change.

References

Chant, S. and McIlwaine, C. (2009) *Development Geographies in the 21st Century*, London: Edward Elgar.

Cowen, M.P. and Shenton, R. (1996) *Doctrines of Development*, London: Routledge.

Hettne, B. (1995) *Development Theory and the Three Worlds: Towards an International Political Economy of Development*, Second Edition, Harlow: Longman.

Kuhn, T. (1962) *The Structure of Scientific Revolutions*, Chicago: University of Chicago Press.

Nederveen Pieterse, J. (2010) *Development Theory*, Second Edition, London: Sage.

Peet, R. and Hartwick, E. (2009) *Theories of Development: Contentions, Arguments, Alternatives*, Second Edition, New York and London: Guilford Press.

Potter, R.B., Binns, T., Elliott, J. and Smith, D. (2008) *Geographies of Development: An Introduction to Development Studies*, Third Edition, Harlow, London and New York: Pearson-Prentice Hall.

Potter, R.B., Conway, D., Evans, R. and Lloyd-Evans, S. (2012) *Key Concepts in Development Geography*, Los Angeles, London, New Delhi, Singapore and Washington DC: Sage.

Preston, P.W. (1996) *Development Theory: An Introduction*, Oxford: Blackwell.

Thirlwall, A.P. (2011) *Economics of Development: Theory and Evidence*, Ninth Edition, London: Palgrave Macmillan.

Chapter 2

The impasse in development studies

Frans J. Schuurman

Introduction

Development studies is a relatively new branch of the social sciences. Coming into being in the late 1960s and early 1970s, it inherited many features of post-Second World War developments within the social sciences. Modernization theory contributed to its developmental orientation and its comparative methodology. From dependency theory it inherited its normative and progressive political character and its interdisciplinary conceptual frameworks.

In the 1970s, with dependency theory denouncing modernization theory as crypto-imperialist and modernization theorists hitting back by accusing dependency authors of being populist pseudo-scientists, development studies found fertile ground and grew into an increasingly accepted new discipline of the social sciences. Universities – often under pressure from leftist professors and students – created Third World Centres. Debates about the nature and impact of development assistance became popular, and the existence of many dictatorial regimes in the South led to numerous solidarity committees in the North. In the 1980s, things started to change for development studies. A number of occurrences in that decade, which will be dealt with in the following paragraphs, led to an increasingly uneasy feeling within the discipline that old certainties were fading away. It was felt that development theories in the sense of a related set of propositions of the 'if . . . then' kind, could ever less adequately explain experiences of development and under-development. Whether it concerned modernization theories or neo-Marxist dependency theories, both sets of development theories were losing out in terms of their explanatory power. From the mid-1980s onwards, the so-called 'impasse in development studies' was talked about. The contours of this impasse were sketched for the first time in a seminal article by David Booth in 1985. In the years that followed other authors continued the discussion, which took on new dimensions with the end of the Cold War and the debate on globalization.

Reasons for the impasse

Three reasons can be held responsible for having changed the panorama for development studies to such an extent that it created this theoretical impasse. Chronologically they were: (i) the failure of development in the South and the growing diversity of (under) development experiences; (ii) the postmodernist critique on the social sciences in general and on the normative characteristics of development studies in particular; and, finally,

DOI: 10.4324/9780429282348-3

(iii) the rise of globalization in its discursive as well as in its ontological appearance. Each of these issues is considered in the account that follows.

The failure of development in the South

Although until the 1980s developing countries realized average improvements in life expectancy, child mortality, and literacy rates, more recent statistics have shown, however, that these improvements were less valid for the poorest of these countries and, more specifically, for the lowest income groups. In fact, in the 1980s there was a reversal in some of the development indicators. It was realized that given the growth rates of that time, it would take another 150 years for Third World countries to achieve even half the per capita income of Western countries. Modernization theories failed to account for these figures and trends. Instead of a self-sustained growth (a much favored concept of modernization), many developing countries were up to their ears in debt, which served to paralyze development initiatives.

Problems such as unemployment, poor housing, human rights offences, poverty, and landlessness were increasing at alarming rates. UNICEF estimated a fall of 10–15 per cent in the income of the poor in the Third World between 1983 and 1987. In 1978, the Third World received 5.5 per cent of the world's income; in 1984 this had fallen to 4.5 per cent. The 'trickle-down' process (another favored concept of modernization) had failed miserably. In 1960, the income ratio between the world's rich and poor countries was 20:1, in 1980 it increased to 46:1 and in 1989, the ratio was as high as 60:1.

Although dependency theory could certainly not be accused of an over-optimistic view concerning the developmental potentials of developing countries, it could not really account for the growing difference between Third World countries, and neither were the developmental experiences of so-called socialist countries particularly enviable. In addition, Marxist and neo-Marxist development theories were dealt a heavy blow when the fall of the Berlin Wall meant the delegitimization of socialism as a political project of solving the problem of underdevelopment.

The postmodernist critique of the social sciences

The 1980s witnessed the advancement of postmodernism within the social sciences, bringing with it a tendency to undermine the 'great narratives' of capitalism, socialism, communism, and so forth. The basic argument was that there is no common reality outside the individual. As such, political alternatives, which always exist by the grace of a minimum of common perception, were maneuvered out of sight. Development theories based on meta-discourses or on the role of a collective emancipatory agency lacked, according to the postmodern logic, a sound basis. The Enlightenment ideal of the emancipation of humanity (shared by modernization and dependency theory alike) had not been achieved nor could it be achieved. In addition, in its quest for hidden metaphors, the postmodern method of deconstruction revealed that the notion of development contained a number of hidden and unwarranted evolutionist, universalist and reductionist dimensions that would definitely lead anyone working with this notion down the wrong path. As such, development studies became a direct target for a wide range of views furthering the notion of 'alternative development'. Under postmodernist or, perhaps better put, anti-modernist pressures, the central object of development studies – unequal access to power, to resources, to a humane existence – became increasingly substituted by some-thing like 'socioeconomic diversity'. Apparently, the notion of diversity was considered to avoid the hidden universalist (read: Western or imperialist) and reductionist dimen-sions that inequality brought with it. At the same time, others considered this switch to a voluntarist and pluralist approach to the development problem not only as anathema but also as inferior to a universalistic emancipation discourse.

Globalization

In the 1990s, the forces that had led to the impasse in development theories were joined by the discourse on globalization. Although the most recent factor, it probably represents the most important positive challenge to development studies. Whether globalization is a real phenomenon (cf. Hirst and Thompson, 1996) or nothing more than a discourse to legitimize neo-liberal market logic, it is undeniable that it has had a major influence on development studies in the 1990s. To understand why this is so, it is important to realize the significance of the (nation-)state for social science theories in general and for develop-ment studies in particular.

It is the declining – or at least changing – position and status of the (nation-)state that has been – and still is – at the core of the literature on globalization. As an interdisci-plinary branch of the social sciences, theories within development studies try to connect economic, political, and cultural aspects of inequality and development trajectories. The connection between these aspects is realized by using the (nation-)state as a linchpin. As such, theories of economic development became focused upon the workings of the national market and on economic relations between countries. In theories of political development the role of the state and the process of nation-building were central objects of study. In more culturalistic development theories, the notion of a national identity was crucial in understanding the differences between development trajectories. This impor-tance of the (nation-)state became visible in modernization theories, in both neo-Marxist and Marxist development theories alike. Globalization changed all that. Many authors writing about globalization agree on the decreasing – or at least the changing – economic, political, and cultural importance of (nation-)states. The central role of the state, it is said, is being hollowed out from above as well as from below. In a political sense there is the increasing importance of international political organisations that interfere politically and also militarily in particular states. In this way, they relegate to the past the Westphal-ian principles about the sovereignty of (nation-)states and their monopoly on the use of institutionalized violence within their borders. The national state is hollowed out from below by the growing phenomenon of decentralization and local government.

Economically, the state is seen as disappearing as an economic actor through priva-tisation supported by deregulation. Also, there is the growing importance of the global financial market where about $1,500 billion is shifted daily around the globe. Culturally, the idea of national identity as the central element in identity construction for individuals or groups is quickly eroding, in favor of cosmopolitanism on the one hand and/or the fortification of ethnic, regional and religious identities on the other.

It is not only that the globalization debate gives reason to suppose that the role of the (nation-)state has been – and still is – declining but also that, as a consequence, the former conjunctive dynamic (i.e., following the same spatial and time paths) of economy, polity, and culture – upon which the interdisciplinary character of many a development theory was based – has been replaced by a disjunctive dynamic (cf. Appadurai, 1990). Development studies has yet to redefine its object and its subject as, in fact, the other social sciences, vis-à-vis globalization, but this quest presents much more of a challenge than the former impasse ever did.

Conclusion

The impasse in development studies can in fact be traced back to a crisis of paradigms. The three reasons that were mentioned as being responsible for the impasse and its deepening – the lack of development and increasing diversity in the South, the postmodernist critique on 'grand narratives', and globalization – challenged, respectively, three post-Second World War developmental paradigms. These were:

1 The essentialization of the Third World and its inhabitants as homogeneous entities.
2 The unconditional belief in the enlightenment concepts of progress and the 'makeability' of society.
3 The importance of the (nation-)state as an analytical frame of reference and a political and scientific confidence in the state to realize progress.

Each of these paradigms came in for criticism, one after the other. Development theories related to these paradigms (such as modernization and dependency theories) became automatically tainted as well, initiating the so-called impasse in development studies.

However, in spite of this impasse, an important number of authors in the field of development studies have continued their work, some using more grounded theories, others trying to elaborate upon new concepts like civil society, global governance, and global social movements. Many feel that the growing inequality between, as well as within, North and South is enough of a reason to continue with development studies. To fit this effort in with the new reality shaped by globalization presents a new and exciting challenge, and one which relegates the impasse to a past period. Nevertheless, from well into the 21st century looking back upon the impasse, development studies has never been the same since. The aforementioned crisis in the three post-Second World War developmental paradigms served as academic warning flags that since the impasse led to an almost continuous process of rethinking that which needs to be explained and the various explanations that have been used in development studies.

Further reading

Corbridge, S. (1989) 'Marxism, Post-Marxism and the Geography of Development', in R. Peet and N. Thrift (eds) *New Models in Geography*, Vol. 1, London: Unwin Hyman, pp. 224–254, identifies and elaborates upon three dimensions in Booth's critique of neo-Marxist development theories, i.e. essentialism, economism and epistemology.

Edwards, M. (1989) 'The Irrelevance of Development Studies', *Third World Quarterly* 11(1): 116–136, approaches development theories from the point of view of the practitioner.

Leys, C. (1996) *The Rise and Fall of Development Theory*, London: James Currey, and Bloomington: Indiana University Press, especially chapter 1 (pp. 3–45) provides a seminal overview of the post-WWII development theories.

Munck, R. and O'Hearn, D. (eds) (1999) *Critical Development Theory. Contributions to a New Paradigm*, London: Zed Books.

Schuurman, F.J. (ed) (1993) *Beyond the Impasse: New Directions in Development Theory*, London: Zed Books, provides a general overview of the dimensions of the impasse and the attempts to develop new theories as well as the problems and possibilities of these attempts.

Simon, D. and Närman, A. (eds) (1999) *Development as Theory and Practice. Current Perspectives on Development and Development Co-Operation*, Harlow: Longman.

Vandergeest, P. and Buttel, F. (1998) 'Marx, Weber, and Development Sociology: Beyond the Impasse', *World Development* 16(6): 683–695, focuses upon Booth's critique upon the underlying meta-theoretical assumptions of Marxism, pointing out the necessity of looking within the heterogeneity of developing countries for common denominators.

Willis, K. (2005) *Theories and Practices of Development*, London: Routledge.

References

Appadurai, A. (1990) 'Disjuncture and Difference in the Global Cultural Economy', in M. Featherstone (ed) *Global Culture, Nationalism, Globalization and Modernity*, London: Sage, pp. 295–311.

Hirst, P. and Thompson, G. (1996) *Globalization in Question*, Cambridge: Polity Press.

Chapter 3

Dependency theories

From ECLA to André Gunder Frank and beyond

Dennis Conway and Nik Heynen

Dependency Theory, more than a theoretical construct, is a way of understanding historically embedded, political-economic relations of peripheral capitalist countries, especially Latin American countries, within the broader context of the global economy. It is, essentially, a *critique* of the development paths, policies, and strategies followed in Latin America and elsewhere in the peripheral Global South. *Dependency Theory* emerged as a critical lens through which the history of Latin American development, marginalised as it was by Western hegemony, could be better understood; the 'development of under-development', no less. The initial theorization was a *structuralist* perspective by economists who were associated with the United Nations Economic Commission for Latin America (ECLA). This was soon transformed – and informed – by more critical *dependency* notions and the spread of Marxist and neo-Marxist critiques of imperialism (Chilcote, 1984).

Perhaps, one of *Dependency Theory's* most important characteristics is that it was a product of Latin American scholarship (much of it written in Spanish) rather than Western or North American/European scholars. These authorities theorized on the Latin American condition as 'insiders', as erstwhile, often passionate native sons. This gave rise to a more informed – and more involved – appreciation of the reasons for Latin American underdevelopment as *Dependista*s dealt with the context of various countries' specific national circumstances, and theorized about Latin America's structures of social organisation and localised behaviors. More widely, it was the publication of the writings of André Gunder Frank (and the collection and translation of other Latin American original contributions by North American Latin Americanists) that brought the *Dependency School's* ideas to the notice of North American and European development studies.

Prior to World War II, Latin American countries' economic strategies primarily revolved around a development path based on the export of natural resources and primary commodities to core countries. Many, including Argentinian Raúl Prebisch, Brazilians Paul Singer and Celso Furtado, and Chilean Osvaldo Sunkel, felt that Latin America's historical marginalisation and resultant underdevelopment were perpetuated by such unequal commercial arrangements. While free-market notions of 'comparative advantage' might suggest Latin America should benefit from providing their primary goods to the industrialized countries, Prebisch (1950) posited there were short-term fluctuations in the terms of trade in Latin America countries, deteriorations in the long-term and improved terms of trade in the advanced countries. Such *structuralist* assessments had core countries, particularly Britain and the United States, benefitting at Latin America's expense.

DOI: 10.4324/9780429282348-4

Consequently, Prebisch and other ECLA *structuralists* felt that major structural changes in development policy were needed to improve Latin America's economic situation. They proposed structural changes that favored switching to more domestic production under tariff protection as a means of replacing industrial imports. In line with this strategy, capital goods, intermediate products, and energy would be purchased with national income revenues from primary exports, and technology transfer would be negotiated with transnational corporations. This development strategy – often referred to as import substitution industrialization (ISI) – became widely practiced throughout Latin America, the Caribbean, and the Third World/Global South in general.

Although the ECLA *structuralist* analyses recognised some of the problems underlying Latin American underdevelopment, the proposed import-substitution industrialization (ISI) remedies brought other, more problematic, forms of dependency. Multi-national and transnational corporate power and authority over technology transfer and capital investment emerged as a new form of neo-colonial dependency. Fernando Henrique Cardoso (1979) pointed this out in his assessments of power and authority in Brazil, and he preferred to characterise the situation in such peripheral economies as *associated dependent development*. Indeed, Cardoso felt that the dependent capitalist process of 'industry-by-invitation' occurred mostly under authoritarian regimes and, further, that state policies would favor multinational capital at the expense of labour.

Prebisch's identification of core-periphery relations as the global historical heritage behind unequal development meant Latin America continued to face a formidable structural reality. Imperialism and colonialism were to be challenged, more rigorously. Capitalism, more especially peripheral capitalism, was not the answer for Latin American development. Accordingly, alternative critical commentary, more deeply rooted within Marxist and neo-Marxist ideologies, emerged to better explain Latin America's subordinate place within the global economy and better understand the processes that led to such exploitive *and dependent* relations. ECLA *structuralism* was re-cast in *dependencia* terms.

Baran's influential (1957) *Political Economy of Growth* described the reasons for Latin America's underdevelopment within a Marxist framework as being a consequence of advanced nations' forming special partnerships with powerful elite classes in less-developed or pre-capitalist countries of the Global South. Such alliances were of course detrimental to the capitalist development of such 'backward' economies since they benefitted the minority class of Latin American elites rather than advancing economic development at large. Such 'partnerships', according to Baran, perpetuated the ability of core countries to maintain traditional systems of surplus extraction, thereby making domestic resources continuously available and making the economic development of Latin American counties unlikely, since any surplus generated was appropriated by the elites. Thus, the imperial core countries would keep Latin America subordinate – and maintain their monopoly-power – to ensure a steady outflow of cheap primary resources.

André Gunder Frank further developed Baran's ideas by focusing upon the dependent character of peripheral Latin American economies. In Frank's (1966) prognosis, the 'development of underdevelopment' was the concept that best characterised the capitalist dynamics that both developed the core countries and at the same time caused greater levels of underdevelopment and dependency within Latin American countries. Frank used this conceptual framework to explain the dualistic capitalist relations that had occurred and that he felt would continue to occur between Latin American and core counties, as a result of the continued domination of Latin America by the core.

Although there was a popular perception that Third World countries regained some sense of self-determination following decolonisation, Frank argued this was a fallacy. Exploitation of many Third World/Global South countries by colonial and neo-colonial core countries intensified following their achievement of political 'independence', further contributing to greater unequal relations. Thus, given the class-based stratification of Latin American society, which Baran blamed for the development of ties between Latin American elites and capitalist and political leaders from core countries, revolutionary action to remove such elites from power would be needed to forge a reformulation of international capitalist relations. Frank (1979) suggested this was only possible through revolutionary action that strove to install socialist ideals within the political systems of the dependent countries.

Besides arguing that the dependent core-periphery relationship was best articulated at the national scale, Frank also posited that a similar metropolis-satellite relationship occurred at smaller (regional) scales. In particular, he described similar dependant circumstances occurring between cities in Latin American countries and colonies and their non-urban peripheries. He illustrates this relationship within the context of the privilege that has always existed for colonial Latin American cities. As the place of administration for colonial powers, the city has always been the power-base from which the expansion of capitalism has spread. Within this more localised scenario, the city and its peripheral hinterland becomes increasingly polarised as a result of the capitalist relations between them, namely the metropolis exploiting its satellites. Given the localised nature of this relationship, dense networks of metropolis-satellite combinations form what Frank referred to as 'constellations across national space'.

As an explanation for Latin America's peripheral position in terms of modern versus traditional structures, Frank contended that this dualist perspective failed to truly comprehend the historical significance and transformative impact of capitalism's penetration of the continent's economic, political, and social structures. The dependent relationship Frank posited as a counter explanation to such dualist notions drew sharp criticism from many, however. Laclau's (1971) analysis is perhaps the most notable.

Laclau asserted that Frank's analytical method has significant shortcomings because it was based on an erroneous characterisation of Marx's notion of modes of production. Instead of basing the construction of a mode of production on social or class relations, as Marx did, Laclau claims that Frank's reliance on market relations as the defining quality of the processes under which production occurs is inherently flawed. As a consequence, Laclau faults Frank for constructing a circular concept of capitalism that is inherently imbalanced. Laclau concludes that as a result of the flawed interpretation of the mode of production, Frank's analysis offers little more than an account of a history that is well reported; in effect, he contributes nothing to theoretical explanation in terms of determining conditions.

The resultant tensions within Frank's analytical framework as a result of arguably incorrect – or less than accurate – usage of Marxist ideology, led the way to other neo-Marxist investigations of the linkages and possible reconciliation between *Dependency Theory* and Marxism. Seeking to 'resolve the debate', Chilcote (1984) effectively situated the various capitalist and socialist approaches to the 'development of underdevelopment' – *structuralism, dependencia, internal colonialism, neo-Marxism,* even *Trotskyism* – as a full set of alternative theories and perspectives on development and underdevelopment. He also found a place for Wallerstein's more worldly focus in this collection of alternatives.

Indeed, Wallerstein (1974, 1980) adapted *dependency* notions to comment on the commercial relations between the core countries and Latin America and examined world historiography in terms of the dominant and subordinate relations that successive emerging cores, their peripheries and semi-peripheries experienced. This account started with the 'long sixteenth century', passing through successive eras of capitalism to the present neoliberal era of globalization (the post-1980s). Wallerstein's 'world systems theory' complements and expands upon Frank's ideas, providing a more comprehensive global stage appropriate for understanding the wider reach and more diverse spatial realignments of commercial capitalist relations in contemporary times. More recent 'world systems' explanations of geo-political eras detail the transformations of world's hegemonic relationships of core-periphery relationships to the present global era that continues into the second decade of the 21st century (Conway and Heynen, 2006). Amin (2003), too, offers a much more critical view of contemporary geo-political times than Wallerstein.

Ghosh (2001) further provided a contemporary critical appraisal and overview of contemporary thoughts on the full set of alternative dependency theories, pointing out the significant 'inter-temporal paradigm shifts' in the theory's wider application in our rapidly globalizing world. As Ghosh (2001: 133) reminds us:

> There are indeed many issues and areas of development where dependency plays a major role. Some of these are; aid dependency, technological dependency, dependency for foreign capital investment, trade dependency, dependency for better human capital formation and so forth.

There are obvious connections between the divergent trajectories of capitalism's expansion in the Global North as opposed to the Global South. Equally obvious, 'unequal competition' remains an extremely powerful, dependency relationship in globalisation's transformative, disciplinary, and destructive influences (Conway and Heynen, 2006). Just as the imperialism of old imposed colonialism fostered dependency and underdevelopment, modern globalization of the post-1980s has several salient features that are de facto, neoliberal successors to these imperial mechanisms. They represent: i) a program of binding individuals, institutions, and nations into a common set of market relationships; ii) a calculated economic strategy of the capitalist economies, corporations, and international financial institutional systems to encourage and stimulate capitalist growth for 'winners' – core and emerging markets – not the 'losers' with no comparative advantages, weak or failed states, or the corruption-weakened and iii) a means of extracting surplus through the exploitation of cheap labour, high-quality manpower, and resources of the Global South (Ghosh, 2001: 158).

'Dependency-thinking' has come a long way since its initial Latin American interpretations, but even in today's 'globalizing world' the geo-political and geo-economic struggles underway in Latin America are anything but predictable and can no longer be so easily framed in the centuries-old structural terms of core US hegemony and Latin American dependency. The evolving world system of core-periphery relationships has entered a new advanced phase of 'modernity' in which there are new dependency relationships, ecological uncertainty, rapid technological change, and a multiplicity of cross-cutting flows of information, cultural messages, and knowledge exchange. They occur at multiple scales and scopes of influential power and authority – ranging from the global to the local, from the exceptional to the ordinary and from the elites to the bourgeoisie and working classes.

Furthermore, and as a concluding recommendation, 'dependency thinking' today requires us to confront the power hierarchies of the recent past (and present) using much more informed critical perspectives on the geo-economic power of transnational corporations, the emergence of BRICs in the Global South and the comparative declines of traditional cores' hegemonic authority in global forum – the G20, the UN, and such (Amin, 2003). Marxist theory may no longer be sufficient in and of itself to explain contemporary processes of Global North and Global South interdependencies, but the derivative critical perspectives drawn from such structural, neo-Marxist analysis by the likes of Amin, Wallerstein, and Frank can still help in the formulation of progressive alternatives for developing societies.

Guide to further reading

Blomstrom, M. and Hettne, B. (1984) *Development Theory in Transition: The Dependency Debate & Beyond-Third World Responses*, London: Zed Books.

Chilcote, R. H. (1984) *Theories of Development and Underdevelopment*, Boulder and London: Westview Press.

Ghosh, B. N. (2001) *Dependency Theory Revisited*, Aldershot and Burlington VT: Ashgate.

Kay, C. (1989) *Latin American Theories of Development and Underdevelopment*, London and New York: Routledge.

References

Amin, S. (2003) *Obsolescent Capitalism: Contemporary Politics and Global Disorder*, London and New York: Zed Books.

Baran, P. (1957) *The Political Economy of Growth*, New York: Monthly Review Press.

Cardoso, F. H. (1979) "On the Characterization of Authoritarian Regimes in Latin America." In D. Collier (ed) *The New Authoritarianism in Latin America*, Princeton: Princeton University Press, pp. 34–57.

Conway, D. and Heynen, N. (2006) *Globalization's Contradictions: Geographies of Discipline, Destruction and Transformation*, Abingdon and New York: Routledge.

Frank, A. G. (1966) "The Development of Underdevelopment." *Monthly Review* 18(4): 17–31.

Frank, A. G. (1979) *Dependent Accumulation and Underdevelopment*, New York: Monthly Review Press.

Laclau, E. (2018). Feudalism and capitalism in Latin America. In *Promise of Development* (pp. 166–190). Routledge.

Prebisch, R. (1950) *The Economic Development of Latin America and Its Principal Problems*, New York: United Nations Department of Economic Affairs, Economic Commission for Latin America (ECLA).

Wallerstein, I. (1974) *The Modern World System, Volume 1. Capitalist Agriculture and the Origins of the European World-Economy in the Sixteenth Century*, New York: Academic Press.

Wallerstein, I. (1980) *The Modern World System, Volume 2. Mercantilism and Consolidation of the European World-Economy, 1600–1750*, New York: Academic Press.

Chapter 4

The new world group of dependency scholars

Reflections of a Caribbean Avant-garde movement

Don D. Marshall

Introduction: post-new world intellectual currents

Forty-five years after the emergence of the New World movement, it might be reasonable to expect that gathering forces in the international system – shaped by the imperatives of globalisation – would present the spectre of the emergence once more of vital new political forces. Then, as now, the region was thrown back into contemplation on the relevance of its development strategy. With the benefit of the backward glance, the 'New World' was first founded in Georgetown towards the end of 1962 against the backdrop of a long general strike and growing racial conflict between African-Guyanese and Indian-Guyanese. The early founders aspired to invent an indigenous view of the region, convinced that the modernisation ideologies very much in vogue neither inhered a strategy for real, independent development nor understanding of the political economy legacy of the Caribbean, of which more later on.

Currently, Caribbean intellectuals in the main, particularly its social scientists, take on the colour of their historical environs: if neoliberal capitalism cannot be successfully challenged, then to all intents and purposes it does not exist; all that remains is the challenge of massaging a link between market liberalisation and populist-statism. To be sure, this concern among Caribbean scholars and commentators does not preclude expression of despair in some quarters over the sustainability of the island-national project of the Caribbean. This forecast is based on an understanding of the export-impetus girding contemporary capitalism and the difficulties associated with making the transition in political economies dominated by merchant capital.

Decolonisation and the rise of new world

The New World movement in the Anglophone Caribbean was marked by an optimism of will and intellect. Newly independent governments were seen to be in pursuit of development guideposts to chart a self-reliant future. At the popular level, claims for social equality through redistribution became intensely salient as an expression of justice. And knowledge producers both within the academic and literary community, no longer under the heel of colonial power, focused energies either on transformative or ameliorative development agendas. Social dialogue and action seemed governed by an impulse towards West Indian self-definition manifested in discussions on race, class, culture and the question of ownership and control of the region's resources. The general decolonisation horizon within which such mood and thought moved was also marked

DOI: 10.4324/9780429282348-5

by raging debates occurring in the academic world between modernisation theorists and neo-Marxist scholars. The New World group of largely historians and social scientists would come to draw from – and intervene in – these debates, combining serious inquiry into the development possibilities under capitalism, with integrative, normative, and programmatic thinking on nation-building.

Considered by their pragmatic counterparts in government, media, and academy as 'radicals', this cluster of writers and commentators across the Caribbean came to be known as the New World Group (NWG). Their thoughts and ideas on socialism, national self-determination, and the delimiting horizons of capitalism reached a West Indian mass audience through public lecture series, various national fora, and newspapers and newsletters of their creation. The *New World*, a Jamaica-based magazine, first appeared in 1963 and was published fortnightly under the editorship of Lloyd Best with assistance from a host of University of the West Indies (UWI, Mona Campus) scholars, George Beckford, Owen Jefferson, Roy Augier, Derek Gordon, Don Robotham, and Trevor Munroe, to list a few. From 1965, *New World* was published as a quarterly. Bearing the imprint of the UWI, the *New World* would serve as a loose association attaching its name to anti-imperialist consciousness-raising activity across the region. Indeed NWGs were said to be formed in St. Vincent, St. Lucia, Washington D.C., Montreal, St. Kitts, Trinidad, Barbados, Anguilla, Jamaica, and Guyana. Other publications that appeared either as complements to or refinements of *New World's* mission included *Moko* and *Tapia*, Trinidad-based weekly newspapers appearing in 1969, Abeng, a Jamaican newsletter launched in the same year as well as the 1970 St. Lucian-based *Forum*.

The first issue of NWQ focused on Guyana's development dilemma. The analysis therein moved beyond conventional state-centric explanations about the country's savings gap, low technologies, unskilled, undifferentiated labour markets, and inadequate infrastructure. Guyana's and indeed the Caribbean's limited development, it was argued, were a function of the region's structural dependent linkages with Europe in terms of its value system and its economic relations. This point of view resonated with the dependency perspective first advanced by Paul Baran and subsequently extended by others who specialised in Latin American area studies. It was certainly a more assimilable 'angle' for Norman Girvan and Owen Jefferson to deploy in their doctoral theses explaining Jamaican underdevelopment (circa 1972) than the market-deficiency arguments of neoclassical proponents. As Girvan and Jefferson saw it, the move towards self-government and independence could not arrest the process of underdevelopment so long as the domestic economies remained dependent on foreign capital and terms of trade set under colonial rule.

Principally, the path of resistance for New World associates was forged out of opposition to Arthur Lewis' (1955) import-substitution-industrialisation (ISI) model, favoured by Caribbean governments in the 1960s and 1970s. Briefly, the ISI programme required state provision of incentives to transnational enterprises in order to attract offshore industrial operations. The various budgetary and fiscal preparatory statements placed emphasis on the prospects for increased employment and technology transfer and stimulated markets for local inputs.

Beckford (1972), and Best and Levitt (1968), levelled a critique of Lewis' model that was representative of the dominant positions New World associates adopted on the question of Caribbean capitalist development. With epistemic insights drawn from orthodox Marxists and Latin American structuralists, their research fitted the growing canon of work seeking to establish dependency as the source of persistent underdevelopment. Beckford and others in the NWG would enrich this stock argument by anchoring the dependency concept within the plantation experience of Caribbean societies.

Dependency theory and plantation economy

Beckford's (1972) *Persistent Poverty* defined the historic plantation slave economy as a quintessential dependent economy, the units of which included Caribbean land, African unfree labour, and European capital. This is Best and Levitt's (1968) 'pure plantation economy' as no other economic activity occurred outside the sugar plantation. Beckford's work was as much a repudiation of Caribbean development strategies, as it was a paradigmatic challenge to the liberal fallacy of 'progress'. For him, the mode of accumulation in the region remained a modified plantation economy variant, as dependent investment and aid ties with London and other metropolitan cities persisted. After lamenting the dis-articulation between branch-plant production and the rest of the host economy and the general mono-product character of local economies, Beckford and, later, Best and Levitt, outlined other structural features of plantation economy which generated underdevelopment:

1 Land requirements of plantation production tended to restrict domestic food production.
2 Terms of trade often deteriorated as rising food and other imports presented balance of payments difficulties.
3 Stagnant educational levels tended to foreclose on product diversification options and improvements.

Havelock Brewster (1973), seized by the plantation economy argument, argued that foreign capital could not possibly champion industrialisation in accordance with common needs and the utilisation of the internal market. This was so, he surmised, because the gridlocked nature of a plantation economy with its lack of an internal dynamic, its reliance on outdated technologies, and hierarchical management practices guaranteed for the region a subordinate role in its relationships with core firms and countries.

From the foregoing we may gather that unlike their dependency counterparts in Latin America, most New World associates relied less on external-determinist explanations to explain Caribbean underdevelopment. They focused on the *internal* workings of Caribbean economies to account for the region's structural dependency, even as they were careful to note that the characteristics of these economies extended back to colonial relations between Britain and the West Indies. *Dependentistas* and structuralists, on the other hand, placed the centre-periphery relations they depict within the context of macro-historical forces intent on locking peripheral societies into an unyielding spiral of exploitation and poverty.

Interestingly enough, Walter Rodney, a Guyanese historian, and Jamaican political scientist Trevor Munroe could be said to have framed Caribbean development in such deterministic terms except that they singled out the social legacy of the plantation experience as especially debilitating for non-white races. Both were inspired by Marx's historical materialist method but Rodney was inclined to argue that nation-building in the region had to be about renewing spirits, constructing grounds for black liberation and pursuing self-reliance. Trevor Munroe's perspective was expressed in more classical but nuanced terms as he was mindful of the plantation slavery experience. As he would frame it, underdevelopment in the region was the predictable outcome of undeveloped class formation – itself partly perpetuated by those mix of domestic policies which threw the territories back on traditional activity and on traditional metropolitan dependence. The extent of the lag in technological, market, infrastructural, and resource development will pose a challenge to aspirant Caribbean societies committed to constructing a capitalist economy.

Of the New World group, however, Best's dependency perspective evinced a deep-seated ambivalence towards Western discourses on development. Perhaps he was self-conscious of the postcolonial scholar's place in such literary transactions, of the dangers of succumbing to the neoclassical association between open economies and automatic economic growth. In the context of plantation economies, such assumptions muddled an already complex situation, Best argued. His dependency perspective was consistently embedded in extended and detailed analyses of ruling circles. Apart from providing address to the aforementioned features of neo-colonial dependency in the region, he singled out the shared outlook of Caribbean elites and Western development planners as a major brake against effecting meaningful socio-economic transformation. Not surprisingly, his appeal was for a shift in the register of social consciousness on the part of the ruling elite. The colonial hangover apart, Best failed to draw sufficient attention to the degree of class conflict decolonisation inheres as new class forces move to reorient the social system and the values that define that system.

The demise of new world

As the 1970s dawned, the New World movement shuffled to a halt as division arose over strategies, tactics, and modes of resistance to neo-colonialism. By this time, Best was especially critical of the group, decrying what he saw as New World's fatal attraction for governments and a tendency to substitute policy-oriented research for contemplative scholarship. Increasingly, such knowledge products, he argued, amounted to exercises in self-justification and as such were quite explicit disclosures of governmental discourse in action. He was also resistant to the idea that New World could move towards the formation of a political party or organisation contending for power. In a polemic entitled, *Whither New World*, Best (1968) spoke of the tensions of the group offering the following observation: 'There is among us, much unwitting intolerance, little cool formulation, hardly any attentive listening and even less effective communication'. Munroe would come to lament their facetious pursuit of class unity and vowed to distance himself from what he termed the 'bourgeois idealism' of New World.

The disintegration of the New World group was in part a result of the attention given by many to the immediate realms of the policy process. Mona-based economists, in particular, played key advisory roles in the Michael Manley Administration of the 1970s, while others across the region responded to appeals from governments for technical and project management assistance. But there are some scholars that instead place emphasis on the internal arguments between Best and others on the question of New World's relevance and its activist orientation. Their analysis, in my view, falls short precisely because they insufficiently recognise that New World, as any avant-garde movement, became compromised not so much by bourgeois acceptance *as by absorption into the intelligentsia*. Attendance to career, administration, and public service would spawn a culture marked by keynote address, cocktail attendance, and doctoral authority. Consequently, the new radicals were to be found on the outskirts of Black Power movements, drawn less to its ideology as to the struggle for worker freedom and justice.

On a wider intellectual plane, the New World could be said to suffer the slump it did largely because the dependency concept itself lacked lasting explanatory power. Overall, there was a circularity in the dependency argument: dependent countries are those which lack the capacity for autonomous growth and they lack this because their structures are dependent ones. Other scholars have also made the point about development in the world economy being in fact *dependent* development pointing to foreign investment relationships between core states and firms. By the late 1970s, the emphasis among neo-Marxists shifted away from an independent weight placed on 'dependency' as undesirable, towards either a normative condemnation of state capitalism or an appeal to Third World states to *negotiate* the scope of their dependency.

Summary: back to the future

If we posit that openings for dissent are as necessary to democracy as securing of consent, then Caribbean civil and uncivil society can continue to offer sites for objection and challenge. But there has been no New World equivalent emerging out of the tensions of the present neoliberal period. True the rise and influence of non-governmental organisations (NGOs), particularly women's organisations, trade unions, and the galvanising work of the Caribbean Policy Development Centre along with that of critical scholars have served to exert pressure on increasing public transparency and inclusion. To be sure it is not at all clear that NGOs constitute an intrinsically virtuous force for the collective good. These can run a similar course to that of the New World. Beyond a certain point NGOs may lose the critical element that caused them into existence as they render services to governance agencies, take funds from them, or 'cross over' to work for government institutions and organisations that they previously challenged. Currently, market-mentalities predominate in government bureaucracies, business firms, and academia. From various nostrums, academicians from the UWI, particularly social scientists, are exhorted by media, business, and government commentators to give advice and attention to the technicality of social control or constitutional and other reforms. In most issue-spaces, ruling discourses of technocratic expertise seem to arbitrarily suppress alternative perspectives. The UWI's role in this is not entirely surprising as the University's struggle for relevance and its sensitivity to budget efficiency do make for a climate where conformity to the prevailing common sense seem the best course for research programming. Hegemony-affirming research thus continues to triumph. Political and intellectual challenges are foreclosed in the prevailing environment where priority of survival continues to be asserted both as an operating principle and as a rationale for the absence of radical critique. This is the 'bourgeois villainy' Best would speak of when the case was hardly self-evident among intellectuals of New World. The associates then at least managed a discussion of Caribbean dependency that was enriched by site characteristics of plantation production relations. This added colour to parallel debates in Latin America. For New World associates, the dependency concept had operative power; it encouraged an interesting entry-point for challenging the colonial mode of accumulation. It also fashioned an intellectual *cachet* of dissent in the region, illuminating history and social fact as economic paradigms came under challenge.

References and Further Reading

Beckford, G. (1972) *Persistent Poverty: Underdevelopment in Plantation Economies of the Third World*, Morant Bay and London: Maroon Publishing House and Zed Press.

Best, L. (1968) 'Forum: Whither New World', in *New World*, Vol. IV, No. 1. pp. 1–6.

Best, L. and K. Levitt (1968) 'Outlines of a Model of Pure Plantation Economy', in *Social and Economic Studies*, Vol. 17, No. 3, September. pp. 283–326.

Blomstrom, M. and B. Hettne (1984) *Development Theory in Transition*, London: Zed Books Ltd.

Brewster, H. (1973) 'Economic Dependence: A Quantitative Interpretation', in *Social and Economic Studies*, Vol. 22, No. 1. pp. 90–95.

Dookeran, W. (ed.) (1996) *Choices and Change: Reflections on the Caribbean*, Washington D.C.: Inter-American Bank.

Hendriks, C. M. (2006) 'Integrated Deliberation: Reconciling Civil Society's Dual Role in Deliberative Democracy', in *Political Studies*, Vol. 54. pp. 486–508.

Lewis, R. C. (1998) *Walter Rodney's Intellectual and Political Thought*, Kingston and Detroit: The Press University of the West Indies and Wayne State University Press.

Lewis, W. A. (1955) *The Theory of Economic Growth*, London: Allen and Unwin.

Marshall, D. D. (2000) 'Academic Travails and a Crisis-of-Mission of UWI Social Sciences: From History and Critique to Anti-Politics (Chapter Four)', in G. Howe (ed.) *Higher Education in the Anglophone Caribbean: Past, Present, and Future Directions*, Mona: University of the West Indies Press, pp. 59–84.

Munroe, T. (1990) *Jamaican Politics: A Marxist Perspective in Transition*, Kingston and Boulder, Colorado: Heinemann Publishers (Caribbean Limited) and Lynne Rienner Publishers.

Rodney, W. (1972) *How Europe Underdeveloped Africa*, London: Penguin Books.

Chapter 5

World-systems theory

Core, semi-peripheral, and peripheral regions

Thomas Klak

Definition

World-systems theory (WST) argues that any country's development conditions and prospects are primarily shaped by economic processes, commodities chains, divisions of labour, and geopolitical relationships operating at the global scale. World-systems theorists posit the existence of a single global economic system since at least the outset of European industrialisation around 1780–1790. According to WST doyen Immanuel Wallerstein and others, the global system dates back even further, to at least 1450, when international trade began to grow, and when Europe embarked on its 'age of discovery' and colonisation (Frank and Gills, 1993). Contrary to much social science thinking, WST stresses the futility of a 'statist orientation' – i.e., the attempt to analyse or generate development by focusing at the level of individual countries, each of which is profoundly shaped by world-system opportunities and constraints (Bair, 2005).

WST has identified a number of regularly occurring historical cycles associated with the level and quality of global business activity. These cycles account for economic booms and busts of various durations. The main economic periods for WST are *Kondratieff cycles*, named after the Russian economist who discovered them in the 1920s. Each cycle or *long wave* lasts about 50–60 years and represents a qualitatively different phase of global capitalism, not just a modification of the previous cycle. Kondratieff cycles are themselves divided into a period of expansion and stagnation. There is first an A-phase of upswing, economic expansion, and quasi-monopolistic profitability, fuelled by technological innovations and organised by new asymmetrical institutional rules. Price inflation increases during the A-phase. This then leads into a B-phase of increased competition, profit decline, economic slowdown, and price deflation. The profit squeeze towards the end of the B-phase motivates capitalists and policy makers to create new and innovative ways to accumulate capital. They work to shift investment out of established economic sectors, regulated environments, and production locations and thereby create the conditions for a new Kondratieff cycle (Knox et al., 2003).

The previous Kondratieff cycle began in the 1940s, expanded until 1967–1973 (A-phase), and then contracted through the 1980s (B-phase). Each cycle's organising institutions and rules are both economic and political. For that Cold War cycle, key economic rules and structures included the US dollar as the global currency and supranational bodies such as the World Bank, the IMF, and the G-7. Political structures included the UN and the geopolitical divisions brokered at the Yalta conference. It divided Europe into US- and Russian-dominated zones, pitted global capitalism against Russian-led state

DOI: 10.4324/9780429282348-6

socialism (communism), and presented the Third World as ideologically contested turf. The early 21st century found the world in a cycle shaped by the WTO, neoliberal free trade, and global financial liberalisation aimed at ensuring quasi-monopolistic profitability and global power for core countries. As in the Cold War cycle, the United States remains the preeminent core (and thus global) power, but its hegemony is now contested by other strengthening core countries and semiperipheral countries, notably China. The global financial crisis since 2008 may signal the dawn of new long-wave shaped by such factors as information technologies, resource scarcity, and climate change (Moody and Nogrady, 2010).

Scholars and disciplines influencing – and influenced by – WST

WST is deeply linked to its principal architect, Immanuel Wallerstein (born 1930). Indeed, few influential theoretical perspectives are so intertwined with one contemporary scholar. WST's conceptual roots are largely in Marxism. Wallerstein (1979) says that WST follows 'the spirit of Marx if not the letter'. Evidence of Marx's spirit includes WST's emphasis on class, the state, imperialism, and control over the means of production and labour power. WST's objections to classical Marxism include concern over a theoretical component known as *developmentalism*. This is the idea that societies move sequentially through feudalism, capitalism, and socialism to communism and that they can be analysed and transformed individually and separately from the world system. WST's alternative view – that there has been for centuries but one world economy driven by capital accumulation – employs a concept of mode of production closer to that of Karl Polanyi than to Marx.

WST has much interdisciplinary relevance and has therefore attracted both supporters and detractors from across the social sciences. WST complements political-economic analysis rooted in the traditions of *dependency theory* (Cardoso and Faletto, 1979), *uneven development* (Smith, 1984), and *dependent development* (Evans, 1979). A conceptually overlapping but perhaps less economistic and highly influential alternative to WST is *the regulation school*. Usually applied at a more local level than WST (i.e., to national or subnational systems), regulation theory seeks to identify phases of capitalism of variable length based on relations between a particular prevailing method of accumulating capital and an associated 'mode of regulation', i.e., a set of state regulations and behavioural norms (O'Hara, 2003).

The geography of WST: three groups of nation-states

WST's temporal cycles of systemic integration, order, turbulence, transition, and reconstitution of the global economy play out variably across geographical space. The world-system is very unequal. Despite (or, world-system theorists argue, *because of*) several centuries of worldwide economic integration and trade and more than 65 years of World Bank-led international development, global inequalities continue to rise, and at an increasing pace. The difference in per capita income separating the richest and poorest countries was 3:1 in 1820, 35:1 in 1950, 72:1 in 1992, 108:1 in 2004, and 384:1 in 2011 (UNDP, 1999, 2006, 2011). Within this highly unequal world order are place-specific dynamics. At times, regions can rise and fall in terms of power, development, and economic potential. WST describes this globally differentiated space with reference to nation-states, regional groupings thereof, and regions within nation-states. These fall into three categories (see figure).

Scholars disagree over which variables best define a country's positions in the world-system. With this caveat in mind, general geographical features can be described. Countries of the *core* or *centre* are the sites of global economic (and especially industrial) control and wealth and the associated political and military strength and influence. Core countries feature higher-skill, capital-intensive production. Politically, they collectively establish and enforce the rules of the global order and through these advantages appropriate surplus from non-core countries. The *semi-periphery* is positioned between the core's strengths and periphery's weaknesses. It mixes characteristics of the core (for example, industry, export power, prosperity) and the periphery (for example, poverty, primary product reliance, vulnerability to core decision-making). The semi-periphery is the most turbulent category, in that its members most frequently rise or fall in the global hierarchy. In semi-peripheral countries, there is much hope for development and joining the core countries and narrow windows of opportunity to do so. But there are also intense interactions with core countries bent on fostering their own capital accumulation by maintaining the hierarchical status quo. The *periphery* is the backwater of the world system. It provides low-skill production and raw materials for industries elsewhere. It has poor living conditions and bleak development prospects. The semi-periphery versus periphery distinction for non-core regions is important. It avoids grouping such a heterogeneous set of countries with respect to development, industrialisation, trade, resource control, and geopolitics. Still, putting the world's 200 countries into just three groups inevitably glosses over much intra-group heterogeneity. Note the regional clustering of countries in the three categories in the figure. At present the core is mainly North America, Western Europe, and Japan. The semi-periphery is essentially East Asia, Latin America's larger countries and most of the former Soviet realm. The periphery is everything else, particularly Africa (Wallerstein's empirical focus).

A nation-state's position in the world system is historically *path dependent*, but not deterministically so. Nation-states can move between categories over time, depending on their accumulation regimes, development strategies, and international aid and alliances. Indeed, WST is quite useful for analysing the upward and downward movement of countries over time. There is not agreement over each country's categorisation, depending on the defining characteristics and their interpretation. In addition, relative positions *within* each of the three categories can also shift over time.

East Asia illustrates the semi-periphery's potential and turbulence. Following massive US aid and industrial export growth in recent decades, South Korea has recently been knocking on the core's gate, although it was set back considerably by the 1997 Asian financial crisis. Indonesia has traditionally been peripheral, but in recent decades it has arguably joined the semi-periphery. Its increased clout derives from economic growth based on industrial exports for Nike and others, large resource endowments including oil exports, and its status as the world's fourth most populous country (see figure). China's industrial export boom and associated capital accumulation since the 1980s drove it into the semi-periphery. Now many Japanese and US leaders fear China's global resource hunger and ambitions as a soon-to-be core country (Zweig and Jianhai, 2005).

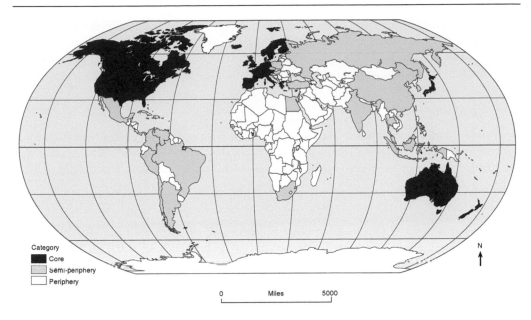

Figure 5.1 The world system early in the 21st century

Criticisms of WST

One capitalist world economy, divided by Kondratieff cycles, since at least 1450?

Need we subscribe to WST's totalising global history to employ it effectively to understand recent development? Compared to Wallerstein, few writers employing a WST framework are as deeply historical, and few treat economic activities during previous centuries in such a globally holistic way. Much work, for example, has been done to identify the evolving features of capitalism associated with five Kondratieff cycles extending back only to 1789. Many other WST-influenced scholars focus on the dynamics of contemporary capitalism. WST purists may reject these approaches as insufficiently historical.

While Kondratieff cycles have considerable historical and empirical support (Mandel, 1980), they remain controversial. Others have assembled evidence to cast doubt on the existence and significance of long waves and to instead suggest that capitalism moves through phases of differing lengths, problems, and features (e.g., Maddison, 1991). As mentioned earlier, the regulation school is one alternative conceptualisation of contemporary capitalist dynamics.

Metatheory?

Beyond the considerable empirical analysis of Kondratieff cycles and their associated production and technological features, many WST claims remain untested and are perhaps untestable. Most WST-influenced scholarship focuses on the contemporary global political economy and the lack of time series data limits testing. Further, how could the simple three-category spatial division of the world system be tested? WST-inspired writing tends to read like an open-ended analysis of unfolding world events. Critics can claim that this method allows one to find and fit the data anecdotally to the theory. Better to think of a world-system *approach*, *analysis*, or *perspective* than a world-system *theory*.

Neglect of the local?

Operating mainly at the global level and concerned with economic cycles over decades if not centuries, WST is too holistic to account for local dynamics. Indeed, WST underplays the generative role of local activities, initiatives, social movements, and people.

Conclusion

World-systems theory, with its keen sense of historical, cyclical, technological, and geographical patterns, has undoubtedly deepened our understanding of the global political economy. It is a satisfying antidote to the reductionism, ahistoricism, and superficiality in most popular interpretations of economic change. WST's historical and holistic perspective and level-headedness serves to counter the recent hyperbole about the uniqueness of globalisation and inevitability of neoliberalism.

In practice, many scholars employing a WST perspective downplay the details and measurement of the cycles of upswing and downswing in the global economy. They instead focus primarily on contemporary trends and adopt a qualitative approach to understanding business cycles, global systemic change, and the associated realignments of geopolitical and economic power, constraints, and potential. Many economists and some WST purists would judge a more qualitative version of WST to be insufficiently rigorous and therefore theoretically deficient. WST defenders would counter that a more qualitative approach is suitable, given their aim to see the 'big picture' and to decipher and rectify contemporary economic and political institutions and options.

Further reading

Knox, P., Agnew, J., and McCarthy, L. (2008) *The Geography of the World Economy* (5th edn), London: Hodder Education. Couples WST with economic geography to explore the workings of the contemporary global economy.
Shannon, T.R. (1996) *An Introduction to the World-System Perspective* (2nd edn), Boulder: Westview Press. Useful overview of WST, endorsed by Wallerstein.
Wallerstein, I. (ed.) (2004a) *The Modern World System in the Longue Durée*, Boulder, CO: Paradigm Publishers. Wide-ranging chapters by leading world-system scholars providing a useful overview of WST.
Wallerstein, I. (2004b) *World-System Analysis: An Introduction*. Durham: Duke University Press. Wallerstein's own introduction to the field.

References

Bair, J. (2005) 'Global Capitalism and Commodity Chains: Looking Back, Going Forward.' *Competition and Change*, 9: 153–180.
Cardoso, F. and Faletto, E. (1979) *Dependency and Development in Latin America*, Berkeley: University of California Press.
Evans, P. (1979) *Dependent Development: The Alliance of Multinational, State, and Local Capital in Brazil*, Princeton: Princeton University Press.

Frank, A.G. and Gills, B. (eds.) (1993) *The World System: Five Hundred Years or Five Thousand?* London: Routledge.

Maddison, A. (1991) *Dynamic Forces in Capitalist Development*, Oxford: Oxford University Press.

Mandel, E. (1980) *Long Waves of Capitalist Development*, Cambridge: Cambridge University Press.

Moody, J. and Nogrady, B. (2010) *Welcome to The Sixth Wave*, North Sydney: Random House Australia.

O'Hara, P. (2003) 'Deep Recession and Financial Instability or a New Long Wave of Economic Growth for U.S. Capitalism? A Regulation School Approach.' *Review of Radical Political Economics*, 35: 18–43.

Smith, N. (1984) *Uneven Development: Nature, Capital, and the Production of Space*, New York: Blackwell.

UNDP (1999; 2006; 2011) *Human Development Report*, New York: Oxford University Press.

Wallerstein, I. (1979) *The Capitalist World-Economy*, Cambridge: Cambridge University Press.

Zweig, D. and Jianhai, B (2005) 'China's Global Hunt for Energy.' *Foreign Affairs*, September/October.

Chapter 6

Neoliberalism
Different paths within a global project

Elizabeth Humphrys

In the 1970s, accepted economic orthodoxy around ideas of developmentalism and Keynesianism became unstuck in a protracted economic crisis. The belief that governments were inefficient and that the best way to allocate resources was through competitive markets became increasingly popular among economists and governments. In the decades that followed, a radical economic transformation took place and it had profound social and political consequences. This period is often referred to as the neoliberal era. Policy making was centred on markets and prices (of commodities and labour) as the primary way to shape social life. The idea of *homo economicus* – where economic modelling presumes self-interested and utility maximising individuals – became hegemonic.

Neoliberalism has been criticised because it has increased inequality and shifted significant wealth into private – and fewer – hands. 'We are the 99%' was the slogan of the Occupy movement, underlining that a disproportionate amount of wealth is in the hand of the richest 1 per cent globally. The world was vastly unequal before the neoliberal period, as a result of who has historically controlled wealth and property in society as well as the subjugation of the Global South through colonialism. Changes in the neoliberal era have re-enforced and extended these inequalities, continuing the racialised and gendered nature of global capitalism.

Neoliberalism's advance

The late 18th and early 19th centuries were dominated by the ideas of the classical political economists, such as Adam Smith, David Ricardo, and John Stuart Mill. These ideas legitimated private property and the central role of the 'free' market, although these scholars believed there was a need for the state to provide for the common good. The stock market crash of 1929 – and the Great depression that followed – delivered widespread unemployment and hardship and shifted focus to the policies espoused by John Maynard Keynes. Keynes advocated for macroeconomic management and a strong role for government, where state policy would ensure steady economic growth, robust investment, and efforts to achieve full employment. Keynes argued this approach would stabilise capitalism's crisis tendencies and allow citizens to flourish.

World War II (1939–1945) and the post-war reconstruction involved enormous financial expenditure on the part of nations, which had a similar effect to Keynesian ideas of using fiscal policy (government expenditure) to drive growth. In this period ideas of developmentalism began to take hold in many countries in the Global South, including those that were newly independent of colonial rulers. Developmentalism focuses on

DOI: 10.4324/9780429282348-7

government led economic development of public infrastructure, import substitution, and the construction of welfare systems. This period, between 1945 and the early 1970s, is often referred to as the golden age of capitalism, where there was rapid growth and low unemployment across the Global North and sections of the Global South – in particular the Southern Cone of South America.

Things changed rapidly in the early and mid-1970s, with protracted economic crisis and rolling recessions. Key events and factors undermined stability in the global economy, including: the 1971 devaluation of the US dollar and the collapse of the Bretton Woods system, and the oil crisis in 1973 and a global imbalance of trade relations. Unemployment and inflation were rising concurrently in a phenomenon called stagflation. Stagflation acted contrary to usual economic situations where unemployment and inflation acted in an inverse relation – if one goes up the other was supposed to come down. Many argued that stagflation discredited Keynesianism and developmentalism, and policy makers began to search for ways to address it.

Ideas of trade liberalisation, fiscal discipline (austerity), and privatisation and deregulation of state-run sectors become popular and eventually hegemonic. In some quarters this was referred to as the 'Washington Consensus'. It was argued this was because Washington power brokers had developed a consensus on a new economic and development model and agreed it should be implemented primarily in almost all Latin American countries, but also more broadly. A similar agenda was promoted and mandated by bodies like the World Bank and International Monetary Fund (IMF) in the Global South as part of loan conditions. These approaches were also endorsed by Global North Governments and instigated in their home countries.

What is neoliberalism?

Most scholars agree 'that neoliberalism can be broadly defined as the extension and installation of competitive markets into all areas of the economy, politics and society' (Birch, 2015: 571). Neoliberalism was not a process of 'deregulation' or setting markets 'free' so they would be in some sort of 'natural' state but a conscious decision to reshape society in the interests of certain sections. New laws and institutions were introduced in a process of 're-regulation' of global capitalism.

In the hands of various authors, neoliberalism can refer to: a set of ideas or ideology; a distinct phase of capitalism; or a governing rationality or a form of regulation or a type of state. At times authors refer to a 'neoliberal era', signposting the term as the period from the 1970s until now. How we define and understand neoliberalism is contested and, ideologically, neoliberal theory is a hybrid. It is informed by theoretical traditions including liberalism and the ideas of classical political economy, neoclassical theory, monetarism, and ordoliberalism. It is particularly associated with the ideas of Friedrich August von Hayek, Milton Friedman, James Buchanan, and others involved in the Mont Pelerin Society.

The sorts of policies that neoliberalism favours include:

- A deregulated financial and banking sector, including floating exchange rates.
- Reduced trade protections like tariffs on imported goods.
- Privatisation of state-owned industries and utilities.
- Flexible labour markets and a reduced role for trade unions.

- Individual over collective rights.
- Austerity and changes to welfare spending, including privatised pension systems.
- User pays healthcare and education and policies such as vouchers.
- Indirect and regressive taxation.
- Monetary policy to keep inflation low, even if this means unemployment increases.

From the neoliberal era onwards, policy making in relation to development was increasingly based in neoclassical economic theory and less pluralist (Hirschman, 2013). No list is complete, but these are the policies that have been widely deployed and proved enduring. Importantly, neoliberalism is more than an ideology or suite of policies, as people have to bring them into effect. In this way, it can be useful to think of neoliberalism as a state-centred hegemonic political project remaking capitalist production and social reproduction after the end of the long boom. David Harvey argued, in the best-selling *A Brief History of Neoliberalism* (2005), that the way in which nations responded to these problems was to pursue a project to restore class power.

Neoliberalism has also included a process of financialisation, where profits were increasingly generated in the financial sector (compared to commodity production), which reshaped social and political life (Adkins, 2018). Financialisation has resulted in households being absorbed into financial markets in new ways, such as through the speculation of retirement income and increasing household debt.

Neoliberalism has philosophical underpinnings, which inform how proponents understand society – particularly around the sovereignty of the market and individualised notions of freedom. However, ideas centred on property rights, the rights of individuals, and the concept that we are all equal before the law and engage in fair exchange in the marketplace can ignore fundamental inequalities (Whyte, 2019). Focussing on the individual in this way can neglect the historical and racialised context of colonialism, which was fundamental to the accumulation of wealth by advanced industrialised countries. Scholars point out that despite positing neoliberalism as a project that would expand freedom, the effect has been de-democratising (Brown, 2015). In this way, 'actually existing neoliberalism' can be a departure from the economic and philosophical ideas who are its supposed inspiration (Cahill, 2014).

More recently, especially since the 2008 global economic crisis and its aftermath, whether neoliberalism continues or has faded has been debated as recessions in many countries resulted in significant government spending. Alternatively, some argue we have entered a new period of populist or authoritarian neoliberalism. Others emphasise that new processes of 'neo-developmentalism' are not a return to old forms of economic management and that new approaches have failed to shed the core elements of neoliberal ideology. At the same time, some argue that the concept of neoliberalism is now used by analysts so broadly its usefulness has diminished.

Global project, many paths

Neoliberalism is a process of geographic variegation and involves 'systemic production of geoinstitutional differentiation' (Brenner et al., 2010: 184, emphasis in original). This diversity results from national or regional historical contexts, what policies were adopted, and how neoliberalism was implemented and resisted (i.e., the balance of social forces and class power in society). Sometimes neoliberalism was imposed by dictators in a context of extreme force, like in Chile under Augusto Pinochet in the 1970–1980s. On other occasions it was implemented democratically – but by directly confronting trade union power, like in the United Kingdom under Margaret Thatcher and in the US under Ronald Reagan in the 1980s. Occasionally it has been implemented more consensually – with the cooperation of labour unions in social pacts – like in Australia in 1980s and in Finland recently. Neoliberalism was implemented exceptionally rapidly in the post-Communist European countries after 1989, resulting in great hardship for residents.

National contexts to neoliberalism are critical, but they exist alongside global changes. Since the 1970s, globalisation has altered how capitalism operates, as national economic systems have become more interdependent through international financial and trade systems. While local decisions about neoliberal policy are central, the global context shapes those decisions. Similarly, there is no one neoliberal path and no fixed end point. At the same time, we can recognise similarities such as the necessary subduing of the social power of labour and trade unions.

The commonly accepted idea that neoliberalism originated in the Global North and was exported to the Global South has been challenged by Raewyn Connell and Nour Dados (2014). They argue that by positing neoliberalism as emanating from US and European examples, analysts place the Global North at the centre of the concept's development and eschew the specific and varied experience of the Global South. While they argue a single 'Southern narrative' is not possible, they contend there can be an 'enriched understanding of neoliberalism . . . when the social experience and intellectual production of the Global South are prioritised' (2014: 132). In examining neoliberalism as a global phenomenon, we need to be attentive to connection and difference across nations and regions and in whose story of neoliberalism we prioritise.

Although neoliberal ideas and policy become hegemonic amongst global institutions and governments, these changes were resisted in various ways – in particular by what was called the 'movement of movements' or the Global Justice Movement in the 1990s and 2000s (McNally, 2006). Large scale mobilisations and movements included: the indigenous Zapatista uprising in Mexico; the global *Via Campesina* peasants and farmers movement; the Bolivian Water Wars (1999–2000) against World Bank mandated privatisation; the Battle of Seattle against the World Trade Organisation in 1999 and the 200,000 strong Genoa protests against the G8 Summit. More recently, the Occupy movement, anti-austerity squares protests, and mobilisations such as the Spanish 15M movement – which were all prominent in 2011 and 2012 – have called for an alternative to neoliberalism.

References

Adkins L. (2018) *The Time of Money*. Stanford, California: Stanford University Press.

Birch K. (2015) Neoliberalism: The Whys and Wherefores . . . and Future Directions. *Sociology Compass* 9(7): 571–584.

Brenner N., Peck J. and Theodore N. (2010) Variegated Neoliberalisation: Geographies, Modalities, Pathways. *Global Networks* 10(2): 182–222.

Brown W. (2015) *Undoing the Demos: Neoliberalism's Stealth Revolution*. New York: Zone Books.

Cahill D. (2014) *The End of Laissez Faire? On the Durability of Embedded Neoliberalism*. Cheltenham: Edward Elgar.

Connell R. and Dados N. (2014) Where in the World Does Neoliberalism Come From? *Theory and Society* 43: 117–138.

Harvey D. (2005) *A Brief History of Neoliberalism*. Oxford: Oxford University Press.

Hirschman A.O. (2013) *The Essential Hirschman*. Adelman J. (ed.). Princeton: Princeton University Press.

McNally D. (2006) *Another World is Possible: Globalisation & Anti-Capitalism*. Winnipeg: Arbeiter Ring Publishing.

Whyte J. (2019) *The Morals of the Market: Human Rights and the Rise of Neoliberalism*. London: Verso.

Suggested further reading

Fine B. (2009) Neoliberalism as Financialisation. In: Yalman G.L. and Saad-Filho A. (eds.). *Economic Transitions to Neoliberalism in Middle-Income Countries: Policy Dilemmas, Economic Crises, Forms of Resistance*. London: Routledge, pp. xv, 251.

Foucault M. (2004) *The Birth of Biopolitics: Lectures at the Collège de France, 1978–79*. Hampshire: Palgrave Macmillan.

Gezmiş H. (2018) From Neoliberalism to Neo-Developmentalism? The Political Economy of Post-Crisis Argentina (2002–2015). *New Political Economy* 23(1): 66–87.

Jessop B. (2019) Authoritarian Neoliberalism: Periodization and Critique. *South Atlantic Quarterly* 118(2): 343–361.

Mirowski P. and Plehwe D. (2009) *The Road from Mont Pèlerin: The Making of the Neoliberal Thought Collective*. Cambridge, MA: Harvard University Press.

Audio-visual material

Barnard Centre for Research on Women: 'What is Neoliberalism?' https://www.youtube.com/watch?v=7kL4p3llmHk

PBS Website with Materials about the Documentary. *The Commanding Heights: The Battle for the World Economy*. http://www.pbs.org/wgbh/commandingheights/lo/index.html

Chapter 7

Development as freedom

Patricia Northover

Amartya Sen and the development imaginary

The discourse on 'Development' – a central motif of narratives of modernity – addresses a highly contentious problem field and offers neither stable signifiers nor much common ground on the nature, tendencies, ethics, and politics of *a process* of development in our contemporary world. Nevertheless, certain strands of thinking have been able to exercise a dominant influence on this discourse and have thus led the way in delineating the possibilities for imagining and even encountering Development Escobar (1995).

Development as Freedom (1999), the seminal text authored by the 1998 Nobel Laureate economist Amartya Kumar Sen, expresses one such emergent dominant narrative that has found a global resonance in the 21st century. For example, the revolutionary uprisings in the Middle East in 2011 and the #BlackLivesMatter global upsurge against racial violence and vulnerability within the context of the COVID-19 pandemic in 2020 rest on themes strongly embedded in the work of Sen – namely, aspirations for freedom, democracy, and social justice. In *Development as Freedom*, Sen argues for development as the expansion of the 'real freedoms' that people may enjoy and have reason to value. In the text, he rehearses his longstanding challenges to competing views on the meaning of social welfare and development and offers his work as a better analytical frame, than, for example, Utilitarian, Rawlsian, Marxist, or Hayekian Libertarian standpoints on development. Indeed, Sen posits that this framework of 'Development as Freedom' *should* act as a foundational and universal principle for all peoples in order to better facilitate the aspirations of the multifaceted forms of social contestation, protest, and resistance against inequality – from race and gender struggles for equal freedoms, to de-colonial and popular uprisings for local and global social justice.

Practically and politically, Sen's conceptual and methodological approaches for assessing and valuing human well-being and his message of 'Development as Freedom' have had an enormous impact on the Development industry. His work, for example, has helped to underpin the rise of new development indicators, such as the Human Development Index, the Human Poverty Index and the Gender Empowerment Index, championed and developed through the United Nations Development Program (UNDP) and promoted through their Human Development reports. Sen's work has also infused the theoretical study of gender and development, poverty, famine, social justice, and democracy. The idea of 'Development as Freedom', moreover, lends its weight to the transition to the Post Washington Consensus from the so-called Washington Consensus aid regime. The latter was a package of neo-liberal aid reforms focused on 'getting the prices right'

DOI: 10.4324/9780429282348-8

through a narrowly focused set of economic stabilisation and structural adjustment policies introduced, within the context of economic crises, to largely developing economies, and it intended to return these countries to economic growth and neo-liberal economic market equilibria. In the Post Washington consensus era – which focuses on a pro-poor and participatory approach to development, as well as on the institutions for 'good governance' – Sen's text, *Development as Freedom*, could easily become the new bible for guiding the current agenda for aid and development policy. Such is the power of this text.

'Development as capability expansion' – the road to *Development as Freedom*

Sen's influential work thus cuts across theoretical, philosophical, and practical development domains to espouse the powerful message of 'development as freedom'. This freedom is, in part, to be indexed by a capability approach to human development, where capabilities are more than skill sets but rather speak to 'valuable beings and doings' that agents have reason to value and, most critically, are free to choose. This focus on capabilities or 'capability freedom' is represented as the key departure from utilitarian thought, and neo-classical strands of economic thought steeped in this tradition, where the focus is placed on either subjective desires and mental states of being, such as happiness, pleasure, and desire fulfillment in the assessment of well-being. The capability approach to human development is also distinguished from early growth and modernisation theories for development and Rawls' theory of justice, since these other approaches to the development problematic tend to focus more on the commodities (that can be marshaled through a process of market driven growth) needed to support either consumption desires and needs or baskets of primary goods and resources. However, as Sen has continuously emphasised, income is not necessarily correlated with well-being; the latter being dependent on the *quality of life* enjoyed through a specific set of *valued agential states and activities* for *individual agents*.

In particular, for Sen, *human* well-being must rely on both a more objective and more personal criteria of welfare. In order to achieve this he draws on the Aristotelian notion of 'functionings'. Functionings describe various doings and beings that a person actually experiences or realizes. A specific functioning vector indicates which *actual* states of doing (e.g., reading or eating) and being (for example, being well-nourished, being literate, being part of a community, and appearing without shame) have been realised. As Sen emphasises, functionings can thus vary from elementary states to more complex ones. However, Sen has been reluctant, as Alkire (2002: 29) and Nussbaum (2003) have highlighted, to establish a list of human capability priorities or even to systematically defend a set of basic functionings or capabilities aligned with basic needs (Alkire, 2002: 157). This allows him to side-step the thorny issue of judging amongst valuable states of being and doing but leaves his approach open to charges of policy impracticability, since no clear guidelines beyond an appeal for dialogue and participation are offered. Nevertheless, there exists a growing field of application of the capability approach in vital areas such as health and disability studies (Ruger and Mitra, 2015; Parey, 2022).

Sen's emphasis on Aristotelian functionings as the foundation for assessing well-being is, however, deliberately (and controversially) set apart from a 'basic needs' strategy for development (Alkire, 2002: 166–174). Thus, rather than encouraging a focus on resources for needs to be met, attention is instead to be placed on a person's '*capability*' defined as 'a set of vectors of functioning, reflecting a person's freedom to live one type of life or another' (Sen, 1992: 40). Capabilities reflect then 'effective possibilities'; they describe what people *could* do and be *even if* they are not actually chosen. Paying attention to human capabilities, or a person's capability set, helps one to distinguish between the un-freedom and deprivation of a starving child and the liberty of the fasting monk, since despite the similar states of experienced functioning, (i.e., a starving hunger), they reflect different states of capability and relatedly the presence/absence of choice.

A capability approach, however, also allows one to better and more equitably attune resources to support desired functioning vectors, as simply allocating the same basket of primary goods or resources to heterogeneous persons would not be equivalent to supporting equal well-being. This is because persons will differ in their ability to translate resources into functionings. The pregnant woman and disabled person would thus need different stocks of resources to attain their valued functionings, in comparison to the able-bodied man. Sen's approach thus accommodates a 'basic capabilities' rather than a 'basic needs' approach to well-being (Robeyns, 2017). Sen moreover sharply illustrates the critical power of this capability approach to human development when he demonstrates that there are around 60–100 million *missing women* in the world due to systematic gender discrimination (sex-selective abortion, neglect for female health and nutrition, especially during early childhood) resulting in severe capability deprivation for women (Sen, 1999: 104–107).

Agency and power in *Development as Freedom*

In elaborating upon his ideas for promoting human and social well-being which culminated in the text *Development as Freedom,* Sen is also keen to establish what he considers to be foundational analytical distinctions in how we may approach an understanding of well-being *through a valuing of individual agency.* He sets out his four elementary concepts as: well-being achievement, well-being freedom, agency achievement, and agency freedom, which are themselves derivative from the ideas of the 'well-being aspect' and the 'agency aspect' of a person (Sen, 1992: 56–57; Alkire, 2002: 9).

For well-being achievement, one looks *only* at the *actual states of valued functionings* that one realises, while for well-being freedom one assesses the *ontological* or *real context of power* in which the achievement *regarding one's own well-being* is carried out. In other words, in assessing a state of well-being freedom, the relevant question to ask would be: was the agent operating in a *context of freedom*, so that their choice reflects a real power that the agent can do, and could have done otherwise? If so, we have well-being freedom, or, a 'real opportunity to accomplish what we value' as wellness of being (Sen, 1992: 31). Consider here, again, the distinction between the starving child and the fasting monk; the monk has exercised a well-being freedom whereas the child has exercised no such freedom, despite the similarity of well-being states. For grasping well-being freedom, one has to analyse then *well-being per se*, as well as *agency* – the ability of an agent to pursue and bring about certain goals and finally the real *context of power* that actions are embedded in and dependent on.

In contrast to well-being achievement, agency achievement is to be assessed on a wider set of objectives than *personal* well-being and so can include other goals, such as seeking the independence of one's or another's country, that may conflict with *individual welfare.* In addition, like well-being freedom, agency freedom, which Sen defines as 'one's ability to promote goals that one has reason to promote' (1992: 60), is also to be assessed by referring to the *context of power* in which one's agency is exercised. In this case, however, the context of power, needed here for agency freedom to be recognised as something distinct from mere agency achievements, seems synonymous with the libertarian view of human freedom as a space of liberty. Agency freedom appears then to be calibrated by a liberal context that affords both negative liberty – absence from interference – and positive liberty, *one's own power* to achieve a desired goal or end state. This affinity between agency freedom and liberty reinforces Sen's persistent promotion of libertarian political ideals; ideals that he seeks to ground in something more than just a focus on rights, by including a concern for more objective moral considerations and consequences, such as individual advantage or well-being and the inequalities in the space of 'capability freedoms'. This is an important step as it allows Sen to avoid the complaint that neoliberal libertarian priorities on rights ignores substantive inequalities in the command over resources and feasible well-being achievements. As Sen highlights in his discussion of famines, a concern with negative rights alone is insufficient to prevent human catastrophes (1999: 65–66). Nonetheless, Sen remains firmly within methodological individualist traditions in his analysis of famines and shies away from underscoring the role of structural dynamics of power.

In line with these conceptual and philosophical orientations, Sen thus emphasises what he refers to as instrumental freedoms, five in all, that are vital to the project of promoting *Development as Freedom*. These instrumental freedoms are i) political freedoms, ii) economic facilities, iii) social opportunities, iv) transparency guarantees, and v) protective security (Sen, 1999: 10, 38–40). These are conceived of as 'distinct sets of rights and opportunities' that help to 'advance the general capability of a person' (1999: 10). All in all then, Sen advances his thesis of *Development as Freedom* by advocating that the expansion of freedom is not only i) the primary end but also ii) the principal means of development (36). Development thus 'consists in the removal of various types of unfreedoms that leave people with little choice and little opportunity to exercise their reasoned agency' (1999: xii).

Critiques and limit points

Sen insists that the impulse to freedom and for freedom is not a peculiarly Western tradition, which is being copied by other peoples across time and space. And while Sen wishes to distance himself from the previous standard of 'Development as Modernization' that informed programs of development especially in the post-war period, he still seems rather wedded to the institutional scaffolding of Western liberal democracy and its plural economic freedoms entrenched in a capitalist market society. Yet, is Sen correct in making an empirical claim that Development is invariably twinned with Liberal Freedoms in a capitalist market society? Sen seems to think so and argues the case based, in part, on his examination of the correlation he detects between famines and democratic freedom, and in particular the presence of a 'free press' (Sen, 1999: 16, chapter 7). In contrast, others such as Corbridge (2002), Selwyn (2011), and Northover (2012), arguing from rather different perspectives, suggest that development and freedom may *not* be so well linked.

Why not? Well, first, as Corbridge (2002) highlights, because the evidence suggests that substantive well-being and structural transformations are often made under authoritarian systems – such as those in East Asia, with China being the most striking case in point. Should such transformational experiences concomitant with the expansion of opportunities *not count* as development? Should we *devalue* these development trajectories in so far as they have hinged on the presence of what is now commonly referred to now as the 'developmental state'? That is, a state that has the capacity to exercise its hegemonic authority to resolve and manage conflicts and direct the use of resources to promote broad development objectives.

Second, from the Marxian standpoint as embraced by Selwyn (2011), it may be argued that under capitalism one can only maintain a façade of liberty or insecure substantive freedoms in the context of commodity fetishism, labour exploitation, and alienation, given the dialectical contradictions shaping the social relations of production. As such capitalism itself is seen as a real paradoxical force that undermines the real freedoms possible in the present system of liberties due to the nature of capitalist relations of power. Indeed, rather than being linked to the absence of democratic institutions, famines are more associated with the historical conditions of structural inequality attendant with the real dis-possessions needed for the creation and continued reproduction of a class of wage-laborers with the hegemonic rise of a capitalist world system.

Finally, from a post-structuralist inflected critique of Sen, Northover (2012) posits that Sen's liberal humanist thesis of 'Development as Freedom' has been politically and historically conditioned by a 'racial philosophy of place'. In particular, she argues that imaginaries of 'agency freedom' have been sustained through a politics of abjection that leaves modern freedoms dependent on an 'abject blackness'; that is, on processes of othering, disavowal, and natal alienation in social and political relations. This politics of abjection is evident *not only* in the links between capitalism and slavery – where property rights and self-determination were determined by racial and gendered signifiers, *but also* in the substantive 'afterlife of slavery' – where blackness remains marked by 'social death' and dying, given complex racialisation processes (Barchiesi and Jackson, 2019). This racially differentiated risk of death given ongoing states of insidious capability inequalities, violent deprivations, and precarious freedoms has been tragically highlighted by the COVID-19 pandemic (Liebman et al., 2020).

From this last standpoint then, more attention needs to be placed on the *intrinsic relations between freedom and unfreedom in liberal humanism's development processes* in order to speak to the racial power effects shaping both labour and market processes (Tilly and Shilliam, 2018), as well as new existential threats, such as climate change and migration vulnerabilities (Gonzalez, 2021). It is not surprising then, in the context of the intersecting power relations among race, gender, and class that the message coming from some feminist critiques is that Sen has placed an undue emphasis on the abstract qualities of freedom. For example, as Des Gasper and van Stavaren (2003) have argued a focus on abstract qualities such as 'freedom of the people' and 'agency freedom' – Liberalism's prized liberties – is hard to justify in the face of the ongoing, growing and complex contradictions troubling modern 'freedom's charms'.

In the end, although Sen's message to the world and 'development community' is generally very attractive – especially in the present time of post-colonial and contemporary world crises – his message may ring hollow for the many who see Development twinned with violence and a new imperial order. Perhaps then, despite Sen's intense and prolific efforts in popularizing this current narrative of Development, his *Development as Freedom* may not hold the key to a brave new world, given the problematic history of freedom and its ongoing violent contexts of power.

Guide to further reading

For a collection of Sen's foundational works on economic development issues, see Sen, A. (1987) *Resources, Values and Development*, Cambridge, Massachusetts: Harvard University Press. This collection gathers his path breaking writings on institutions, social investment, ethics and well-being. For a selection of critical perspectives on Sen and his relevance to feminist economics, see the (2003) special issue of *Feminist Economics*, edited by Bina Agarwal, Jane Humphries and Ingrid Robeyns. For a useful collection of readings on the Human Development and Capability Approach, largely based on the Human Development Reports of the last decade, see Fukuda-Parr, S. and A. Kumar (eds.) (2005) *Readings in Human Development: Concepts, Measures and Policies for a Development Paradigm*, New York and Oxford: Oxford University Press. Finally, for an analytical 'helicopter overview' of the Capabilities approach, see Robeyns, I. (2017) *Wellbeing, Freedom and Social Justice: The Capability Approach Re-Examined*, Cambridge, UK: Open Book Publishers.

Useful website

http://www.ophi.org.uk/: The Oxford Poverty and Human Development Initiative (OPHI), established in 2007, is an economic research centre within the Oxford Department of International Development, University of Oxford.

References

Alkire, S. (2002) *Valuing Freedoms: Sen's Capability Approach and Poverty Reduction*, Oxford: Oxford University Press.

Barchiesi, F. and S. Jackson (eds.) (2019) 'Blackness and Labor in the Afterlives of Racial Slavery', Introduction, *International Labor and Working Class History*, (96), pp. 1–16.

Corbridge, S. (2002) 'Development as Freedom: The Spaces of Amartya Sen', *Progress in Development Studies*, 2 (3), pp. 183–217.

Escobar, A. (1995) *Encountering Development: The Making and Unmaking of the Third World*, Princeton, NJ: Princeton University Press.

Gasper, D. and I. van Staveren (2003) 'Development as Freedom and as What Else?', *Feminist Economics*, 9 (2–3), pp. 137–161.

Gonzalez, C. G. (2021) 'Racial Capitalism, Climate Justice, and Climate Displacement', *Oñati Socio-Legal Series*, 11 (1), pp. 108–147. https://opo.iisj.net/index.php/osls/article/view/1214

Liebman, A., K. Rhiney and R. Wallace (2020) 'To Die a Thousand Deaths: COVID-19, Racial Capitalism and Anti-Blackness', *Human Geography*, (0), pp. 1–5.

Northover, P. (2012) 'Abject Blackness, Hauntologies of Development, and the Demand for Authenticity: A Critique of Sen's Development as Freedom', *Global South*, 6 (1), pp. 66–86.

Nussbaum, M. (2003) 'Capabilities as Fundamental Entitlements: Sen and Social Justice', *Feminist Economics*, 9 (2–3), pp. 33–59.

Parey, B. (2022) 'Well-Being Among Older Persons with Disabilities in Trinidad', *Health & Social Care in the Community*, 30, pp. e1101–e1111. https://doi.org/10.1111/hsc.13518

Robeyns, I. (2017) *Wellbeing, Freedom and Social Justice: The Capability Approach Re-Examined*, Cambridge, UK: Open Book Publishers.

Ruger, J. P. and S. Mitra (2015) 'Health, Disability and the Capability Approach: An Introduction', *Journal of Human Development and Capabilities*, 16 (4), pp. 473–482.

Selwyn, B. (2011) 'Liberty Limited? A Sympathetic Re-Engagement with Amartya Sen's Development as Freedom', *Economic and Political Weekly*, 46 (10), pp. 68–76.

Sen, A. (1992) *Inequality Reexamined*, Cambridge, Massachusetts: Harvard University Press.

Sen, A. (1999) *Development as Freedom*, New York: Anchor Books.

Tilly, L. and R. Shilliam (2018) 'Raced Markets: An Introduction', *New Political Economy*, 23 (5), pp. 534–543. https://doi.org/10.1080/13563467.2017.1417366

Chapter 8

Postcolonialism

Cheryl McEwan

What is postcolonialism?

Postcolonialism is a difficult and contested term not least because it is far from clear that colonialism has been relegated to the past. Its meaning is not limited to 'after-colonialism' but refers to ways of criticizing the material and discursive legacies of colonialism (Radcliffe, 1999: 84). Broadly speaking, therefore, postcolonial perspectives can be said to be *anti-colonial*. They have become increasingly important across a range of disciplines in recent decades.

A number of core issues underpin postcolonial approaches. First, they stress the need to destabilize the dominant discourses (e.g., history, philosophy, linguistics, and 'development'), which are unconsciously ethnocentric, rooted in European cultures and reflective of a dominant Western worldview. Postcolonialism problematizes the very ways in which the world is known, challenging the unacknowledged and unexamined assumptions at the heart of Western disciplines that are profoundly insensitive to the meanings, values, and practices of other cultures.

Second, postcolonial critiques challenge the experiences of speaking and writing by which dominant discourses come into being. For example, a term such as 'the Third World' homogenizes peoples and countries and carries other associations – economic backwardness, the failure to develop economic and political order, and connotations of binaries between 'us' and 'them', 'self' and 'other' – which are often inscribed in development writings. These practices of naming are not innocent. Rather they are part of the process of 'worlding' (Spivak, 1990) or setting apart certain parts of the world from others. Said (1978) has shown how knowledge is a form of power and, by implication, violence; it gives authority to the possessor of knowledge. Knowledge has been – and to large extent still is – controlled and produced in 'the West'. Global economic power might be starting to shift, but the power to name, represent, and theorize is still located in 'the West'. Postcolonialism seeks to disrupt this power.

Third, postcolonialism critiques the spatial metaphors and temporality employed in Western discourses. Whereas previous designations of the Third World signalled both spatial and temporal distance – 'out there' and 'back there' – a postcolonial perspective insists that the 'other' world is 'in here'. The Global South is integral to what 'the West' refers to as 'modernity' and 'progress'. From the beginnings of colonialism it has contributed directly to the economic wealth of Western countries through its labour and economic exploitation. In addition, the modalities and aesthetics of the Global South have partially constituted Western languages and cultures. Postcolonialism, therefore,

DOI: 10.4324/9780429282348-9

attempts to rewrite the hegemonic accounting of time (history) and the spatial distribution of knowledge (power) that constructs the 'Third World'.

Finally, postcolonialism attempts to recover the lost historical and contemporary voices of marginalised and oppressed peoples through a radical reconstruction of history and knowledge production. Postcolonial theory has developed this radical edge through the works of scholars such as Spivak, Said, and Escobar who, in various ways, have sought to recover the agency and resistance of peoples subjugated by both colonialism and neo-colonialism.

These core issues form the fabric of the diverse field of inquiry of postcolonial studies, based in the historical fact of European colonialism and the diverse material effects to which this phenomenon has given rise.

Postcolonialism and development

The possibility of producing decolonized, postcolonial knowledge in development studies became a subject of debate during the 1990s, culminating in dialogue that continues today. In theoretical terms, postcolonialism has been greatly influenced by Marxism and poststructuralism, drawing on the political-economy approaches of the former and the cultural and linguistic analyses of the latter. The politics of postcolonialism diverge sharply from other approaches and, although it shares similarities with dependency theories, its radicalism rejects established agendas and accustomed ways of seeing. Consequently, postcolonialism is a powerful critique of 'development' and an important challenge to dominant ways of apprehending North–South relations.

Critiquing discourses of development

Postcolonialism challenges the very meaning of development as rooted in colonial discourse depicting the North as advanced and progressive and the South as backward, degenerate and primitive. Early postcolonial writers, such as van der Post, challenged this assumption by referring to hunter-gatherers as the first affluent peoples and postcolonial theory has prompted questions about whether the pursuit of capitalism, with its emphasis on individual wealth and incorporation into the global economy, is more advantageous to peoples of the South than indigenous systems of equity, reciprocity, and communalism. The superiority of modern industrialization and technological progress is increasingly questioned by alternative knowledges reshaping perceptions of non-Western societies and their environments.

Critics argue that to subject development to postcolonial critique is a form of intellectual faddism; while ever there are pressing material issues such as poverty in the world, concerns with the *language* of development are esoteric. However, language is fundamental to the way we order, understand, intervene and justify those interventions (Escobar, 1995). Postcolonialism offers new ways of understanding what development is and does and why it is so difficult to think beyond it. The texts of development are written in a representational language – metaphors, images, allusion, fantasy, and rhetoric – the imagined worlds bearing little resemblance to the real world (Escobar, 1995). Development writing often produces and reproduces misrepresentation. Postcolonial approaches seek to remove negative stereotypes about people and places from such discourses. They challenges us to rethink categories such as 'Third World' and 'Third World women' and to understand how location, economic role, social dimensions of identity, and the global political economy differentiate between groups and their opportunities for development.

The texts of development are strategic and tactical, promoting and justifying certain interventions and delegitimizing and excluding others (*ibid.*). Power relations are clearly implied in this process – certain forms of knowledge are dominant, others are excluded, and it is important to ask whose knowledge is excluded and why. Ideas about development are not produced in a social, institutional, or literary vacuum and a postcolonial approach to development literature can reveal the apparatuses of power and domination in which those texts are produced, circulated, and consumed. Development discourse promotes and justifies real interventions with material consequences. It is, therefore, imperative to explore the links between the words, practices, and institutional expressions of development and between the relations of power that order the world and the words and images that represent the world. By doing so, postcolonial approaches can effect change.

Agency in development

Postcolonial approaches challenge the notion of a single path to development and demand acknowledgement of a diversity of perspectives and priorities. The politics of defining and satisfying needs is a crucial dimension of current development thought, to which the concept of agency is central. Postcolonial approaches question who voices the development concern, what power relations are played out, how participants' identities and structural roles in local and global societies shape their priorities, and which voices are excluded. They attempt to overcome inequality by opening spaces for the enactment of agency by non-Western peoples, despite poverty and a lack of access to technology making this increasingly difficult.

Postcolonial critiques also question authorization and authority in development. By what right and on whose authority does a development 'expert' claim to speak on behalf of others? On whose terms is space created in which Southern actors are allowed to speak? Are we merely trying to incorporate and subsume Southern voices into our own canons? It is no longer feasible to represent the peoples of the Global South as passive, helpless victims. They are making their voices heard and their ideas are increasingly being incorporated into grassroots and even mainstream development policies. Within gender and development, they have forced a move away from totalizing discourses and a singular feminism (based upon the vantage point of white, middle-class Western feminists that failed to acknowledge differences between women) towards the creation of spaces where the voices of black women and women from the South can be heard (McEwan, 2001; Mohanty, 2002). Postcolonial feminisms allow for competing and disparate voices among women, rather than reproducing colonial power relations in which knowledge is produced and received in the West, and white, middle-class women have the power to speak for their 'silenced sisters' in the South.

New dialogues and approaches in development

One of the major criticisms of postcolonialism has been that it is too theoretical and divorced from material concerns: its emphasis on discourse detracts from the material ways in which colonial power relations persist and it is ignorant of the real problems characterising everyday life in the Global South. However, recent work at the interface of postcolonialism and development refutes these criticisms. Postcolonial critiques of economic development challenge its amnesia about (neo)colonialism and question its blind loyalty to scientific progress and universal economic prescriptions (Kapoor, 2008; McEwan, 2018). Fundamental questions are being asked about how capitalism reproduces inequality in the name of development and how it is that the deepening of capitalist social relations comes to be understood *as* development (Wainwright, 2008). Postcolonialism does not concede the space of materiality – the provisioning of livelihoods, tangible constraints on life, relations of production and distribution – to economics (Pollard et al., 2011). Rather, it suggests radically different ways of understanding and responding to these issues. Economists, economic models and dominant/orthodox notions of development erase the richness of human agency and experience in response to economic and other crises through their drive to produce 'development aggregates', which then often fail to produce adequate responses to these crises. In contrast, postcolonial approaches emphasise the need to understand development through the eyes of local people who are making daily livelihood decisions in situations of poverty, conflict, uncertainty, ambivalence, hope and resistance (Sylvester, 2011). This different approach is producing innovation in the sources and methods used within development studies, including using novels (Sylvester, 2011) and poetry (Madge and Eshun, 2013) to understand the thoughts and actions of those subject to development interventions.

Criticisms of postcolonialism have crystallized recently in the emergence of decolonial theory which, while sharing common ground with postcolonial thinking, differs significantly. Decolonization as a material concern (e.g., the return to indigenous peoples sovereignty over the lands from which they were dispossessed) does not equate with the more metaphorical concerns of postcolonial theory to decolonize knowledge (Tuck and Yang, 2012). Despite this, engaging postcolonialism with ideas of development and with decolonial theory from Global South contexts is important in ensuring its relevance in critical theories that seek alternatives to social, economic and political injustice. Postcolonial theory enables those marginalised by relationships of power the means to narrate their own lives, and it demands that those positioned more favourably within relationships of power be prepared to listen and learn. This provides new imaginings of what has previously been taken for granted, including development. It is a different political project to decolonization, but may provide a foundation for a deeper engagement with decolonization.

Postcolonial thinking is a significant advancement in development studies. It demonstrates how the production of Western knowledge forms is inseparable from the exercise of Western power. It also attempts to loosen the power of Western knowledge and reassert the value of alternative experiences and ways of knowing (wa Thiong'o, 1986). It articulates some difficult questions about writing the history of 'development', about neo-colonial representations and discourses concerning 'the Third World' and about the institutional practices of development itself. It has the potential to turn critique of conventional development into productive 're-learning to see and reassess the reality' of the Global South (Escobar, 2001: 153). It has been an important stimulus to alternative formulations such as 'indigenous' and 'alternative modernities' and rights-based approaches to development. And, precisely because of their divergent traditions, dialogue between postcolonialism and development studies offers new ways of conceptualizing and doing development (Sylvester 2006).

Postcolonialism has an expansive understanding of the potentialities of agency. It shares a social optimism with other critiques, such as those concerning gender and sexuality in Western countries that have generated substantial changes in political practice. Dialogue between postcolonial and development studies have the potential to engage postcolonial theory in considering questions of inequality of power and control of resources, climate change, human rights, global exploitation of labour, child prostitution and genocide, helping to translate the postcolonial theory into action on the ground and a means of tackling the power imbalances between North and South. They might also inspire critical reshaping of postcolonial futures and counter new forms of Orientalism that continue to disadvantage the developing world. The challenge remains, as Simon (2006) argues, to link postcolonial concerns with local identities, practices and agendas to broader campaigns and projects for progressive and radical change that are substantively decolonial and critically developmental. This is beginning to emerge with new North-South alliances, critical analysis of the role of agencies and institutions and experiments with new economic forms such as guaranteed minimum incomes. Despite the seeming impossibility of transforming North-South relations by the politics of difference and agency alone, postcolonialism is a much-needed corrective to the Eurocentrism and conservatism of much writing on development. It is playing an important role in re-imagining critical development studies, generating new dialogue and action, and opening possibilities for Global North countries learning from the Global South. Through its focus on the politics of knowledge production and problematizing power relations between different actors engaged in the development nexus, it is also reviving within development studies a longstanding concern with the moral imperatives underpinning development research, the ethics of research and an ethos of solidarity with others (McEwan, 2018).

References

Escobar, A. (1995) 'Imagining a post-development era', in J. Crush (ed.) *Power of Development*, London: Routledge, pp. 211–227.

Escobar, A. (2001) 'Culture sits in places: Reflections on globalism and subaltern strategies of localization', *Political Geography*, 20: 139–174.

Kapoor, I. (2008) *The Postcolonial Politics of Development*, London: Routledge.

Madge, C. and Eshun, G. (2013) ' "Now let me share this with you": Exploring poetry for postcolonial geography research', *Antipode* (forthcoming).

McEwan, C. (2001) 'Postcolonialism, feminism and development: Intersections and dilemmas', *Progress in Developing Studies*, 1(2): 93–111.

McEwan, C. (2018) *Postcolonialism, Decoloniality and Development*, London: Routledge. A comprehensive account of the significance of postcolonial theory within development theory and practice.

Mohanty, C. (2002) ' "Under Western eyes" revisited: Feminist solidarity through anticapitalist struggles', *Signs*, 28(2): 499–535.

Pollard, J., McEwan, C. and Hughes, A. (eds.) (2011) *Postcolonial Economies*, London: Zed.

Radcliffe, S. (1999) 'Re-thinking development', in P. Cloke, P. Crang and M. Goodwin (eds.) *Introducing Human Geographies*, London: Arnold, pp. 84–91.

Said, E. (1978) *Orientalism*, London: Routledge and Kegan Paul.

Simon, D. (2006) 'Separated by common ground? Bringing (post)development and (post)colonialism together', *The Geographical Journal*, 172(1): 10–21.

Spivak, G. (1990) *The Postcolonial Critic: Interviews, Strategies, Dialogue*, London: Routledge.

Sylvester, C. (2006) 'Bare life as a development/postcolonial problematic', *The Geographical Journal*, 172(1): 66–77.

Sylvester, C. (2011) 'Development and postcolonial takes on biopolitics and economy', in J. Pollard et al. (eds.) *Postcolonial Economies*, London: Zed, pp. 185–204.

Tuck, E. and Yang, W. (2012) 'Decolonization is not a metaphor', *Decolonization: Indigeneity, Education, & Society*, 1(1): 1–40.

wa Thiong'o, N. (1986) *Decolonising the Mind*, London: James Curry.

Wainwright, J. (2008) *Decolonizing Development*, Oxford: Blackwell.

Additional resources

Comaroff, J. and Comaroff, J.L. (2012) 'Theory from the South: Or, how Euro-America is evolving toward Africa', *Anthropological Forum*, 22(2): 113–131.

A provocative article that challenges the Eurocentric idea of linear development, argues that the economies of the Global North are becoming like those of the Global South and opens the possibilities of the Global South as a site from which development theory emerges.

Lewis, D., Rodgers, D. and Woolcock, M. (2005) *The Fiction of Development: Knowledge, Authority and Representation*, London: LSE Research Online. http://eprints.lse.ac.uk/archive/00000379/

A concise paper that extends this chapter's argument about the power of representation in development. It discusses how novels from around the world can provide different representations of the Global South. The list of recommended reading on literary works (in English) on development is a highly recommended resource.

Mbembe, A. (2016) 'Africa in the New Century', *Africa is a Country*, June 29. https://africasacountry.com/2016/06/africa-in-the-new-century/

A short, accessible account from a foremost African philosopher and scholar of Afro-futurism, which challenges Eurocentric ideas of Africa's 'backwardness' by arguing that the continent will play a key role in 21st-century global affairs. The argument informs postcolonial critiques of representation in development and asserts the agency of African people.

An introduction to the core ideas of a research collective on decoloniality, organized by Walter Mignolo and Arturo Escobar, which brings together scholars of Latin American/European origin working in universities in the United States and Latin America and interested in ideas of dependency theory, colonialism, gender, and critical theory. https://globalsocialtheory.org/topics/decoloniality/

An excellent Ted Talk in which Nigerian novelist Chimamanda Adichie tells the story of how she found her authentic cultural voice and warns that, if we hear only a single story about another person or country, we risk a critical misunderstanding. https://www.ted.com/talks/chimamanda_adichie_the_danger_of_a_single_story

Chapter 9

Postmodernism and development

David Simon

Postmodernism: panacea, placebo, or perversity?

Postmodernism became a major social scientific theoretical paradigm during the 1980s and 1990s, although its prominence has now waned. In development studies it gained prominence as one of the routes for transcending the so-called theoretical impasse that emerged in the mid- to late 1980s. However, the concept assumed diverse meanings, a factor contributing substantially to the often heated but unenlightening debates over its usefulness in the context of development.

The raft of new development textbooks appearing since the late 1990s, when such debate was at its zenith, have devoted surprisingly little attention to postmodernism and other post-structural perspectives. Some made no mention of it and other 'post-' or 'anti-developmentalist' approaches at all, while others included only a few pages or a single chapter, almost as an afterthought. Very few give fuller coverage – the revised edition of *Encountering Development* (Escobar, 2012) and third edition of *Theories of Development* (Peet and Hardwick, 2015) being welcome late exceptions – with the result that most current students continue to have little exposure to these debates.

Postmodernism first emerged in art, architecture and literature in the mid-1970s. The concepts of ideals, absolutes, order and harmonization, which had given rise to increasing alienation of the individual, were challenged, and the objective became a celebration of diverse forms and sharp contrasts, in order to rupture conventional expectations. This is generally achieved through the juxtaposition of radically different styles in street façades.

In Latin America, writers like Gabriel García Márquez and Carlos Fuentes pioneered a literary style that broke with the established tradition of a single, chronological flow to novels and replaced it with multiple, cross-cutting strands, flashbacks, forward leaps and previews in structurally much more complex forms. It won the authors prizes but has not proven a durable literary form.

In the social sciences, postmodernism gained a foothold as part of the ferment in discourse that included post-structuralist rejection of modernist meta-theories and grand narratives of a single mode of explanation or 'truth'. Ahluwalia (2010) demonstrates that post-structuralism derives from a complex blend of colonial, anti-colonial and post-colonial roots, strongly linked to the Algerian liberation struggle. The work of Michel Foucault, Henri Lefebvre, Jacques Derrida, Jean Baudrillard and Frederic Jameson looms large in the foundations of postmodernism. Among the most widely reputed social scientific writers on postmodernism are Jean-François Lyotard, David Harvey (1989) and Ed Soja. They situate it explicitly within the (cultural) logic of late capitalism, as part of the

DOI: 10.4324/9780429282348-10

search for profit accumulation in a context of globalized production and consumption. As such they are critical and see it as having limited social explanatory value, serving mainly to justify conspicuous self-expression, rather than representing a profound new paradigm.

Significantly, several leading advocates of postmodernism and postcolonialism in cultural studies, sociology and allied disciplines, including Homi Bhabha, Trin Minha, Gyatri Spivak and the late Edward Said, hail from the Global South, even though generally later working in Northern universities. Among geographers, sociologists and development specialists, especially those working in Latin America, some of the most trenchant critics of conventional development espoused postmodernism as the way forward during the 1990s – in particular, Santiago Colás, Arturo Escobar, Gustavo Esteva and David Slater. Colás (1994) interprets postmodern developments in different spheres of Argentinian society, while Esteva and Prakash (1998) provide one of the very few detailed expositions of regional and local-level postmodernism in practice as social action. Slater (1992, 1997) advanced geopolitical and development debates across the Global North–South divide.

Nevertheless, most social scientists working in – or concerned with – poor regions of the world have tended to ignore postmodernism or to dismiss it as irrelevant on one or both of the following grounds:

- Postmodernism literally means 'after the modern'; however, in the Global South, the majority of people are still poor and struggling to meet basic needs and to enjoy the fruits of modernization so powerfully held up to them as the outcome of development. In such situations, modernity has yet to be widely achieved, so that which follows on from the modern can have little relevance.
- Postmodernism was merely a temporarily fashionable Northern paradigm, which found expression mainly in aesthetic/architectural terms and as playful, leisured heterodoxes and new forms of consumption centered on individualism which can best be described as self-indulgence by the well-off. Such preoccupations are seen as irrelevant to the Global South, if not actually harmful in terms of distracting attention from the urgent priorities of the poor majority, namely survival, the ability to meet basic needs and related 'development' agendas, as well as the broader structural forces and processes that impact upon them.

A conceptual schema

The following typology distinguishes the different connotations of postmodernism/-ity in order to facilitate understanding. Postmodernity describes the 'condition' or manifestation, while postmodernism is the ideology or intellectual practice. At least three broad interpretations of the postmodern can be distinguished in the vast and multidisciplinary literatures:

The chronological approach

This is the most literal interpretation, in terms of which the postmodern necessarily follows the modern. In practice, however, no clean break between eras can be distinguished: there was no dramatic event to act as signifier, and there has been no agreement on the basis of transition. At best, one might be able to conceptualize a transitional phase of some years' duration.

In terms of globalization and mass consumption, for instance, the traditional mass-market air package holiday would be modern, whereas the more differentiated and personalized small-group luxury tour, complete with ecotourist credentials and/or sanitized versions of conflictual local histories in distant countries for the benefit of international tourists, might be conceived of as postmodern.

The aesthetic approach

The second basic understanding of postmodernism is as a form of expression in the creative and aesthetic disciplines like art, architecture and literature. This perspective reflects the considerations – and is exemplified by the authors and buildings – cited in the introduction. Inevitably, perhaps, most such attention has been centered on elite and middle-class consumption, especially in terms of leisure activities but also increasingly in the working environment and public spaces. Terms most frequently associated with this movement include pastiche, mélange, playfulness, commodity-signs, imaginaries and spectacles. Theme parks, pleasure domes and other purpose built leisure complexes which offer decontextualized time–space representations of various places and experiences (Watson and Gibson, 1995), often in sanitized form, are characteristic of this approach in much the same way as great exhibitions of global exploration, scientific discoveries and industrial achievements were hallmarks of Victorian modernity.

Postmodernism as intellectual practice

The approach of postmodernism as *problematique* or intellectual practice is the most relevant from the perspective of development studies. Here, the postmodern supposedly represents fundamentally different ways of seeing, knowing and representing the world. The modern approach, rooted in Enlightenment thinking about rationality, is concerned with a search for universal truths, linked to positivist scientific methodology and neo-classical economics. Such universalizing, globalizing approaches are referred to as meta-narratives.

Postmodern practice rejects such singular explanations in favor of multiple, divergent and overlapping interpretations and views, in terms of which these different accounts are all accorded legitimacy. The privileging of official and formal discourses should be replaced by approaches lending credence to both the official and unofficial, formal and informal, dominant and subordinate, central and marginal groups and to their discourses and agendas. Top-down development, so closely associated with official national and international agendas of modernization, has been discredited over a long period; instead bottom-up approaches or some hybrid of the two should be encouraged.

Hence, postmodernism represents a potentially fruitful approach for addressing the conflictual and divergent agendas of social groups, be it in relation to access to productive resources and/or the bases for accumulating social power, mediating the impacts of large development schemes, evolving complementary medical services that harness the most appropriate elements from both Western and indigenous systems or addressing long-standing conflicts between statutory and customary legal systems (Esteva and Prakash, 1998; Simon, 1998). Empowerment of the poor and powerless should be the objective.

Such discourses had much in common with earlier, liberal pluralism, basic needs and grassroots development paradigms, although the emphasis on co-existence and multiple modes of explanation is different. Equally, there are considerable areas of overlap with some strands of postcolonialism, which is centrally concerned with the cultural politics and identities of previously subordinated groups. Extreme postmodernism or other form of relativism means that everyone's views are equally valid; without some decision-making rules, any social action not gaining unanimity or consensus becomes impossible.

Accordingly and because of the emphasis by some authors on playful, leisured self-fulfillment, postmodernism was sometimes criticized as a conservative ideology embedded within late capitalism – and hence rejected. Somewhat prematurely, Ley (2003) wrote its epitaph because of its perceived fad status and because of problems operationalizing it in meaningful distinction from the modern, meaning recent or present-day. Postmodernism has lost much of its social scientific prominence since the early 2000s in favor of the closely related umbrella term of postcolonialism, but it contained much of value, especially in terms of understanding contemporary social dynamics and complexities. Gabardi's (2001) articulation of 'critical postmodernism' 'not only as a product of the modern-postmodern debate, but also as a theoretical and ideological response to our current late modern/postmodern transition and a practical tool for negotiating this transition' (p.xxi) remains one of the most detailed and spirited articulations of that perspective.

Conclusion

Postmodern discourses arose within changing intellectual and geopolitical circumstances. Its multiple uses and meanings contributed to confusion and misinterpretation. It found favor among some leading artistic and intellectual voices, who contributed greatly to its refinement and prominence. However, it is also true that many Northern writers linking postmodernism to globalization and other politico-economic changes simply assumed, either implicitly or explicitly, that their Northern research and epistemological arguments had global relevance.

'Moderate' forms of postmodern intellectual practice did and arguably still do have global relevance in the cause of problem analysis, development promotion and empowerment. They lend legitimacy to different social groups and their voices rather than merely seeking a compatible mouthpiece to support external interventions in the name of 'development'. This may help to transcend the shortcomings of discredited official rapid 'modernization as development' and to facilitate local communities to develop according to their own conceptions, governed by acceptable rules of conduct. Such processes are not free from conflict, nor can nostalgia for long-dead traditions and heritages substitute for tackling present-day problems imaginatively in relation to today's dynamic realities.

Extreme postmodernism was unduly relativistic and permissive and in practice often precluded any social contract or practical action – which rather devalued its application in development contexts apart from arguing for a radically post-structural form of post-development.

Guide to further reading

Escobar, A. (2012) *Encountering Development: The making and unmaking of the Third World.* 2nd edn. Princeton: Princeton University Press.

Esteva, G. and Prakash, M.S. (1998) *Grassroots Postmodernism: Remaking the Soil of Cultures.* London: Zed Books.

Gabardi, W. (2001) *Negotiating Postmodernism.* Minneapolis: Minnesota University Press.

Harvey, D. (1989) *The Condition of Postmodernity.* Oxford: Blackwell.

Peet, R. and Hartwick, E. (2015) *Theories of Development; Contentions, arguments, alternatives.* 3rd edn. New York: Guilford.

Simon, D. (1998) 'Rethinking (post)modernism, postcolonialism and post traditionalism: South–North perspectives', *Environment and Planning D: Society and Space* 16(2): 219–245.

Slater, D. (1992) 'Theories of development and politics of the post-modern – exploring a border zone', *Development and Change* 23(3): 283–319.

Slater, D. (1997) 'Spatialities of power and postmodern ethics – rethinking geopolitical encounters', *Environment and Planning D: Society and Space* 15(1): 55–72.

Watson, S. and Gibson, K. (eds) (1995) *Postmodern Cities and Spaces.* Oxford: Blackwell. An important collection of essays, addressing different approaches to post-modernism in Northern and Southern urban contexts.

References

Ahluwalia, P. (2010) *Out of Africa: Post-Structuralism's Colonial Roots.* London: Routledge.

Colás, S. (1994) *Postmodernity in Latin America: The Argentine Paradigm.* Durham, NC: Duke University Press.

Ley, D. (2003) 'Forgetting postmodernism? Recuperating a social history of local knowledge', *Progress in Human Geography* 27(5): 537–560.

Watson, S. and Gibson, K. (eds) (1995) *Postmodern Cities and Spaces.* Oxford: Blackwell.

Chapter 10

Post-development

James D. Sidaway

Instead of the kingdom of abundance promised by theorists and politicians in the 1950s, the discourse and strategy of development produced its opposite: massive underdevelopment and impoverishment, untold exploitation and repression. The debt crisis, the Sahelian famine, increasing poverty, malnutrition, and violence are only the most pathetic signs of the failure of forty years of development.

(Escobar, 1995: 4)

Development occupies the centre of an incredibly powerful semantic constellation . . . at the same time, very few words are as feeble, as fragile and as incapable of giving substance and meaning to thought and behavior.

(Esteva, 1992: 8)

Along with 'anti-development' and 'beyond development', post-development is a radical reaction to the dilemmas of development. Perplexity and extreme dissatisfaction with business-as-usual and standard development rhetoric and practice, and disillusionment with alternative development are keynotes of this perspective. Development is rejected because it is the 'new religion of the West . . . it is the imposition of science as power . . . it does not work . . . it means cultural Westernisation and homogenisation . . . and it brings environmental destruction. It is rejected not merely on account of its results but because of its intentions, its world-view and mindset. The economic mindset implies a reductionist view of existence. Thus, according to Sachs, 'it is not the failure of development which has to be feared, but its success' (1992: 3).

(Nederveen Pieterse, 2000: 175)

Jan Nederveen Pieterse goes on to explain how from these critical perspectives, 'development' has often required the *loss* of 'indigenous' culture or the destruction of environmentally and psychologically rich and rewarding modes of life. Development is also criticized as a particular vision that is neither benign nor innocent. It reworks – but is never entirely beyond – prior colonial discourses (see Kothari, 2005). Development comprises a set of knowledges, interventions and world-views (in short a 'discourse') and hence powers – to intervene, to transform and to rule. It embodies a geopolitics (see Slater, 1993) in that its origins are bound up with Western power and strategy for the Third World, enacted and implemented through local elites. Hence American power and the Cold War containment of communism structured the meaning of development in the second half of the 20th century; associated with ideas of modernization (see Engerman et al., 2003). Such modernization strategies have deeper roots in American prototypes from earlier in the century (see Ekbladh, 2010; Sewell, 2010). Western agencies, charities and consultants

DOI: 10.4324/9780429282348-11

long dominated development's agendas (see Jackson, 2005; Stirrat, 2000, 2008). Development hence came to the fore during the Cold War as a powerful combination of policy, action and understanding. Related to concepts of anti-development and post-colonial criticisms, post-development arose in the 1990s as a critique of the standard assumptions about progress, who possesses the keys to it and how it may be implemented. Such critique also proved suggestive for the those studying the politics of local and regional development in Western Europe (see Donaldson, 2006).

Of course, as a number of people have pointed out, many of the critiques associated with postdevelopment reformulate skepticisms about and calls for alternatives that have long been evident. According to Marshall Berman (1983) an example is the myth of Faust, which crops up repeatedly in European cultures. Faust is a man who would develop the world and himself but must also destroy all that lies in his path to this goal and all who would resist. The myth of Faust, who sells his soul for the earthly power to develop, bears witness to a very long history of critics of progress and modernity. Throughout the 20th century, populist ideas of self-reliance and fulfilling 'basic needs' have also been skeptical of many of the claims of development, particularly when the latter takes the forms of industrialization and urbanization (see Kitching, 1989). Subsequently, the history of ideas of dependency has been, in part, a rejection of Western claims of development as a universal panacea to be implemented in a grateful Third World. From Latin American roots (see Kay, 1989), dependency ideas were disseminated very widely and sometimes took the form of a rejection of Western modernization/development as corrupting and destructive (see Blomstrom and Hettne, 1984; Leys, 1996; Rist, 1997) or as a continuation of colonial forms of domination (Rodney, 1972). In particular, writers from predominately Islamic countries (most notably Iran) saw the obsession with development as part of a misplaced 'intoxification' with the West (see Dabashi, 1993). Either way, the Third World was not simply a passive recipient of development but became a project and site for liberation and struggle (Prahshad, 2007; Sidaway, 2007). Likewise, more conventional Marxist accounts had long pointed to the 'combined and uneven' character of development and its highly contradictory consequences (see Lowy, 1980). Feminist writings had also criticized the ways in which the so-called Third World woman is represented as needing 'development' and Western-style 'liberation' (Mohanty, 1988) and opened up alternative ways of conceptualizing the economic and social change of 'development'.

Some critics have therefore complained that 'post-development' was never really beyond, outside or subsequent to development discourse. In this view, 'post-development' was merely the latest version of a set of criticisms that have long been evident *within* writing and thinking about development (Curry, 2003; Kiely, 1999). Development has always been about choices, with losers, winners, dilemmas and destruction as well as creative possibility. Gavin Kitching (1989: 195), who is concerned to put post-Second World War debates about development into a longer historical perspective (stressing how they also reproduce even older narratives from the 19th century), argues:

It is my view that the hardest and clearest thinking about development always reveals that there are no easy answers, no panaceas whether these be 'de-linking', 'industrialization', 'rural development', 'appropriate technology', 'popular participation', 'basic needs', 'socialism' or whatever. As I have had occasion to say repeatedly in speaking on and about this book, development is an awful process. It varies only, and importantly, in its awfulness. And that is perhaps why my most indulgent judgements are reserved for those, whether they be Marxist-Leninists, Korean generals, or IMF officials, who, whatever else they may do, recognise this and are prepared to accept its moral implications. My most critical reflections are reserved for those, whether they be western liberal-radicals or African bureaucratic elites, who do not, and therefore avoid or evade such implications and with them their own responsibilities.

In this sense, perhaps post-development's skepticism towards grand narratives about development is less original than the theoretical frames (the analysis of discourse) it brings to bear in problematizing these. Yet according to some post-development writers, not only are there 'no easy answers', but the whole question of 'development' should be problematized and/or rejected.

There are a number of more fundamental objections to post-development. The first is that it overstates the case. Such arguments usually accept that development is contradictory (that it has winners and losers) but refuse to reject all that goes under its name. For, to reject all development is arguably a rejection of the possibility for progressive transformation. Or it is to ignore the tangible improvements in life chances, health, wealth and material well-being evident in some places, notably the 'developmental states' of East or Southeast Asia (Rigg, 2003). The changing global map of production, consumption and finance in recent decades is also remaking the map of more and less developed spaces and of global wealth and power (see Sidaway, 2012; Bräutigam and Xiayang, 2012). Moreover, development itself has long been so varied and carried so many meanings (see Williams, 1976) that critiques need to be specific about what they mean when they claim to be anti- or 'post-development'.

In this context, Escobar's (1995) work, in particular, was often criticized. One objection has been that he understated the potential for change within development discourse (see Brown, 1996). Escobar's work reflects his experiences as an anthropologist in Columbia. As an account based on experiences of 20th-century Columbia, Escobar's critique of development could seem suggestive. Columbia has experienced periods of brutal civil war and foreign intervention. It became a major source of cocaine, connected to often violent smuggling networks extending northwards into the United States and Europe. Yet there is a risk that Escobar's text obscures the diversity of experiences of development, not all of which are as troubled as the Colombian experience.

The second objection involves rejecting post-development as yet another intellectual fad, of limited (or no) relevance to the poor in the Third World. Sometimes this objection draws attention to the fact that many of those who write about or disseminate post-development ideas live precisely the cosmopolitan, middle-class, relatively affluent lives that development promises to deliver. Such questions parallel the critique of *post-colonialism* as an intellectual fashion most useful to the careers of Western-based intellectuals.

However a few counter-points are in order here. First a whole set of writings and ideas are grouped together under the rubric of post-development. Michael Watts (2000: 170) explained that:

> There is of course a polyphony of voices within this post-development community – Vandana Shiva, Wolfgang Sachs, Arturo Escobar, Gustavo Esteva and Ashish Nandy, for example, occupy quite different intellectual and political locations. But it is striking how intellectuals, activists, practitioners and academics within this diverse community participated in a global debate.

Moreover, it is important to point out that for Escobar (1995) and others exploring the (geo)politics of development, to criticize development is not necessarily to reject change and possibility. Rather, it is to make us aware of the consequences of framing this as 'development'. It stresses that development is (for good and bad) always about power. Hence, as Clive Gaby (2012: 1249) notes, projects such as the Millennium Development Goals involve 'a logic of ambitious social, cultural and spatial engineering'. Moreover, alternative visions considering, for example, democracy, popular culture, resourcefulness and environmental impacts would transform the imagined map of more or less developed countries. Recognition that development is but one way of seeing the world (and one that carries certain consequences and assumptions) can open up other perspectives. What happens, for example, to the perception of Africa when it is seen as *rich* in cultures and lives whose diversity, wealth and worth are not adequately captured by being imagined as more or less *developed*? Alternatively, why are poverty and deprivation (or for that matter, excessive consumption amongst the affluent) in countries like the United States, New Zealand or the United Kingdom not issues of 'development' (see Jones, 2000; Kurian and Munshi, 2012)? What is taken for granted when the term 'development' is used? For it often seems that, in Escobar's (1995: 39) words, development has 'created a space in which only certain things could be said or even imagined'. Post-development literatures teach us not take this 'space' and its contours for granted.

Guide to further reading

Crush, J. (ed.) (1995) *Power of Development*, London and New York: Routledge.
Escobar, A. (1995) *Encountering Development: The Making and Unmaking of the Third World*, Princeton, NJ: Princeton University Press.
Ferguson, J. (1990) *The Anti-Politics Machine: 'Development', Depoliticization and Bureaucratic Power in Lesotho*, Cambridge: Cambridge University Press.
Rahnema, M. and Bawtree, V. (eds.) (1997) *The Post-Development Reader*, London: Zed Books.
Sauders, K. (ed.) (2002) *Feminist Post-Development Thought: Rethinking Modernity, Post-Colonialism and Representation*, London: Zed Books.
Sidaway, J. D. (2007) 'Spaces of postdevelopment', *Progress in Human Geography* 31 (3): 345–361.
Ziai, A. (ed.) (2007) *Exploring Post-Development: Theory, Practice, Problems and Perspectives*, London and New York: Routledge.

References

Berman, M. (1983) *All That is Solid Melts into Air: The Experience of Modernity*, London: Verso.
Blomstrom, H. and Hettne, B. (1984) *Development Theory in Transition: The Dependency Debate and Beyond: Third World Responses*, London: Zed Books.
Bräutigam, D. and Xiayang, T. (2012) Economic statecraft in China's new overseas special economic zones: soft power, business or resource security? *International Affairs* 88 (4): 799–816.
Brown, E. (1996) 'Deconstructing development: alternative perspectives on the history of an idea', *Journal of Historical Geography* 22 (3): 333–339.
Curry, G. N. (2003) 'Moving beyond postdevelopment: facilitating indigenous alternatives for "development"', *Economic Geography* 79 (4): 405–423.
Dabashi, H. (1993) *Theology of Discontent: The Ideological Foundation of the Islamic Revolution in Iran*, New York and London: New York University Press.
Donaldson, A. (2006) 'Performing regions: territorial development and cultural politics in a Europe of the regions', *Environment and Planning A* 38 (11): 2075–2092.
Ekbladh, D. (2010) *The Great American Mission: Modernization and the Construction of an American World Order*, Princeton: Princeton University Press.

Engerman, D. C., Gilman, N., Haefele, M. H. and Latham, M. E. (eds.) (2003) *Staging Growth: Modernization, Development and the Global Cold War*, Amherst and Boston: University of Massachusetts Press.

Esteva, G. (1992) 'Development', in W. Sachs (ed.) *The Development Dictionary: A Guide to Knowledge as Power*, London: Zed Books.

Gaby, C. (2012) 'The Millenium development goals and ambitious developmental engineering', *Third World Quarterly* 33 (7): 1249–1265.

Jackson, J. T. (2005) *The Globalizers: Development Workers in Action*, Baltimore: John Hopkins University Press.

Jones, P. S. (2000) 'Why is it alright to do development "over there" but not "here"? Changing vocabularies and common strategies of inclusion across the "first" and "third" worlds', *Area* 32 (2): 237–241.

Kay, C. (1989) *Latin American Theories of Development and Underdevelopment*, London and New York: Routledge.

Kiely, R. (1999) 'The last refuge of the noble savage? A critical account of post-development', *European Journal of Development Research* 11 (1): 30–55.

Kitching, G. (1989) *Development and Underdevelopment in Historical Perspective: Populism, Nationalism and Industrialization* (revised edition), London and New York: Routledge.

Kothari, U. (2005) 'From colonial administration to development studies: a post-colonial critique of the history of development studies', in U. Kothari (ed.) *A Radical History of Development Studies: Individuals, Institutions and Ideologies*, London: Zed, pp. 47–66.

Kurian, P. A. and Munshi, D. (2012) 'Denial and distancing in discourses of development: shadow of the 'third world' in New Zealand', *Third World Quarterly* 33 (6): 981–999.

Leys, C. (1996) *The Rise and Fall of Development Theory*, London: James Currey.

Lowy, M. (1980) *The Politics of Combined and Uneven Development: The Theory of Permanent Revolution*, London: New Left Books.

Mohanty, C. P. (1988) 'Under western eyes: feminist scholarship and colonial discourses', *Feminist Review* 30: 61–88.

Nederveen Pieterse, J. (2000) 'After post-development', *Third World Quarterly* 21 (2): 175–191.

Prahshad, V. (2007) *The Darker Nations: A People's History of the Third World*, New York: The New Press.

Rigg, J. (2003) *Southeast Asia: The Human Landscape of Modernization and Development* (second edition), London and New York: Routledge.

Rist, G. (1997) *The History of Development: from Western Origins to Global Faith*, London: Zed.

Rodney, W. (1972) *How Europe Underdeveloped Africa*, London: Bogle L'Ouveture.

Sewell, B. (2010) 'Early modernisation theory? The Eisenhower administration and the foreign policy of development in Brazil', *English Historical Review* 125: 1449–1480.

Sidaway, J. D. (2007) 'Spaces of postdevelopment', *Progress in Human Geography* 31 (3): 345–361.

Sidaway, J. D. (2012) 'Geographies of development: new maps, new visions?', *The Professional Geographer* 64 (1): 49–62.

Slater, D. (1993) 'The geopolitical imagination and the enframing of development theory', *Transactions of the Institute of British Geographers NS* 18: 419–437.

Stirrat, R. L. (2000) 'Cultures of consultancy', *Critique of Anthropology* 20 (1): 31–46.

Stirrat, R. L. (2008) 'Mercenaries, missionaries and misfits: representations of development personnel', *Critique of Anthropology* 28 (4): 406–425.

Watts, M. (2000) 'Development', in R. J. Johnson, D. Gregory, G. Pratt and M. Watts (eds.) *The Dictionary of Human Geography* (fourth edition), Oxford: Blackwell, pp. 167–171.

Williams, R. (1976) *Keywords*, London: Fontana.

Chapter 11

Clarifying confusion between development as 'change' and 'intention'

David Lewis

Introduction

The term 'development' can be used in several different ways: to describe broad processes of economic and social transformation, to refer to 'levels' of progress in terms of economic, political and social conditions found in different countries or regions, and as a descriptor for the broad field of activity undertaken by actors in the international aid system. Development is therefore a word that is made to work extremely hard. Two particular areas of difficulty can arise from confusion between the use of development to refer to societal change and its consequences and its use to describe the intentions and activities of institutional actors designed to bring it about.

Currie-Alder et al.'s (2014) simple definition is that development refers to 'how societies change over time', but this raises important questions about how and why they may change. For Heinz Arndt (1981) the two main meanings boil down to development as *something that is done* (involving intention and choice) and development as *something that happens* (an outcome based on some kind of ex ante logic). How can we make sense of these two distinct meanings, and can they be fitted together?

Exploring unfolding change versus intention

Making this distinction can clarify and add precision to discussions of development. Two main approaches to this issue have become influential. The first is the distinction between 'intentional' and 'immanent' development coined by historians Michael Cowen and Bob Shenton (1996). They distinguished two main ways of thinking about how societies change: i) faith in the unfolding progress through which societies advance and bring improvements to people's lives, and ii) the reformist idea of state action that can minimise the undesirable consequences of capitalism and bring order to disruptive processes of economic change. They argue that the modern idea of development principally arose from the state's need to manage instability within Europe's rapidly industrialising societies: 'it was in Europe that development was first meant to create order out of the social disorder of rapid urbanisation, poverty and unemployment'. From these origins, they suggest, the idea of development became associated with Europe's expansion and extended from the domestic realm into the colonies. Colonial powers came to see themselves as sufficiently 'developed' such that 'they could act to determine the process of development for others deemed 'less-developed'. This was achieved through a paternalistic notion of 'trusteeship', based on the idea that it was the state's job to manage problems of poverty, urban

DOI: 10.4324/9780429282348-12

migration and unemployment, both at home and, increasingly, in colonial territories. This produced a historically specific 'doctrine of development' that linked the growth and expansion of capitalism with the need to control and manage change.

Geographer Gillian Hart (2001) discusses a similar distinction in her work. Hart identifies the post-1945 international aid system, the international financial institutions, the United Nations and non-governmental organisations as a specific historical formation that she calls 'big D Development'. This is contrasted with 'small d development', the unfolding process of capitalist growth and expansion, which produces developmental change, but with uneven outcomes and both winners and losers. This draws on Cowen and Shenton's ideas but adds a number of additional points. The first is the idea that a dialectical relationship exists between these two kinds of development. Indeed, the 'big D' system serves the interests of 'little d' in important ways – both by providing welfare safety nets for those who lose from capitalist change, in order to maintain order and stability and by enabling the progress and expansion of capitalism through promoting policies such as economic liberalisation.

Hart's conceptualisation also opens up the possibility of agency and resistance. A link is made with Karl Polanyi's influential book *The Great Transformation* (1944), which traces the development of a highly commodified market society and pushback from society against this process, which he terms the 'double movement'. Polanyi understood efforts to check the market's expansion dialectically, as arising from potentially productive 'countertendencies' that are contained within capitalism, based on 'the clash of the organising principles of economic liberalism and social protection'. Trade unions, civil society organisations and social movements seek to reshape capitalism. Contestation emerges around the struggle for social protection, such as improved labour standards, equal rights and social justice within harsh processes of industrialisation and modernisation.

Both of these frameworks for thinking about development emerged during the 1990s post-Cold War period, when structural adjustment policies imposed on the Global South were now being accompanied by triumphalist proclamations about globalisation, 'the end of history', and the natural dominance of liberal capitalism. Hart in particular wished to challenge those she called 'neoliberal globalists' by reasserting the importance of the developmental state. She proposed that this could be achieved not by focusing not on the ameliorative terrain of 'big D' but instead by engaging with 'small d' issues of class and power and as a more profound way to engage with increasing levels of inequality, exclusion, and injustice.

How are these distinctions useful?

Distinguishing development as unfolding change from deliberate intervention can be useful in both analytical and practical ways. First, as we have seen, it provides more background and depth to discussion of development, which is a field that is sometimes criticised for its ahistoricism. The principle of trusteeship for example has echoes in the assumptions and practices found in many development agencies working in the Global South today. There is a paternalist element in the discourses of 'capacity building' and 'women's empowerment' that reflects roots in the idea of trusteeship.

Hart (2010) also shows that the complex relationship between 'big D' and 'little d' is not static but dynamic. For example, during the first few decades after World War Two, the state was seen as central to the process of modernisation and capitalist change, while the 1980s ushered in a neoliberal ideology in which market liberalisation and state withdrawal were put centre stage.

Second, the distinction can speak to debates about the sometimes problematic relationship between development theory and practice. For example, Thomas (2000) has argued that development should not be defined narrowly as 'what development agencies do' or that 'the development of capitalism accords to some kind of natural historical law'. Hart's work for example argues that 'the conditions for global capitalism have constantly to be promoted by those political forces which favour them'. Framing this relationship helps us to overcome the regular periods of 'impasse' in development studies (e.g., Booth, 1993) when scholars and practitioners find themselves unable to communicate effectively with each other.

Third, the distinction challenges an increasing emphasis on purely technical approaches to development by insisting on the need to bring in politics. A focus only on 'big D' produces a depoliticised conception of development intervention that focuses on managing institutional capacity building, service delivery, infrastructure and public administration reform. Such reformist interventions do not address underlying processes of capitalist (or indeed socialist) development, paying insufficient attention to alternative ways of organising political, economic and social relationships.

For example, one useful example of the way the distinction has been deployed usefully in theory/practice discussions is in relation to the role of non-governmental development organisations (NGOs). For example, Bebbington et al. (2008) use the 'little d/big D' distinction in their discussion of the different roles taken by NGOs in development, comparing service delivery work with advocacy and rights based approaches. While NGOs delivering services within project or contract settings can have positive implications for citizens, especially in the short term, they argue that NGOs that move beyond the 'big D' realm to engage with 'little d' development through lobbying to change the structural causes of poverty and mobilising against injustice may achieve more transformative effects in the longer term.

The example of Bangladesh's garment based export-driven development model can be used to briefly explore the strengths and weaknesses of the distinction. Since the 1970s, the country has manufactured and exported ready-made garments. This process of industrialisation has become the backbone of the economy and made a major contribution to the country's recent graduation to lower middle income country status. The garment factories have created employment for women, generated rural to urban migration, and led to social change (Lewis, 2011). Lives have been transformed, even empowered. But poor working conditions, regular industrial accidents, and labour unrest also illustrate both the transformative and disruptive power of raw capitalism as a driver of change – as described by the 'little d' formulation set out by Hart, and is reminiscent of the dislocation described by Cowen and Shenton's account of 19th-century Europe.

At the same, the agencies of 'big D' development seek points of contact with the garment industry, with donors supporting initiatives to improve accountability in international supply chains, and NGOs both providing services to workers and factories such as childcare facilities and campaigning to improve workers' rights. In this way 'big D' can be seen to serve 'little d' development through its ameliorative role in welfare service provision, but it may also challenge it by engaging in efforts to open up new spaces for transformation – in line with Polanyi's 'double movement' idea.

The example also illustrates the problem of making a clear separation between the two ideas of development at the level of social process and policy. At the international level, international trade and investment governance arrangements in the multi-fibre agreement (MFA) made it possible through a quota system for capital to flow to Bangladesh during the 1980s in the early stages of the ready-made garment industry. At the national level, the developmental state has also played a key role in providing and managing the policy and regulatory framework that has made the expansion of the country's garment industry. How does this state action fit within our distinction? Where does 'small d' end and 'big D' begin?

Here Jonathan Rigg's (2004) modification of the D/d distinction from the perspective of south east Asia is helpful. In addition to the ideas of 'big D' as managed change ('funds, technologies and expert advice, much of it from the rich world'), and 'little d' (as the 'historical progress of capitalism'), he proposes two further categories. These are additions to the 'big D' category, in the form of 'superscript development' (D^d) and 'subscript development' (D_d). The first draws attention to the role of developmental states that direct development processes, and the second concerns the role of NGOs, movements, and community groups that seek to promote 'alternative' forms of development. He writes that '[t]he point of drawing out development typologies in this way is that it helps to contextualise – intellectually – different critiques of development'. It also allows us to better distinguish the entanglements of state, market, and civil society. For example, the case of the Bangladesh garment sector discussed earlier illustrates the ways that understanding 'big D' also requires us to engage with the role of the developmental state in formulating policies to attract and manage investment in export garment sector and with the diverse range of civil society responses to this industrial policy.

Conclusion

Distinguishing development's two main meanings, as unfolding change and as purposeful intervention, can be useful in three main ways. First, it helps to clarify the different ways in which the word development is used, and, hopefully to sharpen debate. Second, it historicises ideas about development in ways that challenge technical and apolitical assumptions that have become common in development policy and practice today. Cowen and Shenton's work usefully highlights the longer term origins of development ideas – such as early 20th-century British colonial welfare policies – long before the post WW2 moment that many associate with the birth of modern development ideas. Hart's analysis identifies the post-1945 creation of a Western international development system as a distinct historical moment in the 20th century and makes it possible to examine the relationship between this system and the wider capitalist system. Third, it is useful in discussions of development policy and practice, where it can inform debates about reformism and welfare versus structural change and whether and how sites of resistance can usefully be opened up by activists. But some may also feel that there are unresolved contradictions in the distinction as well, around the role of policy actors beyond the aid system (such as non-funded activist organisations and movements) and the precise nature of the dialectic relationship between change and intention. Rigg's revised formulation begins to usefully to address some of this.

What of the future? Following the post-war period first of state-led modernisation and later of market-based neoliberalism, what happens to 'big D' within the rise of new forms of financialised and business-centred ideas of development? As the aid system has become more closely aligned with finance, Western bilateral aid has declined, and there are new Southern actors reshaping the landscapes of development. Does the distinction continue to offer insights into understanding contemporary worlds of development, or does it become less useful as the Western post-1945 order is challenged by new realities?

References

Arndt, H. (1981) 'Economic Development: A Semantic History', *Economic Development and Cultural Change* 29, no. 3: 457–466. doi:10.1086/451266.

Bebbington, A., S. Hickey, and D. C. Mitlin (2008) 'Introduction: Can NGOs Make a Difference: The Challenge of Development Alternatives', in *Can NGOs Make a Difference? The Challenge of Development Alternatives*, edited by A. Bebbington, S. Hickey, and D. C. Mitlin, pp. 3–37. London: Zed Books.

Booth, D. (1993) 'Chapter 2 Development Research: From Impasse to a New Agenda', in *Beyond the Impasse: New Directions in Development Theory*, edited by F. J. Schuurman, pp. 49–69. London: Zed Books.

Cowen, M., and R. W. Shenton (1996) *Doctrines of Development*. London: Routledge.

Currie-Alder, B., Kanbur, R., Malone, D. M. and Medhora, R. eds., 2014. *International development: ideas, experience, and prospects*. OUP Oxford. p. 13.

Hart, G. (2001) 'Development Critiques in the 1990s: Culs de Sac and Promising Paths', *Progress in Human Geography* 25, no. 4: 649–658. doi:10.1191/030913201682689002.

Hart, G. (2010) 'D/developments after the Meltdown', *Antipode* 41, no. S1: 117–141. doi:10.1111/j.1467-8330.2009.00719.x.

Lewis, D. (2011) *Bangladesh: Politics, Economy and Civil Society*. Cambridge: Cambridge University Press.

Polanyi, K. (1944) *The Great Transformation*. New York. Farrar & Rinehart.

Rigg, J. (2004) *Southeast Asia: The Human Landscape of Modernization and Development*. London: Routledge.

Thomas, A. (2000) 'Development as Practice in a Liberal Capitalist World', *Journal of International Development* 12, no. 6: 773–787. doi:10.1002/1099–1328(200008)12:6<773::AID-JID716>3.0.CO;2–7.

Chapter 12

Culture and development

Susanne Schech

Culture is so intrinsic to economic, political, and social processes that it is difficult to make sense of development without taking 'the cultural' into account. Culture and development emerged as a distinctive approach in the 1990s at a time of crisis in development thinking. Existing Marxist, neoliberal, and modernization theories seemed unable to fully explain how, when, and why countries develop; understand what development means to people in different places; and provide development interventions to improve their lived experience. One key contribution of the 'cultural turn' in development was to identify the narrow conceptualization of culture in the mainstream development literature. Second, it turned attention to the non-economic aspects of people's lived experiences and how they are shaped by – and connect to – cultural formations such as religion, gender, ethnicity, and colonialism. Third, drawing on cultural, postcolonial, and critical development studies, scholars started to examine development as a concept shaped by ideas, norms, and values, and hence as a cultural construct that is deeply contested, both intellectually and in everyday struggles over development.

Culture matters – but how?

Development scholars have argued for many decades that cultural values, attitudes, orientations, and opinions are a key variable in determining economic progress. According to modernization theory, which became popular in the 1950s, development is a culturally neutral universal solvent that can be facilitated or hindered by culture and can hinder or facilitate development. Key cultural markers are traditional and modern, whereby traditional cultural traits and practices are destined to be cast aside and replaced through more-or-less well-meaning policy interventions that usher in modernity (Schech and Haggis, 2000: 18–19). In modernization studies, 'traditional man' was typically contrasted with 'modern man' in terms of values, attitudes, and behaviors – one is oriented towards the past, frugal in life-style, rooted in place, and suspicious of innovation, while the other is dreaming of a different, more expansive world, ever-changing and full of possibilities (Lerner, 1958). In this view, culture is traditional, bounded, and static, like a box that has been passed down from one generation to the next until it must be cast aside to make way for modernization. Traditional societies exist outside of history, and any society that resists modernizetion and clings to tradition will remain underdeveloped. Only societies willing to give up their traditional values, institutions, and cultural practices – or societies that happen to possess cultural traits favorable to modernization – will succeed in their quest for development.

DOI: 10.4324/9780429282348-13

One important contribution of the cultural turn in the 1990s was to challenge the idea that culture is bounded and static. Rather, culture does matter, but it is constantly changing. Any attempt to tie it down and determine it is futile; as Sen (2004: 43) puts it, it is as hopeless as 'trying to fix the cultural anchor on a rapidly moving boat'. Furthermore, cultures do not evolve in isolation as separate boxes but always through interaction with other cultures. Cultural interconnections through migration, conquest, trade, exploration, and pilgrimage can be traced back far into history (Nederveen Pieterse, 2001). In the current era of globalization, the boundaries of people's lived experience are more permeable than ever before, informed by an awareness of other circumstances, experiences, images, and ways of living. Arjun Appadurai (1990) described these interconnections as global cultural flows of people, machinery, money, images, and ideas that simultaneously produce sameness and difference. Through these uneven processes of global capitalism, culture is constantly reworked, reproduced, and contested. As Pred (1992: 109) puts it, 'culture is embodied and lived, actively produced and expressed, through all social practices, through all that is concrete and everyday, through all that is enmeshed in power relations and their associated discourses, their associated representations and rhetorics'.

If culture is neither fixed nor internally homogeneous, it makes no sense to portray the cultures of the Global South as obstacles to change. This opens up a space for culture to be conceived a resource to be tapped for development. In this view, cultural practices can generate goods for global markets, such as Indian Bollywood films or African 'world music'. Cultural difference can be used to position a country in a competitive market place, as the Indian government has done with the 'Imagine India' campaign. UNESCO, the United Nations organisation tasked with protecting and promoting the diversity of cultural expression around the globe, sees culture, arts, and heritage as a resource in the production of cultural goods and services for viable and competitive cultural industries, using the transformative power of creativity to bring about desirable development outcomes (Isar and Pyykkönen, 2015).

On the other hand, culture can also be a resource to resist and formulate alternatives to capitalist development. For example, Indian scientist-activist Vandana Shiva evokes indigenous cultural knowledge as a powerful resource in local struggles to protect water and genetic plant material and resist their privatisation and commodification (Opel and Shiva, 2008). Post-development scholars have been criticized for romanticizing indigenous knowledge, portraying these knowledge systems as somehow outside of history and ignoring the often very oppressive class and caste relations within which subsistence lifestyle and culture is placed. To understand how local cultural practices are selectively commodified or why indigenous communities resist pressures to market their culture is to recognise culture as a site of struggle rather than as essentialized difference (Nederveen Pieterse, 2001). In summary, the shift from a reductive understanding of culture in modernization theory has revealed a tension between two different views of culture – culture as a total and distinctive way of living of a social group that must be preserved and malleable, ever-changing culture that is interwoven with every aspect of economic, political, social, and environmental activity. This is seen in the literature on culture as a resource for desirable development that assumes that development is a socio-economic and political process that people can join or resist by utilizing their culture.

Development as a cultural construct

The picture changes radically when development itself is viewed as culturally constructed. According to Pieterse (2001), development is an intercultural transaction where the aim is to develop cultural 'others'. This view of culture and development pays close attention to representation and power. From a cultural studies perspective, European colonization involved not only economic and political domination but also cultural domination by bringing European cultural categories, languages, images, and ideas into the New World in order to describe and represent it' (Hall, 1992: 293–294). In relation to the Middle East, Edward Said (1978) analyses Orientalism as not just a Western way of knowing the Orient but also a Western discourse – or style of dominating, restructuring, and having authority over the Orient. Said draws attention to the connection between discourse and the power to determine what can be said with authority by whom and where. Colonial representations have long lasting effects that continue to frame how the world is seen and legitimize contemporary economic and geopolitical interventions.

Arturo Escobar (1995) brought postcolonial and poststructuralist theories to development discussions in his seminal critique of development as a Western discourse that is globalized and reproduced through powerful institutions. Taking the World Bank as an example, he analyses the ways in which it exercises power by producing the (Western) knowledge and interventions that it claims will address poverty and underdevelopment in the rest of the world. Escobar shows how the ever-expanding array of statistics through countries are categorised from 'very high income' to 'low income' and 'least developed' are a colonizing move. These categories mark least developed countries as lacking and in need of intervention in order to catch up with high income Western countries that present themselves as the desirable endpoint of development. In this construction of underdevelopment, according to Escobar, Third World citizens are represented as helpless victims living in ignorance and unable to do anything for themselves, and their governments are portrayed as incapable, fragile, corrupt, and requiring reform.

New directions

Two decades on, research into myriad interconnections between culture and development has been taken into various directions. Three examples are briefly introduced here to demonstrate the continuing impact of the cultural turn in development studies: development intervention as governmentality (Li, 2007), pluriversal perspectives on development (Escobar, 2015), and the social and cultural context of climate change (Sheller, 2018).

Tania Li's ethnographic research of World Bank-funded community development programs in Indonesia reveals how development is culturally constructed from above. Poor communities are treated as natural spaces for development interventions and poor people as subjects who must be empowered to take responsibility for their own improvement. Li shows how development experts render poverty as a technical issue that can be solved through participatory planning and better governance. Capitalist enterprise, in its diverse forms that include micro-credit and social enterprises, is presented 'only as a solution to poverty, not as a cause' (Li, 2007: 267), leaving the unequal relations of production and appropriation outside the frame of view and intervention.

Escobar (2015) draws on the work of Latin-American intellectuals and indigenous grassroots movements that challenge the cultural binaries underpinning the Western model of development – such as tradition/modernity; nature/culture; North/South – and posit alternatives. In the Andean cultural-political project of *Buen Vivir* (good life), economic objectives are subordinated to collective well-being, culturally appropriateness, and the rights of nature. For Escobar, this is one movement among many others working towards a new cultural and economic paradigm in which pluriversal perspectives and many words can co-exist.

Building on these ideas, Mimi Sheller (2018) brings critical cultural analysis to climate change and how disaster-prone islands in the Caribbean encounter them. If climate change is a shorthand for the worsening ecological conditions across the planet, Sheller argues that it also brings into focus the neo-colonial, gendered, racialized, sexualized processes that have shaped the vulnerability of Caribbean people. Their long history of local struggles against 'global capitalist exploitation of nature in a world-spanning system of vast inequity and injustice' can serve as a guide into a post-development future (Sheller, 2018: 979). Taken together, these studies show that new visions can emerge from local struggles over development and culture, although a sustained engagement with these visions will be necessary to break with conventional development (Wilson, 2017).

Additional resources

Escobar, A. (2015) 'Degrowth, postdevelopment, and transitions: a preliminary conversation', *Sustainability Science*, 10: 451–462.

Isar, Y. R., & Pyykkönen, M. (2015) 'Confusing culture, polysemous diversity: "culture" and "cultural diversity" in and after the Convention', in D. D. Beukelaer, M. Pyykkönen, & J. P. Singh (eds.), *Globalization, Culture, and Development*. Houndmills: Palgrave Macmillan, pp. 13–28.

Li, T. M. (2007) *The Will to Improve. Governmentality, Development, and the Practice of Politics*. Durham: Duke University Press.

Sheller, M. (2018) 'Caribbean futures in the offshore Anthropocene: debt, disaster, and duration', *Society and Space*, 36(6): 971–986.

Wilson, K. (2017) 'Worlds beyond the political? Post-development approaches in practices of transnational solidarity activism', *Third World Quarterly*, 38(12): 2684–2702.

References

Appadurai, A. (1990) 'Disjuncture and difference in the global cultural economy', *Theory, Culture & Society*, 7: 295–310.

Escobar, A. (1995) *Encountering Development: The Making and Unmaking of the Third World*. Princeton: Princeton University Press.

Hall, S. (1992) 'The West and the rest: discourse and power', in S. Hall, & B. Gieben (eds.), *Formations of Modernity*. Cambridge: Polity Press/Open University Press, pp. 274–311.

Lerner, D. (1958) *The Passing of Traditional Society: Modernizing the Middle East*. New York: Free Press.

Nederveen Pieterse, J. (2001) *Development Theory*. Thousand Oaks: Sage.

Opel, A., & Shiva, V. (2008) 'From water crisis to water culture', *Cultural Studies*, 22(3–4): 498–509.

Pred, A. (1992) 'Capitalisms, crises, and cultures II: notes on local transformation and everyday cultural struggles', in A. Pred, & M. J. Watts (eds.), *Reworking Modernity: Capitalisms and Symbolic Discontent*. Brunswick: Rutgers University Press.

Said, E. W. (1978) *Orientalism: Western Conceptions of the Orient*. New York: Pantheon Books.

Schech, S., & Haggis, J. (2000) *Culture and Development: A Critical Introduction*. Oxford: Blackwell.

Sen, A. (2004) 'How does culture matter?', in V. Rao, & M. Walton (eds.), *Culture and Public Action*. Stanford: Stanford University Press, pp. 37–58.

Chapter 13

Development ethics

Des Gasper

The field of development ethics explores questions and debates concerning what is good development of societies and of the world, and good development for individual persons. Generations of experience suggest the inadequacy of the assumption that societal, world or personal development can be simply equated to economic growth and wealth. That assumption neglects issues of equitable processes and distribution, security, personal relationships, physical and mental health and well-being, natural environment, identity, culture and meaningfulness. Not least, equating national development to national economic growth neglects the welfare and rights of many groups of already disadvantaged people. Over ten million people a year, for example, are displaced from their homes due to economic expansion, frequently with little or no compensation (Penz et al., 2011). An important alternative conception of development is 'human development', meaning achievement with respect to a wide range of well-reasoned values – not only those measured in money – and advancement of people's ability to achieve such well-reasoned values (Haq, 1999; Nussbaum, 2011). Development ethics tries to identify and systematically reflect on values and value-choices present in – or relevant to – cases and processes in the development of societies, persons, regions, and the globe.

Topics in development ethics include, amongst others: meanings and evolution of the idea of 'development' and the values that these meanings can contain about what is acceptable and desirable; concepts and evidence about human well-being and ill-being; assumptions and gaps in conventional economic evaluation; meanings and varieties of 'equity' and how equity can be neglected; the significance of various types of human vulnerability and insecurity (Goulet, 1971) and their relationships to economic growth. Practically oriented development ethics looks at how and which values are or can be incorporated in systems of policy, laws, social routines, and public and individual actions.

Ethics of development: why?

'Development' sounds self-evidently desirable. Why did a field of ethics of development arise? First, because of persistent undeserved removable poverty, sickness, insecurity and unhappiness, despite economic growth. Rise of average incomes does not necessarily benefit ordinary and especially poor people. Despite enormous growth in human powers and economic turnover, hundreds of millions of people have remained undernourished, leading to their physical and mental stunting, illness and premature death. As of 2004 UNICEF estimated that almost 30,000 babies and children under 5 died every day from poverty-related causes; a third of the world's people lacked basic sanitation, almost a

DOI: 10.4324/9780429282348-14

billion adults were illiterate, and 170 million children were engaged in hazardous work. Most of the people affected – including the babies and children – were not to blame for their own situation and had little unaided response-ability. As Nussbaum (e.g., 2011) asks: how far should the chance of being born in one nation rather than another determine the life chances of a baby? Huge progress has been made in the past 20 years (including during the Millennium Development Goals period, 2000–2015), though very much remains to be done. The estimated number of children dying per year before the age of five fell from 12.6 million in 1990 and still near 10 million in 2004, to 5.3 million in 2018 (UNICEF, 2017; WHO, 2019).

Second, many people become harmed even within processes of economic development: they are made to bear costs without sharing in benefits, as for example in uncompensated displacement from their homes and livelihoods. The people displaced are nearly always poor, moved in order to clear the way for projects that very largely benefit people who are already better off. So, not only do some groups not share in benefits; they may be deliberately harmed, as were generations of slaves and many other workers. Even when not deliberately sacrificed, many suffer through increased marginalisation and exclusion. Issues of distribution and harm involve also future generations, notably in relation to damage to the natural environment, especially through human-induced climate change.

Development ethics looks at implications of the interconnections and in-built conflicts in socio-economic development, within countries and internationally and over time. Increased carbon emissions, a core feature of modern development, indirectly eventually damage people in vulnerable environments around the world. Investment for future generations can be at the expense of present-day poor people; construction of infrastructure for some people's benefit results in displacement of others, typically with major economic, social and psychological disruption and within markets, the increased wealth of some people competes away resources from poorer people, by forcing up the prices of goods such as land, housing and food. Famines and malnutrition have often been caused less by shortage of supply than by this mechanism, whereby wealth draws in resources from around the world (Davis, 2001). Increased pressures of most sorts typically affect women especially, for they are the main care-givers in a society, the 'shock absorbers' that get noticed only when broken. The disproportionate concentration of costs of development upon some groups has been used as a mechanism of transformation, defended as 'breaking eggs in order to make omelets'. Thus industrialization and the transformation of agriculture have typically partly occurred through processes by which many small agricultural, industrial and artisanal producers are forced out of business. Since the mid-1990s around 10 to 15,000 Indian farmers a year are estimated to have committed suicide due to accumulation of unrepayable debts. This is a case not of exclusion from modernization processes but of vulnerable people who are induced and/or choose to participate in types of economic modernization and who then sometimes suffer severely when the risks and the 'small-print' prove to be more than they can cope with.

Third, the gains in well-being through policy approaches that equate development to economic growth are sometimes very questionable. When and how far does acquisition of and preoccupation with material comforts and conveniences bring – or jeopardies – a fulfilling and meaningful life? (Gasper, 2007). The French economist Louis-Joseph Lebret (1897–1966), who helped to found the subject of development ethics, spoke thus not only of 'development for all persons' but of 'development of all the person' and of 'putting the economy at the service of man'.

Fourth, besides outcomes, major questions arise about democratic participation in processes of decision-making (e.g., Chambers, 1997; Ellerman, 2005) and about responsibilities in relation to harm and undeserved suffering: who has responsibilities and to do what – to help, prevent, refrain, compensate – including in light of past injustices (such as slavery) and their consequences for present-day undeserved advantages and disadvantages.

Issues of development ethics become relevant because of arguments that better alternatives are possible compared to what has happened and that real choices exist for the future too. Evidence of such alternatives can lie in the experiences of other countries (Drèze and Sen, 1989) – for example several East Asian countries that combined a rapid climb out of poverty with a relatively high degree of population inclusion and sharing in benefits. Financial and economic calculations are also relevant; for example, that aid transfers equal to eight days' worth of global military budgets would cover the required annual external support for achieving good-quality universal pre-primary, primary and secondary education (UNESCO, 2015: 8).

Ethics in development: what and how?

The agenda of development ethics includes: to explore how the content of the idea of 'development' as societal improvement is value-relative; to highlight who bears the costs of various types of 'development' and to examine the value-choices that are implied and should be considered in development policy, programs and projects; to present well-reasoned alternatives to mainstream habits regarding those choices, in particular to clarify the values behind evaluative and prescriptive arguments from economics, query the narrowness of using only values from the marketplace, introduce other relevant values and query therefore an automatic superior status for economics arguments in relation for example to human rights arguments. The root concerns of development ethics – the insistence on not *equating* societal improvement to economic growth, on identifying and comparing value and strategy alternatives and on not ignoring costs and their distribution – all apply not only to poor countries but with almost equal force in rich countries and for the globe.

In examining the value-choices in development, work in development ethics operates at the interface of ethics, development studies and development policy. It asks: which life-conditions and which effects are both unfair and avoidable? For those, what changes should be made? Who has which responsibilities – including to remedy the damage that they cause and to respect and support local and global public goods (including environmental systems)? How far are national boundaries ethically relevant, in a world that is increasingly unified economically and in other ways, such that people worldwide influence each other? When is international aid justified? – for serving longer-term self-interest, or as praise-worthy but non-obligatory charity or as an obligation, including as an obligation of former colonial powers and beneficiaries from colonialism? What are ethical requirements regarding its conduct?

Some development ethics work deals with basic issues of concepts and theory (e.g., Gasper, 2004); some engage with specific sectors and policies (e.g., Drèze and Sen, 2013). Some instructive examples of the latter type are: i) much work looks at health, and the gross imbalance between health needs and health spending, including spending on research; diseases of the poor have been grossly neglected. Various authors investigate health policy options and related ethical theory (e.g., Stapleton et al., 2014). ii) The Jubilee 2000 debt-relief campaign studied the history of lending and found that all the high-income countries that had long insisted on total repayment of debts by low-income countries, which had vastly escalated since the 1970s oil crises and the later increased interest rates, had themselves had major earlier episodes of debt relief or repudiation (Hanlon, 2000). iii) Penz et al. (2011), building on the work of the World Commission on Dams, propose a detailed, principled approach to assessing, deciding about and compensating for displacement, based on a synthesis of development ethics thinking, including attention to what is justifiable development and what are good procedures for resolving conflicts.

The field of development ethics is a meeting place of theory and practice and of many disciplines and types of knowledge. It needs to use a broad vision, looking at the range of real experiences of human joys and suffering (see e.g., Narayan et al., 2000) and at interconnections besides only those captured by markets and the categories of economics.

References

Chambers, R. 1997. *Whose Reality Counts? Putting the First Last*. London: Intermediate Technology Publications. About equity in participation in defining situations and providing information.

Davis, M. 2001. *Late Victorian Holocausts: El Niño Famines and the Making of the Third World*. London: Verso. A study of the implications and interactions of global power-systems, global markets, and global climate-systems in the era of European global dominance in the late 19th century, especially the disastrous implications for vulnerable population groups.

Drèze, J. and A. Sen. 1989. *Hunger and Public Action*. Oxford: Clarendon. An overview of feasible public action to address basic human needs, under a variety of different political set-ups.

Drèze, J. and A. Sen. 2013. *An Uncertain Glory – India and its Contradictions*. London: Penguin. An illustration of a treatment of development policy evaluation and design that (like Penz et al.) is systematically guided by an ethical standpoint – here, Sen's capability approach.

Ellerman, D. 2005. *Helping People to Help Themselves: From the World Bank to an Alternative Philosophy of Development Assistance*. Michigan: University of Michigan Press.

Gasper, D. 2007. Uncounted or illusory blessings? Competing responses to the Easterlin, Easterbrook and Schwartz Paradoxes of well-being. *Journal of International Development*, 19(4), 473–492.

Goulet, D. 1971. *The Cruel Choice*. New York: Atheneum. An eloquent early classic in this field.

Hanlon, J. 2000. How much debt must be cancelled? *Journal of International Development*, 12(6), 877–901.

Haq, M. 1999. *Reflections on Human Development* (2nd edn.). Delhi: Oxford University Press.

Narayan, D., R. Chambers, M.K. Shah and P. Petesch. 2000. *Voices of the Poor: Crying Out for Change*. New York: Oxford University Press. From over 10,000 interviews around the world.

Penz, P., J. Drydyk and P. Bose. 2011. *Displacement by Development – Ethics, Rights and Responsibilities*. Cambridge: Cambridge University Press. An outstanding synthesis of theory and practical policy analysis.

Stapleton, G., P. Schröder-Bäck, U. Laaser, A. Meershoek and D. Popa. 2014. Global health ethics: An introduction to prominent theories and relevant topics. *Global Health Action*, 2014(7). https://www.ncbi.nlm.nih.gov/pmc/articles/PMC3925811/

UNESCO. 2015. *Pricing the Right to Education: The Cost of Reaching New Targets by 2030*. Paris: UNESCO.

UNICEF. 2017. *Levels and Trends in Child Mortality 2017*. New York: UNICEF.

WHO. 2019. *Factsheet: 'Children – Reducing Mortality'*. Geneva: World Health Organization.

Further reading

Drydyk, J. and L. Keleher, eds. 2019. *Routledge Handbook of Development Ethics*. London and New York: Routledge.

Gasper, D. 2004. *The Ethics of Development*. Edinburgh: Edinburgh University Press.

Gasper, D. and A.L. St. Clair, eds. 2010. *Development Ethics*. Aldershot: Ashgate. A collection of 28 papers from the previous 35 years.

Ingram, D. and T. Derdak. 2019. *The Ethics of Development: An Introduction*. Abingdon: Routledge.

Nussbaum, M.C. 2011. *Creating Capabilities: The Human Development Approach*. Cambridge, MA: Harvard University Press.

Useful websites

Human Development and Capability Association, HDCA. http://www.capabilityapproach.com
International Development Ethics Association, IDEA. http://www.development-ethics.org/

Audio-visual material

- Lecture (42 minutes) on development ethics by Nigel Dower – https://www.youtube.com/watch?v=H-vVEOSc0IY
- Development ethics – a short (4.5 minutes) video of images, compiled by Marta Pedrajas – https://www.youtube.com/watch?v=UVWLl56CtyI
- A set of four short videos of interviews about issues in development ethics – https://www.youtube.com/channel/UCJkjDKGjHpylVj5QtkbnbAQ
- Three speakers on 'development ethics: Past reflections, current happenings, future directions' (2 hours seminar) – https://www.youtube.com/watch?v=tBsHqPftqjQ
- Ethics in action – a series of talks prepared for the online SDG Academy – https://sdgacademy.org/course/ethics-in-action/, introduced at https://www.youtube.com/watch?v=kKPTMNCAYtY
- 'Are you an ethical aid worker?' – panel discussion (47 minutes) – https://www.youtube.com/watch?v=snlYleJpgx8

Part 2

Histories and discourses of development. Editorial introduction

Development studies, as a field, has changed, adapted, and transformed over time in response to ever-changing social, political, and economic contexts that play out within and across different scales of space and time. The chapters in this section have been brought together to provide a view of some of the historical and discursive shifts that frame, organise, and give meaning to development. This section of the Companion brings together the notions of history and discourse with the aim of highlighting a symbiotic relationship between past events, present interpretations, representations, and experiences. This convergence serves to highlight the multifaceted, changing, and contested nature of development, emphasising the importance of understanding its roots to navigate its future.

History is not merely a chronicle of past events; it is a lens through which we can understand the origins of development paradigms, the objectives they espouse, the consequences of past initiatives, and the socio-political contexts that influence developmental trajectories. Recognising the historical underpinnings of development is crucial. It allows us to discern the many social, political, cultural, and economic factors steering its course, equipping us with insights from past experiences to make more informed decisions for the future.

From this position it is imperative to acknowledge the shadows of colonialism that continue to loom large over contemporary development paradigms and practices. Colonial legacies have not only demarcated geographical boundaries but have also deeply influenced socio-political structures, economic systems, and cultural narratives in previously colonised regions. These legacies often manifest in the form of persistent power imbalances, skewed economic relations, and lingering stereotypes that frame the Global South as 'underdeveloped' or 'in need of guidance'. Moreover, many development strategies echo colonial-era practices by imposing external solutions without adequately engaging with local contexts or valuing indigenous knowledge. Recognising and confronting these colonial legacies is crucial. We are thereby challenged to rethink and reframe development not as a mere replication of 'Western' models but as a collaborative endeavour that respects the histories, identities, and aspirations of diverse communities.

Over the years, discourses of development have been contested, revised, and redefined. Early narratives, predominantly from the 'Global North', were steeped in colonial perspectives, portraying development as a linear progression with the colonial 'West' as the epitome of progress. However, contemporary discourses have begun to grapple with the challenges posed by global inequalities and the consequences of past actions, not least

DOI: 10.4324/9780429282348-15

the implications of problematic binaries such as 'developed' and 'developing', 'first' and 'third' world, and the dichotomy of Global North and Global South. Efforts to unpack and see beyond these binaries have contributed to greater attention to perspectives from outside the Global North, shedding light on the power dynamics, biases, and (un)intended consequences inherent in development initiatives. This has led to a transformative shift in the way development is conceptualised, largely influenced by voices and perspectives emanating from the Global South.

These voices have not only challenged long-standing Eurocentric narratives but have also introduced diverse paradigms rooted in indigenous knowledge, local experiences, and context-specific realities. By foregrounding these perspectives, the discourse on development has become more inclusive, recognising the multiplicity of paths to progress and the importance of local ownership in shaping development trajectories. This inclusion underscores the necessity of moving beyond monolithic, top-down approaches, and embracing a more pluralistic, bottom-up understanding of development that respects and integrates diverse worldviews and experiences.

While each chapter in this section offers a unique focus or perspective, certain themes resonate throughout. First, there is the enduring impact of colonial legacies on contemporary development paradigms. These legacies have influenced not only strategies and policies of development but also the very definitions and perceptions of what constitutes development. Second, the chapters underscore the importance of power dynamics – be it in the form of global interdependencies, institutional roles, or historical narratives. These dynamics shape discourse, determining who gets to define development, whose voices are prioritised, and whose narratives are side-lined.

An attention to history in development studies brings with it several implications. It fosters a deeper understanding of the structural inequalities rooted in historical processes and highlights the need for context-specific, historically informed development interventions. It challenges the idea of a one-size-fits-all approach, advocating for strategies that are cognisant of historical legacies and socio-cultural contexts. Furthermore, a historical perspective encourages reflexivity, prompting scholars, policymakers, and practitioners to critically examine their own positions, biases, and assumptions and the effects these might have on others.

This section of the Companion serves as a bridge between the past and the present, offering readers a range of examinations of the histories and discourses that have shaped development as both an arena of policy and practice and field of study.

Chapter 14

Development in a global-historical context

Ruth Craggs

Introduction

Many texts locate the origins of development in the post-1945 era, alongside the emergence of the United State and the Soviet Union as global superpowers, anti-colonial movements and decolonisation in much of the world. Although development as *global project* and *academic discipline* may have begun in this period, many scholars now argue that the ideas and practices that underpinned post-war development had their origins earlier, in the late colonial period. From this perspective, the

> the post-war crusade to end world poverty represented not so much a novel proposal marking the dawn of a new age, as the zenith of decades, if not centuries, of debate over the control and use of the natural and human resources of colonized regions.
>
> (Hodge, 2007: 3)

If the campaign against poverty, so often thought to begin after World War II, in fact has its roots in earlier colonial development projects, then it is valuable to explore how these earlier incarnations of development were practiced and with what effects. As colonialism is often understood as a causal factor in contemporary poverty, inequality and violence, it is crucial to understand the connections between colonial and post-colonial development.

This chapter provides the global-historical context for the development theories and practices explored in the rest of this volume. It explains the relevance of the colonial histories of places and people caught up in the nexus of development – as donors and recipients, ideologues and practitioners – to the ways that development was imagined, funded, practised and received. It begins by exploring colonialism *as* development. It examines the ways in which colonial rule was presented as premised on the idea of developing and modernising colonies. This section highlights the ideologies of trusteeship and modernisation that underlay imperial rule and shaped colonial territories. The second section explores the material legacies of colonialism in the developing world and the continuing influence of colonial thought and practice in post-colonial development. For reasons of brevity, the focus falls on the British Empire (and its decolonisation), though some of the same trends can also be seen in other European contexts.

DOI: 10.4324/9780429282348-16

Colonialism as development

Enlightenment ideas of 'improvement' – or making a more efficient and orderly use of land – accompanied and legitimated colonial rule from at least the 18th century (Hodge, 2007). Potential colonial land was understood as fair quarry for expanding European empires; cast as empty or ill exploited, it was seen as ripe for improvement by those with the expertise through which to make these transformations. Improvement entailed the development of infrastructure (where linked to European trade or settlement), the increase in economic output (benefitting metropolitan interests and markets) and the augmentation of the population with European settlers. In the 19th century, medical research aimed to bolster the colonial system by protecting the health of colonial servants, armies and settler populations and prevent the 'degeneration' caused by tropical climates. Improvements were oriented towards metropolitan interests, rather than towards increasing the quality of life of local communities.

The late 19th and early 20th century period saw a new model of imperialism in the French, British and Dutch colonies which placed more emphasis on development on humanitarian grounds for native colonial communities. This policy became known as 'trusteeship'. As Power (2003: 131) explains, 'Trusteeship in colonial administration was all about the mission to civilise others, to strengthen the weak, to give experience to the 'childlike' colonial peoples who required supervision'. It therefore provided the mandate for European powers to help these territories develop through following a path towards Western modernity. In Britain, official policy was enshrined in the 1929 Colonial Development Act, providing British funding for economic development overseas for the first time, and the 1940 Colonial Development and Welfare Act, which ushered in state-led large-scale development for the purposes of improving welfare. Thus the 1930s onwards witnessed an increasing move towards interventionist development policies in colonial territories, resulting in a dramatic growth in the number, scale and funding of colonial development projects. This trajectory became even more marked in the immediate post-Second World War era, a period which has been called, as a result, the second colonial occupation (Low and Lonsdale, 1976).

Hodge (2007: 8) argues that this new push towards humanitarian development 'helped to reinvigorate and morally rearm the imperial mission in the late colonial epoch', providing continued legitimacy for empire in a rapidly changing geopolitical landscape between 1930 and 1950. This new style of imperialism was linked to an increasing concern over the poor conditions in the colonies but also aimed to support the colonial system. Development hoped to stabilise colonial populations through the creation of an indigenous middle class invested in the colonial state and to sooth growing local unrest during the period of depression. By creating products for European markets and markets for European goods, development also contributed to struggling European economies. As colonial planners began to publicly discuss eventual decolonisation once colonies had 'progressed' enough, development policies became even more important, both as contributor to this colonial progress and to ensure the creation of stable and amenable newly independent states.

Development projects in the late colonial period were closely allied with a belief in modernisation. This involved the linear progress of states towards a developed, modern, (Western) society and economy. Official British colonial films of this era showcase this discourse of development as modernisation (see www.colonialfilm.org.uk). Projects were based on a growing faith in the role of science and technology to combat poverty and disease and focused on infrastructural improvement and the technical enhancement of agriculture, industry and healthcare through new innovations. Higher yielding seeds, new crops and intensive monoculture were encouraged (and indeed enforced), disease eradication programmes were rolled out and new mining technology was introduced (Tilley, 2011). Late colonial housing and building projects drew on new materials such as concrete, scientific construction techniques and modernism in architectural design, imbricating notions of progress into the design of colonial landscapes (Crinson, 2003).

The high modernism of the Kariba Dam project (1955–1960) in what was then the Federation of Rhodesia and Nyasaland (present day Zambia and Zimbabwe) is illustrative of the discourse and practice of late colonial development (Tischler, 2012). A huge project to dam the Zambezi and provide hydroelectric power for the surrounding territories, its construction took 40 per cent of the colonial state's gross national product to complete and entailed the submergence of 57,000 local Gwembe Tonga homes (*ibid.*). Combining grand scale, new technology, a desire for industrialisation, a modern aesthetic and materials and a notion of population crisis (to be solved through technological fix), this project exemplifies development as modernisation. As is clear in the words of Federal Prime Minister Godfrey Huggins, such modernisation often entailed the sacrifice of indigenous lands or ways of life towards the goal of national development:

> it is vital that we have this cheap power so that we can industrialise and employ our rapidly increasing African population. . . . The available land is limited but the African population is not. A permanent solution can only be found by industrialisation.
>
> (quoted in Tischler, 2012: 7)

The Kariba Dam scheme aimed to stabilise the local indigenous population (at a time of increasing African nationalism) and to boost migration from Britain to Central Africa. Providing evidence of the Federation's modernisation, it was hoped the dam might cement the position of the white European settler community in the territory. As with earlier schemes of colonial improvement, the project laid claim to land through the notion of making it productive (Tischler, 2012).

Development became one of the fundamental tenets of colonial policy in the 20th century. It aimed to both support colonialism and to improve the welfare of local populations, although it often failed on both counts, dispossessing people of land and contributing to anti-colonial critiques emerging in both colony and metropole. Although colonial development was imagined as a rational modernisation planned in Europe and put into practice in the colonies, this was never the case. Development policies were shaped by the specificities of colonial locations and reworked in connection to local knowledge and practices (Tilley, 2011). Development discourse and practice were constructed in negotiation in a colonial system.

We now turn to the legacies of colonialism in poorer nations that were once part of European empires and to the legacies of colonialism present in contemporary development discourse and practice.

Legacies of colonialism in development

Issues that many contemporary development policies and programmes attempt to ameliorate have their roots, at least in part, in colonialism. Border disputes where colonial boundaries were pushed through previously united communities, ethnic tensions stoked by policies of 'divide and rule' and trauma from bloody wars of decolonisation are all elements of a legacy of colonialism. Many other issues can be traced back more specifically to colonial development policies, for example, unsustainable and environmentally damaging agricultural systems, polluting industrial sectors and inadequate workers' rights, big infrastructure projects which disrupted communities and ecosystems and arguments over land dispossession.

Less obvious are the colonial legacies that have shaped post-colonial development practice and ideology. The focus of state-led development in late colonialism fed into the post-colonial planning and policies of newly independent governments who often pursued vigorous large-scale state controlled development projects in areas of health, housing, industrial development and power infrastructure. In addition, many post-colonial states relied, for the implementation of their development policies, on the input of colonial experts of various kinds – such as agriculturalists, other technical advisors and colonial district officers (Hodge, 2007; Kothari, 2006a). These professionals were employed by the governments of newly independent states and made up a large proportion of the staff of international organisations formed in the wake of the Second-World War such as the World Health Organisation, the United Nations Development Programme, the World Bank, the Commonwealth Development Corporation and the government departments of former colonial powers, such as the Ministry of Overseas Development in the UK. They were also integral to the formation and staffing of the first development studies departments in UK universities (*ibid.*). They therefore also contributed to the shaping of the discipline of development as an academic subject in the second half of the 20th century.

These continuities in policy and personnel underpinned – and were underpinned by – a continuation in broader development discourse. Crisis narratives about overpopulation and environmental degradation produced by late colonial experts fundamentally shaped post-war development as discipline and practice (Hodge, 2007). Problems continued to be depoliticised and their solutions cast as scientific or technocratic. Ideologies which coded the West as developed and the rest as developing, Western as normal and non-Western as other and which constructed a linear temporal path of development along which the West had travelled further continued (and continue) to hold sway after decolonisation. Ideologies of partnership and responsibility in development have replayed older colonial notions with relationships between donors and recipients which continue to be less than equal (Noxolo, 2006). Power (2003: 131) has gone as far as to argue that 'Colonial humanitarianism has been reinvented after the formal end of colonial and imperial rule' as international development. Moreover, just as within colonialism, 'whiteness and the west provide symbols of authority, expertise and knowledge', in the post-colonial era, expertise continued (and continues) to be coded as Western and white, providing fundamental challenges for the theory and practice of development (Kothari, 2006b: 10).

Finally, although modernisation has now been discredited in much development discourse, it has continuing effects. As Ferguson (1999: 14) notes for the experience of Zambians living on the Copperbelt, 'the breakdown of certain teleological narratives of modernity . . . has occurred not only in the world of theory, but in the lived understandings of those who received such myths as a kind of promise'. Even if modernisation was always a myth, it was one that late colonial academics, policy makers and ordinary people invested in, shaping experiences and imagined futures. Ferguson illustrates the devastating consequences when this idea of modernisation was 'turned upside down, shaken, and shattered' (*ibid.*: 13).

Conclusion

Though conceived and practiced differently within and between European empires, development was central to the colonial project, particularly from the late 19th and early 20th centuries. Many of the ideas, policies and priorities of post-colonial development can trace their genealogies to the colonial era, where they were shaped through metropolitan concerns to maintain and modernise colonies and through contact with the local people, knowledge and conditions. Colonialism therefore not only contributed to the material economic and social conditions in which development takes place today but also fundamentally shaped the project of development itself, through continuities among the ideologies, people and practices of colonial and post-colonial development.

Guide to further reading

Ashton, S.R. and Stockwell, S.E. (eds) (1996) *Imperial Policy and Colonial Practice 1925–1945, Part II: Economic Policy, Social Policies and Colonial Research (British Documents on the End of Empire, Series A, Volume 1)*. London: HMSO.
A selection of primary sources drawn from British official archives which document the evolution and practice of British Colonial Development Policy between 1925 and 1945.

Useful website

www.colonialfilm.org.uk/theme/empire-and-development
Films about the British colonies and development, with 150 available to view online. Many have accompanying critical essays analysing their contribution to understanding British colonialism.

References

Crinson, M. (2003) *Modern Architecture and the End of Empire*. Aldershot: Ashgate.
Ferguson, J. (1999) *Expectations of Modernity: Myths and Meanings of Urban Life on the Zambian Copperbelt*. London: University of California Press.
Hodge, J.M. (2007) *Triumph of the Expert: Agrarian Doctrines of Development and the Legacies of British Colonialism*. Athens: Ohio University Press.
Kothari, U. (2006a) 'From colonialism to development: Reflections of former colonial officers', *Commonwealth and Comparative Politics* 44(1): 118–136.
Kothari, U. (2006b) 'An agenda for thinking about "race" in development', *Progress in Development Studies* 6(1): 9–23.
Low, D. A. and Lonsdale, J. M. (1976) 'Introduction: Towards the new order 1945-1963', in D. A. Low and A. Smith (eds) *History of East Africa, Volume III*. Oxford: Clarendon Press.
Noxolo, P. (2006) 'Claims: A postcolonial geographical critique of "partnership" in Britain's development discourse', *Singapore Journal of Tropical Geography* 27(3): 254–269.
Power, M. (2003) *Rethinking Development Geographies*. London: Routledge.
Tilley, H. (2011) *Africa as Living Laboratory: Empire, Development, and the Problem of Scientific Knowledge, 1870–1950*. Chicago: Chicago University Press.
Tischler, J. (2012, in press) 'Negotiating development: The Kariba Dam scheme in the Central African Federation', in P. Bloom, T. Manuh and S. Miescher (eds) *Revisiting Modernization in Africa*. Bloomington: Indiana University Press.

Chapter 15

Heritage and development

Charlotte Cross and John D. Giblin

Heritage can be defined as the use of the past in the present for contemporary and future needs. Consideration of heritage in development studies is often implicit within broader discussions about the relationship between culture and development, which has been recognised by development organisations since at least the 1990s. However, more recently attempts to mobilise the past for better futures have received greater explicit attention within development and heritage policy and practice. This is evident in support for formalised heritage interventions and institutions (e.g., monuments, museums, architectural and archaeological sites, and memorial performances) as part of development programming, as well as attempts to capitalise upon practices and meanings derived from the past to address contemporary development challenges (e.g., the institutionalisation of traditional and customary practices). A potential role for heritage has been identified in diverse areas of concern within development studies, from economic development through heritage tourism, to environmental conservation, peacebuilding, gender equality, healthcare, and security.

Why heritage for development?

A range of factors have contributed to the growth in interest in cultural heritage as a potential contributor to development, reflecting shifts in development thinking and practice, as well as the objectives and activities of heritage stakeholders and particularly UNESCO. Although often perceived as an obstacle to development prior to the 1980s, the increasing popularity of heritage tourism led governments and international organisations, such as the World Bank, to revalorise heritage as a potential economic asset (Lafrenz Samuels, 2018). Shifting understandings of development towards an emphasis on human development and well-being, rather than purely economic growth, have also prompted reflection on the broader value of heritage and its centrality for many to understandings of what constitutes a good life (Basu and Modest, 2014). The importance of ideas about 'sustainable development', enshrined in the Sustainable Development Goals (SDGs) adopted by the United Nations in 2015, has also prompted greater consideration of heritage. Sustainable development and heritage share a concern with preservation for future generations, and the environments in which people live often have an important relationship to cultural practices, beliefs, and material culture.

Heritage has also received greater attention due to critiques of top-down and standardised approaches to development, as part of shifts towards participatory or

DOI: 10.4324/9780429282348-17

community-based development since the 1990s, as well as more recent interest in forms of 'hybrid' governance, involving non-state institutions. Although, such initiatives do not typically explicitly refer to heritage, it is hoped that by harnessing established practices, often described in terms of 'tradition' or 'custom', development interventions will be more locally legitimate, more sustainable, and more cost-effective.

UNESCO and other heritage stakeholders have sought to position heritage as relevant to the global development agenda and associated funding streams (Labadi, 2019). A key concern has been the need to generate evidence of culture's 'impact' on development in order to support advocacy, and UNESCO has, for example, produced indicators to measure the contribution of culture to development in general and specifically to achieving the SDGs.

Finally, beyond its instrumentalisation in doing development 'better', more radical reimaginings of positive futures, which reject the imposition of an assumed universalised trajectory towards 'progress', also point to heritage as a source of alternative ideas of what a better life might look like and how it could be achieved. 'Pluriversal' deconstructions of development emphasise the transformative potential of philosophies, values, and strategies informed by diverse cultural heritages (Kothari et al., 2019).

The politics of heritage for development

Both heritage and development are subject to multiple, often conflictual, interpretations and are implicated in political contestation over how we imagine and value the past and the future. Harrison (2013: 2) reminds us that heritage is

> not a passive process of simply preserving things from the past that remain, but an active process of assembling a series of objects, places and practices that we choose to hold up as a mirror to the present, associated with a particular set of values that we wish to take with us into the future.

Similarly, as described by Mosse (2013: 230), 'the meaning, direction, and control of development are at the heart of contentious politics, bound up with identity, place and belonging'.

Nevertheless, within both fields, critical scholars have identified dominant discourses, which serve to *depoliticise* heritage and development, framing them as technocratic and universal and thus something that can best be realised through the application of 'expertise' (Ferguson, 1990; Smith, 2006). Positioning complex and political questions about which pasts and futures are valued as neutral technical challenges amenable to universal technical solutions obscures the power relations that shape processes of heritage and development and their outcomes. Of particular importance to both heritage and development as professionalised international sectors, due to their emergence alongside and implication in European colonialism, are enduring inequalities in knowledge production between the Global North and the Global South. 'Expert' knowledge about heritage and development remains often largely produced in the North about the South. The institutions through which heritage and development are practised retain substantial 'colonial baggage' (Basu and Modest, 2014: 21) and both fields can be said to be characterised by 'colonial amnesia', which occludes the historical role of today's donors in producing the problems they now purport to address (Rutazibwa, 2019: 165).

Mobilising heritage for development is intended in part to address these challenges as part of a move away from 'top down' and standardised models and towards programming that is better attuned to context and capitalises upon local knowledge. However, in practice calls to use heritage for development and initiatives that seek to do so are not always informed by the critical perspectives outlined earlier, despite these critiques being widely acknowledged within the disciplines of Heritage Studies and Development Studies respectively (Cross and Giblin, 2023). Rather than contested and dissonant, heritage and development are often framed as valued objectives about which there is consensus. The entanglement of heritage and development in programming has produced new specialised forms of expertise to be transferred through 'capacity building' (Lafrenz Samuels, 2018) which informs the circulation of policy models between different countries and contexts. Thus, articulations of heritage for development can perpetuate the depoliticisation of these complex and contested ideas and processes, contradicting the ostensible aims of localising and diversifying development.

In seeking to instrumentalise the past to do development 'better', heritage is often equated to a form of capital; a static and immutable attribute of 'communities' or places that can be turned into development outcomes, such as financial gains or social cohesion. Here, considering heritage ostensibly offers a means to make implementation of development plans easier, adding value to a project in terms of legitimacy and sustainability but also, often, a more cost-effective solution. In practice, rather than being a public good, heritage is subject to multiple interpretations and implicated in politics at different scales, meaning that, like other development initiatives, projects that draw on heritage often produce unevenly distributed costs and benefits.

Conversely, within approaches that are primarily concerned with heritage, development can appear as self-evident, uncontested, and always desired. In particular, the SDGs have been adopted by many heritage practitioners and scholars as a shorthand for development, and thus types of heritage, particular approaches, or individual projects are often assessed in terms of their potential contribution to achieving the SDGs. The objective is to demonstrate the impact heritage can have on a pre-determined version of development, rather than considering more fundamentally how heritage may inform different aspirations for better futures and ideas about how to achieve them (Cross and Giblin, 2023).

Rethinking heritage for development

The challenges associated with instrumentalising the past for better futures outlined earlier do not make heritage unimportant in understanding and pursuing development. Indeed, all around the world people frequently draw on elements of the past in responding to challenges and opportunities or imagining a better society. We can see this in government policies, the livelihood strategies pursued by groups or individuals, in people's everyday attempts to protect the environments in which they live, and in claims for restitution and reparation for past wrongs. Thus, instead of asking *whether* heritage can contribute to development, it is instead valuable to consider *how* the past is selectively used to pursue 'progress' and whose pasts and whose futures are at stake.

Rather than being a form of capital to be invoked to make development more effective, heritage may thus better be understood as being articulated and contested as part of a dynamic process through which we work out what kind of development is desired and how it should be achieved. In this way, while development can be reimagined '*as a cultural project, and particularly as a culturally context-specific project*' (Basu and Modest, 2014: 26), so can heritage be rethought as a culturally context-specific development project (Cross and Giblin, 2023). Crucially, it is important to recognise such processes as political. Understanding who has the power to articulate authoritative visions of the past and the future and the relationship between them – and how these are contested and negotiated at different scales – is central to a better understanding of how development happens – or does not happen – and who benefits and who does not.

References

Basu, P. and Modest, W. (2014) 'Museums, heritage and international development: a critical conversation', in Basu, P. and Modest, W. (eds) *Museums, heritage and international development*. New York: Routledge, pp. 1–32.

Cross, C. and Giblin, J.D. (2023) 'Heritage for development: practising the past in the pursuit of progress', in Cross, C. and Giblin, J.D. (eds) *Critical approaches to heritage for development*. Abingdon: Routledge, pp. 1–34.

Ferguson, J. (1990) *The anti-politics machine: development, depoliticization, and bureaucratic power in Lesotho*. Cambridge: Cambridge University Press.

Harrison, R. (2013) *Heritage: critical approaches*. Abingdon: Routledge.

Kothari, A., Salleh, A., Escobar, A., Demaria, F. and Acosta, A. (eds) (2019) *Pluriverse: a post-development dictionary*. New Delhi: Tulika Books.

Labadi, S. (2019) 'UNESCO, culture, aid and development in the new millennium', in Labadi, S. (ed) *The cultural turn in international aid: impacts and challenges for heritage and the creative industries*. Abingdon: Routledge, pp. 73–88.

Lafrenz Samuels, K. (2018) *Mobilising heritage: anthropological practice and transnational prospects*. Gainesville: University Press of Florida.

Mosse, D. (2013) 'The anthropology of international development', *Annual Review of Anthropology*, 42, pp. 227–246.

Rutazibwa, O.U. (2019) 'On babies and bathwater: decolonizing international development studies', in de Jong, S., Icaza, R. and Rutazibwa, O.U. (eds) *Decolonization and feminisms in global teaching and learning*. London: Routledge, pp. 158–180.

Smith, L. (2006) *Uses of heritage*. Abingdon: Routledge.

Suggested further reading

Basu, P. and Modest, M. (eds) (2014) *Museums, heritage and international development*. New York: Routledge.

Cross, C. and Giblin, J. (eds) (2023) *Critical perspectives on heritage for development*. London: Routledge.

Labadi, S. (2022) *Rethinking heritage for sustainable development*. London: UCL Press.

Labadi, S. et al. (2021) *Heritage and the sustainable development goals: policy guidance for heritage and development actors*. Paris: ICOMOS. Available at: https://openarchive.icomos.org/id/eprint/2453/1/ICOMOS_SDGs_Policy_Guidance_2021.pdf (Accessed 31 July 2022).

UNESCO (2022) *World heritage and sustainable development*. UNESCO website. Available at: https://whc.unesco.org/en/sustainabledevelopment/

Chapter 16

The changing language of international development

Daniel Hammett

In her 2007 book *The No-Nonsense Guide to International Development*, Maggie Black launched a scathing critique of international development. She argued that it is, at best, a fictional concept given prominence in academia but with little relevance to the daily lives of those it is supposed to be helping, at worst 'adverse[ly] affect[ing] poor people and inflict[ing] poverty on those who were not poor before' (Black, 2007: 12). These concerns led her to ask 'is the concept any longer useful?' (Black, 2007: 10). Despite such questions, the notion of international development remains a prominent concern in policy and practice, notably with the focus on the Sustainable Development Goals (as a replacement to the Millennium Development Goals).

This is not to say that understandings of international development have remained static. Rather, we have seen changes in the terminology and language used, shifts in the geopolitical framings and priorities, critical questions asked about the power to define where, how and to whom 'development' happens, as well as shifting geographies of recipients and donors.

Inventing development

The current era of international development is commonly understood as having emerged through the economic and political responses to the aftermath of World War Two, including the success of the 1946 Marshall Plan to support the rebuilding of Europe (a much longer history of international development should also be recognised, including missionary and colonial endeavours). It was against this backdrop that the US President Harry Truman's inaugural speech in 1949 entrenched the neoliberal ethos of the Marshall Plan as a blueprint for international development policies and promoted a clear set of geopolitical and power relations. Truman's speech positioned American values and Enlightenment thinking at the core of development approaches which reified economic growth, industrial modernisation and the free-market. Truman's speech simultaneously located the US as a beacon of development, duty bound to guide other states to reaching a similar position of democracy and modernity, against the corollary of where 'development' needed to happen, namely outside of North America and Western Europe. As critics such as Arturo Escobar (1999) have argued, this speech simultaneously embodied a (neo)colonial and imperialist approach to international development based upon the 'discovery' of mass poverty *and* created a justification for Western interventionism rooted in assumptions of superiority and the 'white man's burden'.

DOI: 10.4324/9780429282348-18

The emergence of the international development agenda coincided with advent of the Cold War and subsequent polarisation of global politics between the capitalist West and communist East. As tensions between the Soviet Union and the US grew, fuelled by proxy conflicts including the Korean conflict (1950–1953) and Vietnam war (1955–1975), a simple geopolitical imaginary took hold through a tri-partite division of the world. This division utilised the language of the First, Second and Third World to highlight differences linked to the Cold War. The First World referred to the capitalist countries of North America, Western Europe and Scandinavia plus Japan, South Korea and Australia, while the Second World constituted the countries of the Communist bloc. The Third World (a term appropriated from the French idea of the 'third estate' that was used in connection with the marginalised and impoverished commoners in pre-revolutionary France (Payne, 2001)) referred to the rest of the world.

The Third World, comprised of newly decolonised and decolonising states across Asia, Africa and Latin America, was quickly identified by political leaders as both the key focus for both international development policies *and* the geostrategic objectives of the two superpowers of the Cold War. These states were viewed as providing opportunities for extending political influence, expanding trade and economic opportunities and ensuring geopolitical and geostrategic advantages (such as access to resources, control over/access to key trade or transport routes, ability to project military power) for the superpowers. Consequently, vast sums of money were spent in support of these geopolitical objectives, often characterised as 'development aid' as both donor and recipient countries frequently manipulated international development budgets and rhetoric for geopolitical, military and individual gain, often at the expense of social justice, human rights and economic growth. Disparities in development spending and interest often reflected differing levels of geostrategic importance and political alignments – witness the US's differing engagements with pro- and anti-communist regimes in Cuba, Chile, Argentina and Brazil during the 1960s and 1970s. This is not to say that Third World countries were powerless in these relations – Ethiopia's leader, Haile Selassie I, successfully played the US and USSR against each other to secure hundreds of millions of dollars in foreign aid from both countries during this period.

While the landscape of international development remained predominantly framed by the language and geopolitics of the global superpowers and their influential allies, resistance to the discursive positioning of the Third World and intensification of the Cold War increased. This resistance was evident in the formation of the Non-Aligned Movement (NAM) in 1956 as an alliance of newly decolonised states who resisted pressures to align with the Cold War super-powers. Founded on five principles of mutual respect for sovereignty, non-aggression, non-interference, equality and peaceful co-existence, the NAM sought to ensure member's independence from the USSR and USA. Alongside the NAM, the Group of 77 was established in 1964 as a coalition of Third World states to promote members' economic interests through the New International Economic Order (NIEO). The NIEO called for a reworking of the global economic order in favour of Third World countries, rather than those benefitting from the Bretton Woods system.

Parallel to these political machinations, critical scholarship also emerged from the Third World, including dependency and post-development theory, which challenged how development was defined and understood, the power-relations which framed understandings of how, where and to whom development occurred. These scholars argued that the language of international development – and development agendas – reinforced Western privilege and perpetuated economic and political inequality as this status quo ensured the economies – and thus, countries – of the Third World were exploited to fuel the economic growth of the West. Despite these progressive efforts, the development landscape throughout the Cold War remained framed by the geopolitical interests of the global superpowers.

Changing terminology and spatiality

In 1980, the Brandt Commission was established to review international development concerns. The Brandt Report's argument was that the world was primarily divided by economics, depicted through the Brandt Line dividing the world between the rich North and the poor South based upon GDP per capita. The report argued that the wealth of the North derived from the profits from manufacturing, while the South relied upon less-profitable exports of primary commodities. Consequently, the Brandt Report called for a redistribution of wealth and resources from the North to the South as part of a broader shift towards a more equitable global political economy. While the language used began to shift the lexicon of development, the recommendations of the report were subsumed by the geopolitical context of the Cold War.

Following the fall of the Soviet Union and end of the Cold War, the terminology of international development began to change. During the 1990s and 2000s there was a shift to modernist-framed language which referred to the 'developed' and 'developing' worlds and then Economically and Less-Economically Developed Countries (EDCs and ELDCs). More recently, we have witnessed further changes in language used, to the terminology of the 'Global North' and 'Global South' and then to 'minority' and 'majority' worlds. While this shifting language has been welcomed, such terminology remains 'premised on a relatively static but increasingly problematic spatial dualism' (Overton et al., 2013: 121) which perpetuates the assumption – in the West – that development (needs to) happen 'out there'. Amidst these changes in terminology, we have also witnessed an effort to shift the spatiality of development within key international development policy and practice. The Millennium Development Goals (MDGs) dominated both development policies and public discourse from 2000–2015, positioning development as a series of challenge and goals to be met by countries in the Global South. The introduction of the Sustainable Development Goals (SDGs) to replace the MDGs in 2015 provides the current dominant policy framework for international development. Crucially, the SDGs overtly position international development as a *global* not *Global South* concern, with development targets positioned as applicable for all countries of the world (Willis, 2016). Layered in to these developments, there has been growing acknowledgement over time of the heterogeneity of countries in the Global South, the interdependence of the global (political) economy and the need to understand development in a holistic manner rather than simply as an economic concern.

(Re)Emerging powers

As the development lexicon has changed and policy landscape has evolved, so too has the role and prominence of different donors – in part reflecting differing economic trajectories of countries, including the emergence of the BRICS (Brazil, Russia, India, China, South

Africa) and, more broadly, middle-income countries as both international development aid donors and recipients (Alonso et al., 2014). This (re)emergence of donors is rooted in longer history of alternative development modalities, including the early manifestations of South-South development cooperation from the 1950s onwards (Mawdsley, 2012). During the Cold War period, a range of relatively modest – but symbolically powerful – forms of South-South development cooperation emerged, including Cuba's health and education cooperation programmes and Brazil's agricultural interventions. Since the 2000s, the economic growth of China, India and other countries have further changed these dynamics, as such countries have emerged as 'new' donors. These donors are often popular with recipient countries as their relations are free from colonial histories and often involve fewer ties and restrictions linked to development aid. However, critical questions are increasingly being asked of the role of new donors, including the extent to which development projects meet local priorities, concerns with competing political and economic purposes and motivations and the potential undermining of the principle of 'non-interference' of the NAM.

Conclusions

The language of international development remains both highly influential and contested. This chapter has offered an initial overview of the invention of development as an arena of policy and scholarship and subsequent shifts in the lexicon used to discuss development. In so doing, it highlights how the entwining of political power and economic and geostrategic interests inform not only how development is talked about but how these discursive practices spatially locate development in particular ways and perpetuate entrenched and inequitable representations of people and places (Hammett, 2019).

It is clear that the landscape of international development remains fluid and changeable. At a time of global geopolitical flux and uncertainty, increasing calls for environmentally sustainable growth and development – and a long-overdue engagement with the need to de-colonise knowledge and power – it is opportune to rethink the ways in which notions of international development are conceptualised, discussed and spatialised.

References

Alonso, J.A., Glennie, J., Sumner, A. 2014. 'Recipients and contributors: Middle-income countries and the future of development cooperation', Department of Economic and Social Affairs Working Paper 135. https://www.un.org/esa/desa/papers/2014/wp135_2014.pdf

Black, M. 2007. *The No-Nonsense Guide to International Development*. Oxford: New Internationalist.

Escobar, A. 1999. 'The invention of development', *Current History* 98: 381–386.

Hammett, D. 2019. 'Whose development? Power and space in international development', *Geography* 104(1): 12–18.

Mawdsley, E. 2012. *From Recipients to Donors: Emerging Powers and the Changing Development Landscape*. London: Zed Books.

Overton, J., Murray, W., McGregor, A. 2013. 'Geographies of aid: A critical research agenda', *Geography Compass* 7(2): 116–127.

Payne, A. 2001. 'The global politics of development: Towards a new research agenda', *Progress in Development Studies* 1(1): 5–19.

Willis, K. 2016. 'Viewpoint: International development planning and the sustainable development goals (SDGs)', *International Development Planning Review* 38(2): 105–111.

Additional readings

Brooks, A. 2017. *The End of Development: A Global History of Poverty and Prosperity*. London: Zed Books.
Carmody, P. 2019. *Development Theory and Practice in a Changing World*. London: Routledge.

Further sources

Kharas, H. 2015. 'The transition from "the developing world" to "a developing world" ', Brooking Institute. https://www.brookings.edu/on-the-record/the-transition-from-the-developing-world-to-a-developing-world/
Khokhar, T., Serajuddin, U. 2015. 'Should we continue to use the term "developing world"?', World Bank. https://blogs.worldbank.org/opendata/should-we-continue-use-term-developing-world
President Truman's Inaugural Speech. 1949. http://www.presidency.ucsb.edu/ws/?pid=13282

Chapter 17

Representing poverty

John Cameron

The ways in which global poverty is represented in text, images, videos and sound by aid agencies, non-governmental organisations, media producers, artists, private businesses and social media users has important effects on how ordinary people in the Global North understand and act in response to global injustice, which in turn has significant impacts on the policies and behaviour of governments, businesses and civil society organisations. Over the past 30 years representations of global poverty have changed significantly and have become much more sophisticated – from depictions of poor people as passive, often hopeless victims in need of Northern assistance to an ever-more diverse range of images and stories that highlight hopefulness using positive imagery, celebrities, sexuality, humour, cause-related marketing and storytelling. The purpose of these representations is not simply to inform Northern audiences but also to provoke emotional responses that will motivate action – from making a donation to signing a petition or buying a product. However, images and stories about global poverty can also have powerful unintended effects, often producing and reinforcing erroneous beliefs that the causes of global poverty and injustice lie entirely in the Global South and the solution is charity from the Global North.

This chapter begins with an analysis of how and why representations of global poverty matter. It then highlights some of the most important theoretical perspectives for thinking critically about representations of poverty. The chapter concentrates on the representation of poverty and development by non-governmental organisations (NGOs) and aid agencies because they have specific mandates to combat poverty, but much of the analysis also applies to the news media and other actors. The central argument of the chapter is that surface-level representations of global poverty have changed significantly over the past 30 years, but the underlying narratives often continue to reinforce perceptions of Southern 'others' as needing assistance and Northern 'selves' as the source of solutions for global problems.

Why and how representations of poverty matter

The representation of global poverty through what Smith and Yanacopulos (2004) call 'the public faces of development' in the Global North plays a very important role in the social construction of public understandings of global poverty and development, which in turn plays an important role in shaping global power relations. As Darnton and Kirk (2011) explain in an important report on UK NGOs, the ways in which ordinary citizens

DOI: 10.4324/9780429282348-19

in the Global North understand global poverty and development has very important practical implications. Public understandings of global poverty shape individual attitudes and behaviour (e.g., welcoming refugees and immigrants, ethical consumption, volunteering, charitable donation, involvement in social justice campaigns), which in turn shapes support for NGOs and governments in the Global North to take action on issues of global poverty, inequality and human rights.

Since the 1960s when development NGOs and aid agencies first established ambitious marketing and fundraising strategies, the prevailing representations of poverty in the Global North have reinforced paternalistic and neo-colonial attitudes towards people and countries in the Global South, simplistic understandings of the causes of global poverty and beliefs that charitable donations are the most appropriate and effective way for citizens in the Global North to make a difference.

The representation of people in the Global South as impoverished victims in desperate need of help from the Global North reached a peak during the Ethiopian famines of the 1980s. Images of emaciated black African children with flies buzzing around their eyes in NGO fundraising materials and news coverage was criticized as the 'pornography of poverty' (Plewes and Stuart, 2006). Almost 20 years later, research by a British NGO found that the effects of those images still shaped the public understanding and attitudes (VSO, 2002). The report, titled *The Live Aid Legacy*, found that '80% of the British public strongly associate the developing world with doom-laden images of famine, disaster and Western aid' and '74% of the British public believe that these countries depend on the money and knowledge of the West to progress' (VSO, 2002: 4). The report also highlighted a prevailing belief in the superiority of the UK public in relation to the Global South and perceptions of the relationship between the UK and the Global South as 'we are powerful, benevolent givers; they are grateful receivers' (VSO, 2002: 4). More recent research shows that public understandings in the UK and other countries of the Global North are not only still stuck in these myths but that they are also becoming even more cynical about the impacts of international development efforts (Darnton and Kirk, 2011). The challenge is for organisations to find ways to communicate to ordinary people in the Global North about global poverty and injustice in ways that motivate them to become more interested and to take action but that do not over-simplify and misrepresent the causes of poverty or possible solutions to it.

Theoretical perspectives on the representation of poverty

Critical analysis of representations of global poverty draws on theoretical insights from many different fields of research but especially from post-colonialism, psychology, political economy and art history. One of the most important insights of post-colonial research on representation is that depictions of 'others' ultimately serve to strengthen the ways that dominant cultures understand themselves. In his book *Orientalism*, Said (1979) argued that European representations of the 'Orient' as uncivilized and backward in the 19th century served the larger goal of strengthening the self-perceptions of Europeans as civilized and modern and thus justifying colonial domination. Similarly, in *White on Black*, Nederveen Pieterse (1995) explained that European representations of Africa from the medieval era to the present tell us more about prevailing thinking in Europe than they do about Africa, highlighting the important message that representations of poverty tell us more about the culture and specific organisation that created the message than they do about the culture or people who are represented.

Research in psychology highlights the role of emotions in human decision-making and can help us to better understand the complex range of emotions that NGOs seek to trigger in their communications, marketing and fundraising material in order to motivate potential donors and activists to change their attitudes and behaviour. While representations characteristic of 'the pornography of poverty' aimed to spark feelings of guilt among Northern audiences as the strategy to motivate charitable giving, NGOs now seek to provoke a much wider range of emotional responses, including hope, pride and joyfulness as well as fear and anger. The key question for critical analysis is: what emotions does a particular representation of poverty aim to provoke and for what strategic purpose?

Research in political economy highlights the political and financial factors behind the strategic decisions of NGOs and other actors about how they represent global poverty to public audiences. For most NGOs the primary motive behind the images, text and videos they create to represent poverty is to motivate audiences to make charitable donations. Northern-based development NGOs face very considerable challenges to raise money to support their work and typically rely on a combination of government funding, grants from philanthropic foundations, individual charitable donations and sometimes revenue from social enterprises such as fair trade shops. Funding from government aid agencies and philanthropic donors like the Gates Foundation typically have many conditions attached, so many NGOs seek to diversify their funding through appeals to individual donors. In a highly competitive market for donations, NGOs need to continuously innovate their fundraising strategies, which means reliance on strategies that are proven to work, such as child sponsorship and appeals for humanitarian disasters but also finding new ways to represent global poverty and development in the hopes of attracting new donors. A growing number of NGOs are collaborating with private corporations in cause-related marketing strategies in which businesses contribute some of their profits to a charitable cause in order to market their products as 'ethical' with the goal of both making consumers feel good about their purchases and to align themselves with a particular brand (see Richey and Ponte, 2011).

The study of art history also offers important insights for critical analysis of representations of poverty, particularly in images and video. In his book *White on Black*, Nederveen Pieterse (1992) highlights the ways in which the positioning of people in an image in relation to each other and in relation to those viewing the image can reinforce or challenge unequal power relations. For example, an image in which the gaze of the viewer is cast down at someone can foster feelings of superiority while images that engage the viewer and the viewed in even eye-to-eye contact tend to evoke feelings of equality.

Recognizing the power of communications and marketing material to shape public understandings, the coalition organisations that represent development NGOs in most countries of the Global North have developed Codes of Ethics on appropriate strategies for representing poverty and development. However, with continual innovation in the strategies of representation used by NGOs these guidelines need to be regularly re-visited and updated to protect and promote the core principles of protecting the dignity of people in the Global South and not reproducing harmful stereotypes about poverty and development or glamorizing the roles of the Global North.

Change and continuity in representations of global poverty

While critical analysis of representations of poverty has often focused on the depiction of poor people as helpless victims in need of outside assistance (i.e. 'the pornography of poverty'), these images are becoming harder to find as NGOs employ new strategies to engage viewers' emotions, with increasing use of positive hopeful images as the path to donations. NGOs are also increasingly using celebrities (Kapoor, 2012), humour (Cameron, 2015), sex-appeal (Cameron and Haanstra, 2008) and cause-related marketing (Richey and Ponte, 2011) – and sometimes the combination of all these strategies – to attract the attention of potential donors or campaign supporters.

While these strategies of representation appear to be new in some ways, they also continue to reinforce some of the same stereotypes and myths as the 'pornography of poverty'. Images of sexy celebrities or cause-related marketing campaigns that feature hip Northern consumers do mark a move away from the depiction of poor people as helpless victims. However, the images of celebrities and use of social media in NGO communications and fundraising still frequently reinforce unequal North-South power relations by highlighting the agency of Northern donors and activists as those who are entitled and empowered to solve global poverty and minimizing the agency of people and institutions in the Global South. It is thus essential for development research to continue to critically analyse the ways in which global poverty and development are represented in the Global North and especially to ask questions about how new strategies of representation may reinforce old myths and stereotypes.

Bibliography

Cameron, John. 2015. "Can Poverty be Funny? The Serious Use of Humour as a Strategy of Public Engagement for Global Justice" *Third World Quarterly* 36, 2: 274–290.

Cameron, John and Anna Haanstra. 2008. "Development Made Sexy: How it Happened and What it Means" *Third World Quarterly* 29, 8: 1475–1489.

Darnton, Andrew and Martin Kirk. 2011. *Finding Frames: New Ways to Engage the UK Public in Global Poverty*. London: Oxfam and UKAid.

Kapoor, Ilan. 2012. *Celebrity Humanitarianism: The Ideology of Global Charity*. London and New York: Routledge.

Pieterse, Jan Nederveen. 1995. *White on Black: Images of Africa and Blacks in Western Popular Culture*. New Haven, CN: Yale University Press.

Plewes, Betty and Ricky Stuart. 2006. "The Pornography of Poverty: A Cautionary Fundraising Tale" in Daniel Bell and Jean Marc Coilaud, eds. *Ethics in Action: The Ethical Challenges of International Human Rights Nongovernmental Organizations*. Cambridge: Cambridge University Press, 23–37.

Richey, Lisa A. and Stefano Ponte. 2011. *Brand aid: Shopping Well to Save the World*. Minneapolis, MN: University of Minnesota Press.

Said, Edward. 1979. *Orientalism*. New York: Vintage Press.

Smith, Matt and Helen Yanacopulos. 2004. "The Public Faces of Development: An Introduction" *Journal of International Development* 16, 5: 657–664.

VSO. 2002. *The Live Aid Legacy: The Developing World Through British Eyes – A Research Report*. www.vso.org.uk/Images/liveaid_legacy_tcm8-784.pdf

Suggested further reading

Chouliaraki, Lily. 2010. *The Ironic Spectator. Solidarity in the Age of Post-Humanitarianism*. Cambridge, UK: Polity Press.

Dóchas. 2014. *The Illustrative Guide to the Dóchas Code of Conduct on Images and Messages*. Dublin: Dóchas. https://dochas.ie/sites/default/files/Illustrative_Guide_to_the_Dochas_Code_of_Conduct_on_Images_and_Messages.pdf

Dogra, Nandita. 2012. *Representations of Global Poverty: Aid, Development and International NGOs*. London and New York: I.B. Tauris.
Hall, Stuart, Jessica Evans and Sean Nixon, eds. 2013. *Representation: Cultural Representations and Signifying Practices*. 2nd ed. London: Sage.

Audio visual material

RadiAid. https://www.radiaid.com/ Features the best and the worst of international development marketing videos plus guidelines on ethical representations of poverty.
The Bang-Bang Club (Feature Film) Portrays the lives of four photojournalists covering protests against the Apartheid regime in South Africa from 1990–1994.

Chapter 18

Global North and Global South

Kamna Patel

In contemporary development studies the terms 'Global North' and 'Global South', sometimes capitalised and sometimes not, roll off the tongue as a familiar, known and knowable concept that translates easily across development audiences. I refer to 'Global North' and 'Global South' as a singular concept because one does not make any sense without the other. Its meaning is derived entirely from being a pair where one part serves as an imaginary counterfoil to the other. As a heuristic device, the usage and utility of Global North/South intends to conjure particular imaginations of difference. Within development studies, a field of scholarship that rests on differences between places and people, I argue the main usage of Global North/South is part of a longer history of racialised sense-making that follows a colonial imprint, and its utility is derived from development scholars and practitioners readily discovering, learning and locating subjects of development.

Imaginative geographies of difference

The Global North/South is an example of an imaginative geography of difference. The concept of 'imaginative geographies' draws heavily on the work of postcolonial scholar Edward Said, who uses it to chart the interconnections among knowledge, power and geography. To Said, an imaginative geography invokes a powerful representation of spatially coded 'otherness'. In *Orientalism* (first published in 1978), Said's seminal work on postcolonial representation, he explores representations of 'the Orient' and 'the Occident' through cultural and linguistic markers that denote spaces of difference between, respectively, 'the West' and 'the East'. He notes that markers – or discourses – such as rational and irrational, modern and non-modern, superiority and inferiority, are dual ideas that work in pairs to essentialise difference in simplistic ways and that they are almost always applied by the West over the East. At their base function, simplistic essentialising allows control over a narrative about the world outside of Europe and beyond Europeans and makes that world and its people visible and knowable to Europeans. For example, identifying patterns of housing different from a European norm and labelling them 'informal' becomes a way to make differences that exist outside of Europe visible and known to Europeans. An example from Said's work further illustrates the point. He explains,

> A group of people living on a few acres of land will set up boundaries between their land and its immediate surroundings and the territory beyond, which they call "the land of the barbarians". . . this universal practice of designating in one's mind a familiar space which is "ours" and an unfamiliar space beyond "ours" which is "theirs" is a

DOI: 10.4324/9780429282348-20

way of making geographical distinctions that *can* be entirely arbitrary. I use the word "arbitrary" here because imaginative geography of the "our land-barbarian land" variety does not require that the barbarians acknowledge the distinction. It is enough for "us" to set up these boundaries in our minds; "they" become "they" accordingly, and their territory and their mentality are designated as different from "ours".

(2003 [1978]: 54)

The power and purpose of imaginative geography is then to universalise and differentiate between an 'us' and 'them', based on a process of sense-making that conveys shared understanding and shared meaning between readers and writers of the term. In Said's example of land-barbarian land, we can also see elements of geographical racial determinism. That is the idea that lands and peoples outside of a Europe that is familiar and known to its inhabitants is bound to be *less than* and its unfamiliarity can only be understood in deleterious terms.

It is too simplistic to say that Said's idea of imaginative geography has produced a discourse of Global North and Global South. Rather, we can trace in the usage of Global North/South a legacy or a continuation of a pattern identified by Said where such dualism and its usage reveals the interconnections of knowledge, power and geography.

Global North/South in development scholarship

The evolution of the term Global North/South in development scholarship follows a tradition of other similar dualisms of difference that variously describe macro political, economic and imperial relations: developed and developing; core and periphery and first and Third World. In around the mid-2000s, preference for Global North/South became more mainstream in development scholarship, displacing developed/developing and its 'not quite *there*' terminology, evocative of a path to Western modernity as the only path to progress and progress as the only means of understanding positive change. As a value-free descriptor, 'Global North/South', offered a greater potential to imagine difference without deficit or hierarchy, but the concept is still not free of the geographies of imagination that Said alerted us to and those relationships of knowledge, power and geography.

In thinking about the Global North/South in relation to imaginative geographies of difference, there are two aspects of the Global North/South within development scholarship worth further interrogation. The first is its literal use as a geographical signifier of places or territories that are remarkably distinct from one another. The second is its metaphorical value as a theoretical provocation or conceptual apparatus to de-centre and dislodge the refraction of knowledge through a Euro-North American gaze and, related to this, as a methodological invitation to learn about the world differently.

Global North/South as literal

As a concept, the Global North/South is not tied to terra firma or territory. Just as there was no territory of the Orient or of the West, there is no territory of the Global North or Global South. Southern hemisphere countries such as Australia are readily imagined as part of the Global North and northern hemisphere countries such as India are readily imagined in the Global South within discourses of development. Yet, despite the fluidity the application of the concept can give over geographical terrains, in development scholarship, the concept is frequently used as geographical shorthand to denote a real place afflicted by poverty, inequality or particular vulnerabilities. From this, it serves as an organising logic for development itself, sorting centres of valuable development knowledge, resource and expertise.

The creation of the Global South as a real place has at least three aspects. The first aspect is material; for example, there is a Global South where development as practice is carried out in country offices with staff and equipment. There is a Global North, which may no longer model the behaviours and structures to be emulated in the Global South but is a material supplier of staff and equipment either directly through development projects or indirectly via aid flows. The second aspect is through a connotation of the Global South in popular imagery through grounded representations of otherness such as slums, barren landscapes and black and brown bodies (Patel, 2022). The third aspect is through the use of the 'Global South' as a label, a proper noun, applied to a particular collection of countries that have variously been named 'developing', 'peripheral', 'backward' and/ or 'the Third World' and where the work to develop others is carried out.

The three aspects described here map the concept of Global North/South onto Earth, applying a material grounding to a fluid concept. That is, the usage of Global North/ South in development scholarship takes on a concreteness, a territory, without ever having to demarcate the borders of the Global North and Global South. This position of a *real* Global South with unarticulated borders is possible because of the power of shared understanding – or a shared imagination – between writers and readers of the term. Its utility in development scholarship reflects historically entrenched ideas of difference and lack, while simultaneously absenting from its lexicon the history of global political and economic systems that create/d conditions for underdevelopment and dispossession.

The literal use of Global North/South in development scholarship is not without challenge. Ananya Roy and Emma Shaw Crane (2015), in their book *Territories of Poverty* (2015), question the organisation of poverty by places and advocate strongly for a re-theorisation of poverty that forces us to rethink the value and usage of Global North/ South in a limited geographical way. Notwithstanding the geopolitics of poverty, wealth creation and dispossession and the spatial analytical value in locating sites of poverty in relation to sites of wealth (beyond and within national borders), Roy in the opening chapter explains that the real value of Global North/South lay in its relationality. She continually questions, 'What is the *finis*, the border, that demarcates global North and global South' (2015: 8), noting that, 'the global South cannot be mapped as a single and stable location' (2015: 16). Rather, the challenge laid down by Roy and Crane is to create different (if not new) positions and theorisations that navigate these bands and treat Global North/South as metaphorical concepts that stretch our analyses and knowledge of poverty. It is to this metaphorical use of Global North/South as theoretical, conceptual and methodological invitation to think differently, to which I now turn.

Global North/South as metaphor

The work of a metaphor is to appeal to the imagination and allow a transcendence from a literal meaning. Thinking of the Global North/South as a metaphor creates an intellectual space and provides a conceptual apparatus to think differently and to know differently by shifting our empirical vantage point. The usage of the term is then mainly and purposefully a heuristic device.

As a metaphor, the questions we ask of the concept shift from where is the Global North/South (in its literal use and utility) to how and to whom is the Global North/South relevant in the work they wish to do. To Caroline Levander and Walter Mignolo (2011: 3), the Global South is 'an entity that has been invented in the struggle and conflicts between imperial global domination and emancipatory and decolonial forces'. In this short sentence, the authors do two important things. First, they identify the Global South as a place made through two antithetical forces – imperialism and decolonisation – with the former locating places for exploitation and the latter, based on generational experiences of being exploited, drawing together forces of anti-imperial struggle including projects to radically rethink ideas of freedom, prosperity and well-being. In this making of the Global South, Levander and Mignolo reinforce the importance of the questions *to whom* and *from where* is the Global North/South conceptually valuable. Secondly, resulting from the first, they attempt to decouple the Global North/South and identify the utility of the concept the 'Global South' as a distinct concept, one that is not valuable as an imaginative device in relation to the Global North but as multiple sites across different parts of the world where there are struggles against oppressive forces. These struggles include practising progress outside of imperialist, Western-centric notions of modernity, which of course has a long association with development as a modernising mission.

In articulations of the Global South as a distinct singular metaphorical concept, the Global North is pushed to the background. This does not make the term and what it represents irrelevant or invisible but decentred and deliberately displaced. This is particularly apparent in development discourses of South-South cooperation, for example, as a particular type of international relationship that is not refracted through Europe or North America. South-South relations are an important intellectual and material flow of ideas of development and their attainment. Scholarship in this area locates the history of development in a different time/space and set of relations to the established canon of development scholarship; the Bandung Conference rather than the Truman Doctrine begins a global exercise in poverty alleviation and action for equity (Ndlovu-Gatsheni, 2019).

The project to decentre the Global North is also a project to engender new theories and ideas of development that purposefully draw on scholarship outside of the established canon of knowledge produced by and housed in institutions of the Global North (see Narayanan, 2020; Oldfield and Parnell, 2014 for examples). The symbolic conceptual decoupling of Global North/South is in many ways a historic-spatial commentary on and response to the values, ideas and norms that are privileged in development scholarship. Viewed in this way, the scale of challenge to realise the potential of the Global South and/ or Global North/South as metaphor becomes apparent. Particularly, when we consider that development – as a discipline largely housed in Euro-North American universities and staffed by scholars educated in the canon and as an established industry with material presence – is entangled in powerful imaginative geographies of difference.

So, do I use Global North/South or not?

Often, critical discussions of the concept of Global North/South can lead us to ask these simple questions: do I use the term or not? And if not, what should I use? To these questions, there is no clear or useful answer. A more salient question is what does the Global North/South offer us to understand questions of poverty and inequality? Every answer to this question engages with the relationships between knowledge, power and geography. For instance, where Global North/South is used literally to denote sites of poverty and underdevelopment, following an established pattern that invokes racialised imaginations of difference, we can clearly see the processes that create the imaginative geographies of Said. Where Global North/South is applied metaphorically there is still a politics of difference that plays out that rests on relationships of knowledge, power and geography, which are embedded in the questions of Levander and Mignolo on *to whom* and *from where* the Global North/South is conceptually valuable.

Whether one chooses to use the concept of Global North/South or just Global South (where the North is made implicit), the choice ought to reflect a careful navigation of knowledge, power and geography from the position of the user. I write as a development scholar who studied, teaches and works from 'the North' with a set of bodily and material engagements with 'the South', and this mediates my encounters with the concept of Global North/South and affects the utility with which I value it. From my vantage point, I elevate the analytical lens of race – which allows me to speak to imaginations of difference – to navigate the relationships of knowledge, power and geography that create Global North/South, including the borders implied by scholars advocating for a conceptually singular Global South. In this chapter, I do not wish to diminish the real power in rethinking traditional languages of development. I rather simply identify the tensions that exist between literal and metaphorical use and the heavy imperial baggage of essentialing dualisms that is explicitly and implicitly present in the meanings of Global North and Global South.

References

Levander, C. and W. Mignolo (2011) 'The Global South and World Dis/Order', *The Global South* 5(1): 1–11.

Narayanan, N.P. (2021) 'Southern Theory without a North: City Conceptualization as the Theoretical Metropolis', *Annals of the American Association of Geographers*, 111(4): 989–1001, DOI: 10.1080/24694452.2020.1791040

Ndlovu-Gatsheni, S.J. (2019) 'Rethinking Development in the Age of Global Coloniality', in *Thinking and Unthinking Development: Perspectives on Inequality and Poverty in South Africa and Zimbabwe*, B. Mpofu and S.J. Ndlovu-Gatsheni (Eds.). New York and Oxford: Berghahn Books, pp. 27–46.

Oldfield, S. and S. Parnell (2014) *The Routledge Handbook on Cities of the Global South*, Oxon, UK: Routledge.

Patel, K. (2022) 'Being Cosmopolitan: Marketing Development Studies in the Neoliberal University', *Progress in Development Studies* 22(3): 222–238.

Roy, A. and Crane, E.S. (2015) *Territories of Poverty*, Athens, Georgia: University of Georgia Press.

Said, E. (2003 [1978]) *Orientalism*, London: Penguin.

Additional resources/further reading

Murrey, A. (2018) 'When Spider Webs Unite They Can Tie Up a Lion: Anti-Racism, Decolonial Options and Theories from the South', in *The Handbook of South-South Relations*, E. Fiddian-Qasmiyeh and R. Daley (Eds.). London: Routledge, Chapter 4.

Ndlovu-Gatsheni, S.J. (2021) 'The cognitive empire, politics of knowledge and African intellectual productions: reflections on struggles for epistemic freedom and resurgence of decolonisation in the twenty-first century', *Third World Quarterly*, 42(5): 882–901, DOI: 10.1080/01436597.2020.1775487

Pailey, R.N. (2020) 'Decentring the "White Gaze" of Development', *Development and Change* 51(3): 729–745.

Chapter 19

The shift to global development

Rory Horner

A new development paradigm?

Global development is increasingly referred to as a paradigm for development studies, as well as in relation to policy and practice. The United Nations' Sustainable Development Goals (SDGs), agreed in 2015, relate to all countries and highlight how change is needed beyond just the Global South for a prosperous and more sustainable planet. A global scope, related to development challenges facing the whole world, potentially marks a significant geographic shift. The practice and study of international development has long been associated with that part of the world variously referred to as Third World, Global South, industrialising, developing etc. and with aid underpinned by charity from Northern countries to those in the South. A classic focus of development policy and study has been how the South could become 'developed' like the Global North. Calls for a global development focus are not entirely new (e.g., Hettne, 1995). Yet the replacement of the 2000–2015 Millennium Development Goals (MDGs), which were overarchingly oriented towards development challenges in lower-income countries, with the SDGs has accelerated such calls (e.g., Gore, 2015; Leach, 2015).

Why global development?

Three key factors emerge as reasons underlying a shift to global development (Horner, 2020) – global interconnectedness, the challenge of sustainable development and blurring North-South boundaries. None are completely new, yet they have attracted increasing salience in the 21st century in highlighting the need to move beyond North-South oriented international development in order to capture the processes, actors and challenges facing our contemporary world.

Global interconnectedness

Contemporary *global interconnectedness* involves underlying causal processes whereby development outcomes in different parts of the world are interlinked, challenging the North-South binary. While connections across the globe have shaped development outcomes for at least a few centuries, the nature and extent of global interconnectedness has accelerated from the late 20th century onwards. Contemporary globalisation is increasingly recognised as multi-polar and includes prominent South-South interactions in, for example, trade and migration. Moving beyond a residualisation of the causes

DOI: 10.4324/9780429282348-21

of underdevelopment in the Global South, attention to global interconnectedness high-lights the relational nature of many development processes which cut across and within North and South. If privilege and marginalisation are understood as products of the same economic and social relations of capitalism, development can no longer just be about changes for lower-income people. Elite populations need to change too.

The interconnected nature of global public goods provides a particular challenge to approaches which locate the major development challenges just within the Global South.

Global public goods (Kaul, 2017) have benefits available on a potentially worldwide basis and cannot be provided by one particular part of the world alone. The environment is a prominent example, but other global public goods which come under attention are taxation cooperation and infectious diseases. COVID-19 spread rapidly through an inter-connected world, providing a health challenge for all countries (Oldekop et al., 2020). The experience of the pandemic demonstrates how the Global North can fall short in facing development challenges and the detrimental consequences for more equitable and sustainable development.

Sustainable development

The issue of sustainable development, especially climate change, warrants discussion in its own right as a factor which fundamentally challenges older framings of development problems as just something needed in the Global South. Awareness of the issue of sus-tainable development has grown considerably from the mid-20th century onwards, but the universal relevance of the SDGs is arguably irreconcilable with the old development logic. While there are valid questions about the extent of influence the SDGs (and previ-ously the MDGs) have, due to their non-binding nature and lack of accountability, they are a framing device whereby development is about transformation and sustainability globally rather than just about poverty and related issues in the Global South.

The SDGs – and dashboards or indictors of progress towards them – clearly demon-strate how high-income countries fall considerably short. Prominent such issues include climate change, conservation and sustainable production and consumption. With the world facing planetary boundaries in the face of biophysical limits (Steffen et al., 2015), high-income populations must make changes.

Blurring North-South boundaries

A further key reason for a shift towards global development is the changing nature of global inequality, including blurring of North-South boundaries. Income inequalities between countries in the Global North and the Global South grew for almost two cen-turies from the 19th century onwards. However, some changes in the pattern of global inequality have emerged more recently, which have been termed 'converging divergence' (Horner and Hulme, 2019). Converging is a trend found in terms of between-country inequalities, while 'divergence' refers to growing within-country inequalities. In terms of income, for example, population-weighted between-country inequality has fallen since 1990, the first such decline in more than two centuries, alongside mostly rising income gaps within many countries.

North-South boundaries are blurring in the 21st century across a wide range of economic and other development indicators. According to the World Bank's World Development

Indicators, low and middle-income countries share of global GDP has increased from 23.0 per cent in 2000 to 38.8 per cent in 2020. The number of economies classified by the World Bank as low-income has fallen from 63 in 2000 to 27 in 2020. Fewer countries are aid dependent now. The gap in life expectancy between low and middle-income countries (LMICs) and high-income countries (HICs) has fallen from 21.3 years in 1960 to 9.7 years in 2019. Changes in disease profile have also emerged, especially with the rise of non-communicable diseases in the Global South. Annual carbon emissions from LMICs have been higher than from HICs since 2005. That said, important limitations to trends of convergence must be acknowledged, being only a direction of change and for a relatively short period of time compared to the extent of between-country inequality. Convergence as an end result has not been achieved.

Challenges for a global development paradigm

The idea of a clear spatial demarcation between Global North and South, with the major development problems facing the world lying in the latter, is undermined. Nevertheless, several challenges emerge for a global development paradigm.

The term global development is deployed in multiple ways and according to various meanings, leading to confusion. At times, the term is used synonymously with international development and can just be a relabelling to appear more fitting for the 21st century. Parallels are present with the case of 'global health', which has largely superseded an earlier emphasis on international health in a change which largely preceded the 'international to global' shift in development (Horner, 2022). The ambiguity of the global can be part of its attractiveness. Old associations can persist – that global development and global health are still about Northern solutions and perspectives to global problems, focus on certain places and involve unequal partnerships.

Related to different interpretations of the term, some (e.g., Bangura, 2019) have questioned whether a focus on global development could overlook national and local development, which are important scales for policy and political processes. This understanding is of global (development) as scale, however, rather than as scope (e.g., Horner, 2020, 2022). The former can be thought of in a vertical sense, referring to actors, organisations and processes at a larger scale than the local or national. The latter refers to the whole world – including both Global North and South. The SDGs, for example, are global in scope. To be delivered, they require multi-scalar action which includes local and national as well as global levels.

A further critique of arguments for a global development paradigm questions empirical claims regarding blurring of North-South boundaries. Economic growth in the Global South over the last three decades has been very uneven, heavily concentrated in East Asia and China particularly (e.g., Ghosh, 2019). Trends of convergence from 1990 or so start from an extremely low baseline point – when many parts of the Global South had been severely, detrimentally impacted by structural adjustment (Fischer, 2019). Threshold effects – whereby small movements result in certain elements being categorised above rather than below a certain level – also affect the numbers who have fallen into poverty, middle-class populations and the classification of countries by income group (Sumner, 2019). However, change is not just limited to China or East Asia, or the economic sphere. While 30–40 countries appear to have been stuck, others have had increases in consumption and GDP per person (Sumner, 2019). Less debate emerges around convergence trends in human development aspects such as life expectancy and mortality rates (both maternal and under five), where 'real global convergence' (Bangura, 2019) has occurred. In terms of the underlying rationale for a global development paradigm, there appears little serious contention that climate change is a challenge beyond the usual development scope of the Global South or to the idea that causal development processes cut across different parts of the world due to contemporary interconnectedness.

Some have countered proposals for a global development paradigm with advocacy of visions of development, such as a focus on structural transformation, which remain located just in the Global South (e.g., Fischer, 2019; Sumner, 2019). This perspective provides an important warning that a global focus could lose sight of long-standing development challenges in lower-income countries. After all, 84 per cent of the world's 7.76 billion population (in 2020) live in what are classified as low- and middle-income countries.

A further concern regards the tension between universality and specificity, and recognises the need to still pay attention to the latter (Mohan, 2021). A global development paradigm should not involve a recentring of the West or the projection of claims and understandings of the world just based on the European or American historical or contemporary experience. Instead, a plurality of voices and ideas, as well as multi-directional learning, are needed to bring genuinely global understandings (Hope et al., 2022; Sims et al., 2022).

A final concern is that the argument 'we're all developing countries now' is used to undermine the need for aid or development cooperation. This is not a logical conclusion. Major inequalities are still present in the world, including between countries. Nevertheless, it is difficult to see how aid could address many of the challenges the world faces today – global cooperation for sustainable development will need to go well beyond the traditional development policy system (Klingebiel and Gonsior, 2020).

Conclusion

Our world today faces a different set of challenges from the mid-late 20th century when much of the field and practice of development studies and practice were established. While there is increasing consensus that the old map of development challenges no longer holds, there is still debate over what global development is, as well as regarding its merits and limitations. How a global development paradigm will unfold requires continued and critical attention given the enormous challenge of creating a more inclusive and sustainable world.

References

Bangura, Y. (2019) 'Convergence is not equality', *Development and Change*, 50(2), pp. 394–409.

Fischer, A. (2019) 'Bringing development back into development studies', *Development and Change*, 50(2), pp. 426–444.

Ghosh, J. (2019) 'A brave new world, or the same old story with new characters?', *Development and Change*, 50(2), pp. 373–393.

Gore, C. (2015) 'The post-2015 moment: Towards sustainable development goals and a new global development paradigm', *Journal of International Development*, 27(6), pp. 717–732.

Hettne, B. (1995) *Development Theory and the Three Worlds: Towards an International Political Economy of Development*. Harlow: Longman.

Hope, J., Freeman, C., Maclean, K., Pande, R. and Sou, G. (2022) 'Shifts to global development: Is this a reframing of power, agency, and progress?', *Area*, 54(2), pp. 154–158.

Horner, R. (2020) 'Towards a new paradigm of global development? Beyond the limits of international development', *Progress in Human Geography*, 45(3), pp. 415–436.

Horner, R. (2022) 'Beyond rebranding from international to global: Lessons from geographies of global health for global development', *Area*, 54(2), pp. 159–167.

Horner, R. and Hulme, D. (2019) 'From international to global development: New geographies of 21st century development', *Development and Change*, 50(2), pp. 347–378.

Kaul, I. (2017) 'Making the case for a new global development research agenda', *Forum for Development Studies*, 44(1), pp. 141–148.

Klingebiel, S. and Gonsior, V. (2020) 'Development policy from a systemic perspective: Changes, trends and its future role within a broader framework for transnational co-operation', *Revista Brasileira de Política Internacional*, 63(2).

Leach, M. (2015) 'The Ebola crisis and post-2015 development', *Journal of International Development*, 27(6), pp. 816–834.

Mohan, G. (2021) 'Below the belt? Territory and development in China's international rise', *Development and Change*, 52(1), pp. 54–75.

Oldekop, J. A., et al. (2020) 'COVID-19 and the case for global development', *World Development*, 134, p. 105044.

Sims, K., Banks, N., Engel, S., Hodge, P., Makuwira, J., Nakamura, N., Rigg, J., Salamanca, A. and Yeophantong, P., Eds. (2022) *The Routledge Handbook of Global Development*. London: Routledge.

Steffen, W., et al. (2015) 'Planetary boundaries: Guiding human development on a changing planet', *Science*, 347(6223), pp. 736–746.

Sumner, A. (2019) 'Global poverty and inequality: Continuity and change in late development', *Development and Change*, 50(2), pp. 410–425.

Chapter 20

Enlightenment and the era of modernity

Marcus Power

Introduction: the 'rough and tumble' of early industrialism

Just as light cuts through darkness, the philosophy of the Enlightenment was seen as something that would open the eyes of the world's poor and free them from unjust rule. The 'age of Enlightenment' is most often traced to the 18th century and represented a catalyst for the development of particular styles of social thought in the form of a movement or a programme in which reason was used in order to achieve freedom and progress and during which hostility to religion was omnipresent. In its simplest sense, the Enlightenment was the creation of a new framework of ideas and secure 'truths' about the relationships among humanity, society and nature which sought to challenge traditional worldviews dominated by Christianity. Science – and the scientific approach, – became tools to investigate the world, instead of theological dogmas. According to Gay (1973: 3) at this time educated Europeans experienced

> an expansive sense of power over nature and themselves: the pitiless cycles of epidemics, famines, risky life and early death, devastating war and uneasy peace -the treadmill of human existence – seemed to be yielding at last to the application of critical intelligence.

Fear of change began to give way to fear of stagnation. It was a century of commitment to enquiry and criticism, of a decline in mysticism, of growing hope and trust in effort and innovation (Hampson, 1968). One of the primary interests was social reform and the progression and development of societies built around an increasing secularism and a growing willingness to take risks (Gay, 1973).

There is no monolithic 'spirit of the age' that can be discerned however and the Enlightenment does not represent a set of ideas which can be clearly demarcated, extracted and presented as a list of essential definitions. There were however many common threads to this patchwork of Enlightenment thinking: the primacy of reason/rationalism, a belief in empiricism, the concept of universal science and reason, the idea of progress, the championing of new freedoms, the ethic of secularism and the notion of all human beings as essentially the same (Hall and Gieben, 1992: 21–22). Thinkers such as Kant, Voltaire, Montesquieu, Diderot, Hume, Smith, Ferguson, Rousseau and Condorcet found a receptive audience for their 'new style of life' (Hampson, 1968) producing a large collection of novels, plays, books, pamphlets and essays for the consumption of nobles, professionals (especially lawyers), academics and the clergy. New cultural innovations in writing,

DOI: 10.4324/9780429282348-22

painting, printing, music, sculpture and architecture and new technological innovations in warfare, agriculture and manufacture had a major impact on the *philosophes*, the free-thinking intellectuals or 'men of letters' that had brokered this enlightened awakening in France. The *philosophes* sought to redefine what was considered as socially important knowledge, to bring it outside the sphere of religion and to provide it with a new meaning and relevance. For Hall and Gieben (1992: 36) four main areas distinguish the thought of the *philosophes* from earlier intellectual approaches:

• Anti-clericalism.
• A belief in the pre-eminence of empirical, materialist knowledge.
• An enthusiasm for technological and medical progress.
• A desire for legal and constitutional reform.

There is thus clearly a risk in applying the term 'the Enlightenment' too loosely or too widely, as if it had touched every intellectual society and every intellectual elite of this period equally. The Enlightenment is thus best considered as an amorphous, dynamic and variegated entity (Porter, 1990). More than simply a predominantly French movement centred around a small group of *philosophes*, scholars have recently begun to consider the complex spatiality of 'the Enlightenment' as a cosmopolitan process, to view it in its international context (where its key ideas and views were transmitted across borders) and thus to identify a number of different 'Enlightenments'. Reaching its climax in the mid-18th century in Paris and Scotland but with foundations in many countries (including several outside of Europe such as the USA), 'the Enlightenment' was thus a sort of intellectual fashion or 'a tendency towards critical inquiry and the application of reason' (Black, 1990: 208) rather than a singular coherent intellectual movement or institutional project. The *philosophes* of the 18th-century Enlightenment in France, for example, did not act in concert and neither should they be seen as a unified family, for their views were too disparate (Porter, 1990).

It is also important to remember that the new 'style of life' championed by Enlightenment intellectuals was in the main reserved for the fortunate and the articulate – the rural and urban masses had little share. It was not until the eve of the French Revolution in the 1780s that a new social group emerged concerned with popularising Enlightenment ideas. Similarly, though many women played a major part in the development and diffusion of Enlightenment ideas, applying such ideas to their social conditions meant negotiating a number of contradictory positions within patriarchal societies. The emancipatory potential of this knowledge thus turned out to be limited in that it was conceived of as abstract and utilitarian, as a mastery over nature which thus becomes characterised by power. As Doherty (1993: 6) has argued:

> Knowledge is reduced to technology, a technology which enables the *illusion* of power and of domination over nature. It is important to stress that this is an illusion. This kind of knowledge does not give actual power over nature. . . . What it does give in the way of power is, of course, a power over the consciousness of others who may be less fluent in the language of reason. . . . Knowledge thus becomes caught up in a dialectic of mastery and slavery.
>
> (Emphasis in original)

The Enlightenment was also closely linked to the rise of modernity and provided an important crucible for the invention of the modern idea of 'development' which began to emerge 'amidst the throes of early industrial capitalism in Europe' (Cowen and Shenton, 1996: 5). The metaphor of the 'light of reason' shining brightly into all the dark recesses of ignorance and superstition in 'traditional' societies was a powerful and influential one at this time. In Europe, the light that the process of 'development' brought was intended to 'construct order out of the social disorders of rapid urban migration, poverty and unemployment' (Cowen and Shenton, 1996: 5). Many Enlightenment thinkers also viewed the remedy for the disorder brought on by industrialisation as related to the 'capacity' to use land, labour and capital in the interests of society as a whole. Only certain kinds of individuals could be 'entrusted' with such a role (Cowen and Shenton, 1996). Property, for example, needed to be placed in the hands of 'trustees' who would decide where and how society's resources could be most effectively utilised. In 18th-century France, the prevailing social orders were represented as three 'Estates' – Clergy, Nobility and the 'Third Estate', which comprised everyone else, from wealthiest bourgeois to poorest peasant (Hall, 1992). This 'dialectic of mastery and slavery' and this gap between the *philosophes* (who were often members of the second Estate) and the peasantries of European 18th-century societies are both important parts of the historical context of Enlightenment thinking. Although they appeared to represent a threat to the established order, these ideas and writings sought evolutionary rather than revolutionary change, arguing that progress and development could come about within the existing social order through the dissemination of ideas among 'men of influence' (Hall and Gieben, 1992).

'Modernity' and the rise of the social sciences

The influential Economist John Maynard Keynes (1936: 570), once wrote that '[p]ractical men, who believe themselves to be quite exempt from any intellectual influences, are usually the slaves of some defunct economist'. So it is with much development thinking today. A variety of 20th-century movements including neo-classicism (of which Keynes was an important part) and liberalism can trace their origins back to the Enlightenment. The foundations of many modern disciplines (including development studies) were intimately bound up with the Enlightenment's concept of progress and the idea that development could be created through the application of reasoned and empirically based knowledge. The Enlightenment had forged the intellectual conditions in which the application of reason to practical issues could flourish through such 'modern' institutions as the academy, the learned journal and the conference. In turn, a 'modern' audience was constituted for the dissemination of social and political ideas alongside a class of intellectuals that could live from writing about them (Hall, 1992). Through the Enlightenment, state bureaucracies began to use social statistics to provide the evidence necessary for 'rational' choices in the allocation of resources. This process of labelling people was part of a wider intellectual paradigm that considered categorisation, quantification and measurement as integral to rational and objective decision-making. These 'official' labels were – and still are – generally portrayed and accepted as objective facts, though many are rooted in intensely political processes. For example, many conventional racial and group classifications were created in the imperial and colonial periods, when authorities counted, categorised, taxed and deployed slave, servile and forced labour, often over vast geographical areas (IDS, 2006: 1).

The emergence of an idea of 'the West' was also important to the Enlightenment in that it was a very European affair, which put Europe and European intellectuals at the very pinnacle of human achievement. This view sees 'the West' as the result of forces largely internal to Europe's history and formation (Hall, 1992) rather than as a 'global story' involving other cultural worlds. In the making of 19th-century European 'modernity', Europeans had a sense of difference from other worlds (e.g., 'Africa'), which shaped the ways in which they were viewed as distant, uncivilised and immature stages in the progress of humanity. The establishment of modern modes of scientific enquiry, of modern institutions and the modern 'development' of societies in 19th-century Europe thus partly incorporated a contrast with the 'savage' and 'uncivilised' spaces of the non-Western world. The emergence of area studies disciplines in the 20th century can also be traced back to Enlightenment efforts to support theories of human progress by comparing Europe to other regions of the world and in elaborating the contrast between Europe and other areas (Ludden, 2003). This tradition of universal comparison and ranking has also arguably continued to be a feature of 'development thinking' in the 21st century.

Modernist reason was not as inherently good as the 'enlightened' thinkers believed and has been used for a wide variety of purposes. Reason can be imperialist and racist (as in the making of the idea of 'the West'), taking a specific form of consciousness for a universal, a standard that all must aspire to reach. Reason was also a potent weapon in the production of social normativity during 'the Enlightenment', driving people towards conformity with a dominant and centred 'norm' of behaviour (Doherty, 1993). Modernist reason was therefore dependent on the 'othering' of non-conformists, of cultures and societies that were not informed by this reason and social norms and were thus banished to the lower echelons of humanity, defined as 'backward', 'undeveloped' or 'uncivilised'. The emergence of new ideas about social, political and economic development was therefore bound up with these pressures to conform to particular notions of knowledge, reason and progress and with the making of a 'Third Estate' or 'Third World' of non-conformity as the alter ego of a developed 'West'.

Conclusions: completing the enlightenment beyond Europe

Much contemporary development thinking has its roots in the Enlightenment as the 'age of reason', which shaped concepts of progress, growth and social change. Modernist thought also envisaged a process of enlightenment, of becoming more modern and less traditional, and it also saw a group of enlightened Western intellectuals and scientists 'guiding' the paths to progress of distant others. Arturo Escobar (1995: 2–4) has even argued that the post-1945 development project is 'the last and failed attempt to complete the Enlightenment in Asia, Africa and Latin America' (Escobar, 1995: 221). After 1945, modernisation theorists in the US also 'saw their project as the enlightenment writ large' (Gilman, 2003: 8) and even the vision of the modern developed under Soviet Communism (albeit with a very different collectivist, anti-religious and anti-capitalist belief system) was similarly a product of the Enlightenment. 'Development' thus has complex roots in the emergence of 'the Enlightenment', in the dawn of industrial capitalism in Europe and America and in the rise and formation(s) of modernity. It is also important to remember that the self-identification of European and Western countries as 'developed' has partly been produced through a contrasting of modernity with the tradition and backwardness of the 'Third World' as Other.

The work of Enlightenment thinkers like Adam Smith (with his free market economics) remains very relevant to 'international development' today for some observers. Examples of this can be found in some of the key global development institutions like the World Bank that see their (neoclassical) knowledges as potentially enlightening. Consider the following quotation from a speech given by the World Bank President James Wolfensohn in 1996:

> Knowledge is like light. Weightless and intangible, it can easily travel the world, enlightening the lives of people everywhere. Yet billions of people still live in the darkness of poverty – unnecessarily.
>
> (Wolfensohn quoted in Patel, 2001: 2)

Thus the knowledge and expertise of contemporary development practitioners is seen as something almost universal that easily traverses borders extinguishing the darkness of poverty wherever it shines. For some theorists and practitioners of development today, people and places can become 'developed' simply though acquiring scientific and technical knowledges about the 'normal' or correct series of developmental stages. If only it were that simple.

Guide to further reading

For an excellent introduction and overview to early development discourses and ideas see Rist, G. (1997) *History of Development*, London: Routledge.

Cowen, M.P. and Shenton, R.W. (1996) *Doctrines of Development*, London: Routledge, provides an accessible discussion of enlightenment ideas, exploring their bearing on the construction of particular development approaches and doctrines.

Doherty, T. (1993) 'Postmodernism: An Introduction', in T. Doherty (ed.) *Modernism/Postmodernism*, Hemel Hempstead: Harvester Wheatsheaf, pp. 1–31, offers clear and accessible definitions of modernism.

Hall, S. and Gieben, B. (1992) *Formations of Modernity*, Cambridge: Open University/Polity, focuses on the making of modernity in the non-Western world.

References

Black, J. (1990) *Eighteenth-Century Europe 1700–1789*, London: Macmillan.

Cowen, M.P. and Shenton, R.W. (1996) *Doctrines of Development*, London: Routledge.

Doherty, T. (1993) 'Postmodernism: An Introduction', in T. Doherty (ed.) *Modernism/Postmodernism*, Hemel Hempstead: Harvester Wheatsheaf, pp. 1–31.

Escobar, A. (1995) *Encountering Development: The Making and Unmaking of the Third World*, Princeton, NJ: Princeton University Press.

Gay, P. (1973) *The Enlightenment: A Comprehensive Anthology*, New York: Simon and Schuster.

Gilman, N. (2003) *Mandarins of the Future: Modernization Theory in Cold War America*, Baltimore: Johns Hopkins University Press.

Hall, S. (1992) 'The West and the Rest: Discourse and Power', in S. Hall and B. Gieben (eds.) *Formations of Modernity*, Cambridge: Open University/Polity, pp. 275–331, chapter 6.

Hall, S. and Gieben, B. (1992) *Formations of Modernity*, Cambridge: Open University/Polity.

Hampson, N. (1968) *The Enlightenment*, London: Penguin.

Institute of Development Studies (IDS) (2006) 'The Power of Labelling in Development Practice', IDS Policy Briefing, Issue 28 (April).

Keynes, J.M. (1936) *The General Theory of Employment, Interest and Money*, London: Macmillan.

Ludden, D. (2003) 'Why Area Studies?', in A. Mirsepassi, A. Basu and F. Weaver (eds.) *Localizing Knowledge in a Globalizing World: Recasting the Area Studies Debate*, New York: Syracuse University Press, pp. 131–137.

Patel, R. (2001) 'Knowledge, Power, Banking', *Znet Magazine*, July 20th 2001.

Porter, R. (1990) *The Enlightenment*, London: Macmillan.

Chapter 21

The Washington consensus and the post-Washington consensus

Ali Burak Güven

The Washington consensus: A neoliberal turn in development policy

The term 'Washington Consensus' was coined by British economist John Williamson in 1990 to describe the wide range of policy reforms Washington-based institutions such as the **International Monetary Fund** (IMF), the **World Bank** and the US Department of the Treasury deemed desirable across Latin America during the 1980s. Williamson (1990) listed ten items: fiscal discipline, public expenditure reform, tax reform, interest rate liberalisation, competitive exchange rates, trade liberalisation, promotion of foreign direct investment, privatisation, deregulation and secure property rights. The term quickly caught on not because it offered a precise depiction of concrete prescriptions towards a particular region but because it captured the spirit and scope of universal development policy advice coming from the most powerful governments and multilateral organisations in the North at the time.

Two things about the Washington Consensus were extraordinary. First, it represented a cohesive programme that amounted to comprehensive restructuring in the target country. This differed from the analytical and policy pluralism that had characterised development economics in the post-war decades. Here was a relatively new field that had explored how to spur growth from within the accepted constraints of the default mixed-economy setting across the Third World – a setting where intense government intervention was common, planning was considered vital and fledgling private firms coexisted with state-owned enterprises. By contrast, the Washington Consensus reflected a new scholarly current, that is, the ascendant neoclassical thinking in American universities in the 1970s (Harvey, 2005). Its built-in assumption that markets allocated resources more efficiently than governments meant countries were advised to undergo sweeping reforms to replace their state-directed and often inward-oriented strategies with a vision that prioritised the private enterprise.

Second, this policy realignment was promoted in an exceptional historical context that significantly empowered Northern international financial institutions (IFIs). Triggered by turbulence in commodity prices and a sudden hike in interest rates, the international debt crisis of the early 1980s forced scores of countries to turn to official lenders such as the IMF (or, the Fund) and the World Bank (or, the Bank). Note here that the Fund and the Bank, together known as the **Bretton Woods** twins, are very different organisations with different mandates: the Fund makes loans for post-crisis stabilisation and recovery, whereas financing by the Bank is directed at specific projects. Yet the twins usually

DOI: 10.4324/9780429282348-23

think alike on policy matters, and their business model has also been similar in that in exchange for loans they ask borrowers to meet distinct and often policy-related criteria called 'conditionality'. At the time of the Washington Consensus these conditionalities were particularly stringent: some were quantitative (such as lowering fiscal expenditures by a certain amount), while others were qualitative (such as deregulatory moves in a sector). Often, the IFIs also employed a practice called 'cross-conditionality', meaning satisfying conditions of a Bank programme could also be listed as a condition for a Fund programme and vice versa. The scope of these programmes was deeply transformative, to the extent that both organisations used the term 'structural adjustment' to officially label some types of loans.

It was under these extraordinary circumstances that the range of orthodox neoliberal policies associated with the Washington Consensus found a highly favourable climate for widespread implementation. The disintegration of the Soviet Union in 1989–1991 boosted the clientele of the Fund and Bank, according further justification to their market-oriented policy stance in the process. By the early 1990s the Washington Consensus had become so dominant that it could be considered a 'transnational policy paradigm' (Babb, 2012). A generation of economists and policymakers had operated under the primacy of the neoclassical economic vision, while international investors and lenders, on which countries in the South had come to rely more intensely due to their continued immersion in the global economy, also expected local economies to be governed by this familiar wisdom. As a result, even when countries were not under the 'objective compulsion' of a Fund or Bank programme, they were inclined to uphold the policy outlook and the reformist ethos of the Washington Consensus.

The post-Washington consensus: poverty reduction and institutional reform

For all its dominance, the Washington Consensus in its original form did not survive long. There were severe downsides. Critics had long highlighted that rapid liberalisation worsened domestic inequality as well as the North-South income gap. Meanwhile fiscal and policy restrictions imposed via IFI programmes narrowed down the scope of development strategies available to Southern governments. Yet what truly made mainstream economists and policy insiders begin questioning the orthodox neoliberal wisdom was its poor record in terms of growth and macro stability. The 1980s and 1990s proved to be 'lost decades' for many developing countries. Despite accelerated growth in Asia, per capita income growth in 1980–1998 in the developing world as a whole averaged a dismal 0.0 per cent (Easterly, 2001). Financial instability was another pernicious by-product. A combination of premature capital account liberalisation and poorly regulated financial sectors exposed countries to the vagaries of international financial markets, leading to a string of devastating crises in the semi-periphery from the mid-1990s onwards as exemplified in Mexico and Turkey, Brazil and Russia and most famously in the Asian Crisis. Given these widely recognised policy failures, the Washington Consensus was universally discredited by the late 1990s (Gore, 2000).

The post-Washington Consensus (PWC) refers to the establishment reformulation of neoliberal principles in response to these widely recognised failures of the Washington Consensus. The term was coined in 1998 in a lecture by Joseph Stiglitz, the World Bank's

Chief Economist at the time (Stiglitz, 1998). The title of his talk, 'More Instruments and Broader Goals', captured the essence of this paradigm revision appropriately: rather than negate the market-oriented outlook of the original consensus, the PWC expanded the IFIs' prescriptive focus to include an enhanced set of objectives via a wider range of tools. Specifically, it emphasised not mere growth, but high-quality growth, with fisco-financial stability and social sustainability as key objectives. Attaining such a growth path required not a narrow set of policies but an appropriate institutional environment, which meant a wider and more challenging reform agenda. As such, perhaps a more accurate depiction of the PWC can be found in Dani Rodrik's (2006: 978) notion of 'Augmented Washington Consensus', which adds ten other items to Williams' original list: corporate governance; anti-corruption; flexible labour markets; WTO agreements; financial codes and standards; 'prudent' capital-account opening; non-intermediate exchange rate regimes; independent central banks/inflation targeting; social safety nets and targeted poverty reduction. This expanded prescriptive scope inspired changes to the IFIs' lending instruments, as in the Fund's Poverty Reduction and Growth Facility (PRGF) that stressed social sustainability in concessional loans to low-income members and the Bank's Development Policy Loans (DPLs) that enabled bundling a variety of policy and institutional reforms in one large package.

While the PWC set a corrective tone, it was also criticised intensely from the outset. One major point was that, like its predecessor, the PWC too ignored the structural inequalities in the global political economy, putting the onus of development firmly on domestic reforms in countries in the South (Öniş and Şenses, 2005). Commentators also considered the comprehensive reforms proposed under the PWC too distant from the core competencies of the Fund and Bank and too arduous for borrowers to implement given capacity problems afflicting developing countries. Of particular concern was the ambitious good governance and institutions agenda, which was attacked for sidestepping questions of compatibility by advocating transplantation of Anglo-American institutional blueprints in the very different and varied socio-historical contexts across the South.

The decline of grand paradigms? Development policy after the global crisis

It is important to note that the PWC was not implemented as widely or forcefully as its predecessor. By the time it was consolidated, that is, in the first half of the 2000s, the world economy was booming and fewer countries were knocking on the door of the IFIs. Still, PWC-style social and regulatory reforms, from poverty reduction programmes and anti-corruption schemes to banking reforms, were common across the South, and some major developing countries such as Brazil, Indonesia and Turkey relied on World Bank loans to implement wide-ranging institutional reform programmes that reflected the PWC focus on social protection, good governance and fisco-financial stability.

The global economic crisis of 2008–2009 changed this picture dramatically. While the crisis led to a resurgence of IFI loans to developing countries and while PWC items remained within the policy repertoire of the Fund and Bank, over time both organisations have selectively disengaged from this comprehensive agenda. On the Fund side, PWC-related policy conditionalities are increasingly rare, with the exception of the organisation's existing focus on fiscal and financial reforms; on the Bank side, the politically cumbersome good governance reforms are less prominent, although the social sustainability agenda remains fully entrenched (Güven, 2018). Changes within IFI practice is one reason behind these trends. Some elements of the PWC agenda were long met with scepticism amongst Fund and Bank staff and received criticism from the twins' internal evaluation departments, while reforms to lending framework in both organisations complicate the imposition of structural reforms from outside via harsh conditionality.

More crucial is the ongoing transformation of the international development regime as part of wider shifts in the global balance of power. Manifestations of this transformation include the rise of Southern donors in development assistance, growing South-South trade and investment links, the emergence of South-led multilateral institutions such as the Asian Infrastructure Investment Bank and the ascendance of Southern narratives and practices in development cooperation, all leading to 'an increasingly polycentric development landscape' (Mawdsley, 2018: 182). In this evolving context, Northern organisations such as the Fund and the Bank are less able and less willing than before to champion PWC-style grand paradigms originating in the North and dispensed from above via conditional lending. This is not to say that individual policy elements that comprise the Washington or the post-Washington Consensus are being completely abandoned. Rather, rigid policy packages with a claim to universal applicability are gradually replaced with a more flexible and pragmatic approach, in most cases expanding the policy and institutional choices available to developing economies.

References

Babb, S. (2012) 'The Washington Consensus as transnational policy paradigm: its origins, trajectory and likely successor.' *Review of International Political Economy* 20(2): 268–297.

Easterly, W. (2001) 'The lost decades: developing countries' stagnation in spite of policy reform, 1980–1998.' *Journal of Economic Growth* 6: 135–157.

Gore, C. (2000) 'The rise and fall of the Washington Consensus as a paradigm for developing countries.' *World Development* 28(5): 789–804.

Güven, A. B. (2018) 'Whither the post-Washington Consensus? International financial institutions and development policy before and after the crisis.' *Review of International Political Economy* 25(3): 392–417.

Harvey, D. (2005) *A Brief History of Neoliberalism.* Oxford: Oxford University Press.

Mawdsley, E. (2018) 'The "southernisation" of development?' *Asia Pacific Viewpoint* 59(2): 173–185.

Öniş, Z. and F. Şenses (2005) 'Rethinking the emerging post-Washington Consensus.' *Development and Change* 36(2): 263–290.

Rodrik, D. (2006) 'Goodbye Washington Consensus, hello Washington confusion? A review of the World Bank's Economic Growth in the 1990s: Learning from a Decade of Reform.' *Journal of Economic Literature* 44(4): 973–987.

Stiglitz, J. (1998) 'More instruments and broader goals: moving toward the post-Washington Consensus.' *WIDER Annual Lecture*, Helsinki, 7 January.

Williamson, J. (1990) 'What Washington means by policy reform.' In J. Williamson (ed.) *Latin American Adjustment: How Much Has Happened?* Washington, DC: Institute for International Economics.

Further reading

Bretton Woods Project: Critical Voices on the World Bank and IMF. https://www.brettonwoodsproject.org/

Park, S. and A. Vetterlein, eds. (2010) *Owning Development: Creating Policy Norms in the IMF and the World Bank.* Cambridge: Cambridge University Press.

Peet, R. (2009) *Unholy Trinity: The IMF, World Bank and WTO*, Second Edition. London: Zed.

Serra, N. and J. Stiglitz, eds. (2008) *The Washington Consensus Reconsidered: Towards a New Global Governance*. New York: Oxford University Press.

Woods, N. (2005) *The Globalizers: The IMF, the World Bank and Their Borrowers*. Ithaca: Cornell University Press.

Chapter 22

Concepts and measures of development

Beyond GDP

Jakob Dirksen

Introduction

Not everything that can be counted counts, and not all that counts can be counted

The *development of what* counts – or ought to count – and how to count or measure it are foundational and traditionally controversial questions in development studies and practice. On the one hand, judgements about what counts in, for, and as development necessarily invokes debates about what the term development means, its normative and ethical implications. On the other hand, there is the crucial question of how best to measure that which counts, – if it is countable, at all.

Measurement is used *descriptively,* to summarise states of affairs and offer information, for example, regarding the questions if, where, and why developmental objectives are or are not being met. But measurement is also – and increasingly so – used *prescriptively*, for example to inform the design and implementation of evidence-based, targeted development policies, programmes, and interventions. There is hence a risk for measures to describe and suggest pursuit of the wrong concepts or to poorly approximate and prescribe actions towards the right ones. As Nobel laureates Joseph Stiglitz and Amartya Sen with Jean-Paul Fitoussi have argued (2010: xviii),

> In an increasingly performance-oriented society, metrics matter. What we measure affects what we do. If we have the wrong metrics, we will strive for the wrong things. In the quest to increase GDP, we may end up with a society in which citizens are worse off. Too often, we confuse ends with means.

Based on such considerations, also summarised in the adage quoted at the start of this chapter, attention has increasingly shifted away from earlier concepts and metrics of development as growth, measured in terms of material gain or productivity as Gross Domestic Product (GDP) or Gross National Income (GNI).

Re-conceptualising development

Perhaps the most prominent advocate of the need to rethink concepts and measures of development over the last half century was economist-philosopher and Nobel Laureate Amartya Sen. He argued for a refocussing of attention of concepts and measures of development on its *ends* – the lives humans can and do lead – rather than *means* – resources

DOI: 10.4324/9780429282348-24

such as money, expressed by metrics of purchasing power or aggregate economic productivity. Instead, Sen et al. advocated for a focus on the real opportunities (capabilities) to be or do (functionings) what one has reason to value – *Development as Freedom* (Sen, 1999). Welfare is here expressed in terms of opportunity or 'achievement sets' (or combinations thereof) and poverty conceptualised as unfreedom or deprivation of *basic* capabilities and functionings. Similarly, the basic needs approach prioritised as *end* of development the eradication or alleviation of unmet basic needs necessary for human survival.

The emphasis on human development and capabilities thereby goes one step further than basic needs in its insistence on human freedoms – beyond humans as *needy* and passive recipients, as mere objects, rather than subjects and agents, of development. Extended across time and generations, it then also found re-expression in the concept of *sustainable* human development as a matter of intergenerational equity and justice (Anand and Sen, 2000).

Through these lenses, there are at least four things that are problematic with an exclusive focus on aggregates and averages such as GDP/GNI or GDP/GNI per capita when operationalising and measuring associated concepts of development.

First, as crude aggregates or averages, they are entirely distribution-insensitive. That is, they conceal possibly stark and problematic inequalities.

Second, as measures of overall consumption or spending, they do not capture the outcomes that, arguably, are of direct concern – such as the lives people can and do actually lead.

Third, they are unable to accurately account for the many dimensions of human development and well-being (or shortfall thereof) – and so cannot measure or compare how well people's lives are going *many-things-considered*.

Fourth, by excluding intergenerational, and possibly non-anthropocentric components of development, GDP-guided development might not only be exclusive and unequitable but also *unsustainable*.

Metrics of development

Measurement approaches that seek to overcome these problems are best understood as the results of an interplay of several interdependent factors, including conceptual shifts, changing motivations for measurement, as well as new and arguably better methods of data collection, processing, and analysis. See for example Deaton's (2018) contribution on the crucial role that new and better household survey data played thereby.

Beyond aggregates and averages

In going beyond crude aggregates and averages, the application of existing and the development of new, inequality indices is noteworthy. These include the widespread application of Lorenz curves and Gini coefficients for the measurement and analysis of income inequalities and more refined measures of 'generalised entropy'. These satisfy many desirable axioms and subsume well-known measures such Theil's and Atkinsons's indices of inequality. Among other things, measures from this family can operationalise the ethical prioritisation of those who are least well-off, by putting more weight on welfare and transfers at the bottom end of the distribution.

Moreover, the distribution-insensitivity of aggregates and averages is of concern because it conceals not only possibly stark inequalities but also possibly widespread and severe poverty that may exist, even when aggregate wealth or income of a population appears to be high. Measures of absolute poverty, focused on those who are insufficiently well-off (below a poverty line) have increasingly been used by analysists and policy makers alike.

The most widespread family of such poverty measures, the Foster-Greer-Thorbecke class, mainstreamed the effective use of the so-called 'three I's of poverty': incidence (how many are poor), intensity (how poor – i.e. how far below poverty line – are they, on average), and inequality (among the poor). Many of these and other applications, their motivation, and (dis)advantages are summarised and discussed in Sen's and Foster's (1997) *On Economic Inequality*.

Last, there have also been efforts to link distributional concerns to traditional metrics such as GDP. This has been done by, e.g., putting special emphasis on poverty reducing, pro-poor, or 'inclusive growth'. One of the World Bank's 'Twin Goals' – alongside ending extreme poverty – for example, is the goal of 'shared prosperity': growth should benefit the bottom 40 per cent of the distribution.

The measures mentioned thus far do, however, all fall into the realm of unidimensional and, more specifically, usually monetary metrics. Even broader concepts, such as unmet basic needs, are frequently translated into so-called expenditure baskets thought to express the purchasing power necessary to secure a minimum nutritional food intake or meet additional non-nutritional basic needs. This idea underlies the most widespread monetary poverty statistics, including many official poverty lines, the World Bank's international poverty lines, and the work of many UN Agencies and other development organisations to date.

Measuring ends, not means of development

The first prominent measure that broadened the informational basis and evaluative space of development metrics in line with the call to refocus attention on ends, not means of development, was the Human Development Index (HDI), developed by Mahbuq ul-Haq together with Amartya Sen and published by UNDP since 1990. Many other indices of this kind have followed since. In addition to the economic dimension of human development, (GNI/capita), the HDI also included an educational or knowledge dimension (expected and mean years of schooling), and the dimension of health and longevity (life expectancy at birth). The original HDI itself is distribution-insensitive within countries and within and across dimensions, but since 2010 UNDP has also published the Inequality-Adjusted HDI (IHDI) that discounts each dimensional achievement by the extent of inequality observed within that dimension.

From composite to multidimensional measurement

Measures such as the HDI can be referred to as 'composite indices'. Whilst measuring ends of development across multiple dimensions, they remain in the realm of population-level aggregates and averages (even if discounted for within-dimension inequality).

Composite indices first summarise across people, i.e. into population-level dimensional aggregates – and then across dimensions. The HDI, for example, summarises information on everyone's education and then combines this with summaries of population health and GDP/capita, rather than considering how well each person is doing across these dimensions.

Even if adjusted for within-dimension inequality, such composite indices thus remain insensitive to inequalities in the distribution of dimensional achievements (or deprivations) across dimensions and individuals. That is, they can only make visible how unequally achievements are distributed in each dimension *separately*. But they do not make visible who is at the upper or lower end of a multidimensional distribution, all things (i.e. all dimensions) considered. This falls short of the ethical individualism of development concepts and measurement motivations, and it prevents interpersonal comparisons and welfare evaluations within populations.

To address this limitation, multidimensional measurement methods invert the order of aggregation. Enabled through household survey data and new computational capacities, they summarise first across dimensions – measuring multidimensional human development, well-being, or poverty at the individual or household-level – in order then to aggregate across people into a population-level measure. This allows for the identification of those who are better or worse-off as those affected by what Wolff and de-Shalit (2007) call *clustered (dis)advantages,* the coincidence of achievements, deprivations, or (un)freedoms (Dirksen and Alkire, 2022).

Sustainable (human) development

Addressing the fourth point of criticism, measurement exercises have sought to incorporate intertemporal and intergenerational components by accounting for planetary boundaries and the pollutive and depletive costs of development. Various ways of 'greening' the HDI for the Anthropocene, e.g., by including ecological footprints and capital stocks left for future generations, have been proposed, for example. But how exactly to measure sustainable human development as development that ensures future generations' ability to meet their own needs or the capability to lead worthwhile lives (Anand and Sen, 2000) remains an important area for further research.

Limitations and critiques

Criticisms of composite and multidimensional measures are not necessarily based on a rejection of underlying concepts. Rather, what is often called into question is the measures' ability to accurately approximate and capture refined multidimensional concepts of development. This is often based on scepticism regarding the choices involved when aggregating across dimensions, including trade-offs between different dimensions, such as health, education, material living standard, etc. Ravallion (2012), for example, thus speaks of, 'Mashup indices of development' when referring to measures that aggregate across dimensions. Instead, sceptics have often argued in favour of so-called dashboards that provide information on the many dimensions of human development one-at-a-time. Without the aggregation across dimensions, however, dashboards do not straightforwardly allow for multidimensional interpersonal comparisons and hence cannot identify who is all-things-considered worse (worst) off for either descriptive or prescriptive measurement and policy-purposes.

Conclusion

This chapter described metrics 'beyond GDP' – measures that mitigate or overcome shortcomings associated with the use of GDP/GNI as goalpost indicators of development. Beyond GDP measures does not necessarily mean GDP substitutes. They often still leave room – and a role – for GDP as well as other aggregates and means-focused metrics. But, with a conceptual shift towards human development as a multidimensional phenomenon, such measures can only ever be imperfect approximations and indicators for *some* aspects of development. Similarly, multidimensional metrics can also incorporate or complement a 'developmental' focus on institutional capacity or economic integration, to name just two additional, traditionally influential concepts of development.

The chapter also showed that conceptualising and measuring development thus, the focal idea of concern is often equivalent to ideas motivating social welfare measurement and policies more broadly – equitably advancing welfare across its dimensions. More often than not, associated policies may still be referred to as development policies and interventions when located in the Global South – particularly when involving so-called overseas development assistance. But it may well be that this questionable terminological heritage will merge into a broader discourse on well-being, distributive justice, and social progress before long.

References

Anand, S. and Sen, A. (2000). 'Human Development and Economic Sustainability', *World Development,* Vol. 28, No. 12: 2029–2049.

Deaton, A. (2018). *The Analysis of Household Surveys. A Microeconometric Approach to Development Policy. Reissue Edition with a New Preface.* Washington, DC: The World Bank.

Dirksen, J. and Alkire, S. (2022, forthcoming). 'Multidimensional Poverty. Measurement, Analysis, Applications', in: Zimmermann, K.F. (ed.) *Handbook of Labor, Human Resource and Population Economics.* New York: Springer.

Ravallion, M. (2012). 'Mashup Indices of Development', *The World Bank Research Observer,* Vol. 27, No. 1: 1–32.

Sen, A. (1999). *Development as Freedom.* Oxford: Oxford University Press.

Sen, A. and Foster, J.E. (1997). *On Economic Inequality. Expanded Edition with a Substantial Annexe by James. E. Foster and Amartya Sen.* Oxford: Oxford University Press.

Stiglitz, J.E., Sen, A.K., and Fitoussi, J.P. (2010). *Mismeasuring Our Lives: Why GDP Doesn't Add Up.* New York: The New Press.

Wolff, J. and de-Shalit, A. (2007). *Disadvantage.* Oxford: Oxford University Press.

Further readings and additional resources

European statistical office (EUROSTAT)

The European Union's official poverty measure, published by EUROSTAT, is the so-called indicator At Risk of Poverty or Social Exclusion (AROPE) (Website). AROPE is Europe's first official measure of *multidimensional* poverty and is also reported against Sustainable Development Goal 1 (No Poverty) by all EU members and some additional countries.

Gross National Happiness Commission (GNHC), Bhutan

Bhutan's Index of Gross National Happiness (GNH) and the Royal Governments pledge to grow not GDP but GNH are among the most prominent among the new generation of development policy-goalpost indicators that aggregate not expenditure and productivity, but people's achievements across the valuable domains of life (or happiness). See the website of Bhutan's Gross National Happiness Commission (GNHC) for more details.

United Nations Development Programme (UNDP)

Human Development Reports Office (HDRO) at the United Nations Development Programme (UNDP) (Website). In addition to the HDI and IHDI, UNDP also publishes the Gender Development Index (GDI) that compares HDI between men and women, the Gender Inequality Index (GII), and the global Multidimensional Poverty Index (MPI; Technical Notes). Together with the Oxford Poverty and Human Development Initiative (OPHI) at the University of Oxford, UNDP is also offering a Massive Open Online Course (MOOC) on 'Designing a Multidimensional Poverty Index'.

Organisation for Economic Cooperation and Development (OECD)

The OECD 'Better Life Index' (Website) and their work on 'Measuring Well-being and Progress' (Website) are two prominent examples of beyond-GDP well-being and social progress measurement.

Oxford Poverty and Human Development Initiative (OPHI), University of Oxford

The Oxford Poverty and Human Development Initiative (OPHI), at the University of Oxford (Website) and the Multidimensional Poverty Peer Network (MPPN; Website) – of which OPHI is the secretariat – have pioneered the development of multidimensional metrics of poverty, vulnerability, and well-being as official governmental statistics and public policy tools.

World Bank

The World Bank hosts a number of large databases and interactive analysis tools for development indicators. These include the World Development Indicators (WDI) Databank, the online poverty analysis tool PovCalnet, its household surveys Microdata Library (including the World Banks' Living Standards Measurement Study (LSMS) surveys), and the Global Database of Shared Prosperity to monitor inclusive growth.

Chapter 23

Global economic inequality, the great divergence, and the legacies of colonialism and enslavement

Alan Shipman, Julia Ngozi Chukwuma, and Emil Dauncey

In his famous book entitled *The Wretched of the Earth* (1961), Frantz Fanon, a prominent anticolonial intellectual, born in the 1920s on the Caribbean Island of Martinique, wrote:

> [The] European opulence is literally scandalous, for it has been founded on slavery. . . . The wealth of the imperial countries is our wealth too. . . . For in a very concrete way Europe has stuffed herself inordinately with the gold and raw materials of the colonial countries: Latin America, China, and Africa. From all these continents, under whose eyes Europe today raises up her tower of opulence, there has flowed out for centuries toward that same Europe diamonds and oil, silk and cotton, wood and exotic products. Europe is literally the creation of the Third World. The wealth which smothers her is that which was stolen from the underdeveloped peoples.
>
> (Fanon, 2001: 76, 81)

While Fanon critically emphasised that Europe's development was only possible because of its exploitation of its colonies, economic historians from the Global North had rarely voiced such an indictment. Many emphasised instead the internal, self-generating factors that may have promoted Europe's (and later North America's) early industrial dynamism: free enterprise, free speech and thought, accountable and democratic law-making, faster generation and application of knowledge, thrift and deferral of gratification, or a superior work ethic. This enabled them to focus on the supposedly benign effects of colonisation: spreading new practices, ideas, technologies, and forms of social, cultural, economic, and institutional organisation that removed constraints on their hosts' 'development', even if while so doing they also spread devastating diseases and religions whose appeal at home would soon decline.

But the European tendency to seize and settle the territories they 'discovered' in the Americas, Asia, and Africa and reshape their economies around extraction of raw commodities that depended on slave or low-paid labour makes Fanon's charge of immorality and injustice hard to escape. These actions become harder to consign to distant history if they have an enduring legacy, in the form of extractions that permanently advantaged the plunderers over the dispossessed and structural constraints that prevent the later developers from catching up.

As Fanon highlights, disparities in economic development between Global South and Global North are not natural phenomena but are rooted in colonialism and enslavement.

DOI: 10.4324/9780429282348-25

These gave rise to enduring structures that not only transferred financial resources from colonies to their imperial centres through the exploitation of people and natural resources but also perpetuate flows of finance, people, and traded goods in ways that disadvantage the economies of former colonies in the present.

Present day global economic inequality and the great divergence

Whilst gross domestic product (GDP) is a crude indicator of economic development, concealing other dimensions of human well-being, it continues to be used extensively because its widespread collection and degree of standardisation allows for relatively straightforward cross-country comparison. For international comparisons, GDPs are converted into one currency (usually international dollars, US$), with adjustments made for different countries' price levels. For comparisons across time, GDP is expressed in US$ for a particular year to remove the effect of price inflation. Table 23.1, although focused on a single year (World Bank, 2021), shows how GDP per capita compared across the world's largest countries (by population), using the 2015 value of the US$.

GDP per capita, in the third column of Table 1, is frequently used as a measure to gauge average living standards in a country. It uses the GDP measure of national income that most countries now compile annually and quarterly and which is backward-compatible with the historic GDP time series compiled by Angus Maddison (2007) and his team at the Groningen Growth and Development Centre. By dividing a country's GDP by its population, it measures the income each person in a country would get if GDP were distributed equally across people. In reality however, GDP is rarely distributed equally across people and living standards tend to differ.

Table 23.1 GDP per capita for world's 12 most populous countries in 2021, plus European comparators

Country	Population 2021 (m)	GDP per capita 2021 (constant 2015 US$)	World Bank income classification[1]	Region/continent
China	1,412.4	11,188	Upper middle	East Asia
India	1,407.6	1,937	Lower middle	South Asia
USA	331.9	61,855	High	North America
Indonesia	273.8	3,892	Lower middle	South-east Asia
Pakistan	231.4	1,474	Lower middle	South Asia
Nigeria	213.4	2,430	Lower middle	West Africa
Brazil	214.3	8,538	Upper middle	South America
Bangladesh	169.4	1,684	Lower middle	South Asia
Russia	143.4	10,216	Upper middle	East Europe/Central Asia
Mexico	126.7	9.525	Upper middle	Central America
Japan	125.7	35,291	High	East Asia
Ethiopia	120.3	835	Low	East Africa
Germany	83.2	42,726	High	Europe
France	67.7	38.056	High	Europe
Poland	37.7	15,850	High	Europe

Source: World bank at www.data.worldbank.org

[1] Using 21.7.22 classification: low < $1,045; lower-middle $1,086–4,255; upper-middle $4,256–13,205; high > $13,205

A chart of this data for the main global regions, shown in Figure 23.1, shows the 'Great Divergence' in their economic fortunes once Europe, followed by the USA, began to industrialise. Until around 1800 all regions had similar per-capita GDPs that grew very slowly. But in the past 200 years Western Europe and North America experienced a sustained and rapid rise in per-capita GDP, while other regions – while doing better than before – grew more slowly, falling steadily further behind. Rapid recent expansion in China and some other Asian countries still leaves them a long way below the Global North. And while many 'emerging' countries have now started on a path to industrialisation and urbanisation, they tend to find the exchange value of manufactures and services significantly reduced by the time they adopt them (Ghodsi and Stehrer, 2018; Emmanuel, 1982). Therefore, one structural constraint that many countries in the Global South face is that their economic development oftentimes appears to be 'dependent' on outsourcing services to – and importing specialised technology and manufactured goods from – their former colonising nations in the Global North (Naseemullah, 2020), locking in the first industrialisers' advantage.

Explaining the great divergence

Europe's quest to transform their colonies into sources of financial gain in order to pay for their investments and military protection propelled them towards intensive plantation-based agriculture for which bonded or slave labour was often the only viable form. In his classic history of slavery, Williams (1944) shows how plantation owners, first in the Caribbean and then the American mainland, turned to Africa and Asia as new sources of slaves after running through the supply of indigenous or displaced European workers. In what is often referred to as the 'triangular trade', people enslaved in Africa

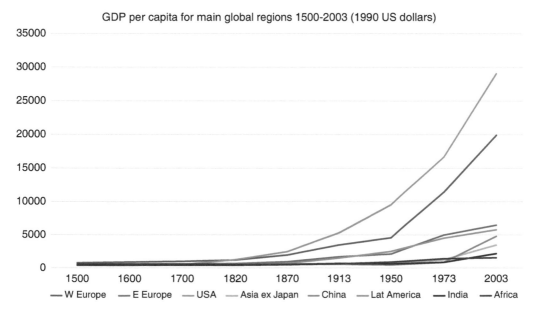

Figure 23.1 GDP per capita for main global regions 1500–2004 (1990 US dollars)

were transported to the Americas, where they produced tobacco, sugar, and other primary commodities, which were then exported to Europe. The goods manufactured from these primary commodities in Europe were subsequently exported throughout the colonies, including arms and ammunition which were traded for slaves, with profits retained in the home countries of the colonial powers (See Rönnbäck, 2018; Inikori, 2020; Nunn, 2020). When liberal economic historians identify investment in industrial and human capital (and the associated willingness to save) as a driving force in Europe's ascendancy, they choose to overlook the scale of capital invested in slaves – whose market value meant that by 1860 they were collectively worth three times the industrial capital stock of the US (Levy, 2021).

Scholars from the Global South such as Fanon (ibid), as well as Walter Rodney (1972), Raul Prebisch (1950) and many others, have convincingly argued that the extraction of resources and labour from colonised regions of Africa, Asia, and the Americas by European colonisers lead to their economic underdevelopment. Resources that could have been invested in African economies, in health, education, and development more broadly, were instead directed to Europe, improving standards of living for Europeans at the direct expense of Africans. Even after countries formally achieved their independence, features of the colonial system of exploitation have often remained, such that former colonies have tended to remain poor because they are locked into a system of supplying cheap labour and resources to wealthier countries in the Global North, receiving too little in return to invest in meaningfully developing their own economies.

Legacies of colonialism and enslavement

In economic terms, standards of living in former colonies have by and large remained worse than in the countries that had colonised them because independence has seldom, if ever, meant fully escaping from the economic structures left behind by colonial powers. These structures include *extractive trade arrangements*, *extractive labour practices*, and *extractive systems of government* oriented towards the continued exploitation of former colonies for the benefit of former colonial powers in the Global North.

i) Extractive trade arrangements

Colonial powers were able to make great profits from selling manufactured goods to consumers both at home and in the colonies. In contrast, the profits enjoyed by the colonised countries that produced the raw materials were limited. This meant they had far less to invest in developing their own industries. Moreover, colonial governments often made it difficult for them to trade globally, and in some cases colonial governments deliberately stopped them doing so. For example, in India, the world-leading textile industry was destroyed after Britain banned the sale and export of India's higher-quality, lower-cost products to make way for textiles produced in Britain (Tharoor, 2018). India was transformed from a major exporter of high value cotton textiles to predominantly an exporter of primary commodities.

ii) Extractive labour practices

Colonial powers economically shaped their colonies around mining and plantation agriculture, where work required long hours and relatively little skill. Therefore, little was invested in the education and training of the colonised. Positions requiring managerial or technical expertise were reserved for European settlers, with technical expertise seldom passed on to Africans and Asians. Colonised people were often denied rights to land, leaving those who had come to own or appropriate it free to extract surplus from them and often leaving colonised people with nothing to invest in their own enterprises, health, or education.

Extractive governance

Wherever possible the colonists practised indirect rule, recruiting a privileged section of the local population to act as administrators. The loyalty of this elite was secured with a share of the power and wealth that the colonists were extracting. Local beneficiaries of indirect rule often retained political power and concentrated ownership of industry and land after independence, resisting a more equal distribution of income, wealth, and authority (see for example Mamdani, 1996). An extreme example was the system of Apartheid in South Africa, a system of racial segregation developed from independence in 1910 until the early 1990s, which left the white minority controlling political and economic institutions whilst the black majority was systematically deprived of rights and resources.

Conclusion

It would be a mistake to say that the colonised are merely the passive victims of colonialism. People have and continue to resist and bring about change. For example, in 1791, enslaved Africans in Haiti united against their colonisers and repeatedly successfully defeated attempts by the French and the British to bring them back under colonial control (James, 1989). But colonisers' past actions have left a legacy of structural constraints that may hold back a country's economic development even when visible colonial subjection has been removed.

The colonial past persists in present-day inequalities and injustices, not only via differences in standards of living but also growing disparity between countries' cumulative carbon emissions and experience of damage from climate change. This is not to say that other factors, such as availability of natural resources or navigable rivers and coast, have not influenced differences in speed and direction of economic development. But these differences cannot be understood without reference to the legacies of colonialism and enslavement. For this reason, activists such as the African-American campaigner Audley Moore have advocated payment of reparations to victims of slavery and their descendants. Understanding such historical injustices is essential if we are to make sense of and challenge contemporary inequalities and uneven development.

References

Emmanuel, Arghiri (1982) *Appropriate or Underdeveloped Technology?* New York: Wiley.
Fanon, F. (1961 [2001]) *The Wretched of the Earth*. London: Penguin Books.
Ghodsi, Mahdi & Stehrer, Robert (2018) *Avoiding and Escaping the 'Commodity Trap' in Development, UNIDO Inclusive and Sustainable Development Working Paper 09/2018*. Vienna: United Nations Industrial Development Organization. www.unido.org/api/opentext/documents/download/10302377/unido-file-10302377
Inikori, J.E. (2020) Atlantic slavery and the rise of the capitalist global economy. *Current Anthropology*, 61(S22), pp. S159–S171.

James, C.L.R. (1989) *The Black Jacobins: Toussaint L'Ouverture and the San Domingo Revolution*. New York: Vintage.

Levy, Jonathan (2021) *Ages of American Capitalism: A History of the United States*. New York: Random House.

Maddison, A. (2007) *Contours of the World Economy 1–2030 AD: Essays in Macro-Economic History*. Oxford: OUP.

Mamdani, M. (1996) *Citizen and Subject: Contemporary Africa and the Legacy of Late Colonialism*. Kampala: Fountain Publishers.

Naseemullah, Adnan (2020) Dependent development in the twenty-first century. *Third World Quarterly*, 43(9), pp. 2225–2243.

Nunn, N. (2020) The historical roots of economic development. *Science*, 367(6485).

Prebisch, R. (1950) The economic development of Latin America and its principal problems. *Economic Bulletin for Latin America*, 7, pp. 1–12.

Rodney, W. (1972 [2018]) *How Europe Underdeveloped Africa*. London: Verso.

Rönnbäck, K. (2018) On the economic importance of the slave plantation complex to the British economy during the eighteenth century: A value-added approach. *Journal of Global History*, 13(3), pp. 309–327.

Tharoor, S. (2018) *Inglorious Empire: What the British did to India*. London: Penguin UK.

Williams, E. (1944 [2021]) *Capitalism and Slavery*. Chapel Hill: The University of North Carolina Press.

World Bank (2021) *World Bank Open Data*. data.worldbank.org.

Chapter 24

Conflict politics as developmentalism

Raktim Ray

Introduction

The discourse of development is contested. The nature of contestation lies in the way development is conceptualised and practised. After decades of developmental efforts, global inequality and subsequent dispossession of people at the margins are increasing. Murray Li (2000) argues that development is a form of governmentality which allows the state to govern and regulate the conditions of its citizens. For her, development is a 'project of rule' for implementing disciplinary powers (Li, 2000).

The idea of development is embedded in the modernisation discourse of the 'West' which for a substantive period of time considered the postcolonial or the colonised world as 'others'. This 'othering' was imparted by the intertwined concept of modernity and developmentalism to the 'non-Western' world where development needs to be done or development is waiting. The only possibility for the non-Western world is to 'catch up' with the West through a developmental process or to be managed towards a categorisation of progress that is legible to the West (Brett, 2009). This 'otherness' has its roots in the colonial imagination, which undermined contextuality and differences along with the hegemonic appropriation of power (Robinson, 2006). The colonial and postcolonial hegemony has been justified by liberal and neoliberal values of private property and free markets, sustaining the Western supremacy and the subordination of the non-Western world. Hence, development has become a problematic discourse that acts as an apparatus for extension of neo-imperialist power of the West (Escobar, 2000; Ferguson and Gupta, 2002; Cooke and Dar, 2008).

The dilemmas of development discourse have become prominent in recent days with increasing challenges of climate change, refugee crisis and the political-economic transformation of the state. Development being a political project often finds itself at the crossroads of 'development alternatives' and 'alternatives to development'. As a discourse, development is often criticised for reproducing hegemonic political norms through new forms of domination and subordination and underestimation of the 'local' (Escobar, 1992; Ziai, 2004; Gibson-Graham, 2007). This chapter acknowledges the importance of these critiques of development and defines development more as a generative process that takes place through conflict politics. Conflict politics can be defined as the heterogenous political relationship between the state and the political society. Defining developmental politics or developmentalism through the framework of conflict politics democratises development processes and goes beyond any binary conceptualisations and opens possibilities for a more generative politics.

DOI: 10.4324/9780429282348-26

What is conflict politics?

Binary understanding of political relationships between the state and the citizens is often parochial and looks at dominance vs resistance. Resistance is often conceptualised as an act against domination (Sharp et al., 2000). Sharp et al. (2000) encourages one to look beyond this and argues for resistance as an extension of power rather than challenging it. Here, power can be interpreted as the flow of social relations that gets mobilised through various networks (Allen, 2003, 2009). Hence, resistance becomes the diagnostic apparatus for identifying differential power relations (Pile, 1997). For citizens, the way they politically engage with the state is much more complex. It is neither always subordination to the state power nor resisting it. It often uses bargaining, negotiations and patronage as political tactics to engage with the state to receive 'fruits' of development. These other forms of engagements do not undermine the importance of direct antagonism or resistance as a political process that is intrinsic to developmental process but expands to a multi-layered understanding of developmental politics.

Hickey (2009) observes conflict and struggle in politics have always been central to development. The very process of state formation has been characterised by violence and struggle, with states establishing control over territorial space and people through coercion. For Hickey (2009), development related political engagements between the state and citizens are a form of a social contract through which the state establishes norms of social protection and provides developmental outcomes that is at the heart of liberal democracy. Simultaneously, the state also uses this social contract to legitimise its control over citizens (Hickey, 2009). This characteristic of the state is more prevalent in the postcolony where colonisers used a similar tactic to control the colonised groups. Post-independence state formation in the postcolonial world happened through a hybrid process, an intermediate state between the Empire-state and the sovereign-state. This hybridity also helped the postcolonial state to represent itself as a provider of redistributive policies in broader political imaginations. Several conflicting and confusing mechanisms were developed for subtle exploitation along with redistribution. This also made the state more a crisis management apparatus than a welfare provider. A very common example of this can be found when the state legitimises evictions of poor people from informal settlements in the name of 'development' but alongside makes ad hoc arrangements with poor people and rehabilitate them at the margins, which are also informal. This conflicting practice by the state can be seen as an oscillation between a 'rational mode' and a 'magical mode of being' (Das, 2004). By adopting a 'rational mode', the state responds favourably to the market demands and promotes capital accumulation through development. Through a 'magical mode of being' the state maintains an elusive trust among its citizens and ensures its own political survival (Ray, 2020).

Similarly citizens also adopt various mechanisms to access any developmental outcome from the state. These mechanisms can be considered as 'strategic essentialism'. 'Strategic essentialism' can be seen when, irrespective of the difference of conflicting class interest or political identity, people engage in essentialising common interest (Spivak, 2012). Citizens sometimes bargain and negotiate with the state to access resources. Sometimes they resist the state through direct antagonism, and sometimes they co-opt with the state. Examples of these can be cited when citizens in the postcolonial context use bribes to access various services offered by the state. Simultaneously, when it is needed, the same citizen groups protest against the state or curtail any developmental interventions that can be seen in various anti-land acquisition movements around the postcolonial world. These various forms of relationships that the state and citizens maintain are defined as conflict politics. Conflict politics often constitute strategic choices of various options, and they are not impromptu solutions. The outcome of the conflict politics on a particular issue is always dependent on how various actors mobilise resources within the power network.

Conclusion

From the previous discussions, it is evident that development as a 'political project' operates through the state where the relationship between the state and citizens are not always linear. Indeterminacy of the state to practice developmental politics is evident where the state maintains certain ambiguity with its citizens and simultaneously implements dispossession and rehabilitation of the dispossessed. This indeterminacy serves two purposes. In one way, it facilitates capital accumulation through development and in another way it maintains its political stability by offering ambiguous trust to the citizens. As an extension of these relations, citizens also maintains heterogeneity of relationships with the state to access developmental outcomes. Hence, conflict politics as a framework to understand these heterogenous political relationships between the state and citizens helps us understand the nature of developmental politics and offers a nuanced understanding of conflict beyond binary of dominance vs resistance.

Bibliography

Allen, J. (2003) *Lost Geographies of Power*. Oxford: Blackwell Publishing.

Allen, J. (2009) 'Three Spaces of Power: Territory, Networks, Plus a Topological Twist in the Tale of Domination and Authority', *Journal of Power*, 2(2), pp. 197–212.

Brett, E.A. (2009) *Reconstructing Development Theory: International Inequality, Institutional Reform and Social Emancipation*. New York: Macmillan International Higher Education.

Cooke, B. and Dar, S. (2008) 'Introduction: The New Development Management', in Dar, S. and Cooke, B. (eds) *The New Development Management: Critiquing The Dual Modernization*. London and New York: Zed Books, pp. 1–17.

Das, V. (2004) 'The Signature of the State: The Paradox of Illegibility', in Das, V. and Poole, D. (eds) *Anthropology in the Margins of the State*. Oxford: Oxford University Press, pp. 225–252.

Escobar, A. (1992) 'Imagining a Post-Development Era? Critical Thought, Development and Social Movements', *Social Text*, 31(32), pp. 20–56.

Escobar, A. (2000) 'Beyond the Search for a Paradigm? Post-Development and Beyond', *Development*, 43(4), pp. 11–15.

Ferguson, J. and Gupta, A. (2002) 'Spatializing States: Toward an Ethnography of Neoliberal Governmentality', *American Ethnologist*, 29(4), pp. 981–1002.

Gibson-Graham, J.K. (2007) 'Surplus Possibilities: Post-Development and Community Economies', in Ziai, A. (ed) *Exploring Post-Development: Theory and Practice, Problems and Perspectives*. London and New York: Routledge, pp. 145–162.

Hickey, S. (2009) 'The Politics of Protecting the Poorest: Moving Beyond the "Anti-Politics Machine"?', *Political Geography*, 28(8), pp. 473–483. Elsevier Ltd.

Li, T.M. (2000) 'Compromising Power: Development, Culture, and Rule in Indonesia', *Cultural Anthropology*, 13(3), pp. 295–322.

Pile, S. (1997). 'Introduction: Opposition, Political Identities and Spaces of Resistance', in Pile, S. and Keith, M. (eds) *Geographies of Resistance*. London and New York: Routledge.

Ray, R. (2020) *Spatial Adhocism as Practice for Conflict Politics: Theorising Urban Politics in Kolkata*. Milton Keynes: The Open University.

Robinson, J. (2006) *Ordinary Cities: Between Modernity and Development*. London and New York: Routledge.

Sharp, J.P. et al. (2000) 'Entanglements of Power: Geographies of Domination/Resistance', in Sharp, J.P. et al. (eds) *Entanglements of Power: Geographies of Domination/Resistance*. London and New York: Routledge, pp. 1–42.

Spivak, G.C. (2012) *In Other Worlds: Essays In Cultural Politics*. New York and London: Methuen.

Ziai, A. (2004) 'The Ambivalence of Post-Development: Between Reactionary Populism and Radical Democracy', *Third World Quarterly*, 25(6), pp. 1045–1060.

Additional resources

Escobar, A. (2017) 'Designs for the Pluriverse' [online video]. Available at: https://www.youtube.com/watch?v=8Ouy7aN6XPs&ab_channel=TheMonadnock

Featherstone, D. (2008) *Resistance, Space and Political Identities: The Making of Counter-Global Networks*. Chichester: Wiley Blackwell.

Mahadevan, S. and Ijlal, N. (2017) 'Contesting Urban Citizenship: The Urban Poor's Strategies of State Engagement in Chennai, India', *International Development Planning Review*, 39(1), pp. 77–95.

Papaioannou, T. (2014) 'How Inclusive can Innovation and Development be in the Twenty-First Century?', *Innovation and Development*, 4(2), pp. 187–202.

Part 3

Actors and institutions. Editorial introduction

An *institution* can be understood as a system of established and embedded social rules that structure social interactions and relationships. These can be formal – such as legal systems – or informal, such as cultural norms, values, or beliefs. The term *actor* in the context of development studies refers to an entity – be that an individual, a group, or an organisation – that actively participates in, influences, or is affected by development processes. Understanding both actors and institutions is crucial, as their interplay forms the backbone of development studies and its concern with understanding social change.

In the theatre of development, institutions matter as they provide the framework within which actors operate. They shape behaviours, influence outcomes, and can either facilitate or hinder development processes. By setting the 'rules of the game', institutions not only set the stage but also define the incentives, opportunities, and constraints that actors grapple with, thereby steering the course of development.

Actors, with their diverse motivations and capacities, continuously shape and reshape institutions. Whether it is social movements advocating for environmental protections or corporate entities driven by profit motives and stakeholder pressures, actors often possess the agency to challenge, reinforce, or even redefine institutional norms. This dynamic interplay, often underpinned by power imbalances, can either catalyse or impede social change.

Key questions emerge around the influence, effectiveness, legitimacy, accountability, and transparency of certain actors in development. For example, multinational corporations have been scrutinised for prioritising profit over sustainable development, leading to potential conflicts of interest and ethical dilemmas.

The relationship between actors and institutions is not merely one of influence but also of collaboration. Synergies arise when NGOs partner with local governments or when international institutions collaborate with grassroots movements. Yet, these interactions can also give rise to conflicts, especially when interests diverge, underscoring the need for effective communication and mutual respect.

Over time, the cast of actors and institutions of interest to development studies has undergone significant shifts. Non-state actors like NGOs, multinational corporations, and social movements have gained greater prominence in the last few decades. Simultaneously, emerging powers such as Brazil, Russia, India, and China (BRIC) have carved out a more pronounced space in the global arena. Their unique development paths have introduced new and competing perspectives, challenging established paradigms, and the hegemony of 'Western' democracies and their associated institutions.

DOI: 10.4324/9780429282348-27

Historically, development studies has often tended towards states and non-governmental organisations (NGOs) as focal points for analysis. However, there is now a recognition of a larger cast of actors shaping the development landscape. From local, civil society organisations, whose knowledge offers invaluable insights, to international organisations such as the United Nations, which plays a key role in coordinating development initiatives and facilitating international cooperation, development studies has sought to understand the multifaceted roles these entities play in shaping, influencing, and driving development agendas.

Fully recognising the complex cast of actors active in the development arena and the diverse perspectives, experiences, needs, and interests they bring them has helped us understand how social change often emerges from competition and conflict as well as collaboration between diverse actors.

The chapters in this section, while diverse in their focus, are unified by several cross-cutting themes such as the evolving nature of agency in development, the blurring of public and private spheres, the challenges and opportunities presented by market-based solutions, and the persistent quest for genuine empowerment amidst the neo-liberalisation and commodification of development.

Chapter 25

Development and nationalism

David Neilson

Introduction

Marx and Engels (1848[1969]) originally argued that capitalism's global market form that turns nation states into 'rivals' would fade away in a socialist world. However, it remains unclear whether this outcome would occur because democratic socialism would transcend the nation state per se or just its capitalist form. They view industrial capitalism's planet-wide spread under a global market framework as democratic socialism's antagonistic precondition. However, they did not specify the global form of this democratic socialism or its relation with the local or national. There is Marx's (1976) brief micro sketch of the 'democratic association of the producers' in *Capital* Vol. 1, but he makes no connection with the transnational. This contribution first makes the case that the presently prevailing 'neoliberal model of capitalist development' has intensified 'national rivalry'. Second, it contrasts the neoliberal model with a democratic socialist model of development that could deliver a progressively cooperative form of the nation state and correspondingly cooperative forms of nationalism.

Nation states, nationalism, and models of development

The nation state and nationalism that have interacting 'internal' (within the nation state) and 'external' (between nation states) dimensions are causally imbricated (Seymour, 2021). As a set of limited possibilities, the form of the nation state determines whether nationalism tends towards a 'regressive' or 'progressive' form. In turn, since WWII, the prevailing 'model of capitalist development' has strongly influenced the form of the nation state. A 'model of development' refers here to the outcome of the interaction between a national regulatory template – which has internal (domestic regulation) and external (foreign economy interface) dimensions – and the transnational framework created as the aggregate effect of the template's widespread adoption, which in turn sets nation states' operational context. Since WWII, there have been two 'models of development', Keynesian-Fordist (1945–1970s) and neoliberal (1980s to present; Neilson, 2020a).

Resonating somewhat with the Left Keynesian post WWII model of development, an ideal-typical progressive model of development refers to a democratically cooperative global mode of regulation that facilitates locally sustainable self-sufficiency for all countries. Implied by this framework of transnational/global-local/national complementarity is a progressive form of the nation-state and correspondingly a progressive form of nationalism characterised internally and externally by cooperative mutuality, positive

DOI: 10.4324/9780429282348-28

sum collaboration, and inclusive solidarity. As the basis of an internally progressive nationalism, democratically inclusive regulation seeks sustainable material security for all locally residing citizens. As the basis of an externally progressive nationalism, each country cooperates with other countries to maximise each one's locally viable accumulation.

At the regressive pole, resonating closely with the neoliberal model of capitalist development, countries are locked into zero-sum locational competition with each other in order to achieve viable local accumulation. Structurally, the internal material security of each nation state conflicts directly with the external relation each has with other nation states. In short, the needs of the one stand in direct conflict with the needs of the other. Externally exclusive, this nationalism tends towards mutual suspicion and mistrust between nations as they contest with each other in zero-sum competitive struggle. Internally, this nationalism tends towards an exclusive form that distinguishes between deserving citizens and undeserving denizens.

Regressive nationalism and the neoliberal model of development

In a direct reversal of the Keynesian-led post WWII model of development, transnational agents of the neoliberal project have pushed the world's countries to each adopt the neoliberal national template. The internal aspect of this template refers to domestic market led regulation that enables the unfettered movement and prerogative of capital within nation state borders. The external aspect refers to economic regulation that opens a country up to capital's free movement across its national borders. The aggregate transnational effect of the neoliberal national template's widespread adoption is a global market terrain on which 'de-territorialised' capital moves freely.

The neoliberal model of development has integrated the world's formally independent nation states into – and made them dependent on finding a viable place within – this level playing field of global market competition. Countries, fixed in space, are drawn into this 'locational competition' with each other in order to attract a viable amount of this free moving capital. Due to the advanced stage of industrial development amongst the mature capitalist countries now combined with the globalisation of productivity norms including in agriculture, labour's oversupply relative to capital's requirements – or the 'relative surplus population' – grows (Neilson and Stubbs, 2011). Because labour's oversupply is capital's scarcity, locational competition becomes zero sum. Inversely, the global scarcity of highly skilled forms of labour that will go to the more profitable companies who can pay winning high rates also drives locational competition. Externally, in the struggle to attract and keep mobile capital at home, the labouring populations of the less-competitive and non-competitive countries experience loss of employment. The zero sum structure of locational competition to attract and keep capital, the source of labour's employment and security, generates rivalry between competing labouring populations, the basic source of regressive nationalism. That is, national labouring populations in competition with each other for scarce jobs undermines international solidarity and encourages suspicion and mistrust between them.

This external competition flows into internal relations within countries. Neoliberal globalisation breaks down a country's real and virtual borders thus facilitating the destabilising (especially for those countries struggling to compete) inward flow of foreign capital and labour. Free-to-move money capital that nation states must attract and retain becomes the direct determinant of a country's financial viability. A growing quantity of foreign products both real and virtual entering into a pre-existing local economy and culture eliminates local producers and undermines local cultures. Free-to-move 'de-territorialised' companies, investors and individual agents with capital roam around the world buying into or replacing local businesses, dispossessing the peasantry and indigenous cultures and buying local real estate that increases accommodation scarcity and pushes up prices for the local population. This zero sum locational competition drives global wage and work norm races to the bottom and generates a hierarchy of unevenly performing and developing countries. This hierarchy is expressed in the uneven size and desperation of national 'relative surplus populations' and in the intensification of illegal and refugee forms of international migration from uncompetitive to competitive countries. At the same time, the wealthy grow wealthier as capital centralises globally and globally scarce labour's income increases, and thus wealth inequality polarises both within and across countries.

Especially for countries – and within them labour market segmented social groups – struggling to achieve or maintain economic viability in this world of intense locational competition, the neoliberal mode of uneven development feeds regressive nationalist tendencies. It provides a terrain on which amoral competitive advantage strategies – and amoral approaches generally – can be rationalised. Rather than grasping the ideologically mystified structural causes of their growing insecurity, uncertainty and inequality that would encourage solidarity, people's everyday experiential knowledge is of alien forces undermining their pre-existing secure and familiar ways of life. This situation encourages an 'amoral familism' where people withdraw into their core identity groups and perceive those outside of them as the enemy. This response aligns with the Far Right's blaming of others, especially including racialised nationalities and demonised foreigners.

The Far Right weaponises people's insecurity and uncertainty induced withdrawal, disorientation and anxiety into a political movement of blame and anger (Neilson, 2015). In particular, Far Right movements incite xenophobia and racism via narratives of a primordial competitive struggle for survival between nations. Moreover, contemporary social media modes of communication provide powerful vehicles for the rapid global spread of these 'willed' narratives that deepen the postmodern malaise of ideological confusion and political polarisation.

As the neoliberal model of development falls towards recurring and deepening multipronged crises, the zero sum competition between countries becomes more brutal, thus intensifying the drivers of regressive nationalism (Neilson, 2021a). When combined with spectres of past struggles and movements still surviving in political institutions and in the minds of the living, competitively regressive nationalist movements, including struggle between Empire nations, will degenerate towards civil and international wars.

Towards an alternative

Inscribed in the neoliberal model of development is an invasively competitive logic that feeds regressive nationalism by undermining national autonomy and pitting countries against each other. The critical alternative, a democratic socialist model of development, imagines a mode of international cooperation that facilitates universally viable local/national viability thereby correspondingly encouraging progressive forms of nationalism. Putting nation states in charge of their own individual and collective priorities via their international cooperation – or 'social democratic multilateralism' – reverses neoliberal templates and trade agreements which create neoliberal globalisation that subordinates countries to capital's imperatives (Neilson, 2021b). Under the alternative project, a group of countries would agree to a common set of core national template principles regarding domestic production and imported products, such as decent labour and ecological sustainability standards. They also would share knowledge and cooperate to help each other maximise the variety of what they can each produce domestically, while continuing to import products they cannot produce themselves. Transforming the United Nations system in a 'cosmopolitan democratic' direction would be a central pre-condition for promoting and supporting the project of social democratic multilateralism. In the spirit of 'knowledge socialism', the United Nations would also establish a 'World Knowledge Bank' whose first priority would be to help the poorest uncompetitive countries develop their own viable local accumulation projects (Neilson, 2020b). In sum, the democratic socialist model of development that imagines a complementary relation between the needs of local national entities and international cooperation and solidarity would drive a positive sum cooperative nationalism for the countries of the world.

Rather than implying the elimination of nation states *per se*, the proposed democratic socialist model of development would transform nation-states into complementary projects and correspondingly facilitate a world of cooperatively complementary nationalisms. Nonetheless, urgent problems remain. If not reversed, human civilisation's present path towards mutual destruction and the increasing scarcity of unevenly distributed natural endowments is undermining projects of inclusive cooperation and increasing the likelihood of international conflicts. Finally, historically shaped by war and imperialism, the present delineation of nation states, when combined with approaching absolute scarcity, will continue to be a source of ongoing civil and international tension.

References

Marx, K. 1976. *Capital*. Vol. 1. Introduced by E. Mandel. Translated by B. Fouwkes. Harmondsworth: Penguin.

Marx, K., and F. Engels. 1848 [1969]. The Communist Manifesto. In *Marx and Engels: Basic Writings on Politics and Philosophy*, edited by L. S. Feuer. London: Fontana.

Neilson, D. 2015. Class, Precarity, and Anxiety under Neoliberal Global Capitalism: From Denial to Resistance. *Theory and Psychology* 25(4): 185–201.

Neilson, D. 2020a. Bringing in the 'Neoliberal Model of Development'. *Capital & Class* 44(1): 85–108.

Neilson, D. 2020b. The Democratic Socialisation of Knowledge: Integral to an Alternative to the Neoliberal Model of Development, In *Knowledge Socialism. The Rise of Peer Production: Collegiality, Collaboration and Collective Intelligence*, edited by M. Peters, T. Beasley, P. Jandric et al., 134–153. Singapore: Springer.

Neilson, D. 2021a. Beyond Regressive Nationalism and the Neoliberal Model of Development. *Review of Radical Political Economics* 53(1): 17–34.

Neilson, D. 2021b. Reversing the Catastrophe of Neoliberal Led-Capitalism in the Time of Coronavirus: Towards A Democratic Socialist Alternative. *Capital & Class* 45(2): 191–214.

Neilson, D., and T. Stubbs. 2011. Relative Surplus Population and Uneven Development in the Neoliberal Era: Theory and Empirical Application. *Capital and Class* 35(3): 435–454.

Seymour, M. 2021. *Nationalism, Companion to Development Studies*. London: Routledge.

Chapter 26

China-Africa relations in a changing world

Frangton Chiyemura

Introduction

The proliferation of economic, political, diplomatic, military and socio-cultural relations between China and Africa since the early 2000s resembles one of the most significant developments in cooperation for Global South countries in the 21st century. Several waves of journalistic, academic and think tank driven research have sought to understand the scope, significance and impacts of the engagements (Oqubay and Lin, 2019). Some publications have focused on drivers and motivations, while others explore the impacts and outcomes of the renewed relations between China and Africa (Kamoche et al., 2021; Alden and Large, 2019). While the majority of research has attempted to understand how China is seen to be driving this engagement, a new set of publications has emerged, shifting the analytical lens to that focused on how Africans – whether as individuals or collectively are influencing, shaping and co-determining the way the engagement unfolds and eventually the benefits and costs associated with such relations (Mohan and Tan-Mullins, 2019). This chapter contributes to this growing literature and begins by briefly contextualising China-Africa relations, followed by a discussion of drivers and motivations. It proceeds with a discussion of the impact and outcomes, ending with a consideration for the future of China-Africa relations.

Contextualising China-Africa relations

Contemporary engagements between China-Africa relations are traced to the Bandung era characterised by Afro-Asian solidarity in which China supported African liberation movements fighting against colonial governments. Notably, from 1970 to 1975 China financed and constructed the Tanzania-Zambia railway project as a gift to African comrades. In the late 1990s, China needed natural resources to drive growth and sought access to foreign markets. Africa was identified as a perfect source of natural resources as well as a market for Chinese goods and services. This resulted in the deepening of political, diplomatic, economic and socio-cultural engagements and in 2009 China became Africa's largest trading partner (see Figure 26.3). While China's foreign direct investment (see Figure 26.4) and loans (see Figure 26.1) to Africa have grown significantly, Chinese funding for infrastructure development remains the most notable aspect of the relationship – partly because the majority of African nations have inadequate and unreliable infrastructure needed to support economic development.

DOI: 10.4324/9780429282348-29

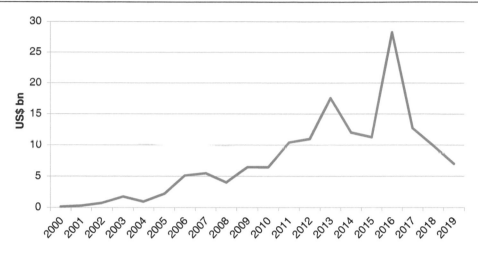

Figure 26.1 Chinese loans to Africa, 2000 to 2021

Against this background, popular discourses about China-Africa relations have been framed by a Chinese-driven analysis of the engagement in which Chinese state-owned businesses, private corporations and state elites were seen to be controlling the development cooperation agenda, (un)wittingly depicting their African counterparts as plaint and subservient. However, new insights from research have shifted the analytical focus to that concerned with how Africans are seen to be significantly influencing their interactions with Chinese actors (Links, 2021; Chiyemura et al., 2022).

Entangled drivers and motivations of China-Africa engagement

Africans and Chinese have interests that are not fixed; they evolve and adjust to policies and popular discourses stemming from within and outside Africa and China's political pressure and priorities. Since the opening of the economy after Xiaoping's reforms and, recently, the Belt and Road Initiative, it became apparent that the growing economy required stable and guaranteed supply of natural resources and for Africa, the rise in demand for these natural resources provided a market opportunity. Africa exports more than 25 per cent of its oil and gas to China, making it the second largest after the Middle East (Vasquez, 2019). In return, China has paid for these natural resources in various ways, key among them being the resource-for-infrastructure model in which a loan for infrastructure construction is 'securitised against the net present value of a future revenue stream from' natural resource extraction (Halland et al., 2014: 1). While China secures natural resources from Africa, Africans secures a market and have the opportunity to finance their infrastructure development, demonstrating the intertwined nature of interests between China and Africa.

Access to foreign markets plays an important role towards economic growth and development. Like any other country, China seeks to access Africa's more than one billion consumers and has been at the forefront of pushing its companies to 'go out' to Africa. Given the domestic competition in China, manufacturers and providers of goods and services turned their eyes to Africa's growing middle class. A 2017 McKinsey report

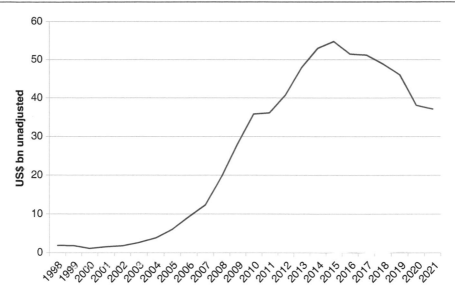

Figure 26.2 Gross annual revenues of Chinese companies' construction projects in Africa

estimates that there were more than 10,000 Chinese-owned firms operating in Africa, and around 90 per cent of these firms are privately owned. In the construction sector, CARI estimate that the gross annual revenues of Chinese companies in Africa was close to $40 billion in 2020 (see Figure 26.2).

Furthermore, Chinese firms have also looked at exploiting Africa's preferential trade agreements with Europe (Everything but Arms) and the United States of America (African Growth and Opportunity Act). This would mean that by relocating to African countries Chinese manufacturers would benefit from reduced customs duties when accessing Western markets. The justification for the relocation is that the cost of production is relatively lower in Africa than in China. As a result, there has been massive influx of both high-end and low-quality Chinese goods and services in Africa, and this has had several impacts which I will discuss in the following section.

As China's economic and business ties with Africa continue to proliferate, scholars argue that aid and development cooperation principles reflect more of China's own development experiences. This means China's involvement in Africa's development landscape is informed by discourses of developing countries solidarity, win-win friendship, non-interference in internal affairs and arguably shaped by principles of justice and equality. Hence some scholars argue that China's engagement with Africa is somehow distinctive. As Power and Mohan (2010: 462) remind us, 'Chinese leaders and strategists believe that China's historical experience and vision of economic development resonates powerfully with African counterparts'. However, it is important to note that China's relations with African countries have evolved dramatically and more emphasis is now on geopolitical and geo-economic interests.

While it is widely undisputed that the market imperatives determine engagements, political drivers also shape the dynamics of the relations. China seeks allies and diplomatic support at various levels of global governance such as the United Nations, and

African governments are eager to be courted by a development partner that treats them as sovereign equals instead of lecturing them about democracy and good governance. China's 'One China' policy, the Hong Kong issue and the Uighurs case have led to Western countries attempting to punish China. African countries have in most cases sided with China. Similarly, China has provided financial and political support for several African regimes with hostile relations with the West, notably Zimbabwe and Sudan.

Against the decline of Western countries' financing of hard infrastructure development in Africa, China provides extensive development assistance that often creates an environment conducive to economic development and friendly political relations. Therefore, African countries generally welcome China's willingness to provide aid without economic conditionalities, which is usually the norm for Western countries' development assistance. However, it is a fallacy to conclude that China does not condition its development assistance. The Chinese prescribe the use of Chinese companies, observance of One China policy, procurement of Chinese technology and services – which is another form of conditionality, albeit different to Western countries and the doners model.

Impact of China's engagement with Africa

Impact and outcomes around China-Africa relations arguably remain highly debated and politicised aspects of this form of South-South development cooperation. This is because it brings to life the big question of 'who benefits'? On the one hand, critical scholars charge that China is an integral part of global capitalism; its interactions with Africa are undesirable because they promote neocolonial tendencies that harm Africa's development policy and practice. Chinese involvement in Africa's development trajectories is seen as entrenching and diversifying African dependency (Taylor, 2014). China is often accused of deliberately indebting Africans in the sense that its 'predatory loan practices and corrupt deals . . . mire nations in debt and undercut their sovereignty, denying them their long-term, self-sustaining growth' in Africa (Tillerson, 2018: 1). China has also been accused of a poor record on environmental standards and of dumping cheap and low-value products (Edwards and Jenkins, 2015) which destructively impact Africa's industrial and manufacturing capabilities. Linked to this, some researchers have focused specifically on political and governance aspects arguing that the 'inter-elite brokerage . . . tends to bypass domestic channels of accountability and so undermines good governance' (Mohan, 2015: 1). Evidently, China has provided economic and political support to African states with poor records on human rights.

Furthermore, trade relations between China and Africa remain highly skewed in favour of China (see Figure 26.3). This shows that China is basically a net importer of African raw materials and unprocessed commodities that do not have added value, while China exports to Africa finished goods and services. Because of these unequal trade patterns, critics conclude that the relations are exploitative, undermine African countries' development and are not so different from Western imperialists.

On the other hand, China has emerged as the largest bilateral creditor to Africa, reaching $160 billion by end of 2020, with the majority spent on infrastructure development (CARI, 2021). However, Chinese foreign direct investment to Africa remains low but steadily rising compared to the USA (see Figure 26.4). While critics argue that China tends to disburse loans to authoritarian regimes and resource-rich countries in Africa, my assessment shows that China's lending practices have much to do with political and

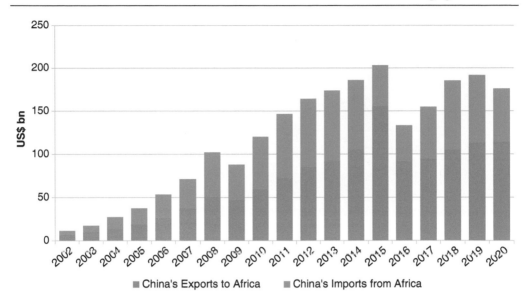

Figure 26.3 China-Africa trade

commercial interests and consider several factors. For example, China does lend to both democratic (South Africa) and authoritarian regimes (Sudan), middle-income (Egypt) and low income (Ethiopia) and resource-rich (Angola) and resource-poor (Ethiopia) African countries.

The relation between China and Africa contribute to the creation of jobs, provision of low-cost capital for infrastructure development, transfer and accumulation of technology, and development of skills essential for Africa's structural and economic transformation (Chiyemura, 2021). My research shows that China provides infrastructure development loans at lower interest rates and with repayment plans more attuned to African countries financial needs (Chiyemura, 2019). By doing so, Chinese firms are investing in productive sectors and creating new ones, contributing to Africa's industrialisation (Calabrese and Xiaoyang, 2022).

These competing understandings of Chinese impacts on Africa highlight several things. First, the impact and outcomes of Chinese involvement in African countries has much to do with the nature of both African and Chinese political and economic interests. This means for African political elites, Chinese investments could therefore be used as an instrument for political legitimacy, and regimes continued stay in power. Second, Western players perceive China's rise as a threat to their supposedly 'zone' of influence in Africa. In fact, the West does consider itself as having a higher moral ground of knowing what is bad or good for Africa. As a result, Western governments, think tanks, and media tend to demonise China's engagement with Africa.

Conclusion: towards the future of China-Africa relations

Academic debates around China-Africa relations are embedded in broader politics of knowledge production and the majority of publications 'seems intent on re-producing

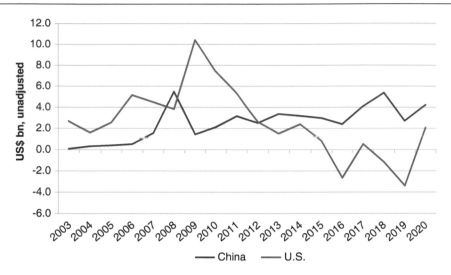

Figure 26.4 Chinese FDI vs. US FDI flow to Africa 2003 to 2020

nationalist-oriented discourses founded on epiphenomena and ideological precepts' (Alden and Large, 2019: 19). Given the depth of relations between China and Africa and the (un)intended consequences of this engagement, there are several issues worth noting about the future of China-Africa relations.

First, both Chinese and African political dynamics play a significant role in shaping the engagement motives, modalities and outcomes which are intertwined, inconstant and undergoing processes of transformations. As new actors with varying interests emerge, new forms of political and economic interdependences would demand thinking differently about the motivations and outcomes of this engagement. Linked to this, the growth of interconnections and increasing complexity between these actors means there is need to move beyond state-based narratives to more plural and dispersed analysis that foregrounds multi-level African engagement and local agency (Benabdall and Large, 2019: 322).

Second, country context and sectorial specificities matters for holistic assessment and understanding of China-Africa relations. There is a need to move away from methodological nationalism, which tends to homogenise Africa or China. To be sure, Africa is not a country but a continent of over 54 countries with over 1.2 billion people and more than 2,000 spoken languages. Similarly, China cannot be treated as monolithic or China Inc.; it is 'a collection of provinces, autonomous regions and municipalities with myriad strategic ties to African countries' (Shen & Fan, 2014: 182) and often characterised by 'competing state agencies, problems of cross-department coordination, and the mismatch between central and local policies' (Su, 2012: 504). Therefore, paying attention to context helps us 'deepen our understanding of the various phenomena which can be observed at the various scales of the multiple relationships that have evolved' between China and Africa (Giese, 2015: 3).

Third, China's engagement with African countries should not be seen in isolation of configurations of global politics of development. There is a need for a comparative lens to understand China's modalities of engagement with other regions and countries, developed or undeveloped. At the same time, it is important to draw comparisons between what the West is doing in Africa and how unique or similar is it to Chinese approaches.

Finally, it is important to pay attention to language used and registers deployed to make sense of this evolving relationship. Oftentimes, we observe 'China in Africa relations', connotations that seem to suggest that African countries are being 'dominated by', 'impacted upon', and 'acted on' by China. We need to frame this as Africa-China relations – in the sense that 'the story of Africa's relations with China is ultimately an African story' (Juma, 2015: 188). This approach returns the focus to Africans – and their agency – which undoubtedly births new perspectives beyond the tired tropes of Africa seen as a victim in its engagement with external powers.

References

Alden, C. & Large, D. (2019). 'Studying Africa and China', in C. Alden & D. Large (eds) *New Directions in Africa-China Studies*. London: Routledge.

Benabdall, L. & Large, D. (2019). 'China and African Security', in C. Alden & D. Large (eds) *New Directions in Africa-China Studies*. London: Routledge.

Calabrese, L., & Tang, X. (2023). Economic transformation in Africa: What is the role of Chinese firms? *Journal of International Development*, 35(1), 43–64. https://doi.org/10.1002/jid.3664

CARI (2021). *China Africa Research Institute*. http://www.sais-cari.org/

Chiyemura, F. (2019). *The Winds of Change in Africa-China Relations? Contextualising African Agency in Ethiopia-China Engagement in Wind Energy Infrastructure Financing and Development*. The Open University. PhD thesis.

Chiyemura, F. (2021). 'Chinese Firms – And African Labor – Are Building Africa's Infrastructure', *The Washington Post*. https://www.washingtonpost.com/politics/2021/04/02/chinese-firms-african-labor-are-building-africas-infrastructure/

Chiyemura, F., Gambino, E. & Zajontz, T. (2022). 'Infrastructure and the Politics of African State Agency: Shaping the Belt and Road Initiative in East Africa', *Chinese Political Science Review*, 8 105–131. https://doi.org/10.1007/s41111-022-00214-8

Edwards, L. & Jenkins, R. (2015). 'The Impact of Chinese Import Penetration on the South African Manufacturing Sector', *The Journal of Development Studies*, 51(4) 447–463.

Giese, K. (2015). 'Adaptation and Learning among Chinese Actors in Africa', *Journal of Current Chinese Affairs*, 44(1) 3–8.

Halland, H., Beardsworth, J., Land, B. & Schmidt, J. (2014). *Resource Financed Infrastructure: A Discussion on a New Form of Infrastructure Financing*. https://elibrary.worldbank.org/doi/epdf/10.1596/978-1-4648-0239-3

Juma, C. (2015). 'Afro-Chinese Cooperation: The Evolution of Diplomatic Agency', in A. Gadzala (ed) *Africa and China: How Africans and Their Governments are Shaping Relations with China*. Lanham: Rowman and Littlefield.

Kamoche, K., Gunessee, S. & Kufuor, N.K. (2021). 'The Africa–China Engagement: Contemporary Developments and Directions for Future Research', *Africa Journal of Management*, 7(4) 447–464.

Links, S. (2021). 'Ascertaining Agency Africa and the Belt and Road Initiative', in F. Schneider (ed) *Global Perspectives on China's Belt and Road Initiative*. Amsterdam: Amsterdam University Press.

Mohan, G. (2015). 'China in Africa: Impacts and Prospects for Accountable Development', in S. Hickey, K. Sen & B. Bukenya (eds) *The Politics of Inclusive Development: Interrogating the Evidence*. Oxford: Oxford University Press.

Mohan, G. & Tan-Mullins, M. (2019). 'The Geopolitics of South-South Infrastructure Development: Chinese-Financed Energy Projects in the Global south', *Urban Studies*, 56(7) 1368–1385.

Oqubay, A. & Lin, J.Y. (ed). (2019). *China-Africa and an Economic Transformation*. Oxford: Oxford University Press.

Power, M. & Mohan, G. (2010). 'Towards a Critical Geopolitics of China's Engagement with African Development', *Geopolitics*, 15(3) 462–495.

Shen, G.C. & Fan, V.Y. (2014). 'China's Provincial Diplomacy to Africa: Applications to Health Cooperation', *Contemporary Politics*, 20(2) 182–208.

Su, X. (2012). 'Rescaling the Chinese State and Regionalization in the Great Mekong Subregion', *Review of International Political Economy*, 19(3) 501–527.

Taylor, I. (2014). *Africa Rising? BRICS – Diversifying Dependency*. Oxford: James Currey.

Tillerson, R.W. (2018). U.S.-Africa Relations: A New Framework. https://sn.usembassy.gov/secretary-264 state-rex-w-tillerson-remarks-george-mason-university-march-6-2018/

Vasquez, P.I. (2019). 'China's Oil and Gas Footprint in Latin America and Africa', *International Development Policy*, 11(1). https://doi.org/10.4000/poldev.3174

Chapter 27

Civil society and civic space

Sarah Peck

The rise of civil society

History is full of examples of human desire for 'collective action in search of the "good" (or better) society' to live in, from the Peasants Revolt of 1381 England to vibrant associational life in 13th-century China (Edwards, 2009: 1, 2014). Varying forms of collective action and associational life have been termed civil society. Conceptualised very broadly as the space between the market, the state and the family, in which people can 'organise, debate and act' (Buyse, 2018: 967), civil society represents one of the most ambiguous terms within political thought (Edwards, 2014).

Civil society – and wider civic space – GDP per Capita for Main Global; Regions 1500–2004 (1990 US dollars) is seen as a key element of global development, as both an arena to be strengthened and organisational bodies capable of delivering development outcomes. Dominant theories of civil society in the development context are heavily influenced by Eurocentric thinking, which places associational life and social capital at the forefront of civil society and subsequent democratic development (De Tocqueville, 1840; Putnam, 2000). Within global development discourse the term civil society often refers to

> formal NGOs and CSOs, often aid- or foreign-funded, involved in service delivery or undertaking a 'watchdog' function by holding government and other actors to account. [However], civil society is properly viewed as a broader category of actors that includes the independent media; human rights defenders; professional associations; academia and thinktanks; and social movements such as land and indigenous people's rights groups, women's and peasant movements, labour organizations, environmental activists, as well as grassroots and community-based organizations.
>
> (Hossein et al., 2019: 9)

Civil society became integral to the World Bank's good governance agenda during its evolution in the 1990s and 2000s. The good governance agenda placed increased emphasis on the role civil society can play in promoting democratic rights and pro-poor development (Banks et al., 2015). Reports stemming from the Bank in the early 1990s outlined a desire to rehabilitate the state following periods of structural adjustment and alongside a revived state; the idea of good governance was able to free and celebrate civil society, accentuating processes such as participation, decentralisation and democratisation to drive development, alongside economic liberalisation and liberal democratisation

DOI: 10.4324/9780429282348-30

(Abrahamsen, 2004; McIlwaine, 1998: 458; Porter, 2003). Civil society was presumed to represent marginalised voices, civil rights and localised interests as a space through which less paternalistic, participatory versions of development could be built and enacted, an arena of social justice and transformation (Banks et al., 2015). Civil society was also thought to provide efficient and effective alternatives to the state in the turn towards market-led development, with civil society organisations seen as innovative, flexible and effective development actors in their own right (Banks et al., 2015; McIlwaine, 1998).

Despite this continued positive emphasis on the role of civil society within the development arena, including recognition within the 2015 Sustainable Development Goals (SDGs), civil society and civil society organisations, also remain the subject of much conjecture. The next section of this chapter will consider how civil society has been critiqued within the global development context.

Critiques of civil society within global development

The promotion of civil society as a crucial development actor, one that can both enhance processes of democratisation and produce alternative, more effective forms of development, has been examined and critiqued from many angles. Conceptual critiques focus on the Eurocentric biases of models of civil society, with these bodies of work questioning the limitations of civil society framings that exclude diverse civil society-state relations and informal and prosaic associational life. Dominant framings of civil society accentuate Eurocentric forms of social relations, leading to concerns about the relevance of these conceptualisations for understanding civil society more globally (Dagher, 2017; Gready and Robins, 2017). These dominant understandings of civil society rely on Eurocentric conceptualiastions of citizenship, individual rights and society-state relations which Chatterjee (2004) contends limits civil society to a narrow section of the urban socio-economic elites in the Global South, ignoring the interconnections between the state and civil society and neglecting other forms of engagement between the two, for example co-operation, bargaining or mutual exchange (Jeffrey, 2007). The idea of civil society may also have limited explanatory power in more global contexts because it does not adequately address the complexities of diverse associational life (Lewis, 2002).

Alongside these conceptual critiques, civil society within the development landscape has also been subjected to multiple practice-based critiques. Optimistic agendas, viewing civil society as the 'magic bullet' (Edwards and Hulme, 1995a: 5), have inevitably led to counter-critiques, with for example Banks et al. (2015: 707) claiming that 'the comparative advantage [of NGOs] was based on ideological grounds rather than evidence'. There has been much debate about whether civil society organisations, NGOs more specifically, are effective development actors, the complexities associated with evaluating their performance and their ability to 'scale up' their work (Edwards and Hulme, 1995a, 1995b). NGOs have however been able to expand services to marginalised groups (Bandyopadhyay, 2013; Banks et al., 2015; Mukute and Taylor, 2013) and studies detail the positive impact they can have on development outcomes, with Townsend and Townsend (2004: 274) concluding that 'the action of NGOs on balance is good in the short term' and Mohan (2002) conceding that NGOs have had micro-level successes.

Much of the literature, however, highlights how civil society has failed to live up to expectations, with two key themes arising, first that civil society is dominated by the elite and second that it has become depoliticised. One significant topic for discussion has been the impact of international aid on NGOs and civil society more widely. Literature details the perceived negative influences of international funding, including the development of paternalistic and dependency relations and the dominance of bureaucratic accountability procedures (Banks et al., 2015). International aid is also connected to the 'NGOisation' and professionalisation of civil society foregrounding formal structures and cultures, with civil society regularly accused of becoming depoliticised, exclusionary, co-opted by the state and democratically limited, solely focused on service delivery – a 'cog in the neo-liberal wheel' (Choudry and Kapoor, 2013; Mercer and Green, 2013: 107; Lebel et al., 2019; Sénit, 2020; Sénit and Biermann, 2021). Perhaps key to understanding civil society is to interrogate, recognise and acknowledge the drivers behind privileging a certain type of civil society and at the same time to search for civil society in its 'actual formation, rather than as a promised agenda for change' (Mamdani, 1996: 19).

Changing civic space

Despite continued emphasis on the role civil society can play in global development, civil society organisations and other civic actors are now coming under greater threat from a number of challenges, including questions about their effectiveness in comparison to other (newer) development actors, changes in aid architecture and constrained civic space, as Green and Pandya (2016) argue on the global media platform openDemocracy:

> Civil society organisations (CSOs) worldwide are under significant pressure as restrictions on foreign funding, barriers to registration, intervention in CSOs' internal affairs, and other forms of harassment have proliferated.

Civic space can be understood as 'the political, legislative, social and economic environment which enables citizens to come together, share their interests and concerns and act individually and collectively to influence and shape their societies' (Civic Space Watch, 2021). Contemporary trends over the last two decades have seen alterations in this space, associated with a narrowing of civic space for more progressive voices and organisations, alongside a rise in right-wing activities and digital engagement (Hossein et al., 2019). The closing down of civic space is a global phenomenon occurring across political regime types, geographies and levels of development, often connected to ethnonationalist and illiberal political discourses (Brechenmacher and Carothers, 2019; Hossein et al., 2019). A variety of processes have narrowed and restricted the spaces in which civil society can operate. For civil society organisations this has included delegitimisation, increased regulation and intimidation (Hossein et al., 2019) as well as restrictions on donor funding, bans on political advocacy, constitutional reforms and restrictions on protests (Carothers and Brechenmacher, 2014; Christensen and Weinstein, 2013; Karim, 2016). Some organisations have closed, some have strategically altered their activities often into a seemingly less political domain and some have reduced their organisational visibility.

Changes in the global economy, alterations in aid architecture, the rise of the private sector actors and the securitisation of development have placed civil society organisations and actors in positions of multiple and increasing vulnerabilities. One concern is the financial landscape in which civil society organisations are now required to operate. Many have experienced decreases in the amount of international donor funding available to them, alongside increasing bureaucratisation and regulation of any funds received.

Connected to this loss of financial support from 'traditional' international donors, civil society groups are now engaging with more diverse financial processes, utilising different forms of funding in their work, including donations from corporate sponsors (Mendonca et al., 2016) and adopting social enterprise models (Cieslik, 2016; Hailey and Salway, 2016).

This narrowing of civic space within an increasingly illiberal political climate reinforces the power of political and social elites, reduces the power of civil society and drives modes of development that disregard civic, political and social concerns (Brechenmacher and Carothers, 2019; Hossein et al., 2019; Poppe and Wolff, 2017). This can be seen in the area of extractive land and resource use and in the promotion of large-scale infrastructure projects that occur in a development landscape that does not allow for challenges from the civic arena, as Hossein et al. (2019: 10) articulate: 'Many efforts to silence civil society actors, it seems, do so in order to pave the way for contentious projects to pass without the fear of public scrutiny or effective legal obstacles'. The closing down of civic space then has impacts on the formation and activities of civil society organisations and on the communities, environments and socio-political causes they advocate for. Whilst civic space may be narrowing for more progressive organisations, Hossein et al. (2019: 10) comment civic space may be opening up for 'right wing extremists, supporters of authoritarian rule, and cultural and faith-based groups that may be opposed to individual human rights or equality'. The rise of digital civic space and also the restrictions that can be placed on the digital sphere are increasingly important in the global development context (Kleine, 2018). It is important then to consider the fluidity of civic space, for whom it is closing and opening and the impacts this has. Changes in civic space will play out differently around the world, including how civil society adapts and responds to these changes.

Conclusion

Civil society and wider civic space have long been associated with global development. Dominant models of civil society within the global development context are based on Eurocentric models of associational life, social capital and the 'good' governance. Civil society within global development has however been critiqued both conceptually and practically as it has often failed to live up to expectations as the 'magic bullet' (Edwards and Hulme, 1995a: 5). Civil society and wider civic space remain integral to global development yet are coming under increasing threats from multiple angles, including illiberal political climates, changes to aid architectures and ethnonationalist and populist discourses. Hossein et al. (2019: 3) comment 'civic space is a precondition for achieving the . . . [Sustainable Development Goals] SDGs'. Civil society and wider civic space are important as spaces in which ideas of social and environmental justice and human rights can be nurtured. If civil society operates as a sphere which can represent, engage and nurture the most marginalised, then the increasingly restrictive civic environment has the potential to limit the progress made towards reducing injustices (Hossein et al., 2019). Civil society and civic space are crucial to global development as spaces in which social justice can be fought for, states and the private sector made accountable and lives and well-being improved.

References

Abrahamsen, R. 2004. The World Bank's good governance agenda: issues of power and democracy. In *Ethnographies of aid: exploring development through texts and encounters*, eds. J. Gould & H. Marcussen, 15–44. Roskilde: Roskilde University.

Bandyopadhyay, K. 2013. Civil society at multiple crossroads in Asia. *Development in Practice* 23 (5–6): 644–652. DOI: 10.1080/09614524.2013.800841

Banks, N., Hulme, D. & Edwards, M. 2015. NGOs states and donors revisited: still too close for comfort. *World Development* 66: 707–718. DOI: 10.1016/j.worlddev.2014.09.028

Brechenmacher, S. & Carothers, T. 2019. *Defending civic space: is the international community stuck?* https://carnegieendowment.org/files/WP_Brechenmacker_Carothers_Civil_Space_FINAL.pdf

Buyse, A. 2018. Squeezing civic space: restrictions on civil society organisations and connections with human rights. *International Journal of Human Rights* 22: 1–23. DOI: 10.1080/13642987.2018.1492916

Carothers, T. & Brechenmacher, S. 2014. *Closing space: democracy and human rights support under fire*. Washington: Carnegie Endowment for International Peace.

Chatterjee, P. 2004. *The politics of the governed: reflections on popular politics in most of the world*. New York: Columbia University Press.

Choudry, A. & Kapoor, D. 2013. NGOization: complicity, contradictions, and prospects: introduction. In *NGOization: complicity, contradictions and prospects*, eds. A. Choudry & D. Kapoor, 1–24. London: Zed Books.

Christensen, D. & Weinstein, J. 2013. Defunding dissent: restrictions on aid to NGOs. *Journal of Democracy* 24 (2): 77–91.

Cieslik, K. 2016. Moral economy meets social enterprise: community based green energy project in rural Burundi. *World Development* 83: 12–26. DOI: 10.1016/j.worlddev.2016.03.009

Civic Space Watch. 2021. *About civic space*. https://civicspacewatch.eu/what-is-civic-space/

Dagher, R. 2017. Civil society and development: a reconceptualization. *Canadian Journal of Development Studies/Revue canadienne d'études du développement* 38 (1): 54–71. DOI: 10.1080/02255189.2016.1196167

De Tocqueville, A. 1840. *Democracy in America*. Translated from French by Henry Reeve. Ware: Wordsworth Editions Ltd.

Edwards, M. 2009. *Civil Society* (2nd Edition). Cambridge: Polity Press.

Edwards, M. 2014. *Civil Society* (3rd Edition). Cambridge: Polity Press.

Edwards, M. & Hulme, D. 1995a. NGO performance and accountability: introduction and overview. In *Non-governmental organisations – performance and accountability beyond the magic bullet*, eds. M. Edwards & D. Hulme, 3–17. London: Earthscan.

Edwards, M. & Hulme, D. 1995b. Policy avenue: NGO performance and accountability in the post cold war world. *Journal of International Development* 7 (6): 849–856.

Gready, P. & Robins, S. 2017. Rethinking civil society and transitional justice: lessons from social movements and 'new' civil society. *The International Journal of Human Rights* 21 (7): 956–975. DOI: 10.1080/13642987.2017.1313237

Green, S. & Pandya, A. 2016. *Introducing openGlobalRights' newest debate: 'closing space for civil society' – analyzing the drivers and responses* [online]. openDemocracy. https://www.opendemocracy.net/openglobalrights/shannon-n-green-archana-pandya/introducing-openglobalrights-newest-debate-closing-s

Hailey, J. & Salway, M. 2016. New routes to CSO sustainability: the strategic shift to social enterprise and social investment. *Development in Practice* 26 (5): 580–591. DOI: 10.1080/09614524.2016.1188886

Hossein, N., Khurana, N., Nazneen, S., Oosterom, M., Schröder, P. & Shankland, A. 2019. *Development needs society – the implications of civic space for the sustainable development goals*. https://www.ids.ac.uk/publications/development-needs-society-the-implications-of-civic-space-for-the-sustainable-development-goals/

Jeffrey, A. 2007. The geopolitical framing of localized struggles: NGOs in Bosnia and Herzegovina. *Development and Change* 38 (2): 251–274. DOI: 10.1111/j.1467-7660.2007.00411.x

Karim, L. 2016. Resistance and its pitfalls: analyzing NGO and civil society politics in Bangladesh. In *The SAGE handbook of resistance*, eds. D. Courpasson & S. Vallas, 461–476. London: SAGE.

Kleine, D. 2018. Development. In *Digital geographies*, eds. J. Ash, R. Kitchen & A. Leszcynski, 225–238. London: SAGE.

Lebel, P., Lebel, L., Singphonphrai, D., Duangsuwan, C. & Zhou, Y. 2019. Making space for women: civil society organizations, gender and hydropower development in the Mekong region. *International Journal of Water Resources Development* 35 (2): 305–325. DOI: 10.1080/07900627.2018.1425133

Lewis, D. 2002. Civil society in an African Context: reflections on the usefulness of a concept. *Development and Change* 33 (4); 569–586. DOI: 10.1111/1467-7660.00270

Mamdani, M. 1996. *Citizen and subject: contemporary Africa and the legacy of late colonialism.* Chichester: Princeton University Press.

McIlwaine, C. 1998. Civil society and development geography. *Progress in Human Geography* 22 (3): 415–424. DOI: 10.1191/030913298669927952

Mendonca, P., Alves, M.A. & Nogueira, F. 2016. Civil society organizations and the fight for rights in Brazil: analysis of an evolving context and future challenges. *Development in Practice* 26 (5): 592–605. DOI: 10.1080/09614524.2016.1190318

Mercer, C. & Green, M. 2013. Making civil society work: contracting, cosmopolitanism and community development in Tanzania. *Geoforum* 45: 106–115. DOI: 10.1016/j.geoforum.2012.10.008

Mohan, G. 2002. The disappointment of civil society: the politics of NGO intervention in Northern Ghana. *Political Geography* 21 (1): 125–154. DOI: 10.1016/S0962-6298(01)00072-5

Mukute, M. & Taylor, J. 2013. Struggles for systems that nourish: Southern Africa civil society contributions and challenges in the creation of flourishing societies. *Development in Practice* 23 (5): 609–616. DOI: 10.1080/09614524.2013.800840

Poppe, A.E. & Wolff, J. 2017. The contested spaces of civil society in a plural world: norm contestation in the debate about restrictions on international civil society support. *Contemporary Politics* 23: 469–488. DOI: 10.1080/13569775.2017.1343219

Porter, G. 2003. NGOs and poverty reduction in a globalizing world: perspectives from Ghana. *Progress in Development Studies* 3 (2): 131–145. DOI: 10.1191/1464993403ps057ra

Putnam, R. 2000. *Bowling Alone: the collapse and revival of American community.* New York: Touchstone.

Sénit, C.-A. 2020. Leaving no one behind? The influence of civil society participation on the sustainable development goals. *Environment and Planning C: Politics and Space* 38 (4): 693–712. DOI: 10.1177/2399654419884330

Sénit, C.-A. & Biermann, F. 2021. In whose name are you speaking? The marginalization of the poor in global civil society. *Global Policy* 12(5): 581–591. DOI: 10.1111/1758-5899.12997

Townsend, J. & Townsend, A. 2004. Accountability, motivation and practice: NGOs North and South. *Social & Cultural Geography* 5 (2): 271–284. DOI: 10.1080/14649360410001690259

Chapter 28

Role of non-governmental organisations (NGOs)

Vandana Desai

The growth of the NGO sector

The term NGO is applied to many kinds of organisations, ranging from large Northern-based charities such as Oxfam or Save the Children to local self-help organisations in the South with an aim to improve the quality of life of disadvantaged people. They are mainly private initiatives, involved in development issues on a non-profit basis. The term 'NGO' is understood to refer to those autonomous, non-membership, relatively permanent or institutionalised (but not always voluntary) intermediary organisations, staffed by professionals or the educated elite, which work with grassroots organisations in a supportive capacity. Grassroots organisations (GROs) on the other hand are issue-based, often ephemeral, membership organisations; they may coalesce around particular goals and interests and dissipate once their immediate concerns have been addressed. Non-governmental organisations (NGOs) have become an important and vocal platform for the involvement of civil society in public affairs.

NGO partnerships

Since the 1950s, NGOs have come to play an increasingly important part in the formulation and implementation of development policy, becoming key actors in the political economy of development. There has been increased collaboration both with governments and developmental aid agencies based on a growing belief over the period that the promotion of NGOs could offer an alternative model of development and play a key role in processes of democratisation, decentralisation of policies and participatory development.

NGOs are popular because they are perceived to be administratively flexible, able to access the disadvantaged or marginalised through their work at the grassroots level, innovative in problem solving, adaptable to local context, more cost effective than corresponding state partners and their grassroots representation brings legitimacy and community mobilisation to programmes and projects.

Donor pressure towards structural reform and privatisation in the 1990s underlies the increased interest in NGOs as 'service deliverers' – part of a wider and explicit objective to facilitate productive NGO–state partnership. There is a realisation among donor countries that aid is becoming ever more complex with new instruments and players with increasing criticisms over the effectiveness of aid. Private financial flows increased rapidly in the wake of increased liberalisation, driven by market reforms, rise in global trade, lowering investment barriers in developing countries and the fall in communications and

DOI: 10.4324/9780429282348-31

transport costs. This also led to the expansion in partnership between Northern and Southern NGOs in changing attitudes and educating constituencies (aim to change public opinion) in the Global North towards development interventions and the underlying causes of poverty, drawing people into active lobbying (attempting to change policy) and campaigning for change in the Global South. A view took hold that merely transferring resources in the form of tools or funds was not an adequate response to alleviating poverty which was rooted in deeper global and local structural problems. Indeed, such transfers could just preserve the situation by creating financial dependency and reproducing inequalities.

Northern NGOs also have a humanitarian function and respond to emergencies, short term relief and long-term rehabilitation, such as for victims of war and of natural or man-made disasters. They raise money in the North, from the general public, private sector and governments to pay for their work and to share as much as possible with their Southern counterparts in building their capacity to support rehabilitation and recovery.

Southern NGOs have the basic responsibility among NGOs for leading the development process in developing countries and the expertise to do so (as outlined earlier). It is expected that relationship between Northern and Southern NGOs is based on an equal partnership incorporating transparency, mutual accountability and risk-sharing. Concerns have been raised in the last few decades on the impact of the unequal relationship among donors, Northern NGOs (see Bebbington, 2005) and Southern NGOs on issues of accountability (see section on accountability)

Roles of NGOs

NGOs play three main roles: service provision, advocacy on behalf of the disadvantaged or marginalised and empowerment i.e. enabling the disadvantaged or marginalised to become advocates for themselves. As service provision agents, NGOs intervene in providing welfare, technical, legal and financial services to the disadvantaged or marginalised or work with community organisations in basic service and infrastructure provision. This is frequently a matter of filling gaps left by the partial service delivery of governments withdrawing from involvement in provision. In the past, governments of developing countries were seen as spearheading the development process. However, such paternalism reached its limits when it became clear that government did not have the financial resources to pay for the essential services of the disadvantaged or marginalised and lacked the organisational expertise to be effective (for example in combatting malaria or HIV/AIDS). In such an environment, the important role for NGOs in the last few decades has been in mitigating the adverse costs of structural adjustment and promoting donor reform packages in offering insurance against a political backlash against harsh adjustment regimes. Such a role raises important questions. Patterns of service delivery through the voluntary sector may lack compatibility and co-ordination. In so far as such efforts rely on government funding, their ultimate sustainability is questioned. At a deeper level, there are worries about the long-term impact of NGO service provision on the sustainability of national health and education systems (rather than programmes) and access to quality services for all. NGOs also constantly face the challenge of the need to adopt 'best fit' approaches within the context of the particular country or local areas and existing institutions rather than to import 'best practices' which may not fit or apply to the local context.

It is also believed that NGOs provide support, in an ad hoc manner but sometimes through more structured programmes. For example, in the past decade NGOs have been very active in promoting local formal savings or microfinance groups in communities supported by or linked to external organisations such as credit unions and banks especially in promoting groups for women as women are least likely to have access to more formal means of saving and borrowing. Women are also considered to be more reliable at repayment than men and to spend the income they control on things that are more likely to benefit family welfare. In addition to their financial objectives, these NGOs may also promote increased collective action by their members, for example investing in community development projects such as construction of wells and schools or lobbying other actors such as larger Northern NGOs to undertake larger projects beyond the means of the community (see Dodman and Mitlin, 2013).

Advocacy: networking and building social movements

In last few decades NGOs' role in policy advocacy has gained importance in seeking social change by influencing attitudes, policy and practice, seeking to reform state services on the basis of NGO experiences and lobbying directly for the policy changes and reform. This is involvement in participatory, public-interest politics, and NGOs engaging in such activity realise the increasing importance of information (which they gather through their own experience of working at the grassroots) as they begin to utilise the power of ideas and information to promote positive change in the wider structures of government and the official aid community. These NGOs often play a catalytic or seeding role – demonstrating the efficacy of a new idea, publicising it, perhaps persuading those with access to greater power and budgets to take notice and then encouraging the widespread adoption by others of the idea. Recently, with the influence of social media, their role as civil society organisation has enhanced in campaigning, either individually or through networks, through partnership between Northern and Southern NGOs in changing attitudes and public opinion both in the North and South towards developmental issues and interventions, inequalities and social justice and drawing people into active lobbying and campaigning for change in perspectives and interventions.

Neither of these roles need exclude the other. Some NGOs naturally progress from filling a gap in service provision to recognising the need to look outward to the wider context in which the need arises and find themselves drawn, possibly through involvement in NGO networks (Bebbington, 2003), into national or global policy advocacy.

NGOs' work with grassroots community organisations often involves disadvantaged and marginalised groups and has been important in mobilising these large numbers of people against either entrenched elites or state interests, campaigning on their behalf and seeking to influence public policy (e.g., placing climate/environmental on the agenda of G8 summit) through citizen participation. This type of 'bottom-up democracy' has been successful in many instances in drawing the attention of politicians, political parties and the wider public to demand change. NGOs have become key actors in a process of transformatory development. This communicative power of information, research, campaigning, lobbying or social media work, because of their moral authority, creates alliances and networks to place pressure on the state which are growing in their complexities (for example, in the area of human rights, corporate social responsibility, child labour etc.).

Recently advocacy campaigns have taken the role of being transformers of consciousness, for example highlighting the reality of sexual oppression in everyday living (domestic violence, human trafficking, gender pay gap, police brutality, sexual harassment at work to name a few) as well as oppression of race, class, inequalities and social justice. This has been amplified and developed through creative communication and advocacy strategies to influence policy, narratives, hearts and minds. There has been an intersection of art and policy that uses and highlights the influential possibilities of art as a medium to discuss and to challenge perceptions of women's experiences, raise awareness and empower women, while simultaneously championing policy change.

The datafication of the world and the increasing availability of (geo)data opens new opportunities for monitoring and documenting human rights violations, particularly in regions and situations that are difficult to access due to security and safety concerns. These advances are raising new technical and methodological challenges and opportunities whilst also posing fundamental ethical, legal and political questions about data availability and use (see Walker, 2020).

In considering the role of NGOs and their links to social movements, there is a growing set of work which conceptualises two meanings of the term 'development' (Hart, 2001; Mitlin et al., 2007; Banks and Hulme, 2012). One is a historical process of social change, while the other refers to specific interventions, particularly those that fit broadly within the post-World War II project of aid and development. Hart (2001) labels these as 'little d development' and 'big D Development' respectively. These classifications are not independent: NGOs are all acting in interventionist Development, while also being part of 'little d development' (Mitlin et al., 2007).

For Banks and Hulme (2012), service provision activities and advocacy undertaken by NGOs on behalf of the disadvantaged or marginalised remain within the domain of 'Big D' Development and depoliticised approaches. In contrast, approaches where the disadvantaged or marginalised act as advocates for themselves should be seen as part of 'little d' development and a way for NGOs to engage with promoting alternatives to 'Big D' Development. It proposes that NGOs should shift further towards radical, system-changing alternatives and be more aligned with social movements rather than merely seeking reforms within existing systems.

NGOs and the role of civil society

There is an increasing interest in the role of NGOs in promoting democratic development by virtue of their existence as autonomous actors. NGOs are said to pluralise and therefore to strengthen and expand the institutional arena and bring more democratic actors into the political sphere. More civic actors means more opportunities for a wider range of interest groups to have a 'voice', more autonomous organisations to act in a 'watchdog' role (especially by international donors) *vis-à-vis* the state and more opportunities for networking and creating alliances of civic actors to place pressure on the state (e.g., environment and climate change initiative) to channel public opinion into policy making and bring about reform. It is believed that NGOs strengthen the state through their participation in improving the efficiency in government services, acting as strategic partners for reform-oriented ministries, filling in gaps in service provision and helping the government forge ties with the grassroots. It is believed that NGOs have a key role to play during and after democratic transitions. For example, the Arab Spring in Tunisia in 2010 was a social movement that tipped the balance across the Middle East.

NGOs have become inextricably implicated in civil society, promoting democracy, good governance and building of social capital. Clarke (1998a) examines the role of NGOs in the politics (see Devine, 2006) of development across the developing world and opines that the failure to theorise the political impact of NGOs has led to an overly 'inadequate, explicitly normative interpretation of NGO ideology' (1998a: 40). This failure has encouraged a tendency to take NGOs' political role as natural/self-evident. Hence most development aid is aimed at promoting such linkages between civil society and the state has been directed to strengthening central governments. NGOs are a key part of civil society, sometimes to the extent that NGOs are conflated with civil society (Mercer, 2002), a move that appeals to donors since it permits them to fund NGOs and then claim to be promoting civil society. Important developments in the last decade are where ordinary people acting in their own country are considered change agents e.g., the Occupy movement of 2011 triggered by widespread discontent with wealthy financiers who caused the 2008 financial crisis – a force whose currency was social-media-enabled organisational skills more than development funding.

It is also important to understand that NGOs are inherently constrained in the extent to which they can act politically or have power to engage with processes of institutional change, then it makes sense to focus on specific problems where there may be some possibility of small positive steps, even if these are likely to be incremental rather than more extensive reforms. The impact of states upon NGOs is central in defining the role NGOs can play in national development, for it is governments which give NGOs the space and the autonomy to organise, network and campaign (Clarke, 1998a, 1998b). Of course, it is difficult to generalise about state–NGO relations, as local political networks are always diverse.

NGOs and accountability

NGOs are increasingly being funded by states and official aid agencies, raising questions about what impact this trend has on NGO accountability. Concerns are raised by many regarding internal and external accountability of NGOs. NGOs have downward accountability to members and upward accountability to donors/governments (patrons). NGOs are internally accountable to beneficiaries, donors, boards of directors, trustees and advisory committees. They are externally accountable to organisations and actors with which they affiliate, international government organisations, states, and people throughout the world. Accountability is crucial for NGOs as NGOs have only their reputation for credibility on which to base their action. Accountability in NGOs needs to be appropriate for their work, for the needs of the beneficiaries (clients) and the values of the organisation itself. There is *functional accountability* in relation to accounting for resources and their impacts and *strategic accountability*, which relates to the wider implications of an NGO's work (especially in the context of impacts on other organisations or the wider environment in which NGOs operate). These are basically mechanisms for assessing effectiveness, monitoring and evaluating NGOs. Accountability in NGOs is becoming quite complex (see Edwards and Hulme, 1995).

Conclusion

The key issue for the future is whether and how NGOs will adapt to the global changes which are currently under way. NGOs are constantly having to link both local and global agendas if they are to be effective, and they will increasingly be forced to learn from – and adapt to – changing demands and opportunities.

The increased availability of large-scale funding has been one of the primary factors driving NGO growth in the 1980s, encouraging the proliferation of social welfare organisations which often had little or no political agenda. The 'inherent' advantages of the NGOs themselves are gradually worn away by increased funding, professionalisation, bureaucracy and the shifting of objectives away from 'social mobilisation' (which might be less attractive to donors) towards service delivery. This process may lead to a widening rift between well-resourced service providers and poorly funded social mobilisation organisations. This highlights that NGOs exhibit potentially illuminating contrasts in emphasis and packaging of activities, in client groups and organisational style. Considerable diversity exists in relation to how autonomous NGOs are from the influence of funding agencies or management influences of donors. Increasingly questions have been asked. Can NGOs deliver all that is expected from them? Is the glowing image realistic? How effective are NGOs? Attention must also be paid to understanding the role of NGOs in influencing institutional change. This is a reminder of the importance of politics in public services. What are the opportunities and constraints related to influencing policies?

There seems to be more concentration on success stories, and there seems to be a gap emerging between rhetoric and practice, which raises issues of objective monitoring and evaluation of NGOs' projects, effectiveness, legitimacy, performance and accountability. Despite growing interest in evaluation, there is still a lack of reliable evidence on the impact of NGO development projects and programmes. If NGOs want to continue to 'sustain their claim to moral authority', NGOs need to maintain attributes such as impartiality and independence, veracity and reliability, representativeness and accountability and transparency.

References and guide for further reading

Banks, N. & Hulme, D. (2012) *The Role of NGOs and Civil Society in Development and Poverty Reduction*, Manchester, UK: Brooks World Poverty Institute, University of Manchester.

Bebbington, A. (2003) 'Global networks and local developments: agendas for development geography', *Tijdschrift voor Economische en Sociale Geografie* 94(3): 297–309.

Bebbington, A. (2005) 'Donor–NGO relations and representations of livelihood in nongovernmental aid chains', *World Development* 33(6): 937–950.

Bebbington, A., Hickey, S. & Mitlin, D. (eds) (2008) *Can NGOs Make a Difference? The Challenge of Development Alternatives*, London: Zed Books Ltd.

Bond, P. (2006). 'Global governance campaigning and MDGs: from top-down to bottom-up anti-poverty work', *Third World Quarterly* 27(2): 339–354.

Brass, J.N., Longhofer, W., Robinson, R.S. & Schnable, A. (2018) 'NGOs and international development: a review of thirty-five years of scholarship', *World Development* 112: 136–149.

Brown, L.D., Ebrahim, A. & Batliwala, S. (2012) 'Governing international advocacy NGOs', *World Development* 40(6): 1098–1108.

Clarke, G. (1998a) *The Politics of NGOs in South-East Asia: Participation and Protest in the Philippines*, London: Routledge.

Clarke, G. (1998b) 'Non-governmental organisations (NGOs) and politics in the developing world', *Political Studies* XLVI: 36–52.

Devine, J. (2006) 'NGOs, politics and grassroots mobilisation: evidence from Bangladesh', *Journal of South Asian Development* 1(1): 77–99.

Dodman, D. & Mitlin, D (2013) 'Challenges for community-based adaptation: discovering the potential for transformation', *Journal of International Development* 25(5): 640–659.

Edwards, M. & Hulme, D. (eds) (1995) *Beyond the Magic Bullet: NGO Performance and Accountability in the Post-Cold War World*, London: Macmillan.

Hart, G. (2001) 'Development critiques in the 1990s: *culs de sac* and promising paths', *Progress in Human Geography* 25: 649–658.

Hulme, D. & Edwards, M. (eds) (1997) *Too Close for Comfort? NGOs, States and Donors*, London: Macmillan.

Khieng, S. & Dahles, H. (2015) 'Resource dependence and effects of funding diversification strategies among NGOs in Cambodia', *Voluntas* 26: 1412–1437.

Mercer, C. (2002) 'NGOs, civil society and democratization in the developing world: a critical review of the literature', *Progress in Development Studies* 2(1): 5–22.

Mitlin, D., Hickey, S. & Bebbington, A. (2007) 'Reclaiming development? NGOs and the challenge of alternatives', *World Development* 35(10): 1699–1720.

Rugendyke, B. (ed) (2007) *NGOs as Advocates for Development in a Globalising World*, Abingdon and New York: Routledge.

Walker, J.R. (2020) 'Remote sensing for international human rights advocacy: critiques and responses', *Journal of Human Rights* 19(2): 183–200.

Useful websites

Debt Justice – a campaigning organisation that exists to end unjust debt and its root causes, formerly named Jubilee Debt Campaign. https://debtjustice.org.uk/

Global Call to Action Against Poverty (GCAP) – supports people in their struggles for justice and brings individuals and organisations together to challenge the institutions and processes that perpetuate poverty and inequalities. https://gcap.global/

Chapter 29

Philanthropy, private foundations, and global development

Adam Fejerskov

Every culture across ancient and modern history has had its share of philanthropy, literally *love of humanity*, albeit in many different forms and shapes. From private individual benefactors, to private businesses or fundamental societal institutions such as the church, philanthropic giving to the less fortunate has always been valued as a human virtue. But philanthropy has never been a politically neutral and altruistic endeavor, solely made for the good of man. The combination of wealth accumulation and gifting away of funds and resources has always caused upheaval and scrutiny for its ability to shape society. The Roman emperor Septimius Severus withdrew the right to gift property to the church in the second century, as it had grown to a size allowing it to self-confidently challenge the emperor, and today history's greatest private foundation, The Bill and Melinda Gates Foundation, faces continuous scrutiny for the way it distributes billions of dollars every year.

Since the late nineteen-hundreds, Western countries have seen the advent of institutionalized giving, that is the forming of professional organisations and institutions that only engage in philanthropic engagements, from Rockefeller to Ford and the Gates Foundation. Though these private foundations have been involved in development efforts for more than a hundred years, particularly two trends are observable over the past decades that have cemented the influence and importance of these organisations in global development. The first is that many foundations have become increasingly strategic and professionalized, themselves now targeting what areas of impact they wish to pursue and the partners with which they want to work. The second is that with this new strategizing has come an increasing outwards focus, from the local and national scene to the international, reflected in growing funds for development efforts. In this chapter, we will take a brief look back at the historical lines of work for private foundations in development, before we focus on their impact and approaches today and their future influence on global development.

Historical avenues of private foundations and development

Modern-day private foundations emerged in the late 19th century, during the golden age of American philanthropy that saw some of the early billionaires looking for ways to spend their fortunes and appease some of the criticism their massive wealth accumulation had evoked. The Carnegie Corporation, the first of what would become known as the 'big three' of American foundations, was formed in 1911 from the riches of Scottish steel-magnate Andrew Carnegie. Though settling for a life far from the largest foundations of

DOI: 10.4324/9780429282348-32

today, the Carnegie Corporation were engaged in some of the first private foundation activities targeting education in Southern and East Africa during the 1920s.

The Rockefeller Foundation was formed only two years after Carnegie and quickly adopted a 'scientific' approach to philanthropy that shares many traits with the science and technology focus of the Gates Foundation today. Built on dominance, if not monopoly, in the oil business, the foundation would go on to support efforts of disease eradication, not least agricultural technologies and reforms, sometimes considered to having been one of the largest development actors up until the 1940s. The Rockefeller Foundation is not least known for having introduced the first 'Green Revolution', an experimental technological response to food shortage that included the development of new varieties of key crops, leading to increased agricultural production from the 1950s and onwards, transforming agriculture across the Global South, mainly in India, Pakistan, Bangladesh, Indonesia and China. The 'Green Revolution' was not so much a coherent framework for agricultural development as a diverse set of experimental interventions across the developing world that took on many forms – modern, scientific, commercial. The Green Revolution had all the elements of success among policy makers and agricultural scientists. But it also entailed a set of undesired – and to a large extent unforeseen – ecological consequences. The new scientific approach of the green revolution prompted the use of 'technology packages' that have been carried over in new initiatives such as AGRA or Alliance for a Green Revolution in Africa, funded by the Gates Foundation today.

As the last of the Big Three, the Ford Foundation chose a different path, removed from science and technology. A historical focus on civil rights, gender equality and education has led the foundation to narrowing its scope on inequality today where it remains the second largest of the US foundations. The early and mid-20th century was a time of mass-institutionalization of private giving, both greatly increasing the volume but also systematizing professional approaches to philanthropy. During this time, private foundations made their mark on the global landscape with self-proclaimed revolutions in medical science, education and agriculture. During the inter-war period, foundations in the US retained a position of almost governmental importance in the home territories, but after the second world war, these increasingly turned their attention outwards, funding projects, programs and international networks crucial to international politics and e.g., the forming of what is now the World Health Organization (Moran, 2014).

Philanthrocapitalism and the Bill and Melinda Gates Foundation

Today, private foundations constitute a main source of private power in global development and international politics. Alongside the advent of other non-state actors, private foundations have grown massively in volume and in active engagements within the field of global development. This has not least coincided with wealth creation along the same heights as that of the Gilded age, from which the traditional big US foundations emerged. Especially the digital realm has proved able to concentrate immense riches in the hands of a few individuals who are now increasingly taking things to the global scene of philanthropy, including Bill Gates, Mark Zuckerberg and Jeff Bezos.

Private foundations are typically defined as a 'Non-governmental, non-profit organisation having a principal fund of its own, managed by its trustees or directors, and established to maintain or aid social, educational, charitable religious, or other activities to serve the common welfare'. Exactly for what they are formed and what aims they pursue are becoming increasingly blurred, and as such we know them mostly from the 'principal fund of their own' today, allowing them to exercise a high degree of self-determination and independence but also making them among the least accountable institutions in modern society (Fleishman, 2009).

Aside from this somewhat generic description, foundations engaged in development today greatly vary. Some have pronounced in-country presence with local offices and conduct almost diplomatic relations with local governments, whereas others like to work solely from afar through implementing intermediaries, typically NGOs. Some are deeply involved in conceiving and forming projects, whereas others provide space for maneuvering for implementers. Despite some of the new digital frontiers that we will discuss in the next part, many of the new foundations mirror the old block in favoring areas of health, education and science.

The perhaps strongest common trait for the new foundations, compared to the traditional ones, is how their founders appear as strong agents of capitalism. But today, these founders increasingly tie in their business skills with their approach to philanthropy. This has led some to coin the term 'philanthrocapitalism' to describe a new generation of philanthropists who employ their capitalist skills to solve global problems (Bishop and Green, 2008). This new generation emphasises efficiency and measurability as well as business-oriented approaches, their deliberate and strategic approach to philanthropy forming a break from traditional reactive philanthropic practice. They now employ diverse financing tools such as social impact bonds, equity, debt and loans but also non-financial support such as mentoring and networking. This approach to private aid is seen by its proponents to form a contrast to the somewhat top-down and centrally planned aid programs of Western states, being more well positioned to support to innovations and take risks in their funding strategies. This is far from true for most foundations, however, who in fact appear risk averse as they are not least guided by a strong focus on measurable results, making high-risk endeavors improbable.

One foundation in particular has risen above the rest today, dwarfing all other foundations through history, The Bill and Melinda Gates Foundation (see Fejerskov, 2018; McGoey, 2015). Formed around 1999 after continuous public calls for Microsoft founder and billionaire Bill Gates to distribute some of his striking wealth, the Gates Foundation has since evolved into an immensely powerful force in global development and international politics more broadly. The Gates Foundation has historically facilitated a surge in funding to both global health and agriculture, when it ventured into each of these two main areas of work. The foundation has granted more than $50 billion through its inception and today holds an endowment of almost the same amount that it continually invests to strengthen its grant making. Its annual grant making of around $5 billion surpasses many key OECD DAC donors, just as it supports work in more than 138 countries around the world. All of this work is facilitated or monitored by the organisation's almost 1,500 employees, the majority of whom work in its massive Seattle headquarters. Inspired by its founders, who act as co-chairs and devote more or less all their time to the organisation, the Gates Foundation is heavily focused on science and technology in its approach to global development, often favoring and funding innovations in its fields (Schurman, 2018).

Digital technologies and new experiments

From the past and the present, we arrive at some of the future lines of work that private foundations are engaged in, not least spurred on by the Gates Foundation. While foundations today broadly mirror the same areas of engagement that many foundations did a hundred years ago, particularly health, education and agriculture, a new movement among foundations can be seen to further approaches of technological experiments and support for a new digital humanitarianism (Fejerskov, 2017; Burns, 2019).

The Gates Foundation leads a movement that practices global development and developing countries as a form of global laboratory in which they experiment with anything from tangible technologies such as new toilets or condoms, to attempts at changing social norms on e.g., reproduction and family planning. Technology and particularly technology-transfer have always been part and parcel of development cooperation, for the many first decades serving as perhaps the main objective. Today, however, we see a second coming of technology in global development, in which foundations increasingly focus on societal progress through technological innovation and experimentation. Historically, developing countries have been thought to have a relative advantage over developed countries because imitation and adaptation of existing technology is less expensive and risky than creating them. With these new developments this seems to be reversing as the Global South is increasingly articulated and utilized as a live laboratory for technological innovation and testing.

The relation between financial resources and political leverage is inevitable in the field of philanthropy and development. Private foundations have historically played an important role in development efforts, but their power and resources continue to grow, not least centering around the Gates Foundation that forms an immensely powerful actor in international politics today, with a growing number of institutions in global development embedded into foundation engagements and institutional arrangements. This growing role of foundations in development hints at the encroachment of business into all sides of political and social life and the privatisation and commercialization of social action (see Richey and Ponte, 2011).

A new generation of givers and foundations has emerged over the past decades with immense expectations about their potential contributions to the world, driving as a prime guiding line of attack that of disruption, upsetting social systems and challenging the status quo to evoke radical change. Some of these see the social world as a computer system that can easily be hacked. It is not, and we are fortunately also seeing signs that some foundations are moving towards a more holistic approach to development, not least acknowledging that social change is incremental and complex and that a long-term perspective is often necessary.

References

Bishop, M. & Green, M. 2008. *Philanthrocapitalism: How the Rich Can Save the World*. London: Bloomsbury Press.

Burns, R. 2019. New Frontiers of Philanthropy-Capitalism: Digital Technologies and Humanitarianism. *Antipode*, 51(4): 1101–1122.

Fejerskov, A. 2017. The New Technopolitics of Development and the Global South as a Laboratory of Technological Experimentation. *Science, Technology and Human Values*, 42(5).

Fejerskov, A. 2018. *The Gates Foundation's Rise to Power: Private Authority in Global Politics*. London and New York: Routledge.

Fleishman, J. 2009. *The Foundation: A Great American Secre*. New York: Columbia University Press.

McGoey, L. 2015. *No Such Thing as a Free Gift*. London: Verso Books.

Moran, M. 2014. *Private Foundations and Development Partnerships*. London and New York: Routledge.

Richey, L. & Ponte, S. 2011. *Brand Aid*. Minneapolis: Minnesota Press.

Schurman, R. 2018. Micro(soft) Managing a 'Green Revolution' for Africa: The New Donor Culture and International Agricultural Development. *World Development* 112: 180–192.

Chapter 30

For-profit consultants and contractors in development

Emma Mawdsley

Introduction

In 2017, a report by the UK Parliament's International Development Select Committee observed that,

> the amount spent through [private sector] contractors has surged in recent years, both in cash terms and as a share of total bilateral expenditure. In 2010/11 contracts represented 12% (£540 million) of bilateral spend and five years later had increased to 22% (£1.34 billion).[1]

In other words, as well as funding multilateral organisations, NGOs and research institutions, a substantial and growing share of UK foreign aid is being directed to a variety of private sector firms which are contracted to take on a range of international development tasks. This trend is one way in which official development agencies are undergoing changes in their missions, narratives, personnel and architectures, as they re-orient themselves towards different sorts of private sector partnerships, outsourcing and investment. This chapter focuses on one aspect of this trend: profit-seeking firms which provide services to the global development sector. This chapter takes the UK as its main example, but it is relevant across 'Northern' and multilateral actors (see, for example, Hayes and Westrup, 2014; Seabrooke and Sending, 2019). To my knowledge, there has been no study of for-profit service contracting by Southern development actors, and they are not included here.

What are development consultants and contractors, and what do they do?

Private sector service providers (the contractors of the 2017 Select Committee report mentioned earlier) include global consultancies like PricewaterhouseCoopers and McKinsey and Company – for which 'development' is one part of their much wider portfolio of work – and specialist development contractors, such as Palladium and Development Alternative Incorporated (DAI), for which winning development contracts is their core business. In addition to these very large and often multinational businesses, there are a vast number of medium-sized and small/micro firms. Some bid for contracts with donors directly, but they are also an essential part of the for-profit ecology as sub-contractors for the leading firms and consortia, who win the large contracts. These leading firms tend to

DOI: 10.4324/9780429282348-33

have a substantial presence in the capitals of donor countries, most prominently the huge Washington DC 'beltway bandits', superbly analysed by Roberts (2014). Subsidiaries of these firms and sub-contracted firms can be found in both donor and client countries – amongst a multitude of examples of this spectrum are River Path Associates, a small development consultancy registered in the UK;[2] Deloitte India, with 14 offices across India[3] and ILC Africa, with offices in Ghana, Sudan and Uganda.[4] Increasingly, financial intermediaries are also on the aid payroll as service providers, as donors seek to launch and stimulate innovative forms of development investment, such as vaccine bonds (see, for example, Hughes-McLure and Mawdsley, 2022).

Spending on or through private sector companies, including consultants and contractors, is not new in international development (Dimier and Stockwell, 2021). The US, in particular, has been a major contractor for many years, as have the UK and Australia. However, amongst OECD-DAC donors there has been an expansion in the scope and scale of private sector outsourcing – always, of course, in specifically contextualised ways. Bearing diverse geographies in mind, we can observe that for-profit firms are being increasingly contracted to deliver tasks that range from: technical advisory support; implementing projects; administering multimillion-pound programmes; undertaking audit and evaluation services and as intermediaries and actors in public-private partnerships. Moreover but not surprisingly, there is also evidence of a greater active policy engagement and influencing roles. At the PricewaterhouseCoopers annual International Development Conference in 2019, for example, the guests of honour included the Chair of the UK's Select Committee on International Development and the Permanent Secretary to the Department for International Development (DFID).

Different perspectives and assessments

For their supporters, there is much to welcome in this infusion of business expertise and private sector efficiency. McKinsey & Company's International Development webpage promises fresh perspectives, analytical rigour and innovative solutions to bear on the world's most urgent and complex issues. Tangible results are to be delivered quickly. These claims are persuasive – in March 2020, for example, DFID appointed McKinsey to administer the £70 million Invest Africa project. The expertise and inclinations of these consultants and contractors are in alignment with many of the wider trends and agendas of global development in the 2020s. These include the turn from foreign aid as the main form of financing 'Development', to the idea that the role of foreign aid should be to catalyse, unlock and leverage the far larger potential of commercial capital: in the World Bank's mantra, to leap from 'billions to trillions'. Consultants are needed to help do the 'work' required to draw in such investment to huge infrastructure projects and other roles in stimulating production, trade and investment opportunities. But even in the social development sectors, such as health, education, disability and gender, for example, private sector service providers are displacing NGOs, government agencies and other non-profit actors. PwC, for example, is one member of the consortium running the £750 million DFID (now FCDO)-funded 'Girls Education Challenge' project in Uganda and Kenya. The Humanitarian Emergency Response Operations and Stabilisation Programme (HEROS) programme, launched by DfID in 2016 (now managed by the FCDO) is an example of outsourcing humanitarian logistics. The contract states that 'The programme will provide humanitarian emergency response operations management and stabilisation support [to] enable UK to respond rapidly, at scale, to global humanitarian disasters as well as supporting DFID teams working on protracted crises, chronic emergencies and humanitarian reform'.[5] Whereas once the Red Cross or Médecins Sans Frontières might have been expected to lead such efforts, the HEROS programme is primarily delivered under contract by Palladium International Ltd. HEROS is an example of a particular rationale to contract the work outside of government, because of the risks and costs of running programmes in fragile and conflict-afflicted regions and states. Rather than employ UK civil servants in Yemen, say, the government would prefer to contract out to a firm which will manage the risks and costs with local partner organisations. Needless to say, this can be a controversial distribution of risk and reward.

Concerns and critiques within the 'mainstream' are not absent. The International Development Select Committee report mentioned at the start of this chapter provides one example of rather critical scrutiny, while a 2013 review of DFID's use of contractors to deliver aid programmes by the Independent Commission on Aid Impact (2013) also sets out concerns around the poor management of contracts after they have been awarded. Amongst other things, it pointed to little monitoring or independent evaluation of their conduct and achievements; and limited 'joined up' learning about the achievements and failures of the work undertaken by these service providers. Contractors can fall well short of their agreed tasks. One example, although at an unusually large scale, is that of Chemonics International, a US-based behemoth, which was awarded the largest ever USAID contract (worth up to $9.5 billion) for the Global Health Supply Chain-Procurement Supply Management project in 2016. A litany of failures followed, detailed in a USAID memo leaked to Devex, which included everything from mishandling urgent orders, to bad data and IT systems, to non-existent oversight and inaccurate financial reporting and more. Chemonics was accused of overseeing 'systematic challenges in . . . procurement and delivery systems', 'chronic delays in shipping orders from suppliers', and a degree of poor planning that had the project's government clients 'questioning the level and quality of service being provided'.[6]

At least four overlapping critiques emerge from more radical commentators. First, for-profit consultants and contractors represent the ongoing infusion of (neoliberal) capitalist logics into the goals, agendas, practices and narratives of development. From this perspective, a global management consultancy or a Washington DC-based megacorp seems an unlikely ally of the pluriverse or wider development alternatives (e.g., degrowth, traditional agricultural systems, non-capitalist exchanges). Second, C&Cs have encroached substantially on work that might once have fallen to not-for-profits like NGOs working in the development and/or humanitarian spheres. While these have certainly been subjected to very considerable critique of their own, supporters point to a very different sort of value-based mission; (ideally) a stronger and closer set of relationships with direct 'beneficiaries' and specifically poor and marginalised people and, in the case of some NGOs, a willingness to stand up for human rights. All of this is lost when the for-profit sector becomes the vector, translator and intermediary of 'development' spending. Third, consultancy firms (notably the 'Big 4') have done far more to serve fossil fuel interests, resource extraction, dispossession, tax avoidance, wealth concentration and other forms of destructive growth than they have contributed to the crumbs of 'development' offered to the world's poor. Whatever they may or may not achieve in specific development contracts, they are indisputably a part of a deeply destructive and unjust contemporary global order. Finally, there are occasional cases not only of incompetence but outright corruption. The Adam Smith International scandal of 2014 provides one example (see additional resources).

Conclusions

For-profit service providers are now a part of the DNA of development and represent a fascinating, controversial, complex, highly diverse and often hard to track set of actors of increasing influence, importance and impact. They represent new opportunities and challenges for critical development theorists (e.g., how to access information on how aid is spent when it disappears into complex sub-contracting arrangements and behind commercial secrecy walls); for aid managers and policy makers (e.g., how to make sure learning is not silo-ed and lost) and for local communities, rights-based campaigners and advocacy organisations, seeking to hold development projects and programmes accountable and, in some cases, seeking to contest underlying modernist principles (universalism, growth, technology, consumption and so on), as well as a politically and economically uneven and unjust world.

Notes

1 https://publications.parliament.uk/pa/cm201617/cmselect/cmintdev/920/920.pdf
2 https://riverpath.com
3 https://www2.deloitte.com/in/en.html
4 https://www.ilcafrica.com
5 https://devtracker.fcdo.gov.uk/projects/GB-1–205176/documents
6 https://www.devex.com/news/exclusive-chemonics-battles-wave-of-challenges-with-9–5b-health-supply-chain-project-91150

References

Dimier, V. and Stockwell, S., eds. (2021) *The Business of Development in Post-Colonial Africa.* Basingstoke: Springer.

Hayes, N. and Westrup, C. (2014) Consultants as Intermediaries and Mediators in the Construction of Information and Communication Technologies for Development. *Information Technologies & International Development,* 10 (2), 19–32.

Hughes-McLure, S. and Mawdsley, E. (2022) Innovative Finance for Development? Vaccine Bonds and the Hidden Costs of Financialization. *Economic Geography,* 98 (2), 145–169.

Independent Commission on Aid Impact (2013) *DFID's Use of Contractors to Deliver Aid Programmes.* https://icai.independent.gov.uk/wp-content/uploads/ICAI-REPORT-DFIDs-Use-of-Contractors-to-Deliver-Aid-Programmes.pdf. Last accessed 2 August 2022.

Roberts, S. M. (2014) Development Capital: USAID and the Rise of Development Contractors. *Annals of the Association of American Geographers,* 104 (5), 1030–1051.

Seabrooke, L. and Sending, O. J. (2019) Contracting Development: Managerialism and Consultants in Intergovernmental Organizations. *Review of International Political Economy,* 27 (4), 802–827.

Additional resources

Devex is a media platform for the global development community. It reports on contemporary events and trends, including consultants and contractors in action. It is an excellent site for constructive, critical and highly specialised/informed journalism and analysis. Some parts are behind a paywall, but there are publicly available elements. https://www.devex.com

To get a sense of how one major management consultancy, PricewaterhouseCoopers, engages with and constructs its role in international development, see https://www.pwc.co.uk/industries/government-public-sector/international-development.html

A *Guardian* article reporting on a now infamous scandal involving one of DFID's top ten contractors, Adam Smith International. https://www.theguardian.com/politics/2017/mar/02/uk-aid-company-bosses-quit-crackdown-profiteering-adam-smith-international

An interview with an international development consultant, Kaene Disepo, capturing one person's journey into the sector. Interestingly, he once might have described himself as having set up one or more NGOs, but this not his framing or positioning, which instead is of an international development consultant. https://www.youtube.com/watch?v=M0bfoS8kMCo

Chapter 31

Corporate social responsibility
Development on whose terms?

Maha Rafi Atal

Corporate social responsibility (CSR) refers both to the normative idea that companies have responsibility for their impact on society and to the practices that companies undertake in the name of that idea. In the past two decades, these practices have stemmed increasingly from the activities of companies from the Global North operating in the Global South. Global development practitioners have increasingly organised their activities around the Sustainable Development Goals, which incentivize donors and non-profits to work in partnership with private actors (SDG 17), explicitly situating CSR activities as contributions to development. This makes it timely to consider what CSR as both a concept and a practice means for development.

i) Scholarly approaches to corporate social responsibility

Scholarly approaches can be subdivided into three broad strands. The first is an instrumentalist approach, which assumes that 'The social responsibility of business is to increase its profits' (Friedman, 1970). In this view, it is best for society that firms promote wider human welfare only when these activities serve shareholder interests. Development research in this tradition evaluates whether CSR activities can increase the profitability of companies operating in developing markets and whether businesses can, through their pursuit of profits, contribute to economic welfare.

A second approach, 'business and human rights' (BHR), takes for granted that companies should have responsibilities for social welfare and considers how these obligations can be enforced through criminal lawsuits, international treaties and self-regulation initiatives like the UN Guiding Principles. This tradition is often critical of CSR as practiced by companies – viewing 'weak CSR as bad development' (Blowfield and Frynas, 2005) – and calls for companies to accept more responsibility and for states to enforce that responsibility in law.

The third approach, 'Political CSR' (P-CSR), argues that neoliberalism and globalization have reduced the authority of state regulators, leading to 'business firms assuming a state-like role' (Moon et al., 2005) to address governance gaps in the global economy. This approach sees corporate labour and environmental standards imposed by firms in the Global North on their supply chains as a solution to a lack of enforcement by Global South states.

DOI: 10.4324/9780429282348-34

ii) Globalization and re-embedded liberalism

Both the BHR and P-CSR framings situate CSR as a response to globalization. One way of conceptualizing this relationship is through the paradigm of 'embedded liberalism' (Ruggie, 1982). As coined by John Ruggie, and drawing on the work of Karl Polanyi, 'embedded liberalism' describes the post-World War II marriage of free international trade in goods and services with capital controls that protected space for states to engage in interventionist and redistributive policies. The Bretton Woods settlement embedded market forces in social obligation at the *national* level, with empowered national civil society organisations, such as trade unions. Its collapse in the 1970s reduced the policy space for states and the influence of national civil society, resulting in a new dis-embedding of capital. Corporations are increasingly unrestrained at a transnational level – but without a transnational state or society into which they can be re-embedded.

In the 1990s, the imbalance of this order was made particularly visible by a series of corporate scandals, which prompted labour rights and environmental activists in developed countries to turn their attention from campaigning for national-level policy changes to campaigning for higher standards for global corporations (Evans, 2007). This globalization of civil society – and the pressure of public scandals about misconduct – prompted businesses to enter into instrumental bargains with NGOs, with NGOs increasingly advising corporations on labour and environmental standards for their global supply chains.

Seen in this light, corporate social responsibility represents an attempt to 're-embed' capital at the transnational level, with transnational bodies like the United Nations playing the role of states and transnational non-governmental organisations, as international civil society, occupying the role once played by national trade unions in the Bretton Woods order. CSR as re-embedded liberalism aims to secure corporations 'social license to operate' in the global realm. This arrangement has been formalized through UN-led initiatives including the Global Compact and the Sustainable Development Goals, both of which bring together corporations and NGOs. The latter in particular incentivizes corporations to enter formal partnerships with NGOs to achieve development goals as defined by the UN in consultation with both states and businesses.

iii) Implications for development

This new global tripartite settlement thus places power to define the terms of 'socially responsible' business conduct in the hands of global corporations and global NGOs, both headquartered predominantly in the Global North, with intergovernmental organisations, whose principal power brokers are Global North states. For scholars of development, this raises concerns about whether the values that inform the new standards for social responsibility can be applied to developing countries that may differ culturally, politically and economically from the liberal market societies of the developed world, with some scholars critiquing CSR as a form of Western 'economic and cultural imperialism' (Khan and Lund-Thomsen, 2011). At the same time, there are considerable political, economic and cultural differences amongst developing countries themselves. The emerging BRICS economies are now donors themselves, with a distinct model of development that is not easily placed within the global CSR discourse.

Among the values that the CSR paradigm takes for granted is the non-negotiability of corporate profit-making as a normative goal (Blowfield, 2005). This framing views the private sector and markets as essential to development, to be balanced through the 'triple bottom line' against social and environmental objectives. The CSR paradigm can be used to justify the privatisation and liberalisation of social and environmental programs, placing responsibility for achieving wider development objectives in corporate hands. This can limit their scope of CSR to what corporations can reconcile to their profits and reduce the capacity of developing country states to deliver social welfare for their citizens. Moreover, such an approach necessarily excludes alternative framings of corporate purpose and alternative conceptions of development that aim to subordinate corporate profits to social and environmental justice.

Moreover, this approach brackets out consideration of business activities as themselves the cause of social and environmental harms. This includes harms causes by CSR programs themselves, which place workers and communities in developing countries under surveillance by multinational corporations in the name of raising standards (Ponte, 2019). These efforts do not necessarily succeed, with superficial auditing and incentives for suppliers to avoid working with buyers who enforce strict standards, since it is suppliers, not buyers, who shoulder the costs of compliance (Knudsen, 2013). In pushing costs downwards, such social responsibility efforts can in fact prevent developing economies from capturing the gains of globalization.

This underscores a paradox in CSR's relationship to development. On the one hand, CSR helps to make developing countries 'safe' for investors by allowing private companies to rely on their own standards and enforcement systems and bypass concerns about the 'risk' posed by investment in developing regions with 'low' standards. Indeed, CSR can rhetorically exempt corporations from obligations to observe local law by privileging their allegiance to the 'higher' standards of international CSR regimes (Evans, 2007). On the other hand, through its embedding of corporations in partnership with international NGOs and transnational bodies like the UN, it absolves individual companies from direct responsibility for harms.

iv) Defining responsibility from the South

Developing countries and their populations have their own conceptions of what it means for business to be responsible. Developing country governments are eager for foreign direct investment, for example, but they expect investors to pay local tax and oppose multinational corporate efforts to use transfer pricing to reduce their liability (Idemudia, 2011). This issue is not usually considered in global CSR discourse. While CSR discourse does emphasise labour, it centers workplace safety and the eradication of forced labour but excludes demands for higher wages and rights to collective bargaining, which communities in developing countries report as primary goals (Blowfield and Frynas, 2005). While CSR programs promote gender equality, they focus on women's access to paid work and ignore unpaid care labour, a more significant driver of inequality (McCarthy, 2018). Where CSR programs consist of investments in social infrastructure, corporations may choose to invest in one form of infrastructure where local residents would prefer another (Idemudia, 2011). Communities in the Global South also define the groups to whom multinational corporations have a social responsibility more broadly than the corporations, whose interest lies in reducing the number of people who can make claims on their resources (Atal, 2017).

Finally, many developing countries are former colonies, and anti-colonial movements framed development in terms of political empowerment (Tignor, 2016). Corporate social responsibility's view of development places such emancipatory considerations outside of its purview. If 'corporate social responsibility' is to play a role in development, it must begin by empowering communities to decide the form development is to take.

Bibliography

Atal, M.R., 2017. White Capital: Corporate Social Responsibility and the Limits of Transformation in South Africa. *Extractive Industries and Society* 4, 735–743.

Blowfield, M., 2005. Corporate Social Responsibility: Reinventing the Meaning of Development? *International Affairs* 81, 515–524.

Blowfield, M., Frynas, J.G., 2005. Setting New Agendas: Critical Perspectives on Corporate Social Responsibility in the Developing World. *International Affairs* 81, 499–513.

Evans, M.D., 2007. New Collaborations for International Development: Corporate Social Responsibility and Beyond. *International Journal* 62, 311–325.

Friedman, M., 1970. The Social Responsibility of Business is to Increase Its Profits. *The New York Times*.

Idemudia, U., 2011. Corporate Social Responsibility and Developing Countries: Moving the Critical CSR Research Agenda in Africa Forward. *Progress in Development Studies* 11, 1–18. https://doi.org/10.1177/146499341001100101

Khan, F.R., Lund-Thomsen, P., 2011. CSR as Imperialism: Towards a Phenomenological Approach to CSR in the Developing World. *Journal of Change Management* 11, 73–90. https://doi.org/10.1080/14697017.2011.548943

Knudsen, J.S., 2013. The Growth of Private Regulation of Labor Standards in Global Supply Chains: Mission Impossible for Western Small- and Medium-Sized Firms? *Journal of Business Ethics* 117, 387–398.

McCarthy, L., 2018. "There is No Time for Rest": Gendered CSR, Sustainable Development and the Unpaid Care Work Governance Gap. *Business Ethics: A European Review* 27, 337–349. https://doi.org/10.1111/beer.12190

Moon, J., Crane, A., Matten, D., 2005. Can Corporations be Citizens? Corporate Citizenship as a Metaphor for Business Participation in Society. *Business Ethics Quarterly* 15, 429–454.

Ponte, S., 2019. *Business, Power and Sustainability in a World of Global Value Chains*. Zed Books, London, UK.

Ruggie, J.G., 1982. International Regimes, Transactions, and Change: Embedded Liberalism in the Postwar Economic Order. *International Organization* 36, 379–415.

Tignor, R., 2016. *The Lessons Ghana Learned from Kwame Nkrumah's Fallout with His Economic Adviser* [WWW Document]. Quartz Africa. https://qz.com/africa/632155/the-lessons-ghana-learned-from-kwame-nkrumahs-fallout-with-his-economic-adviser/ (accessed 7.25.20).

Further reading

Black, D., O'Bright, B., 2016. International Development and the Private Sector the Ambiguities of "Partnership". *International Journal* 71, 144–166.

de Neve, G., 2009. Power, Inequality and Corporate Social Responsibility: The Politics of Ethical Compliance in the South Indian Garment Industry. *Economic and Political Weekly* 44, 63–71.

Jamali, D., Keshishian, T., 2009. Uneasy Alliances: Lessons Learned from Partnerships Between Businesses and NGOs in the Context of CSR. *Journal of Business Ethics* 84, 277–295. https://doi.org/10.1007/s10551-008-9708-1

Jenkins, R., 2005. Globalization, Corporate Social Responsibility and Poverty. *International Affairs* 81, 525–540.

Thérien, J.-P., Pouliot, V., 2006. The Global Compact: Shifting the Politics of International Development? *Global Governance* 12, 55–75.

Chapter 32

Gender, ethical consumerism, and political participation

Celia Bartlett

Introduction

Ethical consumerism is a means by which consumers utilise their position as market actors in seeking to enact change in processes associated with production and trade. Ethical consumers attempt to influence change in diverse societal areas from worker's rights in supermarket value chains (indepth.oxfam.org.uk) to forest stewardship (fsc.org). According to the Ethical Consumer Markets Report, 2020 saw record ethical spending in the UK, reaching £98 billion (Ethical Consumer Markets Report, 2020: 2), with Fairtrade sales increasing by 13.7 per cent between October 2019 and October 2020 (ibid). Whilst continuing to grow globally, particularly amongst the emerging middle class (Bostrom et al., 2019), ethical consumerism is most prevalent in North America and Northern Europe (ibid) which provides the backdrop to this contribution.

The continued growth of the ethical market combined with the participation of diverse stakeholders, including local government, corporate actors and civil society organisations has contributed to the conceptualisation of ethical consumerism, as a 'form of political participation' (Kyroglou and Henn, 2022: 1192). The mobilisation by civil society actors in this area has been defined as both a 'movement and a market' (Lekakis, 2013: 2) and the ethical marketplace has been conceptualised as part of expanding political repertoires (Theocharis and van Deth, 2018; Stolle and Micheletti, 2013) whereby the 'citizen consumer' is empowered to express desires for social change that reach beyond National boundaries (Stolle and Micheletti, 2013). Recognising inequities in global trade, concerned consumers seek to address governance gaps via their market behaviour (Bostrom et al., 2019). In this way ethical consumerism is framed as means by which citizens in the Global North are able to express their concerns about political and economic inequities between the Global North and Global South (Stolle and Micheletti, 2013).

Gendered political consumerism

Ethical consumerism is gendered in a number of ways. First, empirical evidence has established that more women than men participate in ethical consumerism (Stolle et al., 2005; Stolle and Micheletti, 2013; Hall and Holmes, 2017; Gundelach and Kolte, 2020; Chatzidakis & Maclaren, 2020; Scott, 2020). For example, a recent comparative analysis of political consumerism in Greece and the UK found that women, in both countries, are significantly more likely than men to engage with the practice (Kyroglou, 2020).

DOI: 10.4324/9780429282348-35

Furthermore, civil society actors who support and promote ethical consumerism increasingly draw attention to structural gender inequities in production and trade (Fairtrade International, 2016) and to ethical consumerism as a means to address them. For instance, The International Coffee Organisation estimates that women's labour accounts for 70 per cent of total labour in coffee production, yet only 20–30 percent of coffee farms are operated by women, with women having less access to land than men (International Coffee Organisation, 2018: 31). Ethical consumerism is presented as one way of tackling these inequities. Purchasing products produced by women, such as coffee and chocolate, is often promoted as a means of empowering women producers via the market. The Fairtrade Foundation's Women's Leadership Programme, which supports women's leadership in producer organisations, (Fairtrade.org.uk) is one example of civil society challenging structural gender norms in this way.

Concurrently, ostensive empowerment of women via consumerism more broadly, is a growing phenomenon, an example being Nike's 'Dream Crazier' campaign. Eliciting an emotional response to examples of overcoming patriarchal norms in sport, Nike aligns itself with empowering women (Dhonchak, 2021).

The conflation of market-based empowerment and politics is demonstrated by a Nike partnership with DfID whereby Nike was provided with a £12.9 million grant to launch the Girl Hub initiative, aimed at empowering girls via sport (ICAIR, 2012: 2). Whilst there is much evidence to demonstrate that women and girls can be inspired, gain confidence and become more engaged in social change initiatives through sport, (UNOSDP, 2018), it has been argued that market-based women's empowerment commodifies feminism in such a way that undermines its political momentum (Roberts, 2015; Repo, 2020). Complex political realities and the processes required to change them are often reduced to simplistic messages about both individual agency and individual empowerment, rather than addressing gendered societal inequities. Indeed, as noted in the Independent Commission for Aid Impact Report (2012) 'Girl Hub appears to have struggled to . . . reconcile the power of a simple message with its efforts to tackle a complex social problem' (ICAIR, 2012: 5).

Despite a lively debate about the commodification of feminism (Roberts, 2015; Zeisler, 2016; Repo, 2020; Olufemi, 2020) and the efficacy of the market-based women's empowerment narrative for promoting gender equity (Banet-Weiser, 2020; Olufemi, 2020; Zeisler, 2016), there has been little research into the motivations and perceptions of ethical consumers themselves through a gender lens. Neither has there been substantive research into the ways in which ethical consumerism informs and is informed by gendered identities, practices and the enactment of (political) agency.

Evolving citizenship

As conceptualisations of citizenship evolve, perceptions of the ways in which change is possible influence the ways in which citizenship is understood, how it is accessed (Acik, 2013; Lekakis, 2013) and by whom.

The narrative of the 'citizen consumer' moves away from traditional conceptualisations of citizenship, as being a relationship between the citizen and the state (Clarke et al., 2014). Rather it moves towards a notion of citizenship whereby political consciousness, agency and action is embedded in and emergent from daily or routine practices (Neveu, 2015). From this point of view, given the particular ways in which consumerism is gendered, it is important to consider ethical consumerism as a set of gendered practices engaged in social change. It also highlights a need to locate the motivations and perceptions of the ethical consumer within institutionalised gender rules and norms.

Furthermore, consideration of 'citizenship' as a key motivating word for enacting change (Clarke et al., 2014), makes it ever more important to contemplate the potential connections between gendered practices and the nature of agency being expressed via engagement with ethical consumerism. These connections should not be ignored because they are simultaneously *influenced by* and contain *possibilities to influence* gender inequities within the dominant neoliberal paradigm.

The empowered consumer?

Understanding ethical consumerism as form of political participation therefore, at the very least, requires consideration of ethical consumerism in tandem with gendered social practices. Attention should be given to the ways in which the consumer is constructed in relation to categories including gender and ethnicity, particularly the power dynamics associated with the social organisation of these categories.

It is important to reflect on how experiences of, for example, intrahousehold relations that inform consumption decisions may influence the ways in which consumers seek to enact social change. One way of approaching this is to consider the ways in which ethical consumerism may challenge the public/private dichotomy that has been at the heart of much feminist theorising about patriarchy and access to meaningful citizenship (Molyneux, 2000).

Feminist institutionalist scholars have sought to contextualise power within dominant societal norms that influence worldviews (Mackay, 2014; Mackay et al., 2011; Kenny, 2007). The public/private dichotomy is associated with both the reproduction of masculinity in the public space – in which public action occurs – and the persistent feminisation of the private space, in which reproductive labour takes place. This dichotomy has been identified as permeating traditional political institutions and underpinning the reproduction of gendered power relations (Mackay, 2014). A defining feature of ethical consumerism as an emerging form of political participation is that it blurs the public/private dichotomy (Copeland, 2014; Stolle and Micheletti, 2013; Bostrom et al., 2019) by offering diverse spaces for enacting social change. Via 'individualised responsibility taking' (Stolle and Micheletti, 2013: 43) 'citizen consumers' are invited to enact change from within the private sphere. In line with feminist scholarship, which has long highlighted the private and personal as political, this idea further challenges the public/private dichotomy and thus the notion of empowerment as public political action.

At the same time, a narrative of women's empowerment is increasingly used to market goods (particularly to women). Such marketing is central to the debate about the commodification of feminism (Roberts, 2015; Zeisler, 2016; Repo, 2020; Olufemi, 2020) and whether market objectives are in tension with women's empowerment and feminist political goals more broadly. A consideration of ethical consumerism as a gendered political practice draws our attention to tensions between notions of empowerment and market-based social change.

Acknowledging these tensions helps us consider the ways in which market-based empowerment influences 'perceptions of new possibilities' (Grief and Kingston, 2011: 24) and the political motivations, strategies and goals that underpin ethical consumerism as a form of political participation.

Scholars have raised concerns about a form of 'emotional branding' (Illouz and Alaluf, 2019: 245) which via the construction of 'emotional identity' (ibid) creates economic value. Indeed, Nike's 'Dream Crazier' campaign demonstrates the obvious intention to elicit emotional energy, the imagery of successful female athletes intended to inspire its audience to challenge patriarchal norms in sport. Yet when considered alongside the Commission for Aid's assessment of Nike's partnership with DfID (ICAIR, 2012), questions are raised about the ways in which such campaigns might be considered exploitative, in that they elicit emotional energy for commercial gain, whilst placing responsibility for change with the individual. These considerations are especially important when positive emotional energy is recognised as a powerful driver for repeat engagement in practices (Weenink and Spaargaren, 2016), not least consumption.

The point here is that there is a need to interrogate the diverse social, economic and political practices considered under the rubric of ethical consumerism and locate where they might sit on a spectrum between bringing about meaningful social change on the one hand and undermining such efforts on the other (Prugl, 2015; Repo, 2020; Roberts, 2015).

Clearly, this is not as straightforward as discounting market-based forms of empowerment as simply superficial or exploitative. Possibilities of repurposing 'neoliberal technologies for empowerment' (Prugl, 2015: 626) associated with liberal feminist empowerment messaging have been identified. Whilst noting the failure of liberal feminism to challenge structural gender inequities, Prugl (2015) sees the subsequent challenge for scholars being to better understand context, in order to identify entry points to challenge 'oppressive power relations' (Prugl, 2015: 627). Such pragmatism embraces the challenge of enacting change from with the dominant market oriented paradigm.

Questioning agency

Empirical research has established that political consumers have high levels of internal political efficacy, defined as an individual's belief that they are able to promote change (Spada, 2019). This holds even though they may choose not to engage in traditional political arenas, such as political party membership (Lekakis, 2013; Copeland, 2014; Stolle and Hooghe, 2011).

Yet scholars have highlighted broader tensions inherent in pursuing social change via the market, for example, concerns about increasing influence of market mechanisms over political processes (Lekakis, 2013; Friddel, 2018). Recognising the market as a social institution that influences 'social behaviour and practices' (Hussain, 2012: 111), Hussain cautions against the influence of the market over the social, in the absence of associated democratic process or accountability (Hussain, 2012).

Distinctions have been drawn between authentic agency and 'neoliberal agency' (Waller and Wrenn, 2021: 61). Neoliberal agency, it is argued, maintains a 'gendered dualism' (Waller and Wrenn, 2021: 61) that sustains a public/private dichotomy because the individual is 'superficially empowered' via 'the illusion of autonomous decision making' (Waller and Wrenn, 2021: 61) as a consumer. In contrast, authentic agency sees the individual being aware of the limitations of their power within the 'gendered dualism of neoliberalism' (Waller and Wrenn, 2021: 62). Reflection about neoliberal agency in the context of the debate about the political nature of ethical consumerism, therefore, demonstrates why it is important to be vigilant to the nature of agency being expressed.

Furthermore, consideration of gendered perceptions of political efficacy, via the market, reiterates the importance of being alert to the potential for political objectives being diluted by market engagement. It also points to the need to question how far the blurring of the public/private dichotomy, rooted in the empowerment narrative, has the potential to challenge structural gendered inequities in a way that is sustainable.

Conclusion

Ethical consumerism has been conceptualised as an emerging form of political participation. Via the narrative of the citizen consumer, ethical consumerism is invoked as a means by which the consumer is empowered to enact positive social change and to express political agency.

Despite ethical consumerism being more frequently engaged with by women than men and despite debates about the commodification of feminism, broader considerations of tensions inherent in market based social change have not considered conceptualisations of political consumerism in the context of associated gendered practices.

Yet, ethical consumerism is increasingly posited by relevant civil society actors as a means of challenging gendered structural inequities. As ethical consumerism is defined as blurring the public/private dichotomy and as notions of citizenship evolve, debates persist: can market-based social change meaningfully challenge gendered structural inequities or does it engage with and facilitate the institutionalisation of a form of neoliberal agency that undermines recognition of the need for change at the structural level?

References and Further Reading

Acik, N. (2013) Reducing the Participation Gap on Civic Engagement: Political Consumerism in Europe. *European Sociological Review* 29(6): 1309–1322.

Bale, T., Webb, P. & Poletti, M. (2018) *Grassroots. Britain's Party Members: Who They Are. What They Think and What They Do.* Queen Mary University of London.

Banet-Weiser, S. (2020) *Empowered. Popular Feminism and Popular Misogyny.* Duke University Press.

Bostrom, M., Micheletti, M. & Oosterveer, P. (2019) *Introduction. The Oxford Handbook of Political Consumerism.* Oxford University Press Inc.

Chatzidakis, A. & Maclaren, P. (2020) Gendering Consumer Ethics. *International Journal of Consumer Studies* 44(4): 316–327.

Clarke, J., Coll, K., Dagnino, E. & Neveu, C. (2014) *Disputing Citizenship.* Policy Press.

Copeland, L. (2014) Conceptualizing Political Consumerism: How Citizenship Norms Differentiate Buycotting from boycotting. *Political Studies* 62(s1): 172–186.

de Moor, J. & Verhaegen, S. (2020) Gateway or Getaway? Testing the Link Between Lifestyle Politics and other Modes of Political Participation. *European Political Science Review* 12: 91–111.

Dhonchak. *Nike's Dream Crazier – A New Brand of Self Objectification.* Blogslse.ac.uk (accessed 2021).

Ethical Consumer Markets Report (2020). Ethical Consumerism in the Pandemic. *Ethical Consumer.* https://www.ethicalconsumer.org/sites/default/files/inline-files/Ethical%20Consumer%20Markets%20Report%202020.pdf

Fairtrade International (2016) *Gender Strategy: Transforming Equal Opportunity, Access and Benefits for All 2016–2020.* Fairtrade.org.uk (accessed 2022).

Friddel, G. (2018 [2019]) Conceptualising Political Consumerism as Part of the Global Value Chain. In *The Oxford Handbook of Political Consumerism,* edited by M. Bostrom, M. Micheletti & P. Oosterveer. Oxford University Press.

Grief, A. & Kingston, C. (2011) Institutions: Rules or Equilibria. In *Political Economy of Institutions: Democracy & Voting,* edited by N. Schofield & G. Caballero. Springer.

Gundelach, B. & Kalte, D. (2020) Explaining the Reversed Gender Gap in Political Consumerism. Personality Traits as Significant Mediators. *Swiss Political Science Review* 27(1): 41–60.

Hall, S. & Holmes, H. (2017) Gender and Ethical Consumption: Towards a New Research Agenda. *Journal of Consumer Ethics* 1(2): 2–6.

Hussain, W. (2012) Is Ethical Consumerism an Impermissible Form of Vigilantism? *Philosophy and Public Affairs* 40(2): 111–143.

Illouz, E. & Alaluf, B. (2019) Emotions in Consumer Studies. In *Oxford Handbook of Consumption*. Oxford University Press.

Independent Commission for Aid Impact (2012). *Girl Hub: A DFID and Nike Foundation Initiative Inception Report*, indepth.oxfam.org.uk (accessed 2021).

International Coffee Organisation (2018). *Annual Review 2017/2018*. https://www.ico.org/documents/cy2018-19/annual-review-2017-18-e.pdf

Kenny, M. (2007) Gender Institutions and Power: A Critical Review. *Politics* 27(2): 91–100.

Kyroglou, G. (2020) *Political Consumerism as Political Participation: A Mixed Methods Paired Country Comparison Project with Young People in the UK and Greece.* Nottingham Trent University.

Kyroglou, G. & Henn, M. (2022) On Measuring Political Consumerism: An Exploratory Study Among Young People in the UK and in Greece. *Social Indicators Research* 163: 1191–1220.

Lekakis, E. (2013) *Coffee Activism and the Politics of Fairtrade and Ethical Consumption in the Global North. Political Consumerism and Cultural Citizenship.* Springer.

Mackay, F. (2014) Nested Newness, Institutional Innovation and the Gendered Limits of Change. *Politics and Gender* 10(4): 549–571.

Mackay, F., Kenny, M. & Chappell, L. (2011) New Institutionalism Through a Gender Lens: Towards Feminist Institutionalism. *International Political Science Review* 31(5): 573–588.

Molyneux, M. (2000) Comparative Perspectives on Gender and Citizenship: Latin America and Former Soviet States. In *Towards a Gendered Political Economy*, edited by J. Cook, J. Roberts & G. Waylen. Macmillan Press Ltd.

Neveu, C. (2015) Of Ordinariness and Citizen Processes. *Citizenship Studies: Citizenship Agendas in and Beyond the Nation-State* 19(2): 141–154.

Olufemi, L. (2020) *Feminism Interrupted. Disrupting Power.* Pluto Press.

Prugl, E. (2015) Neoliberalising Feminism. *New Political Economy* 20(4).

Repo, J. (2020) Feminist Commodity Activism. *International Political Sociology* 14(2): 215–323.

Roberts, A. (2015) The Political Economy of Transnational Business Feminism: Problematising the Corporate Led Gender Equality Agenda. *International Feminist Journal of Politics* 17(2): 209–223.

Scott, L. (2020) *The Double X Economy. The Epic Potential of Empowering Women.* Faber & Faber.

Spada, P. (2019) The Impact of Democratic Innovations on Citizen's Efficacy. In *Handbook of Democratic Innovation and Governance*, edited by S. Elstub & O. Escobar. Edward Elgar Publishing.

Stolle, D. & Hooghe, M. (2011) Shifting Inequalities. Patterns of Exclusion and Inclusion in Emerging Forms of Political Participation. *European Societies* 13: 119–142.

Stolle, D., Hooghe, M. & Micheletti, M. (2005 [2006]) The Gender Gap Reversed: Political Consumerism as a Woman Friendly for of Civic and Political Engagement. In *Gender and Social Capital*, edited by B. O'Neil & E. Gidengil. Taylor Francis.

Stolle, D. & Micheletti, M. (2013) *Political Consumerism, Global Responsibility in Action.* Cambridge University Press.

Theocharis, Y. & van Deth, J. (2018) The Continuous Expansion of Citizen Participation: A New Taxonomy. *European Political Science Review* 10(1): 139–163.

United Nations Office on Sport for Development and Peace (2018) *Sport and the Sustainable Development Goals. An Overview Outlining the Contribution of Sport to the SDG's.* Geneva, United Nations.

Waller, W. & Wrenn, M. (2021) Feminist Institutionalism and Neoliberalism. *Feminist Economics* 27(3): 51–76.

Weenink, D. & Spaargaren, G. (2016) Emotional Energy and Practice Theory. In *Practice Theory and Research. Exploring the Dynamics of Social Life*, edited by G. Spaargaren, D. Weenink & M. Lamers. Routledge.

Zeisler, A. (2016) *We Were Feminists Once: From Riot Girls to CoverGirl. The Buying and Selling of a Political Movement.* Public Affairs.

Chapter 33

Environmental defenders and social movements

The violent realities of resisting extractivism

Levi Gahman, Filiberto Penados, and Shelda-Jane Smith

Environmental defence: struggles for land and life

Environmental defenders are people or groups who make the personal-political deci-sion to protect ecosystems and communities from the adverse consequences of, inter alia, land grabs and the enclosure/privatisation of the commons; the extraction of raw materials, natural resources, and fossil fuels; pollution and the disposal of contami-nants and noxious materials; environment-altering development mega-projects (e.g., dams, powerplants, seaports, windfarms, highways, tourist attractions); and indus-trial mining, logging, agribusiness, and plantation operations. While environmental defenders can be found in every corner of the earth, i.e., the Global North and South, most tend to be from Indigenous and peasant communities in the Majority World. The pivotal role environmental defenders play in safeguarding both nature and local communities from harm has meant that even the United Nations acknowledges their importance in preserving biodiversity, ecologies, and human/more-than-human species and relations.

Notably, the designation 'environmental defender' is an imperfect and contested catch-all term that is applicable to wide array of activists, social movement organisers, and village residents from different countries, cultures, and contexts who make use of a vari-ety of strategies to protect territories and communities alike (Menton and Le Billon, 2021). Many have different motivations for engaging in environmental defence, although maintaining the health of people, flora, fauna, and the entire planet are often common goals. In short, environmental defenders compose a diverse constituency and heterogene-ous group of political actors who, although operating in specific geographies and facing unique challenges/threats unto themselves, tend to be driven by a shared desire for jus-tice, dignity, and sustainable futures.

Environmental defenders, along with the communities they come from and movements they are a part of, frequently are concerned with land and territory. Land is viewed as a source of life, heritage, and dignity. The survival of countless rural peasants and Indig-enous groups, who should neither be seen as monolith nor romanticised, depends upon access to land and the resources it provides. For many, their cosmologies, identities, cultures, and ancestral memories are *rooted* in land. In turn, many Indigenous and peas-ant communities have long-standing reciprocal relationships with local habitats, forests, and water sources, which gives deep meaning to the places in which they reside. Indeed, for environmental defenders, land is the foundation of life and imbued with a boundless amount of cultural, material, and spiritual significance. This *relationship* with territory is

DOI: 10.4324/9780429282348-36

Figure 33.1 A mural on a community building in a Zapatista *caracol* (administrative centre) in Chiapas, Mexico. The painting depicts the Zapatista's practice of autonomy and the inextricable links their identity, health, and heritage have with nature. Included are clean flowing waters, fertile fields/forests, communal lands, and villagers engaging in meaningful work

a product of refusing to deem land as merely an object of capital that must be instrumentalised, made 'productive', and fashioned into marketable commodities.

Globally, most land continues to be tended to and governed by communities via customary systems of complex tenure. Whether designated as public or traditionally held land, billions of pastoralists, farmers, fisherfolk, and landworkers depend on communally managed fields, farms, forests, and grasslands for their livelihoods and well-being. Under the capitalist 'free market' system, however, joint holdings of the commons and the collective stewardship of land and resources are seen as barriers to economic development, which generally endorses the primacy of profit, private ownership of property, rent-seeking activities, accumulation of assets, prioritisation of revenue streams, and celebration of personal wealth (Federici, 2019). This clash of worldviews, i.e. private property/capital versus communal tenure/the commons, was one of the key drivers of the dispossession, violence, and pollution that are inherent to colonialism and empire-building (Liboiron, 2021).

At present, the ongoing colonial enterprise that constitutes global capitalism continues via the political mandates, economic policies, and development agendas of dominant geopolitical actors (e.g., G7/G20 countries, UN Security Council permanent members), international finance institutions (e.g., the World Bank, International Monetary Fund, World Trade Organisation), and powerful commercial interests (e.g., multinational corporations, large estate holders, landlords) (Lang and Mokrani, 2013). In turn, viewing land as private property and merchandise to be bought and sold, as well as making it available for financialisaton, investment, and economic 'growth' – which are all decidedly Western-liberal constructs – have become an imperilling 'common sense' yet taken-for-granted status quo.

In nearly every instance of environmental defence, activists and movements are confronting 'extractivism', e.g., encroachments onto communal lands, expropriative claims to ancestral territories, and invasive development projects that are sanctioned by both states and corporations. To elaborate, extractivism is the term that signifies the prevailing industrial mode of extractive practices and processes that exploits nature/humans and alienates land/labour with the express intent of selling commodities on the globalised free market. Whilst extractivism and accumulation tend to be thought of as 'business as usual', its effects and aftermaths are devastating.

Extractivism and violence against environmental defenders

Across the world, both transnational companies and domestic governments are the key drivers of extractivism due to the priority each place on income generating activities and the bottom line. While no geography is immune to extractive operations, the age-old colonial model of targeting the Global South for plunder continues in the present-day. In the contemporary moment, however, political elites and members of the ruling class in Majority World countries are also beneficiaries of transactions arranged by transnational corporations, international banks, and foreign financiers. Because extractivism hinges on access/claims to land, conflicts inevitably emerge given companies and states impinge upon the territories, land rights, and sovereignties of Indigenous people and rural peasants, often without appropriate consultation and while ignoring Free, Prior, and Informed Consent protocols (FPIC) (Gahman et al., 2020). Whilst these processes are rolled out under a pretence of 'development' and the socio-ecological costs are framed as 'externalities', for Indigenous and peasant communities it is violence (Tsosie, 2007).

Subsistence farmers and rural landworkers who reside in agrarian communities are particularly vulnerable to the harmful effects of extractivism induced by commercial interests and authoritarian states given: i) they tend to live in secluded remote hinterlands; ii) their land rights over traditional territories tend to go unrecognised and denied; iii) they typically live in biodiverse areas that are rich in resources, raw materials, fertile soils, or rare earth minerals; and iv) they are beholden to cosmologies and worldviews that are habitually denigrated, deemed archaic, and held in contempt. The negative upshots of extractivism, which are social, ecological, and climatic in nature, include but are not limited to: deforestation, acidification, air pollution, soil leaching, groundwater contamination, the bioaccumulation of pesticides, increasing carbon emissions, land grabs, heritage destruction, displacement/eviction, precarious employment, hazardous working conditions, sexual exploitation, fractures in community cohesion, and rising rates of disease, cancer, and birth defects. In response, environmental defenders recurrently take to the frontlines to resist extractivism, which is an assertion of political agency and act of defiance routinely met with retaliation that is meted out in a variety of ways.

Whilst sobering to consider, the reality that nearly 2,000 environmental defenders have been murdered in less than a decade is indicative of just how much is at stake vis-à-vis extractivist development (Scheidel et al., 2020). This statistic also only refers to documented killings, and numerous international advocacy organisations and grassroots movements suggest the actual body count is much higher given environmental defenders are regularly disappeared and because governments and corporations do not report assassinations they orchestrate. Notably, negatively racialised groups are disparately placed in the crosshairs and Indigenous people are overrepresented in the death toll. Here, it is critical to note that the murder of environmental defenders is taking place within an institutionalised culture of silence and complicity. That is, the plausible deniability and impunity afforded to state actors, corporate firms, and private security services (read: paramilitary death squads) who are responsible for killing environmental defenders is directly linked to corrupt and inept judiciaries, not to mention a convenient product of deals brokered between agents of foreign enterprise and ruling class nationals.

In addition to homicide, incarceration, and the destruction of the places and ecosystems in which they live and rely upon for survival, environmental defenders face a host of other human rights violations, which take a toll physically, emotionally, and psychologically. Such abuses are multi-dimensional and made manifest as intimidation, death threats, smear campaigns, harassment, criminalisation, extrajudicial detention, false imprisonment, stigmatisation, racial abuse, sexual assault, rape, militarisation, surveillance, community infiltration, vandalism, and exclusion from consultations and negotiations on development policies/projects (should these even take place; Birss, 2017). The ways in which environmental defenders are vilified and criminalised are exceptionally debilitating, demoralising, and time/energy/resource-consuming given they must cope with the long-term repercussions of defamation, having their credibility undermined publicly, getting dragged through lengthy court proceedings, receiving official arrest records that will follow them throughout life, and being labelled as extremists and divisive to national unity. A point that must be iterated here is that women environmental defenders are disproportionately exposed to abuses that occur in conflicts related to resource extraction, land grabbing, and development. This is a regrettable result of enduring gendered power asymmetries, unequal divisions of labour, and the seemingly intractable patriarchal norms that continue to plague humanity (Tran et al., 2020). In sum, the injustices environmental defenders experience are as jarring to reflect upon as they are myriad in form. Yet, despite the violence, environmental defenders continue to assert their agency and stand up to the driving forces of accumulation because they feel they have no other choice but to do so given that the life of their communities, cultures, and the health of the planet are on the line.

Resistance and the role of social movements

Environmental defenders are neither docile nor passive in the face of either state-sponsored repression or planetary ruin. Markedly, they are organised, acting collectively as an expression of resistance and often associated with or contributing to social movements. A social movement can be defined as a coalition of actors who are engaged in political struggle and working together towards the goal of transforming the status quo and bringing about some form of lasting social, political, cultural, or economic change. Movements employ a wide array of tactics and approaches, including protest, consciousness-raising, civil disobedience, provocation, culture jamming, policy critique/change, legal advocacy, community service, care-work, and even constructing new social relations, alternative economies, or institutions (Ossome, 2021). For a real-world glimpse of Indigenous and peasant movements and environmental defenders in action, prime examples include the Zapatistas, whose efforts in constructing autonomy are focused on building 'a world where many worlds fit'; La Via Campesina, one of largest global movements in history that coined the term 'food sovereignty'; the Landless Worker's Movement, a decentralised organisation focusing on agrarian reform and grassroots popular education; and Abahlali baseMjondolo (The 'Shack dwellers Movement'), a self-organised coalition of resident-activists mobilising against evictions and xenophobia for decent and dignified housing.

Social movements coalesce around a shared purpose, sense of unity, and common dream and mobilise with the objective of correcting historical wrongs, effecting justice, and fighting alongside oppressed, exploited, abandoned, or targeted populations. They can be characterised by peaceful and/or militant 'praxis' (theory-action-reflection), which usually includes some iteration of participatory decision-making, prefigurative politics, assertions of identity/agency/dignity, and direct action. Ultimately, the primary aim of social movements committed to environmental defence is to resist development aggression and the destruction of the planet, as well as *move* societies forward out of structural violence and alienation into safer – more socially just, sustainable, and lifegiving – realities.

References

Birss, M. 2017. Criminalizing environmental activism. *NACLA Report on the Americas*, *49*(3), 315–322.

Federici, S. 2019. *Re-Enchanting the World: Feminism and the Politics of the Commons*. San Francisco: PM Press.

Gahman, L., Greenidge, A. and Mohamed, A. 2020. Plunder via violation of FPIC: Land grabbing, state negligence, and pathways to peace in Central America and the Caribbean. *Journal of Peacebuilding and Development*, *15*(3), 372–376.

Lang, M. and Mokrani, D. 2013. *Beyond Development: Alternative Visions from Latin America*. Amsterdam: Rosa Luxemburg Foundation and Transnational Institute.

Liboiron, M. 2021. *Pollution is Colonialism*. Durham: Duke University Press.

Menton, M. and Le Billon, P. (Eds.). (2021). *Environmental Defenders: Deadly Struggles for Life and Territory*. Oxford: Routledge.

Ossome, L. 2021. Pedagogies of feminist resistance: Agrarian movements in Africa. *Agrarian South: Journal of Political Economy*, *10*(1), 41–58.

Scheidel, A., Del Bene, D., Liu, J., Navas, G., Mingorría, S., Demaria, F., Avila, S., Roy, B., Ertör, I., Temper, L. and Martínez-Alier, J. 2020. Environmental conflicts and defenders: A global overview. *Global Environmental Change*, *63*, 102104.

Tran, D., Martinez-Alier, J., Navas, G. and Mingorria, S. 2020. Gendered geographies of violence: A multiple case study analysis of murdered women environmental defenders. *Journal of Political Ecology*, *27*(1), 1189–1212.

Tsosie, R. 2007. Indigenous people and environmental justice: The impact of climate change. *University of Colorado Law Review*, *78*, 1625–1677.

Additional resources

Asia-Pacific Forum on Women, Law, and Development: https://apwld.org/

Front Line Defenders: https://www.frontlinedefenders.org/en

Global Witness: https://www.globalwitness.org/en/campaigns/environmental-activists/

La Via Campesina: https://viacampesina.org/en/

The Transnational Institute: https://www.tni.org/en/agrarian-environmental-justice

Chapter 34

Religion

Ben Jones

Religion comes in from the cold

Up until the early 2000s, religion was largely absent from the study of development. The twin narratives of modernisation and secularisation meant that religion was understood to disappear from the public realm as societies evolved. Development was understood to be the product of rational, scientific bureaucracies and state-led transformation. Religious ideas and the way they organised society were meant to disappear from the public realm. Many aid organisations, even those with an ostensibly religious orientation, consciously avoided the topic of religion. Catholic Relief Services, an NGO with a religious orientation, struggled to find a place for religion. USAID only developed policies referring to religion or spirituality in the mid-2000s. It was only in the mid-2000s that leading donors started speaking of their active engagement with faith-based organisations (FBOs). USAID talked of the need to work 'closely with faith-based and community stakeholders' while the European Union, with a slightly chillier embrace, noted their 'cooperation with FBOs in development and humanitarian aid'.

Why did religion come in from the cold? In policy terms, the 9/11 attacks on New York and Washington and the ensuing 'war on terror' made religion and Islam, in particular, integral to the formulation of foreign and security strategies. Western development agencies that had been enlisted in the fight against Communism in the 1960s and 1970s were increasingly implicated in the 'war on terror'. The influence of Christian conservatives in shaping US government policy also helped recast development work in ideological terms at this time. More generally, the turn towards aligning development work with the protection of religious minorities and promoting notions such as 'freedom of religion' helped Western donors relate overseas assistance to foreign policy and security concerns.

It was also less and less clear that the relationship between modernisation and secularisation was fixed (Deneulin, 2013). Ways of organising public life along religious lines, such as post-1979 Iran, no longer looked like deviations from the path and instead pointed to a more mixed, multipolar world (Gifford, 2015). Within the academy the fragmentation of development studies also allowed more space for religion. The growing influence of concepts such as human development, sustainable development and community participation made religion and religious actors a more legitimate field of study. There have also been large research programmes and policy initiatives, such as the UK government's five year funding for a 'Religions and Development' research programme exploring the relationship between religion and development.

DOI: 10.4324/9780429282348-37

Religion never went away

Though something of a novelty for many Western donors, religious organisations are nothing new in the field of development. Almost all religions encourage their followers to take care of the poor. Catholic orders have been important providers of health and education services in Latin America and Europe, while Islamic *zakat* systems have provided relief across the Middle East and North Africa. Volunteers, often retired, working out of small charitable outfits in the Global North have been a constant feature of community-level development efforts in much of the world. Evangelical Christian groups from the US have, for many decades, been a visible presence in Latin America, Africa and Asia. The past two decades has seen rapid growth in the practice of short-term-volunteering and short-term missions, where young people visit projects over their summer holiday, mixing evangelisation with development work. Migrants living abroad also send money home via religious channels, with the South Asian diaspora, for example, building up particular spiritual leaders and their health and education projects back home. These varying forms of assistance form a substrate of lived experience, critical to the lives of many people, though, perhaps, less recognised or reflected on in the wider literature on development.

Religion has also been central to politics and the state. In late 19th- and early 20th-century Africa, for example, European-led administrations were reliant on missionaries for the provision of schools and hospitals. In the years after the Second World War, policies of state-led transformation and nationalisation quietened the role of religious actors, with governments taking over the provision of health and education systems from religious actors. With the liberalisation reforms of the late 1980s and the accompanying retrenchment and restructuring of the public sector, many developing country governments have become more relaxed about blurring the boundary between politics and religion. The part played by religious actors in providing social services, whether the Muslim Brotherhood in the Arab World or Pentecostal churches in Nigeria, has become a central of public life and part of mainstream political debate. Elections in 2016, 2018 and 2019 in the United States, Brazil, Israel and India also suggest the emergence of a sort of semi-authoritarian or hybrid politics where candidates explicitly align national political projects with religious identities.

Religion and the development sector

Faith-based organisations (FBOs) can be organised into three types: congregations affiliated with physical structures (e.g., a particular church, mosque or temple); national networks of congregations and their social services wings (e.g., the Catholic Church of Kenya); and unaligned or freestanding religious organisations, separate from congregations and national networks (e.g., Christian Aid, Islamic Relief). In the world of mainstream development, it is the last category of 'FBO as NGO' that has achieved particular prominence. Many of the largest NGOs – World Vision, the Aga Khan Foundation, Christian Aid, Caritas – are faith-based. The support FBOs receive from governments has increased dramatically in the past 20 years. In the wake of 9/11 the US government, for example, increased funding for FBOs from 10.5 per cent of the aid budget in 2001 to 19.9 per cent in 2005.

It is worth adding that much of the research and policy work sponsored by development agencies emphasises a somewhat instrumental approach, focusing on the positive role played by religious idioms or religious organisations. Religious organisations are seen as useful in carrying out development work because they are felt to be legitimate actors in the eyes of poorer people and also because they are felt to adopt an 'holistic' approach – integrating religious and material concerns. FBOs are described as providing the 'best social and physical infrastructure' and the overall emphasis is on the success and reach of religious organisations. Critical research on the subject suggests a more complicated picture (Bornstein, 2004). The distinction between religious and secular so central to the literature on religion and development is much less categorical than is often claimed (Scherz, 2014). In many developing country contexts local staff have a strong faith identity and do not worry about whether or not the organisation is itself 'faith based'.

In understanding the world of religion and development, there is a useful distinction to be drawn between the orthodoxy or 'right belief' religions of Islam, Christianity and Judaism and the 'right action' religions of Asian societies (Tomalin, 2015, chapter 1). NGOs based in 'right belief' societies, whether secular or religious, typically place a strong emphasis on relating development assistance to a particular transformation. The promotion of 'rights-based' approaches to development – human rights, women's rights, children's rights – can be seen as an example of this, as can the more general belief that beneficiaries should change who they are in order to become properly developed. In China, Japan and other Asian societies there has been more of an emphasis on infrastructural development and providing the material conditions for people to live as they are, albeit in better surroundings (Carrette, 2017). This distinction can sometimes be overdrawn, with South Korean NGOs drawing on the country's Christian tradition, for example.

Development as a religion

While much of the available work on religion and development focuses on trying to make religion legible to development workers and scholars, there is also the possibility of thinking about development itself in religious terms (Quarles van Ufford and Schofeleers, 1988). Religious motivations triggered the emergence of development thinking as a framework for North-South encounters in the 19th and 20th centuries, and it is possible to see current iterations of development as a form of religious discourse and practice. World Bank staff go on 'mission' to Africa, NGO workers 'sensitise' people on children's rights, micro-finance institutions make over the lives of women through gender-based lending. The media campaign surrounding 'The Girl Effect', for example, conveyed, in almost messianic terms, the transformation possible through sending girls to school.

What we might think of as the 'developmentalisation' of religion described in much of this chapter can thus be set against what we might think of as the 'religification' of development. The personality cults that define China's growing philanthropic landscape link back to earlier charismatic movements, while the neo-Confucian language of the development agenda of China's Communist Party can be contrasted with the Christian-inflected wording of much Western development assistance. Ethnographic work across a range of sites undermines the distinction between secular and religious development, which can seem so clear-cut at higher levels. Work in the Vietnamese Highlands, for example, shows that development practices which would seem to be ostensibly secularist, can, in fact, be driven by quasi-religious beliefs (Salemink et al., 2004). Development practitioners preach an unquestioning 'gospel of the market' and the 'promise of wealth'.

Figure 34.1 Church-goers relaxing at a healing prayer service in Kaderun Church (Anglican) eastern Uganda

A final word. There is a tendency in work on religion and development to be somewhat reductive and instrumental in how religion is imagined in poorer parts of the world (Jones and Petersen, 2011). If we think of religion as primarily and fundamentally a matter of belief, then the claim that poorer parts of the world as more 'religious' can too easily lead to a view of people in Africa, Asia or Latin America as less questioning. It is worth remembering that many people who profess a faith are open to the possibility of doubt and scepticism, and many people living in poorer parts of the world do not practice a religion (Green, 2006).

Bibliographic references

Bornstein, E. 2004. *The Spirit of Development: Protestant NGOs, Morality, and Economics in Zimbabwe*. London: Routledge.

Carrette, J. 2017. Hindu and Buddhist NGOs and the United Nations. In Carrette, J. and Miall, H. (eds.), *Religion, NGOs and the United Nations: Visible and Invisible Actors in Power*, pp. 195–216. London: Bloomsbury Publishing.

Deneulin, S. 2013. *Religion in Development: Rewriting the Secular Script*. London: Zed Books.

Gifford, P. 2015. *Christianity, Development and Modernity in Africa*. London: Hurst.

Green, M. 2006. Confronting Categorical Assumptions about the Power of Religion in Africa. *Review of African Political Economy*, 33: 635–650.

Jones, B. and Petersen, M.J. 2011. Instrumental, Narrow, Normative? Reviewing Recent Work on Religion and Development. *Third World Quarterly*, 32(7): 1291–1306.

Quarles van Ufford, P. and Schoffeleers, M. 1988. Toward a Rapprochement of Anthropology and Development Studies. In *Religion and Development: Towards an Integrated Approach*, pp. 1–30. Free University Press Amsterdam.

Salemink, O., van Harskamp, A. and Giri, A.K. (eds.) 2004. *The Development of Religion/The Religion of Development*, pp. 121–130. Chicago: University of Chicago Press.

Scherz, C. 2014. *Having People, Having Heart: Charity, Sustainable Development, and Problems of Dependence in Central Uganda*. Chicago: University of Chicago Press.

Tomalin, E. (ed.) 2015. *The Routledge Handbook of Religions and Global Development*. Oxford: Routledge.

Online resources

Berkeley Centre for Religion, Peace and World Affairs, Georgetown University: https://berk-leycenter.georgetown.edu/topics/religion-and-development
Caritas International, an example of a faith based development organization: https://www.caritas.org/who-we-are/
The Immanent Frame on secularism, religion and the public sphere: https://tif.ssrc.org
LSE's Religion and Global Society Blog: https://blogs.lse.ac.uk/religionglobalsociety/
Single graphic (table or image)

Chapter 35

Social capital and development

Anthony Bebbington and Katherine Foo

Introduction

The concept of social capital relates social norms, rules, and reciprocal obligations to patterns of social and economic action (Woolcock, 1998). For James Coleman, 'Social capital is defined by its function. It is not a single entity but a variety of different entities, with two elements in common: they all consist of some aspect of social structures, and they facilitate certain actions of actors' (Coleman, 1998: S98). Meanwhile, Pierre Bourdieu defined social capital as the aggregate of the actual or potential resources which are linked to possession of a durable network of more or less institutionalized relationships of mutual acquaintance and recognition . . . which provides each of its members with the backing of the collectively-owned capital.

(1986: 21)

Economists have used the concept as a way of describing the 'social something' (Hammer and Pritchett, 2006) that their econometric tools could otherwise not handle: the social relationships through which information is exchanged, risk managed, cooperation made possible, etc. (Hammer and Pritchett, 2006; Durlauf and Fafchamps, 2005; Fafchamps, 2006). Other social scientists have used social capital to explore how social relationships affect governance, democracy, livelihood, and collective action (Woolcock, 2010).

Engaging these differences, Uphoff (1999) distinguishes cognitive and structural definitions of social capital. Cognitive social capital pertains to the domain of values, trust, and perceptions. This conceptualization is apparent, for instance, in attitudinal survey research which gives quantitative measures to levels of trust in society and relates this trust to other indicators, in particular ones of economic performance. A structural conception of social capital leads researchers to focus on social relations, networks, loose associations, and formal organisations. Within development studies this structural conception has gained most attention, with social capital referring to the resources – information, reputations, credit – that flow through and are made available by social networks. Some writers view social capital as the interpersonal relationships that individuals mobilize to enhance their wealth and status (Bourdieu, 1977) while others understand it as properties of social organisation that facilitate coordinated, collective action (Putnam, 1993; Woolcock, 2010). This latter approach also considers how social networks can be shaped so that they are conducive to building more democratic, supportive, and inclusive communities.

DOI: 10.4324/9780429282348-38

Why did social capital become prominent in development?

While there are continuities between the concept of social capital and themes in 19th-century classical sociology (Woolcock, 1998), and although the concept had been deployed in urban planning (Jacobs, 1961), sociology (Bourdieu, 1986), and economics (Loury, 1977), it was Robert Putnam's work in political science that popularized the concept as an independent variable in economic and political development while at the same time giving it a quite particular meaning (Putnam, 1993). In his study of regional government performance in Italy, Putnam argued that, ceteris paribus, Italy's local governments were more effective and responsive to their citizens, and its sub-national economies were more dynamic in those regions exhibiting higher rates of participation in civic associations. Through involvement in these associations people learnt citizenship and developed networks of civic engagement (*social capital*) that, in their aggregate, fostered greater levels of accountability and responsibility in society and more efficiency in the economy. Putnam thus tied social capital to coordination, cooperation, and aggregate development performance, a quite distinct conceptualization from that of prior approaches. This conception proved to be much more intuitively accessible to a range of audiences than was the case for earlier renditions.

The visibility of Putnam's work in academic and popular outlets drove collective debate of his argument in sociology and political science, especially in the US. It also caught the attention of senior figures in the World Bank, where both the economic research and the social development communities began exploring the relevance of social capital for their own understandings of development (Bebbington et al., 2004, 2006). This link to the World Bank is important because, while the concept was set to be widely debated within academic social science and North American community development, its passage into development studies was accelerated and amplified by its usage within the World Bank.

Within the World Bank the concept proved especially helpful to those communities who already questioned the value of formal economic approaches to development. They saw in social capital a means of bringing social organisations, relationships, and empowerment into the institution's narrative on development in a way that would still allow conversations with the Bank's economists. Development became understood as a function of different 'forms of capital' at scales that ranged from the nation to the household. Early statements on sustainability and the 'wealth of nations' (Serageldin and Steer, 1994) argued that the sustainability of development could be understood as a function of the mixes and trade-offs among produced capital, natural capital, human capital, and social capital. A 'weak' concept of sustainability would consider development as sustainable as long as the overall capital stock increased; an 'absurdly strong' notion of sustainability would not allow draw down in any of these forms of capital; and 'sensible' sustainability would hold total capital stock intact and avoid depletion of any capital beyond critical levels.

If national development was a function of capital mixes and substitutions, then it was only a few short steps to using similar approaches to the study of poverty, welfare, and livelihoods at the household and individual levels. Work at the World Bank analysed household poverty as a function of household access to human, social, natural, and financial capital, and social capital was identified as an especially critical determinant (Grootaert, 1999; Narayan and Pritchett, 1999). Other development agencies' approaches to livelihoods followed a similar tack (Carney, 1998). These approaches argued that social capital – understood, broadly, as the networks, organisations, and relations to which the person or household had access – facilitated access to other assets or to the institutions providing those assets and in that way reduced poverty and vulnerability. This argument has been used in micro-financial services literature and practice, in which social capital (in the form of group membership) is taken as a guarantee that loans will be repaid. Another strand in this writing (and also at the World Bank) has seen social capital as an important safety net, a means of reducing vulnerability. Here social relationships (formal or informal) are valued for the role that they can play in helping people recover from or cope with crisis, violence, or other sources of risk and perturbation (Moser, 1998).

Two points merit comment here. The underlying influence on the use of the concept came from neo-classical economic approaches to production functions and, to a lesser extent, ideas in ecological economics about stocks of natural capital. Notably absent was Bourdieu's (1977) notion that the distributions of forms of capital (economic, cultural, symbolic, social) have to be understood as inter-related and in large measure mutually reinforcing. In his conception, for example, social capital serves to consolidate control of economic capital and relationships of power. There was no necessary reason why such conceptions could not have influenced development thinking (Bebbington et al., 2006). Second, even if the broader model at work here was underpinned by frameworks from economics rather than sociology, social development professionals latched onto the idea quickly. This type of asset-based framework allowed a development narrative that saw participatory processes and strong organisational fabrics as assets of equal importance to education, finance, or infrastructure. Capital based approaches to sustainable development offered the prospect of incorporating what had typically been local, idiographic, and operational concerns into wider theories of development in which the social was as important as the economic.

Criticisms and elaborations

A good case can be made that development studies research has tended to overstate the potential that social capital holds as a resource for poor people (Cleaver, 2005) and understate the extent to which local, national, and international political economy structure their ability to accumulate more assets and to get ahead. A social theoretical lens would conceptualize social capital as embedded in multiple historical and geographical scales, as both constituted by and constitutive of wider relations of political and cultural economy. Indeed, social capital has been subject to penetrating critique in both social and political science as well as in development studies (e.g., Fine, 2001). These criticisms have been many and varied. Critics note that conceptualizations of social capital can refer to so many dimensions of social life as to become relatively meaningless (Portes, 1998); do not allow for clear identification of causality; perpetuate romanticized notions of community (Cleaver, 2005; Portes, 1998); facilitate the further colonization of social science by neoliberal economics (Fine, 2001); turn social relations into objects of financial calculation; and ignore questions of political economy, power, and politics. In considerable measure such criticism reflects the extent to which early adoption of the concept was underlain by the production function approaches just noted as well as its association with the World Bank.

Even when some claim that such criticisms have been repetitive (Woolcock, 2010), there are indications that social capital research has recognised and responded to some of the points made. For example, Jamal (2009) highlights the tangled and negative dimensions of associational life in an authoritarian context and the ways in which the forms of social capital they involve are prone towards clientelism and patronage. In other studies, methodological progress has been made in developing multi-level approaches to social capital in order to better address the relationships among social structure, well-being, and health (Kawachi, 2008). Meanwhile studies of social capital in local organisations have sought to combine the insights of experimental economics (to understand the emergence and effects of trust and reciprocity) with those of critical social science (to address the effects of power asymmetries on cooperation; Serra, 2011).

Of course, not all new research has been so self-reflective, and more generally publications on (and citations of) social capital continue to boom (Woolcock, 2010; Serra, 2011; Svendsen and Svendsen, 2009). At the same time, the term has found its way into everyday discussions of development (and not only in the English language). Indeed, Woolcock (2010) argues that one of the great strengths of social capital is that the term facilitates many different conversations – both outside and within academia – among groups who otherwise would be unlikely to talk to each other about the relationships between social organisation, development, and democracy. In these different senses 'social capital' may have some affinity with that other slippery development concept, 'sustainability'. Each manages to bundle into a single term something that is at once conceptual, normative, and intuitive. Perhaps for that very same reason, each appeals across a wide disciplinary and political spectrum and has traction in scholarly, policy, *and* popular debate while at the same time being difficult to pin down with great precision. These qualities are simultaneously sources of great strength and great weakness. They may also prove to assure that both concepts will have a long shelf-life in development studies even when many who use them feel some discomfort in doing so.

References and Further Reading

Bebbington, A. 2007. "Social Capital and Development Studies II: Can Bourdieu Travel to Policy?" *Progress in Development Studies* 7(2): 155–162.

Bebbington, A., Guggenheim, S., Olson, E. and Woolcock, M. 2004. "Exploring Social Capital Debates at the World Bank." *Journal of Development Studies* 40(5): 33–64.

Bebbington, A., Woolcock, M., Guggenheim, S. and Olson, E. (eds.) 2006. *The Search for Empowerment. Social Capital as Idea and Practice at the World Bank*. West Hartford: Kumarian.

Bourdieu, P. 1977. *Outline of a Theory of Practice*. Cambridge: Cambridge University Press.

Bourdieu, P. 1986. "The Forms of Capital." In J. Richardson (ed.) *Handbook of Theory and Research for the Sociology of Education*. Westport, CT: Greenwood, pp. 241–258.

Carney, D. (ed.) 1998. *Sustainable Rural Livelihoods. What Contribution Can We Make?* London: Department for International Development.

Cleaver, F. 2005. "The Inequality of Social Capital and the Reproduction of Chronic Poverty." *World Development* 33(6): 893–906.

Coleman, J. 1998. "Social Capital in the Creation of Human Capital." *The American Journal of Sociology* 94: S95–S120. Supplement: Organizations and Institutions: Sociological and Economic Approaches to the Analysis of Social Structure.

Durlauf, S. and Fafchamps, M. 2005. "Social Capital." In P. Aghion and S. Durlauf (eds.) *Handbook of Economic Growth*. Volume 1B. New York: Elsevier, pp. 1639–1699.

Fafchamps, M. 2006. *Development and Social Capital. Global Poverty Research Group. Working Paper Series 007*. Manchester and Oxford: Global Poverty Research Group.

Fine, B. 2001. *Social Capital versus Social Theory. Political Economy and Social Science at the Turn of the Millennium*. London: Routledge.

Grootaert, C. 1999. *Social Capital, Household Welfare and Poverty in Indonesia. Local Level Institutions Working Paper, No.6.* Washington, DC: World Bank, Social Development Department.

Hammer, J. and Pritchett, L. 2006. "Scenes from a Marriage: World Bank Economists and Social Capital." In A. Bebbington et al. (eds.) *The Search for empowerment: Social as Idea and Practice at the World Bank.* West Hartford: Kumarian, pp. 63–90.

Jacobs, J. 1992 [1961]. *The Death and Life of Great American Cities.* New York City: Vintage Books.

Jamal, A. A. 2009. *Barriers to Democracy: The Other Side of Social Capital in Palestine and the Arab World.* Princeton, NJ: Princeton University Press.

Kawachi, I. 2008. "Social Capital and Health." In C. Bird, P. Conrad, A. Fremont and S. Timmermans (eds.) *Handbook of Medical Sociology.* Sixth Edition. Nashville, TN: Vanderbilt University Press.

Loury, G. 1977. "A Dynamic Theory of Racial Income Differences." In P.A. Wallace and A. Lamond (eds.) *Women, Minorities and Employment Discrimination.* Lexington, MA: Lexington Books, pp. 153–186.

Moser, C. 1998. "The Asset Vulnerability Framework: Reassessing Urban Poverty Reduction Strategies." *World Development* 26(1): 1–19.

Narayan, D. and Pritchett, L. 1999. "Cents and Sociability: Household Income and Social Capital in Rural Tanzania." *Economic Development and Cultural Change* 47(4): 871–897.

Portes, A. 1998. "Social Capital: Its Origins and Applications in Modern Sociology." *Annual Review of Sociology* 24: 1–24.

Putnam, R. 1993. *Making Democracy Work: Civic Traditions in Modern Italy.* Princeton, NJ: Princeton University Press.

Serageldin, I. 1996. *Sustainability and the Wealth of Nations.* Washington, DC: World Bank.

Serageldin, I. and Steer, A. (eds.) 1994. *Making Development Sustainable: From Concepts to Action. Environmentally Sustainable Development, Occasional Paper Series No. 2.* Washington, DC: World Bank.

Serra, R. 2011. "The Promises of a New Social Capital Agenda." *Journal of Development Studies* 47(8): 1109–1127.

Svendsen, G. and Haase Svendsen, G. (eds.) 2009. *Handbook of Social Capital: The Troika of Sociology, Political Science and Economics.* Northampton, MA: Edward Elgar Publishing, Inc.

Uphoff, N. 1999. "Understanding Social Capital: Learning from the Analysis and Experience of Participation." In P. Dasgupta and I. Serageldin (eds.) *Social Capital: A Multifaceted Perspective.* Washington: World Bank.

Woolcock, M. 1998. "Social Capital and Economic Development: Towards a Theoretical Synthesis and Policy Framework." *Theory and Society* 27(2): 151–208.

Woolcock, M. 2010. "The Rise and Routinization of Social Capital, 1988–2008." *Annual Review of Political Science* 13: 469–487.

Chapter 36

Is there a legal right to development?

Radha D'Souza

Introduction

The question forms part of a wider question about the status of socio-economic rights in law. Rights in law entail two aspects: normative standards for society and the institutional mechanisms for enforcement. Before the World Wars the right to property was the only economic right that was recognised as part of basic freedoms. Enlightenment thinkers put property rights on par with conventional human rights that existed in pre-modern societies such as right to life, liberty and conscience. Enlightenment thinkers also argued that state interference in the economy must be minimal. Adam Smith, the father of the discipline of economics, for example, argued in the 18th century that the role of the state must be that of a 'night watchman'. The primacy of property rights and its equal status alongside life, liberty and conscience created economic polarisation, financial crises and social unrest. By the late 19th century the economic and social inequalities that property rights introduced in societies made socialist thinkers like Karl Marx argue that the state was 'the executive committee of the bourgeoisie'. The first half of the 20th century was engulfed in political and social upheaval including the revolutions in Russia and Eastern Europe, the World Wars and the anti-colonial struggles. When the World Wars ended, the victorious Allies proposed the formation of the United Nations. The UN Charter for the first time recognised economic rights beyond property rights.

The UN has two goals: promoting peace (Article 1.1) and cooperation for economic, social and cultural problems (Article 1.3). The UN Charter includes harmonisation as a related goal (Art. 1(4), 13) and allows it to codify international law and develop global standards. Chapters IX and X provide the legal basis for the expanded economic role of the UN that we see today. Chapter IX expands on economic goals to include higher standards of living, health and social progress including education, as well as human rights including non-discrimination. It empowers the UN to sign agreements known as the Specialised Agency Agreements (SAA) with the International Economic Organisations (IEO) like the World Bank, the IMF and standard setting organisations like the WHO, ILO and others. The SAAs define the relationship of each Specialised Agency (SA) to the UN. Most importantly the UN Charter created a new economic organ, the Economic and Social Council (ECOSOC), to enable the UN to perform its economic role. The ECOSOC's main role is coordinating the work of the SAs, the states (through its regional commissions), the international standard setting organisations and the NGOs. International law until the UN Charter was almost entirely about war and peace between states. The vastly expanded economic role for international law and its institutionalisation under

DOI: 10.4324/9780429282348-39

the UN Charter were radical departures by any measure from the League of Nations or international law before the League.

Socio-economic rights

On 6 December 1986 the UN General Assembly (UNGA) adopted GA Resolution 41/128 on the UN Declaration on the Right to Development (DRtD). The DRtD forms a continuum in a series of developments in socio-economic rights following the expanded economic role of the UN in the Charter. The first of these was the Universal Declaration on Human Rights (UDHR) adopted in 1948. The UDHR recognises conventional civil liberties and individual freedoms (Art 1–21) together with property rights (Art 17) alongside international cooperation for the realisation of social security (Art 22), right to work without discrimination (Art 23), fair conditions of work (Art 24), standards of living (Art 25) and education (Art 26). There were differences between the Big Four (US, Britain, USSR and China) on the scope of human rights when the UN Charter was drafted. These differences were by no means resolved in the UDHR.

Briefly summarising, the approach of Euro-American states was grounded in classical liberal philosophy where human rights were primarily personal freedoms, non-discrimination by the state, fair process and guarantees of property rights. The Soviet bloc states argued human freedoms were contingent on the capacities of states to meet basic needs of its populations, i.e.: food, clothing, health, shelter and education. Recognition of socio-economic rights was necessary for human rights to be realised. The Third World states, emerging from colonial rule, inherited institutions that were authoritarian and economies that continued to be dominated by colonial/imperial economic actors. Restricting the role of the state in the Third World would, they argued, make de-colonisation impossible. The priorities for Third World states were rights to self-determination and affirmative economic actions. The UDHR was at best a compromise document that straddled diverse approaches to civil and political rights on the one hand and socio-economic rights on the other. The UDHR remained a non-binding declaration. Unlike treaties and covenants, declarations in international law are not binding on the signatories. They are at best statements of intent that guide international cooperation between states. The UDHR became the tool for engagement between the states throughout the Cold War.

In 1966 the UN adopted two binding covenants relevant to DRtD: the International Covenant on Civil and Political Rights (ICCPR) and the International Covenant on Economic, Social and Cultural Rights (ICESCR). The two covenants share similar preambles and recognise some common rights such as the right to self-determination and non-discrimination. If the UDHR was a compromise document the ICCPR and ICESCR was a breakdown of that compromise. The ICCPR expanded on the Western nations' approaches to human rights as civil liberties, individual freedoms and property rights in Articles 1–21 of the UDHR and the ICESCR expanded on the socio-economic rights in Articles 22–29. Against the backdrop of the Cold War, not surprisingly, the ICCPR was canvassed by Western states and the ICESCR by the Soviet Bloc. Unlike the UDHR both covenants created binding obligations on the signatories.

The Third World states mounted a concerted effort to rewrite international law. They argued that international law, including the UN Charter, was unjust because it was written by colonial and imperial powers (Anghie, 2004). Instead they called for a New International Economic Order. In 1974 the UNGA adopted Resolution 3281 which became a charter of economic rights. The Western states were not supportive of the resolution. The DRtD and its emergence within the Commission on Human Rights must be seen against these attempts by Third World states for economic rights. The DRtD came at the tail end of the three UN Development Decades from 1960 to 1990. By that stage it was apparent that the UN's development agenda was not delivering and there was widespread disillusionment.

The DRtD comprises ten articles that guide state actions. It states RtD is an inalienable human right (Art 1) and the human person is the focus of RtD (Art 2). States must create conditions for the realisation of RtD (Art 3) cooperate to achieve RtD (Art 4–6), ensure peace security and disarmament to achieve RtD (Art 7) and take steps to realise equal opportunities, basic needs and democracy (Art 8). Last it pronounces that RtD was indivisible and interdependent (Art 9) bridging the rift in the ICCPR and ICESCR. There is no enforcement or reporting mechanisms in the DRtD.

Globalisation and DRtD

Three years after the DRtD was adopted the architecture of the post-World War world changed dramatically. Two significant moments in the change were the Washington Consensus and the fall of the Berlin Wall, both in 1989. Both these events set in motion the most comprehensive reform of the international legal order since the end of the World Wars. The main thrust of the change was neoliberal transformation of the international order – i.e. rolling back states and rolling in private actors in the global arena. The newly formed WTO took the lead in initiating neoliberal transformation of International Organisations and interstate relations through its trade regime (D'Souza, 2010). With the fall of the Berlin Wall the walls that separated civil and political rights from economic and social rights and both from RtD also fell. The Vienna Declaration and Programme of Action (VDPA) was adopted in 1993 at the World Conference on Human Rights and endorsed by the UNGA resolution 48/121. The VDPA effectively merged the UDHR, the ICCPR and ICESCR and DRtD into a single programme. The VDPA in the new international legal regime became part of the Good Governance requirements proposed under the Washington Consensus and underpinned by IEOs (Woods, 1999).

Until the neoliberal reforms the IEOs interpreted development primarily as lending for infrastructure, industrial and agricultural production and building capital markets. Managing social, political and human consequences of development projects, they argued, was the legal responsibility of the states and beyond the remit of their Articles of Association. The neoliberal reforms of the international economic regime changed all that. An important part of the VDPA's plan of action was 'mainstreaming' human rights (Part II). The scope of the renamed Human Rights Council (HRC) included an 'institution-building package', Universal Periodic Review Mechanism, a new Advisory Committee, a revised Complaints Procedure, and participation of private experts. Within the overall framework of Good Governance, 'mainstreaming' aligned RtD to the goals of the IEOs. Good Governance programmes require Third World states to remove restrictions on private economic actors, domestic and international. Under the overall remit of the IEOs, compliance with human rights standards is seen as a risk-mitigation strategy by private industries, states and financial institutions (Likosky, 2006). Good Governance programmes envisage wide-ranging law reforms within Third World states. These requirements form part of contractual agreements between the IEOs and states, leading some scholars to argue that IEOs increasingly function as law makers (Alvarez, 2005).

Does that make DRtD enforceable through international contracts with IEOs? The VDPA like the DRtD is also normative. The constitutions of many Third World states recognise socio-economic rights as Directive Principles that ought to guide state actions and policies. They are not justiciable in that they cannot be enforced by courts. Generally courts have taken the view that economic entitlements under RtD are contingent on the states' economic capacity to provide for its population and cannot be enforced through courts. Courts have been more open to enforcing social aspects of DRtD like non-discrimination. In the US the Alien Torts Claim Act has been used by activists to hold US corporations to account with mixed results. In the UK courts have held that tort claims must have a direct nexus between the action and wrongdoer. Responding to criticisms that lending policies breached human rights, the WB set up the Inspection Panel as a forum to address grievances arising from WB-funded projects. The credibility of the mechanism is limited by the in-house procedures and management.

Justiciablity of economic rights eludes modern law. 'Where there is a right, there must be a remedy' is a basic common law principle. If RtD is not justiciable, if a person without food or clothing cannot claim them as a 'right', RtD challenges the commonplace understandings of 'rights'. RtD sits on a conceptual fault line: the meaning of development. If development is what states and IEOs say it is, namely that development is about building dams, railways, industries, and banks, then invariably it is followed by displacement, dispossession and disenfranchisement and winners and losers. If development is understood as human well-being then it challenges the concept of modernisation. The fault line that runs through RtD sits on a deeper rift on which modern law stands. That rift is the idea that property rights are inalienable human rights to be valued on par with life, liberty and conscience. RtD challenges us to think about which of these two ought to be privileged when the two do not work in tandem.

References

Alvarez, J. E. (2005) *International Organizations as Law-Makers*, Oxford: Oxford University Press.

Anghie, A. (2004) *Imperialism, Sovereignty and the Making of International Law*, Cambridge: Cambridge University Press.

D'Souza, R. (2010) Law and Development Discourse About Water: Understanding Agency in Regime Changes. In Cullet, P. et al. (eds) *Water Governance in Motion: Towards Socially and Environmentally Sustainable Water Laws*, New Delhi: Foundation Books, p. 491.

Likosky, M. B. (2006) *Law, Infrastructure, and Human Rights*, Cambridge: Cambridge University Press.

Woods, N. (1999) Good Governance in International Organizations. *Global Governance* 5: 1, p. 39.

Further readings

International Commission of Jurists (2008). *Courts and the Legal Enforcement of Economic, Social and Cultural Rights: Comparative Experiences of Justiciability*, Geneva: ICJ Provides Information on Socio-Economic Law.

A special symposium on Right to Development is published in the California Western International Law Journal 1985, p. 429.

Part 4

Identities and practices. Editorial introduction

The term 'identities' is used to refer to the distinct characteristics, beliefs, values, and expressions that individuals or groups use to define themselves and differentiate from others. Identities can be shaped by factors such as culture, ethnicity, gender, religion, personal desires, and experiences, and they play a crucial role in influencing behaviour, perceptions, and social interactions. This section has been titled identities and practices to reflect the relationship between individual and collective self-conceptions and the actions that embody them.

Identities are intrinsically linked to practices as they both shape and are shaped by the recurrent actions and behaviours individuals or groups engage in, reflecting their underlying values, beliefs, social and cultural norms, needs, and desires. This is not to say that this relationship is straightforward or deterministic. Far from it. Identities seldom remain static, often evolving and adapting, making their interplay with practices uncertain if not complicated. Practices not only reinforce and manifest identity but also serve as a means through which identities are negotiated, contested, and transformed within social contexts.

Identities have long been a focal point in development studies. Initially, they were viewed as markers of socio-economic status, cultural and regional affiliation. Historically, development studies primarily focused on economic growth, infrastructural advancements, and policy implications. However, as the field expanded, there was a growing recognition of the human and social dimensions of development. Identities, with their complex ties to social, cultural, political, and economic relationships, emerged as crucial to understanding the nuances of development.

Identities provided an entry point into the lived experiences of individuals and communities, revealing the disparities, power dynamics, and socio-cultural factors that influence development outcomes and processes. Development studies has examined how identities influence access to resources, opportunities, and social standings. Furthermore, the discipline has examined the role of identities in shaping political mobilisation and group solidarities, exploring how they can empower or marginalise, acting either as catalysts for change or as instruments perpetuating existing power structures. Identities have also been seen to inform how individuals interpret, navigate, and respond to their surroundings, influencing societal norms, decision-making processes, and collective action. Moreover, they shape the formulation, interpretation, and responses to policy decisions and interventions.

Whilst few would contest the importance of identities in development, a degree of caution is required against viewing identities in essentialist terms or as the sole or primary determinant of socio-economic processes or outcomes. Essentialism, in this context, refers

DOI: 10.4324/9780429282348-40

to the idea of viewing identities as having an inherent, unchanging nature, often based on stereotypes or generalised notions. Such a perspective can lead to oversimplifications and fails to recognise the dynamic and socially constructed nature of identities. It's crucial to approach identities with an understanding of their fluidity and the socio-cultural factors that shape them.

Interdisciplinary discussions have reconceptualised identities as dynamic constructs that emerge in response to socio-political and economic change. This shift has led to a growing emphasis on the interplay among identities, associated practices, and broader development processes.

Identities should not be taken to imply inherent characteristics of individuals or groups but neither should they be downplayed as merely personal labels; they intertwine deeply with social, economic, and political structures, influencing experiences of marginalisation, exclusion, and discrimination.

Contemporary perspectives recognise identities as socially constructed, multifaceted, fluid, and contingent, changing throughout an individual's life course and with their changing relationships with others. This shift has significant implications for development studies. As societies transform, so do identities within them, introducing new challenges and opportunities.

While identities might appear coherent at a given moment in time, they are in a state of constant change. They are constructed and redefined through interactions with prevailing discourses and power structures.

Identities are complex combinations of political, social, cultural, and economic elements. They influence an individual's self-perception, how they view others, and their interactions with the world. Identities, in essence, are moulded by and in turn shape our beliefs, values, cultural backgrounds, and social roles.

The concept of intersectionality (Crenshaw, 2000) highlights how multiple identities can merge, diverge, and interact, resulting in unique experiences and challenges. Intersectionality refers to the idea that our identities are multi-dimensional, made up of various factors like race, ethnicity, class, gender, and more. These factors intersect in unique ways, creating distinct experiences and unique forms of discrimination or privileges for different individuals.

Contemporary approaches to identity in development studies recognise their dynamic nature and provide a deeper appreciation for the lived experiences and practices in which they are foregrounded. They emphasise that development is not just about policy frameworks or indicators but deeply intertwined with individuals' daily experiences and self-perceptions.

The chapters in this section, while not all directly focused on identities per se, emphasise the interplay between identities and practices and their role in shaping development outcomes and processes. They illuminate the historical and structural roots of identities, the inequalities framing them and stress the importance of viewing development in broader terms to account for lived experiences, social identities, and socio-political milieus they emerge from.

References

Crenshaw, K. 2000. "Demarginalising the Intersection of Race and Sex: A black Feminist Critique of Antidiscrimination Doctrine, Feminist Theory and Antiracist Politics." In *The Black Feminist Reader* (eds. J. James and T. Sharpley-Whiting). Malden, MA: Blackwell, 208–238.

Chapter 37

Children and development

Kristen E. Cheney

To understand the implications of international development for children, it is necessary to understand how young people have traditionally been framed within development discourse, from innocent objects of development to actors in rights-based participation. Throughout children's entanglement with humanitarianism and development, however, the idealized childhood propounded by the development industry has consistently imposed neoliberalism and has problematically codified idealized Western childhoods as universal.

Children as objects of humanitarian intervention

International development's focus on children has its roots in charity and humanitarian aid – which still largely influences interventions for poor children around the world. In the 18th century, European humanitarianism focused on children through Christian social movements in colonial projects. They also responded to children's needs following natural disasters. In the 19th and 20th centuries, there were many local and national charitable children's organisations in the Global North, whose efforts expanded internationally at the end of the World Wars I and II, when many European children were displaced. Save the Children, for example, was established in 1919 by Eglantyne Jebb, an originator of the Declaration on the Rights of the Child. As their name connotes, Save the Children was founded on the notion that children were in need of 'saving' from a plethora of adverse post-war circumstances. Children were thus roundly defined by humanitarian organisations as victims – and not perpetrators – of violence.

The United Nations Emergency Children's Fund (UNICEF) was similarly founded in 1946 to respond to the needs of European children who were orphaned or displaced in World War II – but later shifted to focusing on children in the newly independent but poor countries of the Global South. In the late 1950s and early 1960s, UNICEF primarily targeted health issues and child survival but gradually shifted its focus to broader issues of poverty reduction as part of the U.N.'s wider movement to end the poverty associated with colonialism and the rise of national independence. Children's issues were thus placed squarely within the context of the new, international development efforts that were also part of the geopolitics of decolonization.

In the 1980s, child-focused international nongovernmental organisations (INGOs) as well as intergovernmental agencies like UNICEF again shifted their objectives to dealing with the disastrous consequences of structural adjustment programs (SAPs). SAPs caused children disproportionate suffering and did massive damage to humanitarian efforts to

DOI: 10.4324/9780429282348-41

eradicate child poverty. In some ways, though, humanitarian efforts to counteract SAPs inadvertently facilitated an even faster retreat of the state from the welfare services that SAPs might otherwise have necessitated; as civil wars and other social upheavals raged across Africa, Latin America, and Southeast Asia, INGOs found their efforts necessarily relegated to emergency relief services as well as the health and welfare services that failing and impoverished states were neglecting (Black, 1996).

These organisations still largely viewed children as hapless victims in need of intervention for their very survival. Although on one level this may be true – particularly in contexts of war, disaster, and persistent structural violence – the approach took children as mere objects of humanitarian intervention; indeed, several scholars have argued that children's innocence is often problematically symbolic of all humanity in ways that patronize developing countries as infantile (Valentin and Meinert, 2009; Malkki, 2010).

The assumptions rooted in colonial domination and native infantilization manifested in mid-20th-century programs that targeted children directly for humanitarian intervention so as to transform their lives in the direction of broader 'progress' and 'modernization'. Poverty reduction in this context was as much about national growth as concern for child well-being, but it had the effect of depoliticizing development efforts and decontextualizing childhood in favor of universalized standards – often based on economic and political liberalisation. Several international instruments have impacted the interaction between international development and children.

The United Nations Convention on the Rights of the Child

Such notions of 'progress' are reflected in children's rights discourses after the 1989 drafting and promulgation of the United Nations Convention on the Rights of the Child (UNCRC). Over 40 years in the making, the UNCRC was the culmination of numerous documents drafted due to a sense, following the promulgation of the Universal Declaration of Human Rights (UDHR) in 1948, that children were in need of special protections not adequately covered by nonbinding declarations such as the UDHR or even the Declaration on the Rights of the Child.

Proposed in 1978, approved in 1989, and ratified by 20 states in 1990, the UNCRC guarantees children's civil, political, economic, social, health, and cultural rights (United Nations, 1989). The UNCRC fast became the most ratified UN human rights convention in history, with every country but the United States having signed on.

The UNCRC has not been without its critics, however: many dissenters view the UNCRC as a Western-centric document that promotes the notion of the ideal 'child' as at once both a sovereign, rational individual of a liberal democratic society and as a dependent who should be quarantined in child-specific spaces (Ennew and Milne, 1990). Although many countries rushed to sign the convention, some regional bodies proceeded quickly to qualify the UNCRC with declarations of their own that they felt more closely reflected their social values. The Organization for African Unity's 1990 *African Charter on the Rights and Welfare of the Child*, for example, closely resembles the UNCRC while at the same time emphasising children's communal responsibilities. In spite of these efforts to assert regional autonomy over discourses of child well-being, the hegemony of the UNCRC persisted in setting the universal standard for a singular, idealized childhood based on 'modern' theories of both social progress and child development to which societies should aspire (Twum-Danso Imoh and Ame, 2012).

Some also saw the promulgation of the UNCRC as an outcome of the broader failures of international development to eradicate global poverty and engender world peace (Pupavac, 2001). This allowed for the naturalization of hegemonic notions of 'progress' in the hands of childhood's future adult subjects, with the figure of the immature child operating as metaphor for postcolonial nations' underdevelopment resonating throughout development discourses.

Furthermore, development discourses placed increasing emphasis on 'culture' as a barrier to socioeconomic progress, citing 'harmful traditional practices' and the failure of parents as duty bearers as primary obstacles to the fulfillment of children's rights. International development accordingly shifted concern towards instigating culture change to remake the ideal childhood in line with neoliberal individualism, drawing attention away from structures of inequality and lack of social provisioning.

Despite critiques, children's rights-based approaches became the standard modus operandi in the 1990s, revealing a deep ambivalence about the place of childhood as both a protected and empowered category – and, indeed, reflecting broader doubts about modern humanitarianism. Such unease has resulted in challenges for the implementation of children's participation in development efforts.

Along with provision and protection, many organisations made participation a stated priority in their rights-based programming. But they found it difficult to implement in practice because of varying cultural and organisational notions of child participation. As a result of this and other political and cultural resistance to the notion of 'rights', many organisations pulled back from their earlier child rights frameworks in the first decade of the 21st century, focusing instead on the Millennium Development Goals and the subsequent Sustainable Development Goals.

Children and the MDGs and SDGs

In 2000, world leaders committed to the Millennium Development Goals (MDGs) – eight targets for reducing poverty by 2015 (United Nations, 2000). According to UNICEF, the MDGs could be achieved only if children's rights to education, health, and protection were met. Out of this came several global initiatives that explicitly targeted children, including Education for All (EFA) to achieve MDG 2: universal primary education.

Under rights-based approaches, education had attained a central role in both individual and national development, but this goal to universalize education took on new urgency under EFA. Though only about a third of all countries had reached every goal of the program, including gender parity and quality improvement in education, by 2015, many critical childhood studies scholars pointed out that the drive for education had the adverse effect of making children more dependent on adults by stigmatizing certain forms of labour and even 'deskilling' young people in developing countries (Katz, 2004).

Moreover, many have observed how the emphasis on formal education has encouraged high aspirations while also causing the devaluation of manual, rural, and unskilled forms of labour in favor of the kinds of white-collar jobs that are reserved for the educated – even as universal education has been difficult to attain and possibilities for young people to obtain skilled employment are still rare in many countries.

As 2015 drew to a close, it became clear that despite great strides, not all MDGs would be universally met. So the international community created an even more ambitious and comprehensive set of interconnected goals for 2030. The Sustainable Development Goals include 17 points to address the social, economic, and environmental dimensions of sustainable development, including new targets for climate change and environmental sustainability, economic inequality, peace, and justice (United Nations, 2015).

Of the 230 SDG indicators, 50 concern children directly, yet some organisations have expressed concern that the SDGs do not explicitly protect and ensure children's rights – thus continuing to treat children as objects of, rather than participants in, development. Again, UNICEF argued that children must be at the center of the SDG agenda if it is to reach its goals by 2030. However, the SDGs continue to rely largely on principles of economic liberalisation, privatisation, and free trade without examining their role in growing income inequality and their deleterious effects on children.

The continued paternalism of international development targeting children

Narratives of innocent childhood suffering and child empowerment continue to sit uncomfortably side-by-side in development efforts on behalf of children. Throughout this history, both local and international organisation have consistently encouraged charitable contributions to children's causes based more on victimhood than empowerment. Such charity on behalf of children therefore fails to be transformative; rather, it upholds status-quo international power dynamics and translations of childhood need while also stripping children of their personhood. Moreover, those same actions often dehumanize – and even demonize – their thoroughly racialized parents.

In this sense, the colonial project to 'parent' the children of the Global South is rejuvenated through development interventions. At times, development practices necessarily reify the child – out of and away from family and community contexts – to make children the singular objects of charity, aid, and development. And yet if there is one lesson childhood and development studies have taught, it is that children's lives are necessarily relational because of the dependence of children on older generations when children are young – but also because of the ways in which younger and older generations become interdependent as children grow and gain agency. International development practice still struggles to embrace this relational interdependence in child-focused programming.

Revised with permission from CHENEY, K. 2019. Children and International Development. In BURMAN, E. & COOK, D. (eds.) *The SAGE Encyclopedia of Children and Childhood Studies*. London: Sage.

See also Children's Rights; Education; Millennium Development Goals (MDGs); Postcolonial Childhoods; Sustainable Development Goals (SDGs); Well-Being, Children's

References

Black, M. 1996. *Children First: The Story of UNICEF, Past and Present*. New York: Oxford University Press.

Ennew, J. & Milne, B. 1990. *The Declaration of the Rights of the Child. The Next Generation: The Lives of Third World Children*. Philadelphia: New Society Publishers.

Katz, C. 2004. *Growing Up Global: Economic Restructuring and Children's Everyday Lives*. Minneapolis: University of Minnesota Press.

Malkki, L. 2010. Children, Humanity, and the Infantilization of Peace. In Feldman, I. & Ticktin, M. (eds.) *In the Name of Humanity: The Government of Threat and Care*. Durham: Duke University Press.

Pupavac, V. 2001. Misanthropy Without Borders: The International Children's Rights Regime. *Disasters*, 25, 95–112.

Twum-Danso Imoh, A. & Ame, R. (eds.) 2012. *Childhoods at the Intersection of the Local and Global*. New York: Palgrave Macmillan.

United Nations. 1989. *United Nations Convention on the Rights of the Child*. New York: United Nations General Assembly, 20 November 1989.

United Nations. 2000. *United Nations Millennium Declaration*. New York: United Nations General Assembly.

United Nations. 2015. *Sustainable Development Goals*. New York: United Nations General Assembly.

Valentin, K. & Meinert, L. 2009. The Adult North and the Young South: Reflections on the Civilizing Mission of Children's Rights. *Anthropology Today*, 25, 23–28.

Further readings

Ansell, N. 2017. *Children, Youth and Development*, 2nd ed. London: Routledge.

Boyden, J. & Dercon, S. 2012. *Child Development and Economic Development: Lessons and Future Challenges. Young Lives*. Oxford: Young Lives, Department of International Development.

Cheney, K. & Sinervo, A. (eds.) 2019. *Disadvantaged Childhoods and Humanitarian Intervention: Processes of Affective Commodification and Objectification*. London: Palgrave Macmillan.

Hanson, K. & Nieuwenhuys, O. (eds.) 2013. *Reconceptualizing Children's Rights in International Development: Living Rights, Social Justice, Translations*. Cambridge: Cambridge University Press.

Myers, W. & Bourdillon, M. 2012. Introduction: Development, Children, and Protection. *Development in Practice*, 22, 437–447.

Chapter 38

Youth

Perspectives and paradigms in global development

Emil Dauncey

On 15 November 2022, the global population reached 8 billion people., taking just 12 years to increase from 7 billion in 2010 (UN DESA, 2022a). This population landmark was in large part the result of advances in medicine, public health, and nutrition that have meant many people live longer lives. It was also an effect of rising fertility rates in many parts of the world. One consequence of these changes is that the absolute number of young people is larger than at any time in human history. Indeed, at the time of writing there were 1.22 billion people aged between 15–24, just over 15 per cent of the global population. There are significant national and regional variations. For example, in Europe young people constitute just over 10 per cent of the population, whereas in Sub-Saharan Africa it is closer to 20 per cent (UN DESA, 2022b).

This chapter provides a brief survey of what the significance of this demographic change might be in the context of global development. It begins with a brief discussion of what the term *youth* means, highlighting its fluidity and ambiguity and some of its social, political, and economic connotations. The next section focuses more firmly on what the implications of a large *youth* population might be, not least for young people themselves. Finally, it describes and critically evaluates three paradigms that have characterised engagement with *youth* in the context of development policy and practice.

The meaning of youth

For statistical purposes the United Nations defines *youth* as a person between the ages of 15 and 24 years (UN DESA, 2022b). However, there is no universally agreed upon definition of youth, with significant variations between different cultures and societies and at different moments in time.

In many societies youth connotes a formal political and legal status differentiated from that of an adult or child. For example, in England, 18 is the age at which a person legally becomes an adult and is given equal treatment with other adults under the law, with a legal right to control their privacy and make decisions without parental consent or oversight. This is also the age at which a person can vote and stand for election to parliament. Whilst 18 is a common age at which such rights are achieved, it is not universal. For example, Scotland, Ethiopia, and Ecuador are just some of the countries where 16-year-olds are able to vote.

The significance of the term *youth* extends beyond that of a formal legal or political status. Rather than being a property or characteristic of an individual, youth is perhaps best thought of as a socially constructed, culturally specific category of the person. This

DOI: 10.4324/9780429282348-42

category may be defined with reference to a range of factors including but not limited to biological age, physical characteristics, employment, economic independence, or parental or marital status, reflecting the values, beliefs, and practices of a particular society. Furthermore, the idea of youth is not immutable but contingent and fluid, meaning that those who identify or are identified with the term are not a homogenous group. Their experiences, perspectives, and sense of self will be shaped by multiple and intersecting identities and relationships, not least those of gender, class, (dis)ability, and ethnicity but also marital and parental status, education, and health status amongst others.

Youth is often defined as a transitional or liminal status between childhood and adulthood (Honwana and de Boeck, 2005). In some societies, youth is a brief and narrowly defined phase, while in others, it may span several decades and encompass a diverse range of experiences and identities. Distinctions between youth and adulthood are seldom clear cut. Transitions are rarely linear or irreversible processes, but multifaceted and contingent, such that people may move between categories in different spaces and at different moments in time (Johnson-Hanks, 2002; Valentine, 2003). Nevertheless, the distinctions drawn among childhood, youth, and adulthood are not without consequence. Youth often do not enjoy the same the protections accorded to children or the social, political, and economic rights and responsibilities of adults.

Scholarship across the social sciences has often highlighted the ways young people experience, navigate, and even contest notions of youth and the subject positions they imply and their role in making and remaking social worlds (See for example, Weiss, 2009; Newell, 2012; Masquelier, 2005). Nevertheless, within the context of policy and practice, there had until recently been a tendency to see young people as a homogenous group, as either a problem to solve or as hapless victims, without agency. This has sometimes meant that the ways that young people themselves go about negotiating and navigating social change – and the positive contributions they can and do make to it – go unrecognised (Ansell, 2017).

Opportunities and challenges: some implications of a growing youth population

The implications of a large population of youth are far reaching. Some authors have highlighted the potential benefits, whilst others have sounded a cautionary note about the challenges it poses. It has been argued that a large youth population constitutes a 'demographic dividend' from which economic benefits can be reaped in the form of a larger working age population, an associated increase in productivity and economic growth, and reduced dependency ratios. Young people can play a crucial role in supporting infants or an aging population, either through direct care and economic support or indirectly through contributions to taxation and social protection systems such as pensions (Bloom et al., 2003).

However, a large youth population also presents challenges. It generates a greater demand for employment, education, health, and other services that some societies are unable or unwilling to meet, such that a demographic dividend is unlikely to be realised. Indeed, according to 2020 figures, 23.3 per cent of youth globally were not in education, employment, or training (Dasgupta, 2022) with evidence to suggest that high levels of youth unemployment have become pervasive across many societies. This has consequences that go far beyond an immediate lack of income to longer term health, education, and employment prospects (Fergusson and Yeates, 2021: 1), as well as a potential diminishment of the social and political rights and statuses that go alongside them. There are many cases in which young people do not have access to the resources and opportunities necessary to fully participate in society.

The term 'waithood' has been used to describe young people's experience of being trapped, suspended, or held back, struggling to achieve the milestones often associated with adulthood, such as securing a job, forming a family, or becoming financially independent (Singerman, 2020; Honwana, 2020; Inhorn and Smith-Hefner, 2020). This is exacerbated by political systems where young people are underrepresented or excluded from decision-making processes and political institutions and that prioritise the interests of older generations, at the expense of younger ones. This puts the well-being and long-term opportunities available to young people at risk. This is evident in the context of the climate crisis, where younger generations are marginalised in formal decision-making processes, despite suffering the consequences of environmental degradation brought about by previous generations.

The frustration associated with waithood, when coupled with political marginalisation, can contribute to wider social and political unrest. For example, in the Arab Spring protests, which swept across several countries in the Middle East and North Africa in 2011, young people mobilised to demand greater political and economic opportunities, political change, and social justice more broadly (Honwana, 2012; Singerman, 2020). In a similar vein, the rise of violent extremist groups, such as Boko Haram and Al-Shabaab, has been linked to the failure of governments to address the needs and expectations of young people. It has been argued that these groups have exploited the frustrations and grievances of young people who feel marginalised and excluded from mainstream society, offering them a sense of belonging and purpose (see for example Chigudu, 2020; Ingiriis, 2020). In other cases, the frustration and anger of young people can manifest in anti-social or personally harmful behaviours such as substance abuse, delinquency, and crime (French et al., 2014).

Whilst the frustrations and struggles of youth might be at play in social upheaval and insecurity, we must be cautious of stereotyping young people as inherently threatening. Indeed, reproducing such tropes, serves to reinforce structures that further marginalise young people, not least the tendency to view youth as a problem, or group in need of guidance, rather than recognising their agency and the positive contributions they make to society.

Youth participation and engagement

Within the context of global development, there is a growing recognition that youth are key actors in shaping the course of social, economic, and political change (Ansell, 2017). Youth are often at the forefront of social movements and activism, challenging existing power structures and advocating for change (French et al., 2014; Honwana, 2020). It is both logical and just that as inheritors of the future, young people can and should play a central role in addressing the complex and pressing global challenges, such as climate change, poverty, and inequality, that are at the centre of global development. Policy and practice has sought to engage young people in a number of ways, which might be broadly defined in terms of three approaches to youth engagement.

Passive approaches

Often described as 'youth friendly' these approaches aim to provide dedicated resources or services to young people, taking into account the opportunities and obstacles they encounter when attempting to access them. One example is the establishment of health clinics designed specifically for young people, ensuring confidentiality and offering services at a reduced cost or free of charge, in situations where young people might ordinarily require parental consent or independent means to access them.

A key feature of passive approaches is that young people are the targets or objects of intervention but often exert little influence over which needs are addressed, which services or resources are provided, and how. Youth are viewed as passive recipients of development interventions rather than active agents of change. This approach can be critiqued for ignoring the voices and needs of young people themselves, failing to address the structural relations that marginalise and disadvantage them, whilst perpetuating unequal power dynamics between youth and other stakeholders in development.

Consultative approaches

Consultative approaches, sometimes referred to as 'youth-focused' or 'youth-centred' approaches, tend to emphasise consultation with young people with a view to designing and delivering interventions on their behalf. Such an approach might ask young people what they perceive to be their needs and then provide resources or services to meet them. These approaches might be critiqued for treating young people as informants, rather than partners in development activities. They seek the input and feedback of young people but do not involve them in decision-making processes or programme or policy design, leaving the overarching objectives to be determined by the funding or implementing organisations.

Transformative approaches

Transformative approaches aim to support the empowerment of young people as agents of change. Rather than merely focusing solely on the immediate needs or interests of young people, these approaches aim to tackle the structural conditions that marginalise and disempower them. At the same time, they recognise the capacity of young people to contribute to change, such that their voices, perspectives, and experiences are valued and heard. In these approaches, young people are not only involved but lead at every stage of an initiative from its inception, through design, implementation, monitoring, and evaluation of outcomes. The emphasis here is on actively supporting or enabling young people as leaders and decision-makers, with others providing guidance, support, and resources as and when young people themselves decide they need them.

The central theme in the above typology is the varying degrees to which young people are involved in different stages of policy or intervention, including design, delivery, evaluation, and governance. This aligns with the concept of 'meaningful' youth engagement, which emphasises not only the right of young people to be heard and listened to but their active participation in determining which issues to address and how.

Transformative approaches are underpinned by the idea that empowering young people holds intrinsic value. However, a gradual shift from passive to transformative approaches has perhaps foremost been driven by the instrumental value of youth empowerment to the achievement of a range of development objectives from economic growth and poverty reduction (Roberts, 2007) to improving health outcomes (Mazur et al., 2018) and tackling climate change (Spajic et al., 2019).

It might be argued that transformative approaches are more effective and sustainable than passive or consultative approaches, as they empower young people to take ownership of development initiatives and ensure that their perspectives are taken into account. However, such approaches can also face challenges, not least a lack of institutional support for youth-led initiatives and the resistance of other stakeholders to youth participation in decision-making (Delgado and Staples, 2007).

Transformative approaches are also subject to questions of sustainability and scalability. Young people sometimes lack the resources, experience, and influence necessary to effectively navigate complex political and economic systems, achieve long-term impact, and scale up their interventions to reach broader populations. Here, young people benefit from the support of allies who can work to foster an enabling environment – engaging the wider social relationships, structures, institutions, norms, values, and beliefs that impinge upon the opportunities for young people to empower themselves.

Conclusion

A large youth population presents both opportunities and challenges, not least for young people themselves. However, the experiences, perspectives, and agency of youth and the positive contributions they make often go unrecognised. They are often marginalised in the decision-making processes and structures that shape their present and the future they stand to inherit. Whilst there remains work to do, approaches to youth engagement in development policy and practice increasingly recognise both the instrumental and intrinsic value of supporting young people to empower themselves.

References

Ansell, N. 2017. *Children, Youth, and Development*. Abingdon: Routledge.

Bloom, D., Canning, D. and Sevilla, J. 2003. *The Demographic Dividend: A New Perspective on the Economic Consequences of Population Change*. Santa monica, California: Rand Corporation.

Chigudu, D. 2020. Dealing with Violent Extremist Organizations in Africa: The Case of Boko Haram in Nigeria. *African Renaissance*, 17(2), pp. 37–88.

Dasgupta, S. 2022. *Global Employment Trends for Youth 2022 Investing in Transforming Futures for Young People (No. 995171991002676)*. Geneva: International Labour Office, Economic and Labour Market Analysis Department.

Delgado, M. and Staples, L. 2007. *Youth-Led Community Organizing: Theory and Action*. Oxford: Oxford University Press.

Fergusson, R. and Yeates, N. 2021. *Global Youth Unemployment: History, Governance and Policy*. Northampton: Edward Elgar Publishing.

French, M., Bhattacharya, S. and Olenik, C. 2014. *Youth Engagement in Development: Effective Approaches and Action-Oriented Recommendations for the Field*. Washington, DC: United States Agency for International Development.

Honwana, A.M. 2012. *The Time of Youth: Work, Social Change, and Politics in Africa*. Sterling: Kumarian Press.

Honwana, A.M. 2020. Youth Struggles: From the Arab Spring to Black Lives Matter & Beyond. *African Studies Review*, 62(1), pp. 8–21.

Honwana, A.M. and de Boeck, F. 2005. *Makers & Breakers: Children & Youth in Postcolonial Africa*. Oxford: James Currey.

Ingiriis, M.H. 2020. The Anthropology of Al-Shabaab: The Salient Factors for the Insurgency Movement's Recruitment Project. *Small Wars & Insurgencies*, 31(2), pp. 359–380.

Inhorn, M.C. and Smith-Hefner, N.J., eds. 2020. *Waithood: Gender, Education, and Global Delays in Marriage and Childbearing* (Vol. 47). New York: Berghahn Books.

Johnson-Hanks, J. 2002. On the Limits of Life Stages in Ethnography: Toward a Theory of Vital Conjunctures. *American Anthropologist*, 104(3), pp. 865–880.

Masquelier, A. 2005. The Scorpion's Sting: Youth, Marriage and the Struggle for Social Maturity in Niger. *Journal of the Royal Anthropological Institute*, 11(1), pp. 59–83.

Mazur, A., Brindis, C.D. and Decker, M.J. 2018. Assessing Youth-Friendly Sexual and Reproductive Health Services: A Systematic Review. *BMC Health Services Research*, 18(1), pp. 1–12.

Newell, S. 2012. *The Modernity Bluff: Crime, Consumption, and Citizenship in Cote d'Ivoire*. Chicago: University of Chicago Press.

Roberts, B. 2007. Youth and Poverty Reduction Strategy Processes in Sub-Saharan Africa: Poverty, Development and Livelihoods. *Africa Insight*, 37(3), pp. 432–453.

Singerman, D. 2020. Youth, Economics, and the Politics of Waithood. *Waithood: Gender, Education, and Global Delays in Marriage and Childbearing*, 47, p. 31.

Spajic, L. et al. 2019. Beyond Tokenism: Meaningful Youth Engagement in Planetary Health. *The Lancet. Planetary Health*, 3(9), pp. e373–e375. https://doi.org/10.1016/S2542-5196(19)30172-X.

UN DESA. 2022a. *A World of 8 Billion*. Policy Brief 140. United Nations Department of Economic and Social Affairs. https://www.un.org/development/desa/dpad/wp-content/uploads/sites/45/publication/PB_140.pdf.

UN DESA. 2022b. *World Population Prospects 2022 Summary of Results*. UN DESA/POP/2022/TR/NO. 3. New York: United Nations Department of Economic and Social Affairs.

Valentine, G. 2003. Boundary Crossings: Transitions from Childhood to Adulthood, Children's Geographies, 1(1), pp. 37–52. https://doi.org/10.1080/14733280302186.

Weiss, B. 2009. *Street Dreams and Hip Hop Barbershops: Global Fantasy in Urban Tanzania*. Bloomington: Indiana University Press.

Further reading

Cooper, V. and Holford, N. 2021. *Exploring Childhood and Youth*. Edited by V. Cooper and N. Holford. Milton Keynes: Open University.

Fletcher, A. 2014. *The Practice of Youth Engagement*. Common Action Publishing.

Hart, R. 1992. *Children's Participation: From Tokenism to Citizenship*. Innocenti Essays No. 4. New York: UNICEF. http://www.unicef-irc.org/publications/pdf/; childrens_participation.pdf.

UN DESA World Youth Reports. https://social.desa.un.org/issues/youth/united-nations-world-youth-report-wyr.

Chapter 39

Ageing and poverty

Vandana Desai

Development initiatives and population policies adopted in the 1960s and 1970s have helped to reduce fertility and mortality rates, with rising life expectancy due to improvements in health in many developing countries. Consequently, many countries are experiencing demographic transition from a young to an increasingly older population (Hussain et al., 2006; Dyson, 2010). It is important to understand how social and economic changes have affected or will affect the well-being and support situation of present or future older people and how older people's needs and position in society are related to development issues and the consequences for policy. This has implications for developing countries that need to recognise ageing and development as part of poverty reduction strategies (see Desai and Tye, 2009; Aboderin and Ferreira, 2008).

Population ageing mainly means the progressive increase in the numbers and the proportions, of older people, compared to working-age adults and children in a total population. The UN definition of 'old age' is 60+ years. This definition is quite useful for comparing internationally but poses questions on its appropriateness when applied in different societies and culture. One of the major differences in the context of Asian developing countries is the pace of population ageing and the sheer numbers of older people in the developing world. This makes ageing a development issue. Globally, there were 727 million persons aged 65 years or over in 2020 and the share of the population aged 65 years or over is expected to increase from 9.3 per cent in 2020 to around 16.0 per cent in 2050 (UNDESA, 2020: 1). In 2050, it is expected that nearly 8 in 10 of the world's older persons will be living in the developing regions; 80 per cent of older people will live in low- and middle-income countries by 2050, see Table 39.1.

The reality of the global demographic situation is that some developing countries will face 'unprecedented ageing' without the social and economic infrastructure or public institutions to support those frail and dependent elderly people who are unable to support themselves economically. Developing countries will contain the most 'frail dependent elders' in more vulnerable situations. This end of the spectrum contains the pool of those living without the support of a 'healthy retirement' but who survive on the bare minimum social insurance and patchy health services. The time has come for proactive planning of appropriate institutional frameworks and policies. The challenge is to create those 'safety nets' that will protect the present generation over the coming decades but also to provide guidance, support, and care for the ageing populations. Questions related to ageing populations have secured very little attention in Global South countries. The elderly face economic and resource deprivation but also face identity-based

DOI: 10.4324/9780429282348-43

Table 39.1 Number of persons aged 65 years or over by geographic region, 2019 and 2050

Region	Number of persons aged 65 or over in 2019 (millions)	Number of persons aged 65 or over in 2050 (millions)	Percentage change between 2019 and 2050
World	**702.9**	**1548.9**	**120**
Sub-Saharan Africa	31.9	101.4	218
Northern Africa and Western Asia	29.4	95.8	226
Central and Southern Asia	119.0	328.1	176
Eastern and South-Eastern Asia	260.6	572.5	120
Latin America and the Caribbean	56.4	144.6	156
Australia and New Zealand	4.8	8.8	84
Oceania, excluding Australia and New Zealand	0.5	1.5	190
Europe and Northern America	200.4	296.2	48

Source: United Nations, Department of Economic and Social Affairs, Population Division (2019). *World Population Prospects 2019.* *Excluding Australia and New Zealand

discrimination. These multiple sources of deprivation create deep-rooted social exclusion which cannot be addressed solely through conventional income poverty reduction programmes.

The emerging global trade pattern, in which the capital and/or technology intensive goods and services constitute the fastest rising segment of exports (e.g., electronics, garments, and IT), tends to reduce employment per unit of exports. Thousands of jobs have been created for younger skilled workers in big metropolitan cities like Mumbai, especially in call centres, the IT sector, tourism, and the garments sector. For example, economic liberalisation in 1991 in India created large scale retrenchments from no-longer competitive industries (e.g., textiles and printing industries) that left older urban poor workers unemployed in the age group of 50s onwards, with no prospect of re-employment in the new emerging sectors due to skills mismatch and no digital literacy among ageing population. They are also likely to experience age discrimination (Vera-Sanso, 2004). There is an increasingly urgent concern to create conditions for inclusive citizenship and participation of those ageing people in wider society. Honouring older people's dignity and rights has received very little general attention and is a matter of social justice, human rights, and an extension of rights to social protection (Lloyd-Sherlock, 2004).

Older people are particularly vulnerable to multi-dimensional poverty, ill-health, and social exclusion, particularly in the context of diminishing employment opportunities and barely any social service and security provision in many of the Asian developing and transitional countries. Elders living in absolute poverty are a considerable challenge for many developing countries.

Intergenerational equity

As the elderly population grows, there is a concern about potential intergenerational conflicts over resources. This includes debates about how to allocate resources fairly between younger and older generations, especially in regions with limited resources.

The living arrangements of older people are an important determinant of their economic well-being as well as their physical and psychosocial health and life satisfaction. There is an assumption by policy makers that the elderly live with and are cared for by their children or extended family (e.g., Vietnam and India), but, with shifts in family structure caused by industrialisation, urbanisation, increased migration, the increased level of education among the young generation, and women entering the labour market and education, the context in which older persons live is being reshaped, including the size and composition of their households and their living arrangements. Traditional three-generational households (comprised of grandparents, parents, and children) are decreasing in number in many Asian countries. The rates of elderly living in their own homes separate from children are increasing. There is widespread concern over the decline of the family, its stability and 'effectiveness' to cater for the needs of the elderly.

Living arrangements explain part of the observed international differences in age patterns of COVID-19 mortality, in particular for older persons whose living arrangements affect the risk of contracting COVID-19. Understanding the interconnections between the living arrangements of older persons and their health and well-being has relevance considering the pledge made by governments in the 2030 Agenda for Sustainable Development that no one will be left behind. In practice, this pledge implies that the Sustainable Development Goals (SDGs) must be achieved for all segments of society and at all ages, with a particular focus on the most vulnerable, including older persons (UNDESA, 2020: 2).

Older people can also find themselves in a vulnerable position – in extreme cases, as victims of abuse – when the balance of power in the mother-in-law/daughter-in-law relationship shifts towards the younger woman. There is a history of conflict between women in extended households, particularly where brides have traditionally moved in with their husbands' parents (very common in Asian countries), which can generate conflict and impact on the welfare of elderly women (i.e. the mother-in-law). Elderly women in such situations are marginalised and can consequently live on their own. In addition, 'the growing number of young married couples who live on their own means that care by the daughter-in-law is no longer automatic' (Wilson, 2000: 120). Older women in Thailand are five times more likely than older men to have been abused by family members (Chompund et al., 2010: 283). Only 17 per cent of sexual violence studies worldwide took women aged over 49 into account, in 2013. Most surveys on violence against women are carried out on women aged 15–49 years and therefore exclude older women (WHO, 2018).

Most of the cost of the support for the elderly falls on the family, which also largely bears the costs of educating children (sandwich generation). This burden on young families may prove unsustainable in the context of high unemployment or under-employment, if job creation fails to keep up with labour force increases, or where incomes are low and uncertain. A trend towards nuclear households, with the elderly living separately from their offspring, may also undermine the family support system, as may out-migration of younger workers from the rural areas, leaving behind the elderly and infirm (as is increasingly occurring in China and India).

The economic contribution of older people in some form of employment either for their own economic security or for the collective economy of the household is not well understood. Old-age dependency ratios are based on questionable assumption that older people do not work, ignoring the 1 in 5 over 65s who remain in employment worldwide, higher in low-income countries (see Lloyd-Sherlock, 2010). There is no doubt, though, that there are high rates of economic participation and diverse strategies being adopted by older men and women until late in their lives mainly because of the limited degree of formal social security coverage in regions like South Asia. The majority of older workers – particularly those in the private and informal sector – continue to work until physical disability or sickness prevents them from further participation in the labour force. Greater still is the economic contribution made to the informal economy, and there is rising evidence of 'income strategies' adopted by older urban people primarily to provide vital income (Desai, 2020).

Older people have naturally assumed roles such as tending to childcare, cooking and other household tasks, and many part-time jobs in the informal sector. Younger adults are 'released' for employment purposes and the cumulative effect is beneficial to the whole household unit. If a longer time is spent in productive employment, ageing challenges can be offset without shifting unacceptable burdens onto the young. Increasing the level of economic activity in the economy provides a proactive policy solution and increases the per capita income of the population. By reducing economic dependency, the state can expand social security protection too.

Old people can find themselves sharing their accommodation, pensions, or limited incomes from informal employment with the younger generation for longer than they may have anticipated or desired. The vicious circles of poverty which these households experience from generation to generation are very prominent in some parts of South Asia and many countries in Africa.

Widowhood

Gender plays a significant role in how aging is experienced in the Global South. There are debates around issues such as caregiving responsibilities and the impact of gender-based discrimination on older women. Since women live longer than men, on average, they comprise the majority of older persons, especially at advanced ages. This, together with a tendency for women to marry partners older than themselves, means that demographic ageing may be increasing the percentage of woman-headed households. The highest proportion of women-headed households occurs where the householder is aged 60 or over, pointing to the significance of widowhood in this context (Varley and Blasco, 2000). In some Global South regions, child marriage is prevalent and young girls become widows at an early age. There are deep-rooted cultural practices (social stigma) and traditions related to widowhood which have an impact on widows' rights (e.g., widow inheritance, remarriage, and autonomy). Women who are unmarried (widowed, divorced, or separated) are at greater risk of economic difficulty and living in isolation and exclusion. Throughout the world older women are more likely than older men to live alone. The demographic and fertility transitions take on particular significance for women in the Asian region which marks gender imbalances arising from deeply entrenched forms of discrimination (see Desai, 2020) such as those arising from son-preferences in parts of East and South Asia. Intersectional identities of widows (gender, age, socio-economic status, disability, etc.) impact the challenges widows face. Accurate data on the prevalence and experiences of widows in the Global South is often limited and there is need for more comprehensive research to inform policy and programming.

Conclusion

The experience of ageing and the impact of ageing population on development are complex, difficult to predict, and highly dependent on context. There is almost total absence of discussion on ageing and poverty in the global development agenda and a comprehensive and multidimensional approach is needed. The knowledge base on well-being, poverty, and vulnerability among older poor and their households is very limited. Traditional cultural values and norms play a significant role in how older adults are perceived and treated. How to balance these cultural perspectives with the need for modern approaches to elderly care and support is an ongoing dilemma for many families in the Global South. The issue of elder abuse, including physically, emotionally, and financially is a growing concern in the Global South. It is important in addressing the root causes of abuse to challenge discriminatory practices and raise awareness along with developing protective measures.

Though many older people are working beyond the age of 60, what should be a realistic retirement age has become the subject of debate in both developed as well as in developing countries. Is there a 'surplus' share of the population at 'productive' ages? What are older people's important 'productive' contributions to families and communities and what are the potential economic 'costs' of failing to cater for older people have become important questions that need further exploration. There is a need to reform social security arrangements, particularly pensions (Willmore, 2007) so that they are fairer and sustainable for longer life expectancy and so that there is provision of basic minimum living standards in old age, especially as very limited state welfare systems (social security and healthcare) are in place for the elderly.

There is a need to adopt a critical perspective in understanding ageing and the preparedness of developing countries to address future challenges of ageing populations, especially those living in poverty in later stages of their lives. How can opportunities be created for economically active ageing populations so that they are not dependent on either the state or their families in the later part of life cycles? Some Asian countries have less time to set in place the institutions needed to cope with population ageing and are disadvantaged by fiscal constraints, poor governance, and weak public agencies. It is important to create a society for all ages in which the changing demographic, economic, and social changes will mean an evolution of policies catering to the elderly in a society that promotes support between generations and at the same time sustains economic growth and development.

References and guide for further reading

Aboderin, I. (2005) 'Changing Family Relationships in Developing Nations', in M.L. Johnson and V.L. Coleman (eds.) *The Cambridge Handbook of Age and Ageing*. Cambridge: Cambridge University Press, pp. 30–46.

Aboderin, I. (2006) *Intergenerational Support and Old Age in Africa*. New Jersey: Transaction Publishers.

Aboderin, I. and Ferreira, M. (2008) 'Linking Ageing to Development Agendas in Sub-Saharan Africa: Challenges and Approaches', *Journal of Population Ageing*, 1(1): 51–73.

Barrientos, A. and Lloyd-Sherlock, P. (2002) 'Older and Poorer? Ageing and Poverty in the South', *Journal of International Development*, 14: 1129–1131.

Chompund, M., Charoenyooth, C. and Palmer, M.H. (2010) 'Prevalence, Associated Factors and Predictors of Elder Abuse in Thailand', *Pacific Rim International Journal of Nursing Research*, 14(1): 283–296.

Croll, E. (2008) 'The Intergenerational Contract in the Changing Asian Family', in R. Goodman and S. Harper (eds.) *Ageing in Asia*. Oxford: Routledge.

Desai, V. (2020) 'Urban Widows: Living and Negotiating Gendered Dispossession in Speculative Slum Housing Markets in Mumbai', *Gender, Place and Culture*, 1–20.

Desai, V. and Tye, M. (2009) 'Critically Understanding Asian Perspectives on Ageing', *Third World Quarterly*, 30(5): 1007–1025.

Dyson, T. (2010) *Population and Development: The Demographic Transition*. London, UK: Zed Books.

Gender and Development (2009, November) 'Special Issue on Ageing, Gender and Development', *An Oxfam Journal*, 17(3).

Glaser, K., Agree, E.M., Costenbader, E., Camargo, A., Trench, B., Natividad, J. and Chuang, Y.L. (2006) 'Fertility Decline, Family Structure, and Support for Older Persons in Latin America and Asia', *Journal of Aging and Health*, 18(2): 259–291.

Hussain, A., Cassen, R. and Dyson, T. (2006) 'Demographic Transition in Asia and Its Consequences', *IDS Bulletin*, 37(3): 79–87.

Lloyd-Sherlock, P. (2000) 'Population Ageing in Developed and Developing Regions: Implications for Health Policy', *Social Science and Medicine*, 51: 887–895.

Lloyd-Sherlock, P. (ed.) (2004) *Living Longer: Ageing, Development and Social Protection*. London: Zed Books.

Lloyd-Sherlock, P. (2010) *Population Ageing and International Development: From Generalization to Evidence*. Bristol: Policy Press.

United Nations, Department of Economic and Social Affairs, Population Division (2019) *World Population Ageing 2019: Highlights (ST/ESA/SER.A/430)*. New York: UNDESA.

United Nations, Department of Economic and Social Affairs, Population Division (2020) *World Population Ageing 2020 Highlights: Living Arrangements of Older Persons (ST/ESA/SER.A/451)*. New York: UNDESA.

Varley, A. and Blasco, M. (2000) 'Intact or in Tatters? Family Care of Older Women and Men in Urban Mexico', *Gender and Development*, 8(2): 47–55.

Vera-Sanso, P. (2004) 'Modelling Intergenerational Relations in South India', *Generations Review*, 14(1): 21–23.

WHO (2018) *Global and Regional Estimates of Violence Against Women: Prevalence and Health Effects of Intimate Partner Violence and Non-Partner Sexual Violence*. Geneva: World Health Organisation.

Willmore, L. (2007) 'Universal Pensions for Developing Countries', *World Development*, 35(1): 24–51.

Wilson, G. (2000) *Understanding Old Age: Critical and Global Perspectives*. London: Sage.

Useful websites

Age International: https://www.ageinternational.org.uk/

Aging in Developing Countries Network: http://adcnet.psc.isr.umich.edu/

HelpAge International: http://www.helpage.org/Home

UN Programme on Ageing: http://www.un.org/ageing/

Violence Against Women Prevalence Estimates, 2018: https://www.who.int/publications/i/item/9789240022256

WHO Ageing: http://www.who.int/ageing/en/

Chapter 40

Disability

Ruth Evans and Yaw Adjei-Amoako

Disability, poverty, and development

The complex links among disability, poverty, and development have been increasingly acknowledged within development discourses and research[1]. Poverty exacerbates and/or intensifies impairment or vulnerability to its effects, while the barriers faced by disabled people result in a greater chance of experiencing intense poverty (Grech, 2016). Although reliable statistical data on the incidence of disability/impairment is not widely available and there is no agreed consensus on how disability should be defined and measured, it is estimated that over a billion people worldwide (15 per cent of the global population) live with a disability (WHO and World Bank, 2011). The majority of the global population of disabled people (80 per cent) live in low- and middle-income countries, which often have limited resources to meet their needs. Furthermore, disabled people are disproportionately represented among the numbers of people living in chronic poverty. The high rates of impairment and preventable illness in the Global South are largely caused by malnutrition, poverty, lack of access to sanitation, safe drinking water, healthcare and other services, hazardous work, landmines, armed conflict, structural violence and inequalities.

Being poor also increases the likelihood of an individual experiencing ill health and becoming disabled (Yeo and Moore, 2003). Childhood impairment is often caused by preventable injuries and illnesses in homes and neighbourhoods that are related to poor living conditions, inadequate access to healthcare and sanitation and accidents among working children (McEwan and Butler, 2007). Many disabled children are denied access to education, due largely to the fact that education systems in the Global South lack the resources and skills to include disabled students within mainstream educational settings and adequately meet the needs of disabled students. This in turn leads to high levels of illiteracy, reduced skills and employment opportunities for disabled people in adulthood, perpetuating the cycle of poverty from one generation to the next (Yeo and Moore, 2003). Disabled women and girls often experience multiple disadvantages, on the basis of their gender and disability, and they are particularly vulnerable to abuse, chronic poverty and exclusion.

It is also increasingly acknowledged that understandings of 'normal' bodies, disability and illness vary according to the economic, geopolitical, socio-cultural and spatial context. In many Sub-Saharan African countries, for example, disability in children is associated with maternal wrongdoing and witchcraft and in contexts of poverty, negative cultural attitudes and a lack of support, families may 'hide' or abandon disabled children who are considered 'abnormal' (Kabzems and Chimedza, 2002). However, impairment

DOI: 10.4324/9780429282348-44

does not always lead to exclusion and many individuals are supported and included within their families and communities.

Improving health systems, infrastructure and the prevention and treatment of diseases are critically important in preventing and reducing the prevalence of disability in the Global South. Fulfilling disabled people's rights and enhancing their well-being, however, also requires efforts to tackle poverty and the socio-cultural, political, economic and climate-related inequalities, vulnerabilities and structural violence that people experience.

From the 'medical' and 'social' models to the 'bio-psycho-social' model of disability

Development approaches based on multidimensional understandings of poverty that aim to enhance human capacities and well-being have been influenced by disability politics and social theories of the body, health and disability that developed from the 1970s onwards. Disability activists in the Global North rejected medical, rehabilitative models of disability which were based on assumptions that disabled people suffer primarily from physical and/or mental abnormalities that medicine can and should treat, cure or at least prevent. Within the dominant 'medical model' approach, disability is perceived as an 'individual misfortune' or 'tragedy'. The disability movement in the Global North instead developed a 'social model' of disability to focus attention on the socio-cultural, economic, political and spatial barriers to participation that disabled people experience. An individual's 'impairment' was seen as separate from the social, attitudinal and environmental dimensions of 'disability' that exclude disabled people. The 'social model' thus focuses on changing society to facilitate the participation and inclusion of disabled people, rather than on efforts to 'rehabilitate' individuals and overcome biological constraints of the body.

These understandings of disability have been crucial to improving accessibility, achieving equality of opportunity and securing disabled people's rights within the public sphere in the Global North. Commentators, however, have questioned the appropriateness of applying Western-centric social models of disability in the Global South (McEwan and Butler, 2007). The wider macro-economic context, resource constraints and limited availability of technical solutions to make environments more accessible constrain the implementation of social model approaches to disability in many low-income countries. Chronic poverty, limited income earning options and restricted access to health and education that many disabled people experience mean that access to basic services is likely to represent a higher priority for disabled people, governments and policymakers, rather than issues of accessibility or assistive technology.

Debates about the need to reconcile both medical and social models of disability in the 1980s led to the establishment of 'community-based rehabilitation' approaches in the Global South and the eventual emergence of the 'bio-psycho-social' model of disability that underpins the United Nations Convention on the Rights of Persons with Disabilities. Community-based rehabilitation approaches aimed to provide rehabilitation through medical intervention and care, as well as promoting the social inclusion and participation of disabled people within their communities. Community-based rehabilitation projects have been criticised, however, for being ill-conceived and lacking sensitivity to local cultures and practices (McEwan and Butler, 2007). Community-based rehabilitation can be seen as reinforcing medical/charitable models of disability that were introduced in the colonial era, perpetuating ideas that disabled people are dependent and need to be supported by charitable fundraising and donations (McEwan and Butler, 2007). Community-based rehabilitation projects have sought to shift towards a more community development approach in recent years and aim to empower disabled people and facilitate their participation in the development process.

The UN Convention on the Rights of Persons with Disabilities (UNCRPD) is broadly informed by the social model of disability but also developed from community-based rehabilitation approaches that paid greater attention to the impact of impairment. This led to what has been termed the 'bio-psycho-social' model of disability. Disability and impairment are not explicitly defined in the UNCRPD, but 'persons with disabilities' include: 'those who have long-term physical, mental, intellectual or sensory impairments which in interaction with various barriers may hinder their full and effective participation in society on an equal basis with others'. Key tenets of the convention are disabled people's rights to participation and inclusion, non-discrimination and accessibility.

Disability politics and research from the Global South are increasingly challenging Northern framings of the disability debate (Grech, 2016). Social model approaches to disability have been criticised for failing to acknowledge the materiality of the body such as the effects of pain and impairment on people's everyday lives and the impacts of structural violence, such as impairments resulting from processes of imperialism and colonisation. Meekosha and Soldatic (2011) argue that a 'politics of impairment' is critical for understanding disability in the Global South, pointing to the example of the Vietnamese Agent-Orange Movement's claims for redistributive justice against the US military. Such political mobilisations to achieve compensation for impairments caused by crimes committed as part of the colonisers' project draw heavily on medical science to make claims for a global resource transfer from the North to the South. Similarly, people living with HIV in Africa and others with chronic illness may identify with others on the basis of their biomedical diagnosis rather than according to a strategic notion of 'disability' and the focus of their activism may be on access to healthcare and medical treatment, which differs from the focus of the disability movement in the Global North (Evans and Atim, 2011).

The separation of 'impairment' from 'disability' that underpins the social model can result in impairment being constructed as 'natural' (as opposed to disability which is viewed as 'social'). However, as Meekosha and Soldatic (2011: 1393) argue, 'impairment is not in fact always natural but the outcome of deeply politicised processes of social dynamics *in* bodies that then become medicalised and then normalised through a raft of moral discursive and real practices'. They call for a 'politics of diversity within unity' as a central strategy of global mobilisation on disability and impairment.

'Mainstreaming' disability in development and contemporary challenges

Disabled people's organisations, led by disabled people in the Global South, have played an important role in collective advocacy for the representation of disabled people in all stages of the development process at the national and international levels. The National Union of Disabled People of Uganda (NUDIPU) lobbied for the inclusion of disabled people at all levels of political administration and disabled people have achieved a higher level of political representation in Uganda than in any other country (McEwan and Butler, 2007). International non-governmental organisations, coalitions and networks, such as Disabled People's International (established in 1981) have helped to strengthen national disabled people's organisations and facilitate collective advocacy for disabled people's rights at the global level.

The adoption and rapid ratification of the UNCRPD by many countries from 2008 onwards has resulted in a high level of state and civil society mobilisation around disability. Disability issues appear to be increasingly mainstreamed within the 'rights-based development' agenda, although the legal rhetoric is often very distant from the lived reality experienced by many disabled people in the Global South (Meekosha and Soldatic, 2011; Kingston, 2017). While the emphasis is on mainstreaming disability into all development activities, such as Poverty Reduction Strategy Papers and Sustainable Development Goals (SDGs), it is recognised by the UNCRPD that disability specific measures may be necessary to 'accelerate or achieve de facto equality of persons with disabilities'.

The Millennium Development Goals were critiqued for not recognising the specific inequalities faced by disabled people. Recent years have seen increasing efforts focused on promoting the inclusion of disability rights within the SDGs framework and their associated targets. While disability is not one of the 17 overarching goals, disability and 'persons with disabilities' are mentioned, for the first time, in relation to targets for five of the goals. It is too early to assess whether the SDGs will have a positive impact (if at all) on disabled people.

While the 'mainstreaming' of disability within rights-based approaches to development has been broadly welcomed by advocates and activists, tensions remain. Concerns focus on the dangers of tokenistic involvement of disabled people and the neglect of their self-determination and equality, in addition to the lack of attention to global structural inequalities and the role of imperialism and colonialism as root causes of violations of human rights, famines, malnutrition, ecological degradation and growing impairment in the Global South (Meekosha and Soldatic, 2011).

The risks and impacts of human-induced climate change and severe natural disasters are increasingly rising up both national and international policy agendas. While the proportion of disabled people among those affected by climate change-related disaster is unknown, it is widely acknowledged that disabled people are particularly vulnerable to its impacts. Disabled people have the least capacity to adapt to changes in their environment and are virtually ignored in discussions and planning about preparation for the risks and impacts of climate change (Wolbring and Leopatra, 2012).

The effects of climate change on disabled people are exacerbated in low- and middle-income countries where high levels of poverty, limited resources and lack of adequate welfare support often combine to reinforce disabled people's vulnerability. Disabled people are more likely to experience natural disasters as human disasters (Priestley and Hemmingway, 2007), due to their disproportionate representation among the poor and their own underlying vulnerabilities, arising from factors such as stigma, inaccessible infrastructure and lack of accurate data on their whereabouts and needs. These issues can be magnified when mass displacement occurs due to rising sea levels, flooding, water shortages and climate-related conflict. The rights and needs of disabled people who are forced to migrate may be ignored, despite their specific requirements for accessible transportation, healthcare, housing and inclusive education and livelihood opportunities. Others may be unable to move and simply left behind when faced with climate shocks and disasters, remaining in precarious circumstances, often with reduced family and community support and increased vulnerability to further economic and climate-related shocks.

The perspectives, needs and rights of disabled people and other vulnerable groups must be included within efforts to prevent and mitigate climate-related risks and impacts. Indeed, we conclude that it is crucial to ensure the participation of disabled people in all aspects of development policy, planning and research on an equal basis with others in efforts to achieve global justice and sustainable development in the 21st century.

Note

1   This chapter is a revised and updated version of a more extensive discussion of health, disability and development in Potter et al.'s 2012 book, *Key Concepts in Development Geography*, London: Sage.

References

Evans, R. and Atim, A. (2011) Care, Disability and HIV in Africa: Diverging or Interconnected Concepts and Practices? *Third World Quarterly*, 32(8): 1437–1454.

Grech, S. (2016) Disability and Poverty: Complex Interactions and Critical Reframings. In S. Grech and K. Soldatic (Eds.), *Disability in the Global South: The Critical Handbook*, pp. 217–235. New York: Springer.

Kabzems, V. and Chimedza, R. (2002) Development Assistance: Disability and Education in Southern Africa. *Disability and Society*, 17(2): 147–157.

Kingston, D. (2017) Can the Implementation of the Sustainable Development Goals (SDGs) be Achieved without Addressing Disability Rights? *Disability in the Global South*, 4(1): 1180–1188.

McEwan, C. and Butler, R. (2007) Disability and Development: Different Models, Different Places. *Geography Compass*, 1(3): 448–466.

Meekosha, H. and Soldatic, K. (2011) Human Rights and the Global South: The Case of Disability. *Third World Quarterly*, 32(8): 1383–1397.

Priestley, M. and Hemmingway, L. (2007) Disability and Disaster Recovery: A Tale of Two Cities? *Journal of Social Work in Disability and Rehabilitation*, 5(3–4): 23–42.

Wolbring, G. and Leopatra, V. (2012) Climate Change, Water, Sanitation and Energy Insecurity: Invisibility of People with Disabilities. *Canadian Journal of Disability Studies*, 1(3): 66–90.

World Health Organisation (WHO) and World Bank (2011) *World Report on Disability*. Geneva: WHO/The World Bank.

Yeo, R. and Moore, K. (2003) Including Disabled People in Poverty Reduction Work: "Nothing About Us, Without Us". *World Development*, 31(3): 571–590.

Further reading

Eide, A. and Ingstad, B. Eds. (2011) *Disability and Poverty: A Global Challenge*. Bristol: The Policy Press. This wide-ranging collection explores the social, cultural and political dimensions of disability and poverty in different contexts in the Global South.

Grech, S. and Soldatic, K. (2016) *Disability in the Global South: The Critical Handbook*. New York: Springer. This edited collection critically examines social understandings of disability in global health, disability studies and international development and explores disabled people's lives and intersecting social inequalities across cultures, space and time.

Groce, N., Kett, M., Lang, R. and Trani, J. (2011) Disability and Poverty: The Need for a More Nuanced Understanding of Implications for Development Policy and Practice. *Third World Quarterly*, 32(8): 1493–1513. This article provides a useful review of existing knowledge and theory regarding the disability–poverty nexus.

The journal, *Disability in the Global South*, focuses on all aspects of disability and disabled people's lives in the Global South.

Third World Quarterly (2011) 'Disability in the global South', special issue, 32(8). This collection of articles discusses conceptualisations of Southern bodies, disability, poverty and human rights.

Useful websites

www.un.org/disabilities
https://sustainabledevelopment.un.org/

Chapter 41

Sexualities and development

Andrea Cornwall and Vanja Hamzić

What would the world be like if we really did have the right to choose our sexuality and pleasurable sexual relations?

Karin Ronge, Women for Women's Human Rights, Turkey

Sexuality has had a place in international development from the early days of colonial and missionary intervention in the countries of the Global South. Late 19th-century European sexual mores were transported to the colonies as part of imperial expansion. With it came an ordering of the world into rigid gender and sexual binaries, with the naturalisation of the categories 'women' and 'men', and the late 19th-century European categories 'heterosexual' and 'homosexual'. Development came to be cast as a *moral* as well as economic endeavour, famously described by Gayatri Spivak (1988), for India, as 'white men rescuing brown women from brown men'. Laws and social policies sought to encourage the formation of nuclear families based on the – still relatively novel in the Victorian period – European model of heterosexual monogamy. Practices considered 'barbaric', such as child marriage and female genital cutting, became a prime focus of colonialism's 'civilising' project. For the same reason, the relative sexual freedoms and non-binary concepts of selfhood enjoyed in certain colonised societies were actively supressed and remodelled to fit forcefully globalised Victorian social mores (Hamzić, 2014). Education for girls and women sought to produce domesticated 'good women' who would maintain hygienic homes and respect their husbands (Hunt et al., 1997); the contradictions of colonial extraction also fostered 'bad women', who left rural homes for lives as sex workers in the city (Jeater, 1993; White, 1990).

Colonial Christian moralities produced a normative categorisation of 'good' and 'bad' women (Hunt et al., 1997) and a notion of the perversity of certain sexual acts that came to be constitutive of certain types of persons. In many countries of the Global South, colonial governments put laws in place that regulated sexuality (Stoler, 1989), proscribing non-reproductive sexual expressions as 'carnal intercourse against the order of nature' (Baudh, 2008), outlawing and disinheriting traditional gender-non-conforming communities (Hamzić, 2014) and reshaping the contours of marital practices. Many of these laws remain in place, notoriously the laws on 'unnatural practices' that have been used to persecute gender-variant people and men who have sex with men, described by Judge Michael Kirby (2011) as 'England's least lovely criminal law export'.

Traditions of rescue and the uses of law, social and economic policy and education in the institutionalisation of normative heterosexuality – with all its contradictions – connect the colonial past and a significant part of contemporary development practice. As

DOI: 10.4324/9780429282348-45

the development industry came to take institutional form in the post-WW2 period, the overt concern of the colonial powers with sexual morality receded as the modernisation agenda brought its technocratic rationality to development practice. The control of reproduction came to take centre stage, as efforts were made to curb rapid population growth in the Global South. Sex came to be represented within international development as a cause of unwanted pregnancies, disease, harm and hazard, rather than a source of pleasure, joy, intimacy and happiness. Such efforts went hand in hand with the newly independent nation-states' statecraft, reliant as it was on producing and maintaining an idealised (rigidly heterosexual) 'man' and 'woman' (see, for example, Arnfred, 2013; Hamzić, 2017). And social and economic development policies came to be premised on a particular form of normative heterosexuality, exemplified in the hegemonic concept of the 'household' with its assumptions about male breadwinners and female dependents, which shunted all other forms of economic, sexual and domestic arrangements out of the frame.

Feminist critiques of mainstream development's 'gender blindness' often tended to reproduce rather than challenge the normative assumptions about sexuality implicit in development policy. The first wave of Women in Development projects naturalised women's roles as wives and mothers. The next wave of feminist engagement with development, Gender and Development (GAD), sought interventions in policy discourse that inscribed an asymmetrical power relation between women and men as constitutive of 'gender relations'. Premised on a very specific 'coital and conjugal' (Ogundipe-Leslie, 1994) relationship, heterosexual marriage, it came to eclipse other kinds of relations of gender and power. Powerful gender myths cast women as heroines or victims and men as shadowy, oppressive figures (Cornwall, 2000). A curious prurience hung over feminist engagement with development that relegated questions of sexuality to the margins. Amidst much talk about paying closer attention to women's lived experiences, when it came to sexuality a guarded silence often prevailed. Attempts to break that silence were met with such reactions as a concern that sexuality was 'frivolous' compared to the *real* issues – as if sexuality had nothing to do with poverty or indeed any other core concern in international development – or plain 'embarrassing' (Cornwall et al., 2009).

Sex and sexuality remained a 'health issue', and in striking contrast to mainstream social and economic development, an explosion of interest in sex and sexuality was provoked by the Aids epidemic. Epidemiological research revealed patterns of sexual networking that profoundly challenged normative beliefs about sexuality. The Aids response made visible a rich seam of sexual practices that departed from – and flouted – these norms. Categories created for epidemiological analysis and HIV prevention efforts, such as Men who have Sex with Men (MSM), came to take on a life of their own as they were appropriated as identity and used to claim rights (Boyce and Khanna, 2010). But in Pakistan, for example, the sudden rise of what local activists for sexual and gender diversity dubbed an 'HIV prevention industry' also exacerbated the existing class divisions and further marginalised certain 'traditional' sex- and gender-non-conforming communities (Hamzić, 2016).

Anthropologist James Scott explores in his book *Seeing Like a State* (1998) the ways in which people come to be made legible to the state through forms of discursive ordering – counting, registering, surveying, categorising. Foucault drew attention to how discourses produce subjects; the very categories through which we come to know ourselves are not natural or fixed but products of particular historical and cultural frames. So too the 'homosexual', a category produced in late 19th-century Europe in a very particular cultural context (Plummer, 1981). Its corollary, which came into being a few years later, 'heterosexual', corresponded with a stage in the growth of capitalism that came to depend on the very regulation of sexuality and production of nuclear families with the heterosexual couple at their heart. Like MSM – or the virtually invisible WSW (women who have sex with women) – the categories 'gay' and 'lesbian' – and indeed that of 'heterosexual' – are less ontological categories with a universal existence than the products of a particular period in Euro-American history. As Tom Boellstorff (2003) has observed in the case of *gay* and *lesbi* in Indonesia, the globalisation of these categories has come to them carrying meanings altogether more complex than their Global Northern cognates.

'Transgender' is of more recent provenance still, originating in the United States in the 1990s (Stryker, 2008). And yet, as currents in the whirlwind of transnational debates on gender and sexuality (Hamzić, 2019), these categories of personhood can sometimes be productively destabilised and localised – especially in the Global South and when intersected with local taxonomies of sexual and gender diversity. For instance, in the context of a trans-in-Asia *and* Asia-in-trans debate, the South Asian gender-variant subjectivity known as *hijra* (in Pakistan: *khwajasara*) has been described as helpful to 'provincialize, decolonize, de-Cold War, and/or decolorize the category and practice of trans' (Chiang et al., 2018: 299).

On the other hand, the circulation of a globalised narrative on sexual identity amongst metropolitan elites has, some would argue, rendered as fixed identities local sexual expressions and arrangements that were practices rather than identifications. Writing of the Arab world, Joseph Massad argues that what he calls 'the Gay International' – Western LGBT (Lesbian, Gay, Bisexual and Transgender) activist organisations engaged in constituency-building in the Global South – operate in a way 'that both produces homosexuals, as well as gays and lesbians, where they do not exist, and represses same-sex desires and practices that refuse to be assimilated into its sexual epistemology' (2002: 363). This, he contends, is '*heterosexualizing* a world that is being forced to be fixed by a Western binary' by 'inciting discourse about homosexuals where none existed before' (2002: 383). And yet, as Nivi Manchanda has argued, 'in his zealous, but not wholly unfounded, preoccupation with the imposition of categories, Massad reifies the notion of "gay" and "homosexual" himself' – disallowing 'those in the West' to think outside of them – while having 'the experience of those in the Arab and Islamic world who identify with those terms' roundly dismissed and disqualified as inauthentic (Manchanda, 2015: 137). The resulting normative position is strangely familiar: 'In holding up a singular image of the Middle Eastern man as often partaking in same-sex desire but never *love*', emphasises Manchanda, 'Massad ironically mirrors the Gay International by setting the definitional and experiential parameters of legitimate male sexual encounters' (Manchanda, 2015: 137).

As Oliver Phillips argues, in a paper on the emergence of the 'global gay' written some years before Massad's, the process of subjectivity-making (and naming) is made more complex by the ways in which these global categories come to be taken up and used:

These 'new' identities are merged into local histories and contexts, so that the end product has signifiers of local significance while simultaneously providing a strategy for either laying claim to international human rights agreements, or enabling more effective AIDS/HIV preventative work, or simply buying into an expanding market of western signifiers of 'modern' and bourgeois status, or serving all of these purposes. *What is clear is that these identities are not simply imposed through an imperialistic cultural discourse or economic dominance, but they are actively assumed and proclaimed from below, by those marginalised in these hegemonic formations.*

(2000: 2, our emphasis)

Through the taking up of identities that Phillips describes, we see activists in the Global South coalescing around such labels as MSM and using them politically to claim rights and recognition; we also see spaces being created across and within social movements, including the feminist movement, for greater recognition of gender and sexual diversity. Yet it does not seem to be this openness to diversity of gender expressions, of sexualities, that is being manifested in the embrace of LGBT rights by US State Department and the UK's Foreign Office. Rather, what we are seeing it seems is 'homonormativity', described by Lisa Duggan as:

A politics that does not contest dominant heteronormative assumptions and institutions, but upholds and sustains them, while promising the possibility of a demobilized gay constituency and a privatized, depoliticized gay culture anchored in domesticity and consumption.

(2002:179)

Indeed, some would charge, this 'recognition' and defence of LGBT rights by prominent Western governments represents not only homonormativity but 'homonationalism' (Puar, 2007): a process through which selective acts of recognition and incorporation of gay and lesbians within normative liberal institutions work to deflect attention from the persistent injustice experienced within – *and perpetrated by* – the very societies that construct themselves as liberal and progressive, through pinkwashing and what Rahul Rao (2015) has called 'homocapitalism'. Agathangelou and colleagues ask:

What bodies, desires, and longings must be criminalized and annihilated to produce the good queer subjects, politics, and desires that are being solidified with the emergence of homonormativity?

(Agathangelou et al., 2005: 124)

Those who are most marginalised by this are typically those whose departure from dominant gender, sexual and social norms make them the most visible dissenters from normative gender and sexual categories. Giuseppe Campuzano (2008) uses Robert Chambers' matrix of poverty to highlight the multiple forms of exclusion experienced by the Peruvian *travesti* community, whose transits of gender flout normative expectations of sexual and gender binaries. From bullying and marginalisation at school making it difficult to remain in education, difficulties gaining and remaining in employment because of pervasive prejudices, prejudice making it hard to find housing and the violence of stigma affecting access to health and social services, Campuzano spells out myriad effects of discrimination.

That the 'global gay' has become an object of interest to the development industry and, with it, categories of persons produced by discourses on sexual risk and hazard such as

MSM, has arguably had the effect of reducing the discussion on sexuality in development to one about political rights and deflecting attention from the glare of social and economic injustice. After all, arguably, what a Peruvian *travesti*, South Asian *hijra*,[1] South African *tommy boy*[2] and Brazilian *puta*[3] have in common is not membership of a category of persons – the 'sexually marginalised' or 'sexual minorities' – but marginalisation and vulnerability to violence that must be understood in intersectional terms, exposing the race and class dimensions of these effects of discrimination and difference. Armas (2007) and Jolly (2010) explore the far-reaching dimensions of these forms of exclusion and discrimination on the basis of sexuality and highlight how international development's disregard for sexual and gender diversity can injure, impoverish and produce ill-being.

It is engagement with precisely these broader issues of justice that animates the struggle for the elimination of all forms of discrimination on the basis of gender and sexuality in which so many sexual rights activists in the Global South are engaged. This is a struggle that is about creating space for sexual and gender diversity and coming together in a common fight against exclusion and socio-economic hierarchies and their dire economic, political, social and psychological effects. And it is about establishing connections that, by reclaiming pleasure as a right, reframe the relationship between sexuality and development. Such work involves opening a space for dialogue about pleasure (Jolly, 2009) that can confront the internalised societal prejudices may otherwise prevent people from exploring what they might have in common, as people experiencing sexual exclusion. An inspiring example of this can be found in the work of Chinese NGO Pink Space, who facilitate conversations between the wives of gay men, female sex workers and lesbians that shatter prejudices and create opportunities for these women to explore their own sexualities, rights and pleasures (He, 2013). Another example comes from the human rights training courses run by Turkish organisation Women for Women's Human Rights, who teach rural women sexual pleasure as a human right (Seral Aksakal, 2013). And such work involves confronting society with its prejudices, like the Indian sex workers' collective VAMP whose slogan 'Save us from Saviours' sends a powerful message to the development industry (Seshu, 2013). It's in these sites and spaces that a very different kind of 'development' is happening, one that renegotiates and reimagines, over and over, sexualities as sites of pleasure, contestation and struggle.

Notes

1 Born with male or intersex bodily characteristics, some undergoing the ritual of emasculation and initiation known as *nirban* or *nirvan*, living in a complex gender, sexual and spiritual identity that is sometimes described as 'neither man nor woman', using 'female' personal pronouns and wearing 'feminine' clothes.
2 'Masculine' women-loving-women, who adopt 'masculine' styles of dress and expression.
3 'Prostitutes' – as the leader of the Brazilian Association of Prostitutes put it, 'don't sanitise us with your labels and call us sex workers, we're prostitutes, call us "prostitutes"'.

References

Agathangelou, Anna, Bassichis, Daniel and Spira, Tamara (2005) 'Intimate Investments: Homonormativity, Global Lockdown, and the Seductions of Empire', *Radical History Review*, 100: 120–143.
Armas, Henry (2007) *Whose Sexuality Counts? Poverty, Participation and Sexual Rights, Working Paper 294*, Brighton: IDS.

Arnfred, Signe (2013) *Sexuality and Gender Politics in Mozambique: Rethinking Gender in Africa*, London: Boydell and Brewer.

Baudh, Sumit (2008) *Human Rights and the Criminalization of Consensual Same-Sex Sexual Acts in the Commonwealth, South and Southeast Asia*, New Delhi: South and Southeast Asia Resource Centre on Sexuality.

Boellstorff, Tom (2003) 'Dubbing Culture: Indonesian "Gay" and "Lesbi" Subjectivities and Ethnography in an Already Globalized World', *American Ethnologist*, 30(2): 225–242.

Boyce, P. and Khanna, A. (2010) 'Rights and Representations: Querying the Male-to-Male Sexual Subject in India', *Culture, Health & Sexuality*, 13: 89–100.

Campuzano, Giuseppe (2008) *Building Identity While Managing Disadvantage: Peruvian Transgender Issues, IDS Working Paper 310*, Brighton: IDS.

Chiang, Howard, Henry, Todd A. and Leung, Helen Hok-Sze (2018) 'Trans-in-Asia, Asia-in-Trans: An Introduction', *Transgender Studies Quarterly*, 5(3): 298–310.

Cornwall, Andrea (2000) 'Missing Men: Reflections on Men, Masculinities and Gender in GAD', *IDS Bulletin*, 31(2): 18–27.

Cornwall, Andrea, Corrêa, Sonia and Jolly, Susie (2009) *Development with a Body: Sexuality, Development and Human Rights*, London: Zed Books.

Duggan, Lisa (2002) 'The New Homonormativity: The Sexual Politics of Neoliberalism', in Russ Castronovo and Dana D. Nelson (eds.), *Materializing Democracy: Toward a Revitalized Cultural Politics*, Durham and London: Duke University Press.

Hamzić, Vanja (2014) 'The (Un)Conscious Pariah: Canine and Gender Outcasts of the British Raj', *Australian Feminist Law Journal*, 40(2): 185–198.

Hamzić, Vanja (2016) *Sexual and Gender Diversity in the Muslim World: History, Law and Vernacular Knowledge*, London: I.B. Tauris (Islamic South Asia Series).

Hamzić, Vanja (2017) 'Selfhood and Archipelago in Indonesia: A Case for Human Polyversality', in Oishik Sircar and Dipika Jain (eds.), *New Intimacies, Old Desires: Law, Culture and Queer Politics in Neoliberal Times*, New Delhi: Zubaan.

Hamzić, Vanja (2019) 'The *Dera* Paradigm: Homecoming of the Gendered Other', *Ethnoscripts*, 21(1): unknown [forthcoming].

He, Xiaopei (2013) 'Building a Movement for Sexual Rights and Pleasure in China', in Susie Jolly, Andrea Cornwall and Kate Hawkins (eds.), *Women, Sexuality and the Political Power of Pleasure*, London: Zed Books.

Hunt, Nancy Rose, Liu, T. and Quataert, T. (eds.) (1997) *Gendered Colonialisms in African History*, London: Blackwell.

Jeater, Diana (1993) *Marriage, Prostitution and Power: Divorce in Colonial Southern Rhodesia*, Oxford: Oxford University Press.

Jolly, Susie (2009) 'Why the Development Industry Should Get Over Its Obsession with Bad Sex and Start Talking about Pleasure', in Suzanne Bergeron and Amy Lind (eds.), *Development, Sexual Rights and Global Governance: Resisting Global Power*, New York: Routledge.

Jolly, Susie (2010) *Poverty and Sexuality: What are the Connections*, Stockholm: Sida.

Kirby, Michael (2011) 'The Sodomy Offence: England's Least Lovely Criminal Law Export?', *Journal of Commonwealth Criminal Law*, 22–43.

Manchanda, Nivi (2015) 'Queering the Pashtun: Afghan Sexuality in the Homo-Nationalist Imaginary', *Third World Quarterly*, 36(1): 130–146.

Massad, Joseph (2002) 'Re-Orienting Desire: The Gay International and the Arab World', *Public Culture*, 14(2): 361–385.

Ogundipe-Leslie, Molara (1994) *Recreating Ourselves: African Women and Critical Transformation*, Trenton and New Jersey: Africa World Press.

Phillips, Oliver (2000) 'Constituting the Global Gay: Issues of Individual Subjectivity and Sexuality in Southern Africa', paper presented to the Queering Development seminar series, IDS, Autumn 1999.

Plummer, Ken (ed.) (1981) *The Making of the Modern Homosexual*, London: Hutchinson.

Puar, Jasbir (2007) *Terrorist Assemblages: Homonationalism in Queer Times*, Durham and London: Duke University Press.

Rao, Rahul (2015) 'Global Homocapitalism', *Radical Philosophy*, 194: 38–49.

Scott, James (1998) *Seeing Like a State: How Certain Schemes to Improve the Human Condition Have Failed*, New Haven: Yale University Press.

Seral Aksakal, Gulsah (2013) 'Sexual Pleasure as a Woman's Human Right: Experiences from a Human Rights Training Programme for Women in Turkey', in Susie Jolly, Andrea Cornwall and Kate Hawkins (eds.), *Women, Sexuality and the Political Power of Pleasure*, London: Zed Books.

Seshu, Meena (2013) 'Sex, Work and Citizenship: The VAMP Sex Workers' Collective in Maharashtra', in Naila Kabeer, Ratna Sudarshan and Kirsty Milward (eds.), *Organizing Women Workers in the Informal Economy: Beyond the Weapons of the Weak*, London: Zed Books.

Spivak, Gayatri (1988) ' "Can the Subaltern Speak?" "Can the Subaltern Speak?" ', in C. Nelson and L. Grossberg (eds.), *Marxism and the Interpretation of Culture*, Chicago: University of Illinois Press, pp. 271–313.

Stoler, Ann (1989) 'Making Empire Respectable: The Politics of Race and Sexual Morality in 20th-Century Colonial Cultures', *American Ethnologist*, 16(4): 634–660.

Stryker, Susan (2008) 'Transgender History, Homonormativity and Disciplinarity', *Radical History Review*, 100: 145–157.

White, Luise (1990) *The Comforts of Home: Prostitution in Colonial Nairobi*, Chicago: Chicago University Press.

Chapter 42

Rethinking gender and empowerment

Jane Parpart

Empowerment, particularly for women and the poor, has become a central issue for development institutions such as the UNDP, the World Bank, Oxfam, government agencies and many smaller non-governmental organisations (NGOs). Initially seen as a key element for challenging and transforming unequal political, economic and social structures, empowerment was regarded as a weapon for the weak – best wielded through participatory, grassroots community-based NGOs. However, empowerment is a flexible concept, and by the mid 1990s mainstream development agencies had begun to adopt the term. The language of participation, partnership and empowerment increasingly entered mainstream development discourse (World Bank, 1995; Elson and Keklik, 2002). While the wording remained the same, meanings varied, and mainstream institutions and their practitioners for the most part envisioned empowerment as a means for enhancing efficiency and productivity within the status quo rather than as a mechanism for social transformation (Parpart et al., 2002).

The issue of empowerment first surfaced in gender and development debates in the work of Caroline Moser (1993) and Gita Sen and Caren Grown (1987). These writings reflected a growing concern that gender equality would not be achieved unless women could challenge patriarchy and global inequality. To achieve this end, Moser argued that women needed to gain self-reliance and internal strength in order 'to determine choices in life and to influence the direction of change, through the ability to gain control over crucial material and non-material resources' (1993: 74–75). Sen and Grown emphasised the need for a collective vision, a set of strategies and new methods for mobilising political will and empowering women (and men) to transform society (1987: 87).

Scholar/activists from the Global South raised crucial questions. Srilatha Batliwala (1994) warned that 'empowerment', which had virtually replaced terms such as poverty alleviation, welfare and community participation, was in danger of losing its transformative edge. She called for a more precise understanding of power and empowerment, asserting that empowerment must be seen as 'the process of challenging existing power relations and of gaining greater control over the sources of power' (1994: 130). It requires political and collective action against cultural as well as national and community power structures that oppress women and some men and, consequently, transformative political action.

Naila Kabeer (1994), who has played a key role in operationalising empowerment as development practice, has placed empowerment at the centre of efforts to achieve gender equality. Criticising the liberal and Marxist emphasis on *power over* resources, institutions and decision-making, Kabeer adopted a more feminist position, which emphasises the transformative potential of *power within*. This power is rooted in self-understanding

DOI: 10.4324/9780429282348-46

that can inspire women (and some men) to recognise and challenge gender inequality in the home and the community (1994: 224–9). Like Batliwala, she highlights collective, grassroots participatory action – the *power to* work *with* others 'to control resources, determine agendas and to make decisions' (1994: 229). Kabeer is particularly concerned with enhancing women's ability to exercise choice (associated with access and claims on resources, agency and achievements; 1999: 437).

Jo Rowlands argues that 'empowerment is more than participation in decision-making; it must also include the processes that lead people to perceive themselves as able and entitled to make decisions' (1997: 14). It is personal, relational and collective and 'involves moving from insight to action' (1997: 15). Drawing on in-depth research in Honduras, she points to the crucial role played by social, political and economic contexts, warning that consciousness and agency are always context specific. Building on these debates, Sarah Mosedale (2005: 252) suggests that women's empowerment is best seen as 'the process by which women redefine and extend what is possible for them to be and do in situations where they have been restricted compared to men'. For her the issue is not simply enhancing choice but extending the limits of the possible.

Increasingly, particularly as top-down approaches to development failed to alleviate poverty in the 1990s, especially among women, empowerment became a central focus for mainstream women and development programs. For example, the Beijing Platform, produced at the 1995 UN global women's conference, stated unequivocally that women's empowerment is 'fundamental for the achievement of equality, development and peace' (United Nations, 1995: para. 13). Official development institutions picked up the language of empowerment, gender equality and gender mainstreaming. The Canadian International Development Agency's (CIDA) 'Policy on Gender Equality' declared women's empowerment one of its guiding principles for development policy (1999). While generally framed within neo-liberal discourses of productivity and efficiency, the official commitment to empowerment and gender equality continues (CIDA, 2011; UN Women, 2010).

However, the adoption of empowerment as a basis for gender and development policy and practice has obscured the difficulties facing those trying to understand, implement and measure women's empowerment projects. Empowerment has been framed as a doable, reachable and measurable goal. Yet it is highly context specific, fluid and messy – all factors which affect attempts to operationalise and evaluate empowerment projects. Kabeer has critiqued empowerment projects for assuming that 'we can somehow predict the nature and direction that change is going to assume'. As she points out, ' In actual fact, human agency is indeterminate and hence unpredictable in a way that is antithetical to requirements of measurement' (1999: 462). These critiques have been picked up by many others. Indeed, even internal evaluations often prove the difficulties facing agencies that have promised to empower women and ensure gender equality (Cornwall et al., 2007; Parpart, 2012).

Critical voices have begun to question whether the gender and empowerment approach can produce fundamental change. Parpart et al. (2002) warned that critical thinking about power and empowerment must be incorporated into empowerment projects and policies. Women's empowerment cannot be regarded simply as the need to bring women into established power structures. This approach ignores the deeply held resistances facing marginalised groups around the world and the subtle attitudinal and structural impediments to collective action (*power with*) and generative *power to* support gender equality. Empowerment requires attention to language and meanings, identities and cultural practices as well as the forces that enhance *power to* act *with* others to fight for change, often in hostile and difficult environments. It also requires moving away from the limits of a preoccupation with difference, which as Rai (2007) points out, has undermined efforts to support feminist activism across borders and cultures.

The shift towards the local has enriched knowledge of local conditions and highlighted the plight of small communities and the gendered crises faced by many women (and some men) in these communities. However, while empowerment is often a local affair, the local is also embedded in the global and the national and vice versa. Power (and empowerment) can only be understood within this fluid, complex and messy context. It requires attention to local struggles and their intersection with broader forces, particularly the increasingly global inequality and its gendered nature. As Sharma points out, women's struggles for empowerment and governance require analysis of local complexities within their regional, national and global context (2008). Understanding local processes requires attention to larger forces as well. Only then will local empowerment projects be able to address local inequalities and injustices.

Empowerment projects and advocates must also pay more attention to the ways institutional structures, material and discursive frameworks shape the possibilities and limits of individual and group agency and choices. This does not undercut the importance of local participation and consultation. It does, however, point to the need to situate individual and group action/agency within the material, political and discursive structures in which they operate. This requires careful, historically situated analyses of women's struggles to gain power in a world often neither of their own making or choosing (Berik et al., 2009).

Finally, empowerment is both a process and an outcome. At times the two are indistinguishable, at others they merge, and sometimes the process is the outcome. While recognising that specific outcomes should (and often can) be measured, measuring empowerment remains elusive. Many subtle and often unexpected strategies have the potential – but not the certainty – of empowerment (Kabeer, 1999). Others, such as international covenants and gender-sensitive laws, seem to guarantee empowerment but fail due to patriarchal cultural practices and structures. Thus, while attempts to measure outcomes can focus the mind and encourage new thinking, an obsession with outcomes and measurement can endanger the very processes most apt to nurture women's empowerment, even if not apparent at the time.

These critiques offer some guidelines for trying to ensure that women's empowerment is more than simply a 'motherhood' term for development agencies. They offer ways of making both the concept and practice of empowerment more rigorous, effective and nuanced. Empowerment, particularly as part of the struggle for gender equality around the world, continues to be a development concern, both for policy and praxis. The inclusion of men and masculinity (Cornwall et al., 2011) as well as attention to cultural differences and economic empowerment is a welcome expansion of the term and key to addressing gender empowerment in an increasingly complex, global and still very patriarchal world.

References

Batliwala, S. (1994) 'The Meaning of Women's Empowerment: New Concepts from Action', in G. Sen, A. Germain and L.C. Chen (eds.) *Population Policies Reconsidered*, Boston: Harvard University Press.

Berik, G., Meulen Rogers, Y. and Zammit, A. (eds.) (2009) *Social Justice and Gender Equality: Rethinking Development Strategies and Macroeconomic Policies*, New York: Routledge.

Canadian International Development Agency (CIDA) (1999) *Policy on Gender Equality*, Ottawa: CIDA.

CIDA (2011) *Gender Equality*, Ottawa: CIDA.

Cornwall, A., Edstrom, J. and Greig, A. (eds.) (2011) *Men and Development: Politicising Masculinities*, London: Zed Books.

Cornwall, A., Harrison, E. and Whitehead, A. (2007) *Feminisms in Development*, London: Zed Books.

Elson, D. and Keklik, H. (2002) *Progress of the World's Women: Gender Equality and the Millenium Development Goals*. New York City: UNIFEM.

Kabeer, N. (1994) *Reversed Realities: Gender Hierarchies in Development Thought*, London: Verso.

Kabeer, N. (1999) 'Resources, Agency, Achievements: Reflections on the Measurement of Women's Empowerment', *Development and Change* 30(3): 435–464.

Mosedale, S. (2005) 'Assessing Women's Empowerment', *Journal of International Development* 17: 243–257.

Moser, C. (1993) *Gender Planning and Development*, London: Routledge.

Parpart, J. (2012) 'Exploring the Transformative Potential of Gender Mainstreaming', *Journal of International Development* (forthcoming).

Parpart, J., Rai, S. and Staudt, K. (eds.) (2002) *Rethinking Empowerment: Gender and Development in a Global/Local World*, London: Routledge.

Rai, S. (2007) *The Gender Politics of Development*, London: Zed Books.

Rowlands, J. (1997) *Questioning Empowerment: Working with Women in Honduras*, Oxford: Oxfam Publications.

Sen, G. and Grown, C. (1987) *Development, Crises, and Alternative Visions*, New York: Monthly Review Press.

Sharma, A. (2008) 'Crossbreeding Institutions, Breeding Struggle: Women's Empowerment, Neoliberal Governmentality, and State (Re)Formation in India', *Cultural Anthropology* 21(1): 60–95.

UN Women and UN Global Compact Office (2010) *Women's Partnership Principles*, New York: United Nations.

United Nations (1995) *Beijing Platform for Action*, New York: United Nations.

World Bank (1995) *World Bank Participation Source Book*, Washington, DC: World Bank Environment Department Papers.

Web sources on empowerment

http://web.worldbank.org. See Alsop, R., Bertslsen, M.F. and Holland, J. (2005) *Empowerment in Practice*, Washington, DC: World Bank.

http://www.undp.org/women See the Resource Guide for Gender Theme Groups, January 2005.

http://www.unifem.org

http://unwomen.org

Chapter 43

Critique of feminism from the South

Madhu Purnima Kishwar

I would like to clarify at the outset that feminism and commitment to strengthening women's rights in society through concrete, demonstrable actions are not always synonymous in the countries of the South. However, the public domain on women's issues is dominated by the discourse made fashionable by international development agencies of developed countries.

Taking the example of India, there is a long tradition of male social reformers dedicating their entire life to battling prejudices against equal participation of women in public life and strengthening their rights within the family. The 19th-century social reform movements and the 20th-century freedom movement led by Mahatma Gandhi bought women's rights to the core of these movements. Even in contemporary India, many radical mass-based women's rights struggles have been led by men. Several outstanding women have emerged in leadership roles in rural movements. But the development analysts invariably present a unidimensional picture of patriarchal oppression.

Many of the rights for which valiant feminists in the West fought long drawn out battles in the face of hostile attacks from men of their societies, especially those in the political establishment, came to Indian women through the efforts of male social reformers and freedom fighters who took the lead in fighting battles along with women. They braved forces of resistance in order to make dignified space for women in public and political life in India. Unlike in the Western democracies, rights of women to education on par with men, to vote, to hold political office and enter male dominated professions, came without a lonely and harsh battle. The right to contraception and abortion have never evoked religious opposition or hostility in India as they did in countries where the Church holds the veto power on these issues (see Kishwar, 1985).

Eurocentric view of development

To illustrate the limitations and inbuilt biases, I would like to use the World Development report from 2012 entitled *Gender Equality and Development* (GED). The very first chapter of GED starts by saying,

> Despite the hardships many women endure in their daily lives, things have changed for the better and at a speed that would not have been expected even two decades ago. . . . Improvements that took 100 years in wealthier countries took just 40 years in some low and middle income countries.

DOI: 10.4324/9780429282348-47

It further reminds us that

> Japan's Equal Employment Opportunity Act of 1985 obliged employers merely to endeavour to treat men and women equally during job recruitment, assignment, and promotion. The mandate for equal treatment came about in 1997. The first domestic violence law was passed in 2001.

This is meant to hammer the point that even a 'developed' Asian country like Japan is decades behind Europe and America – in comparison to 'underdeveloped' countries of Africa and Asia! We are then reminded that it is 'the ratification of CEDAW (Committee on the Elimination of Discrimination against Women) and other international treaties' – all of which emanated from Western countries – that 'established a comprehensive framework to promote equality for women'.

The GED report is supposed to provide us a global overview of the challenges faced by women the world over. However, there is hardly any mention of the problems faced by women in developed countries. The assumption is that they have solved all their problems and have provided benchmarks and a road map for countries of the South.

That is why in almost all the charts and graphs presented in the Report, countries of the South are presented as not yet measuring up to 'developed' countries. Countries of the North are no doubt ahead of most countries of the South in matters of education and employment opportunities for women. But such reports completely overlook the fact that domestic violence is fairly rampant in developed countries, which have very high participation of women in the work force as well as high education levels for females. Sexual violence is also very common in these societies. What does the flourishing pornographic industry and sex trafficking say about the status of women in developed countries? Why are these not used as important indicators for evaluating the status of women?

In this Eurocentric world view, problems of women in developed countries are erased out of existence and the non-European world projected as politically and culturally backward and therefore in need of perpetual guidance by the West. This is not the only 'development report' that displays a total lack of awareness about the fact that many of the 'underdeveloped' countries of the South, such as India, China and other Asian countries, were highly prosperous economies and societies during the period Europe describes as its 'Medieval Dark Ages'. Even in the 18th century, India and China were world leaders in trade, manufacturing, education, medicine, architecture, arts and even engineering.

The era of 'progress' for Europe was the era of colonisation for many of the countries of the South. While Europe prospered, the colonised societies were politically subjugated through brute force and wrecked – economically, socially and culturally. Apart from their legendary wealth, many of the Asian countries had matrilineal family structures. Even in the 18th century, India's large regions in the South and East followed matrilineal inheritance. In patrilineal family structures too Indian women had inalienable rights to family property in the form of *stridhan* (totally different from modern-day dowries) which passed from mother to daughter.

But because in Britain and US the law began changing in favour of women only in 1857 and 1858 respectively, it is assumed that, in the world over, women had lived like helpless dependents. This world view wants us to erase memories of colonial rule in the South which disinherited women by enacting laws which transformed community-owned assets and joint family-owned property into individual property vested only in male hands. For the Victorian-minded British, patriarchal control over property and women was the 'natural' God-given order of things. Matrilineal communities in India were condemned for encouraging 'promiscuous and immoral' conduct among women. Women of matrilineal families were described with the disdain reserved for prostitutes because they could change their partners at will and divorce was not an issue in the way it was in the Christian world. They were legally pushed into giving up matrilineal inheritance systems in favour of patriarchal controls. For examples see Schneider and Gough (1961) and Saradamoni (1999). Today, the same societies are being delivered sermons against the sexual repression of women and advised to be more sexually liberated!

Obsession with 50 percentism

Another problem with the 'Women and Development' framework is its obsession with measuring women's well-being through the '50 per cent' benchmark. The parameters crafted by international development agencies for evaluating women's status in different societies are all centred on counting the percentage of women in various fields of life such as education, professions, labour force and politics. The march to 'equality' means only one thing – how far are women from occupying 50 per cent of the spaces in all domains including jobs? Wherever they are less than 50 per cent it is assumed that they are deprived and excluded and therefore oppressed.

'Percentism' in and of itself makes good sense in some areas such as sex ratio figures since the ratio at birth for males and females is near equal or in favour of baby girls. However, in China and most South Asian countries the growing culture of son preference as well as neglect of women's health and nutritional needs has resulted in an alarming deficit of females in the overall population of these countries. Therefore, '50 percentism' is a valid criterion in this domain.

However, in many other domains, it does not tell the whole story or can lead to erroneous conclusions. For example, there is a great deal of excitement about equal participation of women in parliaments and state legislatures as a means of political empowerment. However, feminists and development experts have paid far less attention to the nature of representative democracy in countries of the South as well as North.

When political parties are themselves under the grip of money and muscle power and institutions of governance are steeped in crime and corruption because they lack accountability and transparency, can the presence of a certain percentage of women by itself make our polity and governance machinery more citizen friendly? India, Pakistan, Bangladesh and the Philippines have seen women rise to the top echelons of power only to outperform the worst of men in crime and corruption. Did American foreign policy become less hegemonic or militarist when Condoleezza Rice or Hillary Clinton became Secretary of State? Did they try to curb the arms manufacturing lobby any more than the macho males who assumed those positions of power?

It is not enough to demand that women be given a 50 per cent share of the available pie without first examining whether the pie is worth eating at all. And if it is rotten, women should know how to bake a new and healthy pie which has enough portions for all.

All this is not to undermine the importance of enabling women to have a decisive say in public policy and political affairs but merely to point out this 'ladies compartment' approach to women's issues which demands that women should have reserved compartments in every train, never mind whether the train is headed for doom and destruction because its engineering is intrinsically faulty and disaster prone.

Similarly, much is made of school enrolment figures for girls as an indicator of high or low status. But the percentage of girl's enrolment in school tells a partial story about the future prospects of these girls or for that matter that of boys. The quality of education being imparted to them has far greater implications on what doors open or remain closed for them in life. In countries like India that have adopted English as the language of elite education, administration and access to job opportunities – the quality of education in most government schools for the poor is abysmal.

There is pride associated with being a master craftswoman or craftsman. There is no pride in a man becoming a street vendor, low-paid assembly line worker, peon or lowly paid clerk after some years of shoddy schooling. The economic and social status of supposedly 'illiterate' women in weaver or potter households was far better and respected than that a woman or who has received poor quality high schooling or even a bachelor's degree which does not equip her for anything better than a lowly clerical or sales girl job involving soulless drudgery. In a traditional weaver or potter household, women were an indispensable part of a very creative production process involving specialised skills. Even if they did not get a separate pay cheque they were not treated as dispensable. If today women of such households come out low on the Human Development Index, it is primarily on account of state policies aimed at marginalising and destroying these traditional crafts and technologists despite the fact that these artisanal groups produce high quality aesthetic products which have ready buyers in the national and international markets. We need policy interventions that enable these traditional home-based production units to earn dignified incomes – not necessarily make factory workers or clerks out of them, unless of course they abandon these occupations out of choice.

Ideological preference for employment in the organised sector

The 'development' paradigm has a strong bias against employment in the self-organised sector. Any country that shows a higher percentage of people employed in the organised sectors of the economy is considered 'developed' and 'advanced' and those with a high percentage in the self-organised sectors are considered 'backward' and 'under developed'. Low percentage of women in the organised sector is also held up as an example of discriminatory policies against women. However, in countries like India, the self-organised sectors of the economy have proved to be a boon and provide far greater avenues of economic advancement and upward mobility than most jobs in the organised sectors of the economy. In most developed economies, over 90 per cent of people are in the organised sector and a relatively small proportion is self-employed. Therefore, the moment there is an economic crisis that leads to job cuts and the retrenchment of workers. However in countries like India where over 92 per cent are in the self-organised sectors of the economy (with no more than 3 per cent in corporate sector and 3 per cent in government employment), most people don't even notice economic downturns. This is an important reason why the economic crisis that hit Europe and US in the last decade did not produce any cataclysmic changes for countries like India or Bangladesh. These economies kept growing through the period of global economic crisis.

Many in the self-employed sector prosper quickly *except* when government policies are outright hostile to them. Take the example of street vendors – many woman hawkers earn much more than a factory worker or an office clerk because profit margins in retail are usually high and business in street hawking brisk since they sell low-value goods. They experience rapid upward mobility because they get an opportunity to develop entrepreneurial skills and explore new opportunities. By contrast, a bank cashier or an office clerk may have relatively greater security but avenues of upward mobility are limited.

If the vast majority of street vendors and other self-employed groups remain trapped in low-income ghettos it is mainly because most governments treat them as illegal encroachers and an unwanted nuisance. They are forced to survive by seeking protection of political mafias and police who siphon off large parts of their incomes by way of bribes. The routine violence during clearance operations unleashed on vendors makes it doubly risky for women to survive in this occupation. That is why in countries that have hostile policies towards street vendors far fewer women are found in this trade. But in vendor-friendly countries like Thailand women thrive and dominate the markets and make good money. Most important of all, it provides free business training to the younger generation with young boys and girls joining their mothers as helping hands after school. No organised sector job allows such hands-on training for family members with flexible timings – all for free.

Similarly, women who set up small tailoring shops or beauty parlours at home are able to combine housework with their small enterprises. Many expand their businesses and start earning good money in no time minus the guilt of neglecting their family responsibilities. But 'development' experts would rather put all these small entrepreneurs in the organised sectors.

Devaluation of women as home makers

Apart from undervaluing the importance of home-based enterprises, development experts have a strong ideological preference for seeing women 'gainfully employed' in paid jobs outside their homes. The flip side of this ideological position is systematic devaluation of women's role as mothers and as homemakers and as nurturers of future generations. But the dominant feminist discourse within the 'development' framework treats women who focus on this role as 'unproductive' an work force which is not contributing to national wealth and economic growth.

We need to respect the fact that a lot of women would rather not have dual jobs – as homemakers and wage earners. For many, going out to labour outside the house is a distress response. Many prefer to devote their full attention to family and children. For many women, withdrawal from wage labour comes as a big relief. Many prefer that their husbands earn enough to support the whole family in comfort because world-wide experience shows that the best of paid child care does not compare favourably with parental and family care. If both husband and wife are in high pressure jobs, it is difficult to do justice to children, unless one has the support of extended family – grandparents, uncles and aunts. But joint family is also anathema to most feminists. They only see it as the site of restrictions and oppression for women. This despite the fact that there is plentiful evidence in countries like India that women who live in supportive extended families where grandparents and others share a large part of the responsibility for child care rise high in their professions because they do not have to take mid-career breaks and can devote time to their professions without worrying about neglecting family.

It is noteworthy that the early battles waged by the working class in Europe were for family wage so that women did not have to work under compulsion. The job of women's rights activists should be limited to ensuring that every woman who wishes to take up a paid job or profession outside the house is not prevented from doing so due to lack of opportunities or discriminatory social practices. But it is presumptuous to insist that every woman must work for a wage and those who don't are socially 'unproductive' and culturally 'backward'. Feminism should be about respecting women's choices, not imposing a pre-set, ideologically determined road map for all. To conclude, a one-size-fits-all approach advocated by development experts doesn't take into account the diverse needs and aspirations of women.

References and further reading

Kishwar, M. (1985) 'Gandhi and Women', *Economic and Political Weekly*, XX(40–41).

Saradamoni, K. (1999) *Matriliny Transformed: Family, Law, and Ideology in Twentieth Century Travancore*, Sage Publications, New Delhi.

Schneider, D.M. & Gough, K. (eds) (1961) *Matrilineal Kinship*, University of California Press, Berkeley.

World Development Report (2012) *Gender Equality and Development (GED)*, World Bank, Washington.

Chapter 44

Identities and intersectionality

Sara de Jong

Several perspectives and concepts are helpful for making sense of identities in development studies. The concept of identity often appears alongside other analytical terms, such as subjectivity, relationality, positionality, and reflexivity (*see index*). The construction of identity is intertwined with the formation of the self as a subject. To be recognised as a subject is to be placed within power relations, which constrain as well as enable. In Nancy Cook's words: 'As subjects gradually begin to think and behave according to a particular set of discursive norms, they undergo the paradox of subject formation; they become subjected to power at the moment they become discernible subjects' (2007: 18).

Of key relevance to analysing identities in relation to power is the concept of 'intersectionality', grounded in the work of feminists of colour (Hill Collins, 1993; Crenshaw, 2000; Brah and Phoenix, 2004), which describes the complex interplay between structures of difference. Avtar Brah and Ann Phoenix's definition of intersectionality 'as signifying the complex, irreducible, varied, and variable effects which ensue when multiple axis of differentiation – economic, political, cultural, psychic, subjective and experiential – intersect in historically specific contexts' (2004: 76) captures neatly the various levels of differentiation and its embeddedness in specific contexts. The innovation of the concept of 'intersectionality' lies in the recognition that the various categories emerging from structures of power and inequality are co-constitutive and inseparable. When analysing identities, race, class, and gender should not be approached as categories which can be isolated and then added up; instead, race, class, and gender 'function as parallel and interlocking systems that shape . . . relationship[s] of domination and subordination' (Hill Collins, 1993: 29).

The concept of intersectionality has been key to foregrounding the experiences of women of colour. As Kimberlé Crenshaw (2000) has argued, women of colour often remain invisible, because the category of race and race struggles are primarily associated with men of colour and gender and feminist struggles with white women. Intersectionality can therefore 'be drawn upon for its normative and political (rather than merely 'analytical') function' (de Jong, 2017: 136), highlighting the relation between inequalities and intersectional identity positions. For instance, an intersectional lens can reveal that humanitarianism's organisational culture and working conditions have an aggravating effect on the well-being of *female* national staff compared to *male* national staff (from so-called developing countries), due to care responsibilities and gender norms (Houdley, 2019). It can further help to understand how the experiences of female *national* staff and female *international* staff differ, despite both being affected by gendered structures, as international staff tend to be more mobile due to privileged passports and access to career

DOI: 10.4324/9780429282348-48

progression (Ibid.). Intersectionality can thus also be employed to transcend binary thinking in which the 'both/and position of simultaneously being oppressed and oppressor becomes conceptually impossible' (Hill Collins, 1993: 28). This is vital for understanding the implications of the growing diversity among the people inhabiting Aidland.

Making sense of identities in development studies

As is highlighted by Barbara Heron (2007) in her study of Canadian development workers, modern subjects tend to consider their identity as comprising a core or essence. To maintain a coherent sense of self, we construct stories about ourselves. In Heron's words: 'Through personal accounting for action and self-description, unitary subjects work to compose non-contradictory identities both publicly and privately . . . draw[ing] on culturally available discourses' (2007: 11–12). This offers some important insights. First, it highlights that while modern identities are constantly created and in process, identity constructions tend to coalesce around an imagined core with a distinguishable logic. Second, it recognises that identities are not constructed in a vacuum. Instead, individual self-understandings are made intelligible in the context of wider structural discourses, for instance about the relation between the so-called developed and developing world, that precede the Self. Discourses provide us with frames that make our experiences intelligible to ourselves and others. Contemporary discourses have historical legacies; for instance, the discursive subject formations of women development workers in modern Pakistan echo that of their female imperial predecessors in colonial India (Cook, 2007). Third, since the stories we tell about ourselves (and others) are an important part of our identity formation, narrative methods are useful for research on identities. For instance, development workers' autobiographies or life story interviews could be analysed with attention to both what is said and what remains unsaid.

Attention to silence is important as reference to the term 'identity' often elicits 'reflection on categories linked to subordination rather than privilege and domination, which remain . . . unnamed and invisible' (de Jong, 2017: 134). Studying the stories that development workers tell about themselves also requires consideration of the (imagined) audience for their narration. Hence, the identities of researchers and their relationship with research participants needs reflection and consideration too. For instance, Barbara Heron, who is a researcher *and* a former development worker, chose to include within her sample an interview with herself, conducted by one of her research participants, and she positions herself as 'simultaneously ethnographer and participant . . . with intimate knowledge of the subject position of white middle-class woman development worker' (2007: 19).

Identities matter in international development

It is easy to lose sight of the people who populate 'Aidland' when engaged in debates over whether the Sustainable Development Goals constitute a paradigmatic shift, the challenges of global poverty, and power imbalances between donor organisations and local partners. But behind projects, log frames, and agendas are development workers who go on field visits, learn new buzzwords, and worry about their careers.

The early 2000s saw a surge in critical development studies research addressing the identities, experiences, and subjectivities of development workers (e.g., Eriksson Baaz,

Figure 44.1 Satirical mini-bio on Barbie Savior's Instagram account

2005; Cook, 2007; Heron, 2007). The semi-autobiographical accounts of development workers-turned-researchers, who drew on their own experiences to ask themselves and colleagues critical questions about international development practices and discourses, constitute an important section of this area of development research (e.g., Heron, 2007). This growing interest in development workers' identities occurred against the backdrop of broader research agendas focussing on the 'everyday', for instance in the writing of colonial history – and on 'positionality', drawing on feminist theory.

In recent years, the emerging interest in the identities of development workers has spread to the public realm. The British newspaper the *Guardian* featured a series of anonymous posts by development workers about their dilemmas, desires, and doubts. The entries by the 'Secret Aidworker' provided visibility to the persons behind the headlines of international development news. Social media has drawn attention to the identities of aid workers, problematising the figure of the 'white saviour', epitomised by the naïve yet privileged white, young volunteer development worker. She is mockingly depicted, for instance, in the 'Barbiesavior' Instagram account. Alongside pictures of Barbie cradling African babies and posing with wildlife, her profile reads: 'Jesus. Adventures. Africa. Two worlds. One love. Babies. Beauty. Not Qualified. Called. 20 years young. It's not about me . . . but it kind of is'. (https://www.instagram.com/barbiesavior/?hl=en). This satirical mini-bio not only presents intersectional identity positions framed by age, religion, and gender but also captures how development volunteering is frequently invested with a desire to project a certain moral identity.

Drawing attention to the identities of development workers can serve the purpose of 'returning the gaze' by placing the 'subjects' rather than the 'objects' of development into the centre. This counterbalances the hypervisibility accorded to (stereotypical and simplified) recipients of development aid, such as the 'average third-world woman [who] leads an essentially truncated life based on her feminine gender (read: sexually constrained) and being 'Third World' (read: ignorant, poor, uneducated, tradition-bound, religious, domesticated, family-oriented, victim' (Mohanty, 1984: 337) and the reticent development beneficiary whose resistance is read as reconfirming colonial images of the passive, 'culturally backward' Other (Baaz, 2005: 77). Research on the identities of development workers shows that development beneficiaries are often used as a foil, as figures against which to contrast oneself, for instance as the 'helper' versus the 'helpless' (Baaz, 2005; de Jong, 2017).

In the edited volume *Inside the Everyday Lives of Development Workers,* Heather Hindman and Anne-Meike Fechter 'call for the insertion of development workers' lives into studies of aid not merely as a call to examine the unexamined but because we feel that they are part and parcel of the theory of development' (Hindman and Fechter, 2011: 3). Considering the identities of development workers and their interplay with development policy, practice, and discourse offers an entry point into macrolevel phenomena such as global inequality and meso phenomena like organisational structures.

For instance, the identities of aid workers vis-à-vis aid 'recipients' reflect macrostructures of inequality such as global border regimes, colonial legacies, and hierarchies of knowledge. Who can be a mobile expat, which language skills have currency in the aid world, and who can claim expertise? How are aid workers from different backgrounds variously affected by the precarious work structures that are characteristic of the global capitalist labour market? Moreover, attention to the identities of development workers can help illuminate broader discussions around ethics and morality. What we understand to be the 'right thing to do' is tightly connected to the idea of development work as a vocation. Listening to development workers whose sense of self is tied to their professional activities problematises the boundaries between public/private, personal/professional.

On a meso-level, the identities of development workers impact, for example, on the missions of development organisations. Attention to identities helps to unpack the dynamic nature of organisational missions as they are negotiated on an everyday basis by staff from different backgrounds and with various degrees of power. As development workers move between organisations over the course of their careers, from large to small, field and headquarters-based, radical and mainstream, their personal identities and organisational identities are mutually constitutive; sometimes intimately intertwined, other times in tension. Hence, a focus on the identities of aid workers can deepen our analysis of issues that are associated with the organisational level. We can, for instance, understand more about the challenges surrounding accountability and safeguarding when we have a better sense of staff's identities. While media exposure of sexual abuse and corruptions scandals might single out specific organisations, by taking identities seriously, the interplay between power structures and privileged staff identities and how these inform perpetrator and bystander roles can be made apparent.

References

Baaz, M. E. 2005. *The Paternalism of Partnership: A Postcolonial Reading of Identity in Development Aid.* London: Zed Books.

Brah, A. and Phoenix, A. 2004. "Ain't I a Woman? Revisiting Intersectionality." *Journal of International Women's Studies,* 5(3): 75–86.

Cook, N. 2007. *Gender, Identity and Imperialism: Women Development Workers in Pakistan.* Houndmills: Palgrave Macmillan.

Crenshaw, K. 2000. "Demarginalising the Intersection of Race and Sex: A black Feminist Critique of Antidiscrimination Doctrine, Feminist Theory and Antiracist Politics." In *The Black Feminist Reader* (eds. J. James and T. Sharpley-Whiting). Malden, MA: Blackwell, 208–238.

De Jong, S. 2017. *Complicit Sisters: Gender and Women's Issues Across North-South Divides.* New York: Oxford University Press.

Heron, B. 2007. *Desire for Development: Whiteness, Gender and the Helping Imperative*. Waterloo: Wilfred Laurier Press.

Hill Collins, P. 1993. "Toward a New Vision: Race, Class, and Gender as Categories of Analysis and Connection." *Race, Sex, and Class*, 1(1): 25–45.

Hindman, H. and Fechter, A.-M. 2011. "Introduction." In *Inside the Everyday Lives of Development Workers: The Challenges and Futures of Aidland*. Sterling, VA: Kumarian Press.

Houdley, G. 2019. "Humanitarian Response and Stress in Kenya: Gendered Problems and Their Implications." *Gender and Development*, 27(2): 337–353.

Mohanty, C. T. 1984. "Under Western Eyes: Feminist Scholarship and Colonial Discourses." *Boundary 2*, 12(3): 333–359.

Further reading

Charlès, L. L. 2007. *Intimate Colonialism: Head, Heart and Body in West African Development Work*. Walnut Creek, CA: Left Coast Press.

Goetz, A. M. 2001. *Women Development Workers: Implementing Rural Credit Programmes in Bangladesh*. New Delhi: Sage.

Kothari, U. 2005. "Authority and Expertise: The Professionalisation of International Development and the Ordering of Dissent." *Singapore Journal of Tropical Geography*, 27(3): 235–253.

McCall, L. 2005. "The Complexity of Intersectionality." *Signs*, 30(3): 103–112.

Roth, S. 2015. *The Paradoxes of Aidwork: Passionate Professionals*. New York: Routledge.

Chapter 45

A K-shaped crisis

COVID-19 and inequalities

Ana Luíza Matos de Oliveira and Magali N. Alloatti

1. Defining a K-shaped crisis

Demertzis (2021) reminds us that the shape of the post-pandemic economic recovery has gone through many 'letter characterisations'. The 'K-shaped crisis' shows that individuals at opposite ends of the income distribution or access to rights end up even more apart from each other than in the beginning, thus increasing previous inequalities. This is the case of the COVID-19 crisis, since those with resources, stable social protection systems, and formal jobs were able to endure the crisis and recover more easily. In the Global South, inequality is considerable higher. For example, Brazil's Gini Index for 2019 is 0.53 and Colombia's 0.51, around double that of European countries, and the region presents the highest percentage in the world of total income appropriated by the richest 1 per cent (ECLAC, 2021a). This makes the 'k-shaped' phenomenon particularly visible during the COVID-19 crisis. In this chapter, we briefly discuss how this pattern is reflected in Latin America and the Caribbean (LAC).

2. Inequality during the crisis in LAC

A frequent measure of inequality is through income. However, there are various ways of accounting for income differences, such as the amount received by the richest, access to social rights, and other forms of multidimensional inequality. Our discussion focuses on the latter. LAC has a history of inequality, sexism, and enslavement which has left profound marks in the region making class, gender, age, and racial/ethnic discriminations pillars of its social configuration. In the region, the Covid-19 crisis has been severely felt in 2020 and 2021, and its K-shaped format reverberates beyond the short-term. Before the pandemic, a higher incidence of poverty was observed among people living in rural areas, children and adolescents, indigenous population, Afro descendants, and those with incomplete primary education. Women, younger, and lower qualified individuals presented a more pronounced drop in occupation levels during the pandemic (ECLAC/OIT, 2021). According to ECLAC (2021b), the direct and indirect impacts of the crisis have widened differences, reconfirming the tendency of disasters to align with existing structural inequalities.

2.1 Class inequality

While at this point we are not able to discuss the concept of class in depth, data show that inequality in terms of class increased during the pandemic. According to ECLAC

DOI: 10.4324/9780429282348-49

(2021b), income inequality measured by the Gini coefficient increased in LAC from 0.460 in 2019 to 0.473 in 2020.

Additionally, while more than 140 million jobs were lost globally, directly affecting those dependent on labour income, global wealth increased 7.4 per cent in 2020 (ECLAC, 2021c). The contraction in employment, concentrated in the most precarious and informal jobs, shows that job losses during the crisis were uneven across different social strata, resulting in a significant drop in the total wage bill. The combination of the reduction in employment and increase in inflation has led to a worsening of social conditions and an increase in poverty, extreme poverty, and hunger. According to ECLAC (2021c), despite income and food support policies during the pandemic, the incidence of moderate or severe food insecurity reached 40.4 per cent of the LAC population in 2020, representing an increase of 6.5 percentage points compared to 2019.

Although poverty and extreme poverty were on the rise in the region after 2015 (ECLAC, 2021b), from 2019 to 2020 8 million people in the region fell to extreme poverty (totaling 78 million in extreme poverty) and 22 million to poverty (totaling 209 million in poverty).

2.2 Gender inequality

In a general way, authors have highlighted how the COVID-19 crisis has increased gender inequalities internationally (Bidegain et al., 2020; Collins et al., 2020) and in Brazil (Matos de Oliveira et al., 2022; Matos de Oliveira and Alloatti, 2021). The main reasons refer to complex and structural features. For LAC, we highlight women's labour market participation and its characteristics.

First, the greater female presence among informal workers and the poor – and in sectors highly affected, such as tourism, care work, and paid domestic care. Sectors at high risk of employment loss concentrate around 56.9 per cent of the employment of women and 40.6 per cent of the employment of men in Latin America. In the Caribbean, 54.3 per cent of female employment and 38.7 per cent of male employment are concentrated in high-risk sectors. In the Caribbean subregion, tourism represents 11.9 per cent of female employment and 5.7 per cent of male employment (ECLAC, 2021d). Furthermore, the over-representation of women in the front line during the pandemic is worthy of mention, especially in the commerce and health sector (ECLAC, 2021c).

Second, women were more affected than men in terms of loss of occupation during the COVID-19 crisis (Collins et al., 2020). Moreover, female unemployment rates are masked by a reduction of female participation in the labour force due to the increased demand of unpaid domestic and care work. The closing of schools and day-care centers (ECLAC, 2021c) led to an enhanced economic dependency on the male partner (Bidegaín et al., 2020). Simultaneously, this contributed to the confinement of a victim with her abuser, in case of domestic abuse (Kabeer, 2020). According to ECLAC/IOL (2021a), the reincorporation of women into the labour force at a slower rate occurs in a context in which many school and care activities have not yet been fully resumed. Due to their greater participation in sectors that have not yet completely recovered, such as paid domestic work and employment in the service and commercial sectors, women have greater difficulties in reintegrating into paid work.

2.3 Age inequality

Regarding age discrimination, ECLAC (2021c) highlights that during the pandemic, children and adolescents are at greater risk of experiencing food insecurity, violence, and increased exposure to child labour as educational institutions were closed. This is particularly salient in developing countries due to the dependence on public institutions for childcare, meals, and social assistance. For the elderly population, social distancing produced isolation in a general fashion, yet with a heavier toll among low-income groups. The latter rely significantly on familiar unpaid care work and regular assistance for shopping, cooking, and dealing with public services.

Specifically for the youth, the pandemic could represent a five-year setback in the reduction of the specific fertility rate in adolescents in LAC (ECLAC, 2021c) hampering educational and development trajectory. As for the labour market, youth in LAC face structural barriers that have worsened during the pandemic. In seven out of eight countries in LAC, the age group most affected by the reduction in employment during the pandemic was young people (ECLAC, 2021c). The pandemic has impacted significantly among young women in terms of job losses and decreased labour market participation (RCWG, 2021).

2.4 Regional inequality

Regional inequality has also increased during the pandemic, as well as inequality among urban and rural settings. According to ECLAC (2021b), urban areas have generally suffered the greatest health impact of COVID-19 (as they are faster-spreading locations). Yet, rural sectors are more vulnerable to its social and economic effects. This derives from limited access to public clinics – characterised by scarce resources and equipment – and social assistance centers and welfare. Moreover, restrictions around good-quality food (distribution and availability) have increased poor health conditions, paired with long-term poverty. Thus, gaps with urban areas regarding labour conditions, poverty, and food security have widened.

2.5. Racial and ethnic inequality

Structural deficits in access to services and in living conditions in LAC intersect and potentiate various axes of the social inequality matrix (Matos de Oliveira et al., 2021). This makes indigenous and Afro-descendants particularly vulnerable, conveying that race and ethnicity are key markers before and during the crisis. Racial and ethnic inequalities are intrinsically intertwined with health, social, and economic effects of the pandemic, as they outline low quality of life, restricted healthcare, and crowded households (Pires et al., 2021). This conveys the importance of intersectionality in analysing the impacts of the crisis, as it can further increase the gaps between groups and the final distance between the 'two tips of the K'.

3. Final remarks

According to ECLAC (2021b), in the absence of sufficient individual and institutional capacities to prevent, mitigate, and cope with exposure to risks, disasters fuel a downward spiral. The expected outcomes are severe limitations in the medium- and long-term for improved living conditions and overcoming poverty and inequality. This is the case of the COVID-19 crisis in LAC.

By examining this region, we have shown how distinctive demographics were affected differently by the crisis. Moreover, we address how the struggle in developing countries with modest assistance, multidimensional inequality, and limited or no recovery plans has augmented the gap between the Global North and South.

We echo recommendations that approach the COVID-19 crisis from a holistic perspective signaling the consequences of policies according to different demographics (RCWG, 2021). We argue that the lack of women in leadership positions affects the ability to develop policies targeting significant gendered aspects and issues (Kabeer, 2016: 304). In this sense, inequality is germane to recognizing and addressing i) gender relations and social reproduction; ii) race, age, and ethnic disparities; iii) job losses, informality, and labour market participation. All these three axes outline the complexity of inequality in LAC and, to a certain extent, in the Global South.

References

Bidegain, N., Scuro, L., & Trigo, I. (2020). Women's economic autonomy during the Covid-19 pandemic. *December 2020 CEPAL Review (N 132)*.

Collins, C., Landivar, L., Ruppanner, L., & Scarborough, W. (2020). COVID-19 and the gender gap in work hours. *Gender, Work & Organization*, 28(S1), 101–112. https://doi.org/10.1111/gwao.12506

de Oliveira, A. L. M. and Alloatti, M. N., (2022). Gendering the crisis: austerity and the Covid-19 pandemic in Brazil. *Economia Politica*, 39(1), pp. 203–224.

Demertzis, Maria (2021). A K-shaped recovery and the role of fiscal policy. *Bruegel-Blogs*. https://go.gale.com/ps/i.do?p=AONE&u=anon~b6e0ddd2&id=GALE|A653672505&v=2.1&it=r&sid=googleScholar&asid=6cdb688b Last accessed 25th November 2022

ECLAC (2021a). *Panorama Fiscal de América Latina y el Caribe, Santiago, 2021*. https://repositorio.cepal.org/bitstream/handle/11362/46808/1/S2100170_es.pdf Last accessed 25th November 2022

ECLAC (2021b). *Disasters and Inequality in a Protracted Crisis. Towards Universal, Comprehensive, Resilient and Sustainable Social Protection Systems in Latin America and the Caribbean*. Fourth session of the regional conference on Social Development in Latin America and the Caribbean. Virtual meeting, 26–28 October. https://repositorio.cepal.org/bitstream/handle/11362/47376/3/S2100467_en.pdf Last accessed 25th November 2022

ECLAC (2021c). La paradoja de la recuperación en América Latina y el Caribe. Crecimiento con persistentes problemas estructurales: desigualdad, pobreza, poca inversión y baja productividad. *Informe especial COVID-19 (N 11)*. https://repositorio.cepal.org/bitstream/handle/11362/47043/5/S2100379_es.pdf Last accessed 25th November 2022

ECLAC (2021d). La autonomía económica de las mujeres en la recuperación sostenible y con igualdad. *Informe especial COVID-19 (N 9)*. https://repositorio.cepal.org/bitstream/handle/11362/46633/5/S2000740_es.pdf Last accessed 25th November 2022

ECLAC/OIT (2021). Políticas de protección de la relación laboral y de subsidios a la contratación durante la pandemia de COVID-19. *November 2021 (N 25)*. https://repositorio.cepal.org/bitstream/handle/11362/47510/3/S2100695_es.pdf Last accessed 25th November 2022

Kabeer, Naila (2016). Gender equality, economic growth, and women's agency: The "endless variety" and "monotonous similarity" of patriarchal constraints. *Feminist Economics*, 22(1), 295–321.

Kabeer, Naila (2020). Labour market inequalities are exacerbated by Covid-19. *LSE COVID-19 Blog*. https://eprints.lse.ac.uk/104456/

Matos de Oliveira, Ana Luíza, & Fares, Lygia Sabbag (2021). Sexual division of labour in Brazil: Interpretations of domestic and care work under COVID-19, *Agenda*, 35(4), 48–59. https://doi.org/10.1080/10130950.2021.2015167

Pires, Luiza Nassif, de Carvalho, Laura Barbosa, & Rawet, Eduardo Lederman (2021). Multi-dimensional inequality and Covid-19 in Brazil. *Investigación Económica*, 80(315), 33–58. https://www.jstor.org/stable/26965502 Last accessed 25th November 2022

Rutgers Center for Women's Global Leadership (2021). *A COVID-19 Feminist Recovery Plan to Achieve Substantive Gender Equality (Issue June)*. https://cwgl.rutgers.edu/blog-details/644-a-covid-19-feminist-recovery-plan-to-achieve-substantive-gender-equality Last accessed 25th November 2022

Additional resources

CEPALSTAT Statistical Databases and Publications. *ECLAC. United Nations*. https://statistics.cepal.org/portal/cepalstat/index.html?lang=en

Economic Survey of Latin America and the Caribbean. Published since 1948, this report examines various aspects of the previous year's macroeconomic situation in the region and makes projections for the coming months. The study also includes country notes that review the performance of the main economic indicators in the period analysed. *ECLAC. United Nations*. https://www.cepal.org/en/taxonomy/term/8122

Montoya, M., Krstikj, A., Rehner, J., & Lemus-Delgado, D. (Org.) (2021). *COVID-19 and Cities: Experiences, Responses, and Uncertainties*. 1st ed, v. 1. Cham: Springer.

Oliveira, A. L. M. (2020). *Save Lives and Guarantee the Rights of the Black Population*. Available in English, Spanish, Portuguese, and French. CLACSO. https://www.clacso.org/salvar-vidas-e-garantir-direitos-da-populacao-negra/

Sempreviva Organização Feminista (SOF) (2020). *SEM PARAR: o trabalho e a vida das mulheres na pandemia*. https://mulheresnapandemia.sof.org.br/

[faded, illegible reference text]

Part 5

People and the planet. Editorial introduction

Though the climate crisis has long demanded urgent attention, like many disciplines, development studies has not always prioritised environmental concerns. However, as the world grapples with the realities of climate change, dwindling natural resources, the loss of biodiversity, and the broader consequences of human actions on the environment, the field of development studies has found itself re-evaluating its priorities. In doing so it confronts debates concerning how best to meet human needs whilst addressing the immediacy of climate change, determining whose needs ought to be prioritised and the differential impacts of and opportunities to adapt to change in the context of global inequalities. In many instances, countries in the Global South bear the brunt of environmental degradation and climate change, despite contributing minimally to the problem relative to countries of the Global North.

The transformation of existing power relations is essential to achieving environmental justice. This means challenging the unequal distribution of power and resources that has led to the disproportionate burden of environmental degradation being borne by hitherto marginalised communities. It also means ensuring that those who are most affected by environmental change have a say in the decisions that are made about how to address it.

The chapters in this section converge on exploring relationships between development and the environment. They examine a range of interdependencies between human needs and actions, environmental sustainability, and the contested and changing goals of development. Whilst environmental concerns remain a major cross-cutting theme throughout this book, we present these chapters together, with the aim of underscoring the growing centrality of environmental concerns in development studies.

Development is in essence about improving human well-being. Yet, this well-being is deeply anchored to the health of our planet. Recognising the dynamics of human-environment interactions and the implications of these interactions is paramount. We can no longer afford to overlook how human endeavours impact the environment and, conversely, how environmental change influences human societies.

Historically, dominant narratives in development, particularly in the Global North, focused on a vision of human progress that often considered environmental concerns as peripheral. Where environmental concerns were addressed, it was mainly through the lens of the effective and efficient use of natural resources to meet human needs. Broader environmental issues, such as loss of biodiversity and global heating, were considered separate and secondary. However, the increasing visibility of the climate crisis has shifted the environment to a more central position in development studies. This has prompted a

DOI: 10.4324/9780429282348-50

re-evaluation of development models, especially those that emphasise economic growth, industrialisation, production, and consumption. Traditional metrics like Gross Domestic Product (GDP) and the Human Development Index, which arguably failed to fully account for the environmental costs of industrialisation and consumption, have been widely critiqued, whilst new frameworks like the circular economy and doughnut economics have emerged to offer more balanced approaches that consider both economic progress and ecological responsibility.

Several themes emerge from the chapters in this section, including the dynamics between localised and global environmental challenges, the socio-economic repercussions of environmental degradation, and the ethical dimensions of development amid the climate emergency. Additionally, the politics of environmental knowledge and its influence on development strategies are important themes not only in this section but also throughout the Companion.

This repositioning of the environment within development studies has profoundly influenced the field. It calls for redefining development objectives to prioritise sustainability and a broader sense of human and planetary well-being over immediate gains. It also highlights the need for international collaboration, as environmental challenges often transcend geopolitical boundaries. On reflecting on chapters in this section, it becomes clear that the success and sustainability of development studies itself hinges on its commitment to environmental concerns.

Chapter 46

Sustainable development

Michael Redclift

Discourses of 'sustainable development'

The expression 'sustainable development' has been used in a variety of ways, particularly within the context of development studies. Today we are confronted with several different discourses of 'sustainable development', some of which are mutually exclusive. For example, campaigners for greater global equality between nations, huge international corporations and local housing associations have all had recourse to the term 'sustainable development' to justify their actions.

Sustainable development was defined by the Brundtland Commission in the following way: 'development that meets the needs of the present without compromising the ability of future generations to meet their own needs' (Brundtland Commission, 1987). This definition has been brought into service in the absence of agreement about a process which almost everybody thinks is desirable. However, the simplicity of this approach is deceptive and obscures underlying complexities and contradictions. It is worth pausing to examine the apparent consensus that reigns over sustainable development.

First, following the Brundtland definition, it is clear that 'needs' themselves change, so it is unlikely (as the definition implies) that those of future generations will be the same as those of the present generation. The question then is, where does 'development' come into the picture? Obviously development itself contributes to 'needs', helping to define them differently for each generation and for different cultures.

This raises the second question, not adequately covered by the definition, of how needs are defined in different cultures. Most of the 'consensus' surrounding sustainable development has involved a syllogism: sustainable development is necessary for all of us, but it may be defined differently in terms of each and every culture. Furthermore, how do we establish which course of action is *more* sustainable? Recourse to the view that societies must decide for themselves is not very helpful. (Who decides? On what basis are the decisions made?) At the same time there are problems in ignoring culturally specific definitions of what is sustainable in the interest of a more inclusive system of knowledge. There is also considerable confusion surrounding *what* is to be sustained. One of the reasons why there are so many contradictory approaches to sustainable development (although not the only reason) is that different people identify the objects of sustainability differently.

DOI: 10.4324/9780429282348-51

What is to be sustained?

For those whose primary interest is in ecological systems and the conservation of natural resources, it is the natural resource base which needs to be sustained. The key question usually posed is the following: how can development activities be designed which help to maintain eco-logical processes, such as soil fertility, the assimilation of wastes and water and nutrient recycling? Another, related issue is the conservation of genetic materials, both in themselves and (perhaps more importantly) as part of complex and vulnerable systems of biodiversity. The natural resource base needs to be conserved because of its intrinsic value.

There are other approaches, however. Some environmental economists argue that the natural stock of resources – or 'critical natural capital' – needs to be given priority over the flows of income which depend upon it (Pearce, 1991). They make the point that human-made capital cannot be an effective substitute for natural capital. If our objective is the sustainable yield of renewable resources, then sustainable development implies the management of these resources in the interest of the natural capital stock. This raises a number of issues which are both political and distributive: who owns and controls genetic materials, and who manages the environment? At what point does the conservation of natural capital unnecessarily inhibit the sustainable flows of resources?

Second, according to what principles are the social institutions governing the use of resources organised? What systems of tenure dictate the ownership and management of the natural resource base? What institutions do we bequeath, together with the environment, to future generations? Far from taking us away from issues of distributive politics and political economy, a concern with sustainable development inevitably raises such issues more forcefully than ever (Redclift, 1987; Redclift and Sage, 1999).

The question 'what is to be sustained?' can also be answered in another way. Some writers argue that it is present (or future) levels of production (or consumption) that need to be sustained. The argument is that the growth of global population will lead to increased demands on the environment, and our definition of sustainable development should incorporate this fact. At the same time, the consumption practices of individuals will change too. Given the choice, most people in India or China might want a television or an automobile of their own, like households in the industrialised North. What prevents them from acquiring one is their poverty, their inability to consume, and the relatively 'undeveloped' infrastructure of poor countries.

Is there anything inherently unsustainable in broadening the market for TV sets or cars? The different discourses of 'sustainable development' have different answers to this question. Many of those who favour the sustainable development of goods and services that we receive through the market – and businesses – would argue that we should broaden the basis of consumption. Others would argue that the production of most of these goods and services today is inherently unsustainable – that we need to 'downsize' or shift our patterns of consumption. In both developed and, increasingly, developing countries, it is frequently suggested that it is impossible to function effectively without computerised information or access to private transport.

The different ways in which 'sustainability' is approached, then, reflect quite different underlying 'social commitments', that is the patterns of everyday behaviour that are seldom questioned. People define their 'needs' in ways which effectively exclude others from meeting theirs and, in the process, can increase the long-term risks for the sustainability of other people's livelihoods. Most importantly, however, the process through which we enlarge our choices – and reduce those of others – is largely invisible to people in their daily lives.

Unless these processes are made more visible, 'sustainable development' discourses beg the question of whether – or how – environmental costs are passed on from one group of people to another, both within societies and between them. The North dumps much of its toxic waste and 'dirty' technology on poorer countries and sources many of its 'needs' – for energy, food and minerals – from the South. At the same time, the elevated lifestyles of many rich and middle-class people in developing countries are dependent on the way in which natural resources are dedicated to meeting their needs. Finally, of course, the inequalities are also intergenerational, as well as intragenerational: we despoil the present at great cost to the future. Discounting the future (as economists call it), valuing the present above the future, is much easier to do in materially poor societies, where survival itself may be at stake for many people.

The final element in the redesign of 'sustainable development' policy was the creation of the 'consumer-citizen', the idea that the individual could best express their preferences for goods and services through their own (and their household's) personal consumption. Parallel with the development of cleaner technology and carbon markets came the concern with sustainable consumption. Partially as a result of their insufficient understanding of the link between social structures and consumer habits and the awkward politics of wealth redistribution, governments came to favour consumer encouragement to live more sustainably and to reduce household 'footprints'. This implied the design of new 'lighter' consumer goods, evocations to act in more environmentally responsible ways, and an accent on 'lifestyle' and the consumer at the expense of livelihoods and citizenship.

From the perspective of those most critical of market-based environmental valuation, the conjunction of newly 'liberated' markets and environmental concern was a necessary contradiction of capitalism seeking a resolution and could in hindsight be seen as a 'managed senescence' if we continue with the biological metaphors of 'development' (Redclift, 2008; Bellamy-Foster, 2010). A more mainstream view, however, would be that they addressed system failures, and it could even lead to a rejuvenated, if scarcely recognisable, type of materials 'light' capitalism (Lovins et al., 2000).

Human rights, democracy, and sustainable development

Finally, since the various discourses of sustainable development began to flourish, it has become evident that another dimension to the problem of diminished sustainability needs to be considered. This is the extent to which, at the beginning of the 21st century, we need to refer to processes of democracy and governance in the context of sustainable development. The Brundtland Report took a highly normative view of both the environment and development, as did the Earth Summit deliberations in 1992. With the second Earth Summit in 2002 in mind, it may be useful to pause to consider whether we can ever achieve 'sustainable development' without increased democratisation at all levels of society. Today questions of sustainability are linked, intellectually and politically, to other issues, such as human rights and 'identity', with which they are connected. But notions of 'rights' and 'identity' are themselves changing. In the era of genetic engineering, the genetic modification of humans, as well as plants and animals, is shaping new senses of 'identity'. As individuals change, so do the groups to which they belong. In the era of globalisation it is sometimes argued that 'sustainable development' may be more difficult to achieve, as economies converge towards shared economic objectives. At the same time it may prove impossible to achieve 'sustainable development' (if it *is* achievable) without acknowledging quite distinctive accounts of human rights in nature – and even the rights *of* nature, which were hitherto ignored. The concept of 'sustainable development', as understood by different people, is contradictory, obscure and illuminating at the same time.

References and guide to further reading

Bellamy-Foster, J. (2010) 'Marx's ecology and its historical significance', in M.R. Redclift and G. Woodgate (eds.) *The International Handbook of Environmental Sociology*, Second Edition. Cheltenham: Edward Elgar.

Brundtland Commission (World Commission on Environment and Development) (1987) *Our Common Future*, Oxford: Oxford University Press. This report led directly to the term 'sustainable development' passing into common use. It was also the first overview of the globe which considered the environmental aspects of development from an economic, social and political perspective; cf. *Man and the Biosphere* (*MAB*) almost a decade earlier. Among the principal omissions was detailed consideration of non-human species, and their 'rights'. The 'Brundtland Report' (named after its Chairperson, the Norwegian Prime Minister at the time) also opened the way for non-governmental organizations (NGOs) to be considered a serious element in environment and development issues.

Lovins, A. and P. Hawken and L. Hunter Lovins. (2000) *Natural Capitalism*. London. Earthscan.

Pearce, D. (1991) *Blueprint 2: Greening the World Economy*, London: Earthscan. Following on from Blueprint One, David Pearce illustrated the applications of economic analysis to environmental problems, in a way that particularly interested policy-makers.

Redclift, M.R. (1987) *Sustainable Development: Exploring the Contradictions*, London: Routledge. This was the first, and probably the best, treatment of 'sustainable development'. The case studies and ethnographic illustrations, combined with the accessible intellectual discussion, make this a 'classic'.

Redclift, M.R. (2008 [2009]) 'The environment and carbon dependence: Landscapes of sustainability and materiality', *Current Sociology* 57: 369–387.

Redclift, M.R. and Sage, C.L. (1999) 'Resources, environmental degradation and inequality', in A. Hurrell and N. Woods (eds.) *Inequality, Globalisation and World Politics*, Oxford: Oxford University Press. A good general overview of the relevance for development of inequality in resource endowments.

Chapter 47

The Sustainable Development Goals (SDGS)

Jonathan Rigg

Introduction: from MDGs to SDGs

In September 2015, world leaders gathered at the UN in New York and signed up to the 2030 Agenda for Sustainable Development, with the overarching aim to 'leave no one behind'. This represented the culmination of a series of activities including the 2012 Rio+20 United Nations Conference on Sustainable Development, the 2015 Sendai Framework for Disaster Risk Reduction, 4.5 million online responses to the MY World Survey, civil society engagements, and multiple high-level consultations with national governments, private sector representatives, and scientists. At the heart of this Agenda are the Sustainable Development Goals or SDGs, replacing the Millennium Development Goals (MDGs) which ran over 15 years, from 2000 to 2015.[1]

The 2030 Agenda for Sustainable Development:

> provides a shared blueprint for peace and prosperity for people and the planet, now and into the future. At its heart are the 17 Sustainable Development Goals (SDGs), which are an urgent call for action by all countries – developed and developing – in a global partnership. They recognize that ending poverty and other deprivations must go hand-in-hand with strategies that improve health and education, reduce inequality, and spur economic growth – all while tackling climate change and working to preserve our oceans and forests.
>
> (https://sustainabledevelopment.un.org/sdgs, accessed: 11.11.19).

The 17 SDGs (Table 47.1) and associated 169 targets are seen to represent a step change from the MDGs – a 'paradigm shift', in the words of the UN General Secretary Ban Ki-moon. In his speech to the UN General Assembly on 1 September 2015 launching the SDGs, he spoke of '[completing] the unfinished business of the MDGs, [lifting] the level of ambition and [tackling] emerging issues and challenges' (https://www.un.org/press/en/2015/sgsm17044.doc.htm). They comprise, he said, 'a to-do list for people and planet, and a blueprint for success'.

A transformative agenda?

A case can be made for the SDGs as transformational – a paradigm shift, in Ban Ki-moon's phrase – in four regards: how they were arrived at; what they seek to achieve; who is involved; and the means by which the SDG agenda is pursued.

DOI: 10.4324/9780429282348-52

Table 47.1 The 17 sustainable development goals

Goal 1. End poverty in all its forms everywhere

Goal 2. End hunger, achieve food security and improved nutrition, and promote sustainable agriculture

Goal 3. Ensure healthy lives and promote well-being for all at all ages

Goal 4. Ensure inclusive and equitable quality education and promote lifelong learning opportunities for all

Goal 5. Achieve gender equality and empower all women and girls

Goal 6. Ensure availability and sustainable management of water and sanitation for all

Goal 7. Ensure access to affordable, reliable, sustainable, and modern energy for all

Goal 8. Promote sustained, inclusive, and sustainable economic growth, full and productive employment, and decent work for all

Goal 9. Build resilient infrastructure, promote inclusive and sustainable industrialisation, and foster innovation

Goal 10. Reduce inequality within and among countries

Goal 11. Make cities and human settlements inclusive, safe, resilient, and sustainable

Goal 12. Ensure sustainable consumption and production patterns

Goal 13. Take urgent action to combat climate change and its impacts*

Goal 14. Conserve and sustainably use the oceans, seas, and marine resources for sustainable development

Goal 15. Protect, restore, and promote sustainable use of terrestrial ecosystems, sustainably manage forests, combat desertification, and halt and reverse land degradation and halt biodiversity loss

Goal 16. Promote peaceful and inclusive societies for sustainable development, provide access to justice for all and build effective, accountable, and inclusive institutions at all levels

Goal 17. Strengthen the means of implementation and revitalise the global partnership for sustainable development

Source: https://sustainabledevelopment.un.org/post2015/transformingourworld

The SDGs emerged out of a broad, inclusive, and participatory consultative process, as noted earlier (Pogge and Sengupta, 2015). A second distinguishing feature is that the SDGs are paraded as truly global. Unlike the MDGs which were aimed at the 'developing' world and focused on human development and, in particular, poverty alleviation, the SDGs are seen to be germane to high-income as well as low-income countries. Third, the SDGs are integrative, encompassing the co-achievement of economic, social, and environmental goals. These goals cannot be pursued separately; it is a unified project where the goals are interdependent. And fourth, the achievement of the SDGs is not down to the actions of national governments and multilateral institutions alone; the private sector and civil society are integral to the sustainable development agenda (Scheyvens et al., 2016).

In the light of such claims, proponents of the SDGs argue that they are of 'unprecedented scope and ambition, applicable to all countries, and go well beyond the MDGs' (WHO, 2015: 3).

Critiques of the SDGs

Given the ambition of the SDGs and the claims that are made of the 2030 Agenda for Sustainable Development of which they are a part, it is not surprising that scholars and development practitioners have critiqued the SDGs from the very start, just as they did the earlier MDGs (Spangenberg, 2017). These critiques call into question whether the SDGs are as radical, global, holistic, or as well defined as their advocates suggest.

To begin with, far from being a 'paradigm shift', the SDGs are seen by some critics to represent (neoliberal) business as usual (Spann, 2017). For radical scholars, the SDGs do not address the structural factors that explain global poverty and inequality and drive climate change. Far from being transformative, they perpetuate the systems that have created and sustain the challenges that the SDGs putatively address, not least growing inequality and the climate emergency.

Second, overlapping with this first critique, are the tensions that exists between the global nature of many of the challenges facing the globe and the continuing tendency for the nation state to be the key development actor (Sexsmith and McMichael, 2015). It is necessary to go beyond methodological nationalism if challenges that cross borders are to be addressed. The SDGs do not, it is claimed, grasp this nettle.[2]

A third area of concern relates to the central role accorded to the private sector – business – in the achievement of the SDGs. It is said, for instance, that business is seen as 'a *per se* benevolent actor for the public good' (Spangenberg, 2017: 316), when the contribution of business to the globe's environmental crisis, for instance, is hard to dispute. Further to this, is the tendency of the private sector to focus on technological and market and growth-centred solutions to the problems that the SDGs highlight. Sustainability with regard to agriculture, for instance, is based firmly on an agri-business and global commodity chain model of development, one that critics (e.g., Spann, 2017) maintain works against the interests of true sustainability.

Fourth, other commentators have focussed on the details of the 169 targets that underpin the SDGs. The International Council for Science (ICSU) and the International Social Science Council (ISSC), for instance, found that fewer than one-third (49) of the targets were 'well defined and based on the latest scientific evidence', while more than two-thirds either 'needed more work' (91) or were 'weak or non-essential' (29; ICSU and ISSC, 2015).

Perhaps most important, and this arguably returns to the inherent contradictions in a framework which is said to represent an 'indivisible and universal whole', is progress towards the SDG targets (while bearing in mind that the SDGs still have some years to run). The *Global Sustainable Development Report 2019* (UN, 2019) observes that 4 of the 17 goals are 'not even moving in the right direction', namely inequalities, climate change, biodiversity loss, and waste from human activity. The report continues:

> Critically, recent analysis suggests that some of those negative trends presage a move towards the crossing of negative tipping points, which would lead to dramatic changes in the conditions of the Earth system in ways that are irreversible on time scales meaningful for society.

The result will be that 'social and natural biophysical systems cannot support the aspirations for universal human well-being embedded in the Sustainable Development Goals' (2019: xx). This highlights the trade-offs and contradictions between the aims of the SDGs, not least between 'spur[ring] economic growth' and 'tackling climate change and working to preserve our oceans and forests', and therefore the impossibility of meeting the 'holistic' vision of a richer, better, and more sustainable world embodied in the exhortation at the start of this entry.

Finally, there is the question of whether the SDGs are truly global and universal. This is, arguably, the key feature that distinguishes the SDGs from the MDGs. Just because this point is endlessly repeated, however, does not give it validity, and intuitively it is hard to sustain. Many low- and middle-income countries take the SDGs seriously, allocating people and resources to their tracking and achievement; for high-income economies, the SDGs are largely ignored.

The SDGs and the global development debate

The SDGs are closely linked to the recasting of 'international development' as 'global development'. The MDGs were shaped by rich countries and targeted at poor countries, with rich – or developed – countries providing financial transfers and technological support to meet the MDGs. This was the era of international development. The SDGs, by contrast, are said to go 'beyond aid' in a new architecture of partnership. They mark, then, the emergence of an era of global development. There are a number of other shifts that echo this distinction between the MDGs and the SDGs: in 2016 the World Bank announced that it was no longer going to distinguish between developed and developing countries; development cooperation is far from being a simple flow of 'aid' from North to South as new actors enter the field of overseas development assistance; the spectacular growth of countries like China has led to the 'rise of the South' and a 'great convergence' in global wealth; while climate change is the global challenge par excellence (Horner and Hulme, 2019). The SDGs play to these trends and transformations. The risk, however, is that in presenting development as global, the deep inequalities that continue to characterise the world – and in some respects are deeper now than they were under the MDGs – will be shaded from sight.

Figure 47.1 The 17 sustainable development goals

Credit: Shutterstock / checy

Source: https://www.un.org/sustainabledevelopment/news/communications-material/

Resources

The SDG website provides an abundance of material on the SDGs, including the annual *Global Sustainable Development Report* (download from https://sustainabledevelopment.un.org/globalsdreport/) as well as annual Sustainable Development Goals Progress Reports (download from https://www.un.org/development/desa/en/news/sustainable/sdg-progress-reports-2019.html).

Notes

1 See Rigg 2014 for a critical appraisal of the MDGs in an earlier edition of this *Companion to Development Studies*.
2 Somewhat at odds with this last point is the contention that in adopting a global perspective the SDGs may overlook the special and particular problems that face individual countries and regions (see, for example, Valensisi and Karingi, 2017 on the case of Africa). There is a risk that the specific challenges facing individual countries may be dissolved in a global initiative.

References

Horner, R. and Hulme, D. (2019). "From international to global development: New geographies of 21st century development." *Development and Change* 50(2): 347–378. https://doi.org/10.1111/dech.12379.

ICSU and ISSC (2015). *Review of Targets for the Sustainable Development Goals: The Science Perspective*. Paris: International Council for Science (ICSU). https://council.science/cms/2017/05/SDG-Report.pdf.

Pogge, T. and Sengupta, M. (2015). "The sustainable development goals (SDGs) as drafted: Nice idea, poor execution." *Washington International Law Journal* 24(3): 571–587. https://digitalcommons.law.uw.edu/wilj/vol24/iss3/8.

Scheyvens, R., Banks, G. and Hughes, E. (2016). "The private sector and the SDGs: The need to move beyond business as usual." *Sustainable Development* 24(6): 371–382.

Sexsmith, K. and McMichael, P. (2015). "Formulating the SDGs: Reproducing or reimagining state-centered development?" *Globalizations* 12(4): 581.

Spangenberg, J. H. (2017). "Hot air or comprehensive progress? A critical assessment of the SDGs." *Sustainable Development* 25(4): 311.

Spann, M. (2017). "Politics of poverty: The post-2015 sustainable development goals and the business of agriculture." *Globalizations* 14(3): 360–378.

UN (2019). *Global Sustainable Development Report 2019: The Future is Now – Science for Achieving Sustainable Development*, New York: United Nations. https://sustainabledevelopment.un.org/content/documents/24797GSDR_report_2019.pdf.

Valensisi, G. and Karingi, S. (2017). "From global goals to regional strategies: Towards an African approach to SDGs." *African Geographical Review* 36(1): 45–60.

WHO (2015). *Health in 2015: From MDGs to SDGs*, Geneva: World Health Organization.

Chapter 48

Transformations to sustainability

Lakshmi Charli-Joseph and Jesús Mario Siqueiros-García

Introduction

The study of global change and pathways towards sustainability involves the analysis of profound social changes at different scales. These changes can be manifested as radical, directional shifts or incremental modifications in regulatory, technical, and cultural aspects (O'Brien, 2012). Such changes may or may not be planned, but they are always the result of complex and multi-scalar processes that involve multiple actors and occur at different spatial dimensions and temporal rates (Pelling et al., 2015). A better understanding of how lasting social changes occur and how they can be promoted is essential for dealing with the pressing challenges of sustainability on the planet.

Transformations, as the most profound type of change, involve fundamental modifications of social-ecological systems (SES) by challenging the status quo and creating new policy response areas, new practices, and new ways of relating to others and the environment (O'Brien, 2012; Olsson et al., 2014; Pelling et al., 2015). Transformations are changes in structure, function, and relations that lead to genuine and qualitative new forms of interactions and outcomes (Patterson et al., 2015) and are required to foster more sustainable social-ecological systems (SES) when these exhibit undesirable or untenable conditions (Redman, 2014). Therefore, transformations towards sustainability imply deep shifts that necessarily alter the constitutive human and environmental interactions and feedback.

O'Brien and Sygna (2013), for example, argue that transformations must be conceived of as taking place in three spheres: the personal, the practical, and the political. Within the personal sphere, reflexivity, confronting one's own and collective worldviews and making values explicit, is part of a necessary process of personal transformation. The practical sphere entails people's goals and the strategies and actions they employ to achieve them. The political sphere includes the structures, institutions, and processes beyond the individual that influence system dynamics. As O'Brien and Sygna (2013) point out, although these spheres are interdependent, the interactions among them are rarely sufficiently explored in transformation efforts.

From a wider perspective, Scoones and colleagues (2020) have suggested three different approaches to deliberate transformations: structural, systemic, and enabling. Structural approaches deal with the foundations upon which politics, economy, and society are sustained. Systemic approaches are those that 'emphasise the need for knowledge on system dynamics: the interdependency of social, ecological, institutional and technological elements that together mark thresholds in system states' (p. 67). Finally, enabling

DOI: 10.4324/9780429282348-53

approaches centre their efforts in human agency. These three approaches are not mutually exclusive and compromise complementary analytical lenses on transformative change.

This chapter delves into certain aspects related to enabling approaches to transformations and the importance to promote agency through processes of reframing. The chapter ends with a reference to Transformation Laboratories, as spaces for exploring and enabling agency. Enabling approaches underscore creating social capacities that can empower people to take meaningful action on their own behalf. As their focus is more on process rather than outcomes, these approaches seek to explore how agents can mobilise towards desired change – from grassroot movements to more radical activist endeavours. Therefore, this entails paying more explicit attention to the diverse values, power asymmetries, ways of relating, and agency of the different individuals and groups that are involved in such processes (Scoones et al., 2020).

Human agency in transformation processes

Human agency is an implicit and sometimes explicit concept in the burgeoning literature on transformation. According to Hewson (2010), agency can be present as individual agency, proxy agency, and collective agency. Individual agency is the most basic and it refers to individuals acting. Proxy agency is when someone acts on behalf of someone else (e.g., when an employee acts on behalf of the employer). Collective agency is when a group of individuals organise in such a way that they act as a group. Collective agency is more than the sum of individual agencies since the intentions and the actions cannot be an attribute of a single individual but of a group structured in certain ways that require a system of values and social cohesion (e.g., firms, states, and social movements; Hewson, 2010).

In the context of transformations, agency in its different forms is considered an essential ingredient for any effective and fundamental change that might be expected from sustainability interventions (O'Brien and Sygna, 2013; Westley et al., 2013; Pesch, 2015; Marshall et al., 2018). Thus, agency can be understood as the capacity of an individual to act in pursuit of goals set by the agent her/himself in accordance with her/his values (Sen, 1999; Alkire and Deneulin, 2009; Siqueiros-García et al., 2022) and is constrained by resources and power relationships but also by motivations, cognition, and emotions (O'Brien, 2012). To act means that the agent can make use of and mobilise the resources available, which may include others as means to an end.

Agency, whether individual, proxy, or collective, is not an essential property but something that is built in interaction with the environment and with others, thus it is context dependent. To build agency it is necessary to bring about conditions that foster an iterative interplay of the individual and the collective to create a space that will enable its emergence at both levels. Such is a cognitive space as it fosters sense-making but also an affective one as it depends on the dynamics of mutual affectation of social interactions (O'Brien and Sygna, 2013; Pesch, 2015; Charli-Joseph et al., 2018, Maiese 2019. Consequently, the notion of human agency, in the context of deliberate transformations more recently typified as transformative agency (e.g., Westley et al., 2013), must be central to the design of any space and process aimed at generating significant transformation.

While agency is a recurrent theme in the evolution of resilience thinking and sustainability science, there is yet no consensus on how to operationalise the concept in social-ecological and transformative research. Most conceptualisations of agency are based on observations of the actions of an entity that embodies agency (i.e., agents do this and that), in which the agent itself is a black box full of intentions. SES dynamics are contingent on human agents' intentions towards the system, which include deliberate efforts to maintain or alter the system's emergent structures and identity (Manuel-Navarrete, 2015). Nevertheless, there is still a lack of empirical evidence on how to identify, foster, and evaluate agency.

Enabling agency through reframing

Placing human agency at the centre of social-ecological transformations highlights the intrinsic involvement of humans in emergent structures of SES, but due to system complexity, invoking the emergence of a desirable transformation requires people to more clearly 'perceive' and 'feel' the system to recognise that they, as agents, are part of it. Thus, self-reflexive processes may challenge, transform and open up the frames of reference through which an individual knows and behaves within the world (Mezirow, 2000). Self-reflexivity targets the transformation of frames of reference from which an individual understands system structures, roles, and problem situations and relates to the world (Scott, 2003). Because transformation is relational it is better facilitated in a collective setting since it is through putting into perspective one's own frame of reference with those of others that new, broadened frames of reference emerge (Scott, 2003). In this sense, processes of both individual and collective reflexivity and reframing are central concerns in the design of methods and spaces for exploring transformations to sustainability.

In this view, particularly when deliberation is unaware of the positionalities and contrasting possible framings, reflexivity is necessary for 'opening up' spaces to constructively challenge such framings of a problem or situation, as well as subjective narratives about preferred pathways of social-ecological change (Manuel-Navarrete and Pelling, 2015). Through its focus on reflexivity but also on self-reflexive processes around examining one's own frames (shaped by beliefs, values, symbolic boundaries, agendas, goals, or methods), this approach contributes to the formalisation of human agency in SES dynamics (Stirling, 2008; Barnett et al., 2014). Human agency includes the creation of framings and narratives about transformative pathways of change and the deliberate interventions that are guided by these framings and narratives (Marshall et al., 2021).

Transformative space-processes aim at facilitating change towards a more sustainable and just future; the implication is that transformation is a collective activity, and designed spaces-processes can facilitate the social interactions necessary to catalyse such change. Because transformation processes necessarily must mediate individual perspectives and subjectivities with the imperative to generate shared meanings and goals, it is crucial to elicit, understand, and share the particular frames of the participants (Eakin et al., 2019). Accordingly, there has been increasing interest in studying the associated cognitive processes (i.e. double-loop and triple-loop learning, reflexivity, and reframing processes) of the actors that participate in such spaces (Pahl-Wostl, 2006; Lyon and Parkins, 2013; Lumosi et al., 2019), as strategies for promoting conceptual and behavioural change.

While agency is a central concept in the literature on transformations to sustainability, little has been said about the connection between reframing and agency. Thus, for agency to be transformative, its emergence needs to entail a reframing of the problem situation and of the roles and positions of the agents in a system. Reframing requires specific tools and methods of engagement to create safe-enough spaces where participants are allowed to be vulnerable and truly open (Ramadier, 2004). Transformative spaces thus potentially can enable the reframing of a situation in ways that allow the emergence of agency.

Transformation laboratories

The concept of 'laboratory' has been mobilised as a means of capturing the idea of a space or suite of collaborative activities designed to support experimental social processes, aimed at tackling complex social and environmental challenges (e.g., *transition labs, food labs, change labs*, etc.). For example, *social innovation laboratories* are defined as 'highly designed and expert-facilitated processes intended to support multi-stakeholder groups in addressing a complex social problem' (Page 7, Social Innovation Lab Guide -SILG). The *Transformation Laboratories* (or T-Labs) approach was adapted from the concept of social innovation labs (Westley and Laban, 2012), with the main goal of developing a space in which participants can test multiple solutions or strategies for change, which together could address a social-ecological challenge. This approach was designed to help to collectively frame a challenge, create early prototypes of 'interventions', and build momentum for action, while strengthening the capacity of change agents to address the challenge at stake and, thus, foster transformations to sustainability.

Furthermore, the transformation aspects for the T-Labs' processes include: i) a dominant domain to be challenged; ii) the existence of diverse perspectives that see the problem in a different way; iii) the enablement of the participants' agency; iv) the institutional context to act and the partnerships or alliances (present and potential); v) the identification of a window of opportunity; and the explicit recognition of the social-ecological component. Accordingly, a T-Lab process should be convened when the following conditions are present: i) the problem is recognised as complex with no clear solution; ii) a 'convener' is present with significant ownership over the problem and strong motivation; iii) there exists dissention over the issue and a sense that current practice is no longer an option; iv) there is an opportunity for transformation; v) there is a collective sense of urgency for change; and vi) there is a history of unsuccessful experiments with alternative solutions (Pathways, 2021; Pereira et al., 2021).

Hence, the emphasis of T-Labs is not just on a multi-stakeholder-based, creative search for novel solutions or interventions. T-Labs might also focus on facilitating collective learning about the nature of a problem or challenge and about different kinds of possible solutions or pathways of possible change. Accordingly, T-Labs aim to identify which actors have transformative agency, helping to create a collective sense of the need for change – within and beyond the stakeholders directly involved, by co-constructing strategies for affecting change (Pereira et al., 2021). By positioning multiple intentions into collectively desirable pathways, T-Labs are places to 'open up' spaces and make plural futures happen. Hence, the T-Labs approach has emerged as a means to provide interactive innovation spaces that allow for experimentation with new social-ecological-technological system configurations and sustainability pathways.

Conclusion

Although there is increasing consensus on the importance of examining and fostering more profound deliberate and non-incremental type of changes in social-ecological systems (e.g., O'Brien, 2012; Olsson et al., 2014; Pelling, 2014; Pelling et al., 2015; Abson et al., 2016), and while there is growing agreement on the crucial role that human agency plays in fostering processes of social-ecological transformation (e.g., Westley et al., 2011; Westley et al., 2013), there is still insufficient empirical evidence and methodological development in the criteria and methods that contribute to building agency to enhance the potential for novel interventions in sustainability challenges (Tschakert et al., 2016; Abson et al., 2016; Charli-Joseph et al., 2018). In particular, there is still a need to further examine and understand how to design and implement spaces that may foster the emergence of agency, as many processes appear to assume its presence or emergence, without specific attention to what approaches and tools might be mobilised to construct spaces amenable to its emergence.

T-Labs seem adequate for enabling approaches since they work as spaces for experimentation about different ways to promote agency. T-Labs are safe-enough spaces for exploring new social interactions, one's own views, and others, while critically assessing them from deep mutual respect. Moreover, power asymmetries can be collectively analysed and re-worked in favour of the group and its individual participants. These ambitions, if explicitly embraced in the design of a T-Lab, can address an existing gap in transformations work, so to speak, 'how to engage with and more fully reflect the lived struggles, aspirations and hopes of diverse groups of people' (Shah et al., 2018: 258), along with how to engage with different understandings of what comprises a desirable future in a specific context.

Additional resources

Blog: *How rethinking local people's agency could help navigate Xochimilco's troubled waters.* https://steps-centre.org/blog/new-forms-agency-help-navigate-xochimilcos-troubled-waters/
Blog: *What 'agency' do researchers have in transformative research projects?* https://steps-centre.org/blog/agency-researchers-transformative-research-projects/
Guide: Pathways Network (2018) *T-Labs: A Practical Guide – Using Transformation Labs (T-Labs) for innovation in social-ecological systems.* https://steps-centre.org/news/new-publication-t-labs-a-practical-guide/
Guide: Ruizpalacios B, Charli-Joseph L, Eakin H, Siqueiros-García JM, Manuel-Navarrete D, Shelton R (2019) *The Transformation Laboratory of the social-ecological system of Xochimilco, Mexico City: Description of the process and methodological guide.* Mexico City, Mexico: LAN-CIS-IE, UNAM.
Manuel-Navarrete D, Charli-Joseph L, Eakin H, & Siqueiros-Garcia JM (2021) Applying technologies of the self in transformation labs to mobilize collective agency. *Social Innovations Journal*, 5. Link.
Video: *From being rooted to uprooted, and from trapped to transformed.* https://steps-centre.org/publication/from-being-rooted-to-uprooted-and-from-trapped-to-transformed-voices-of-xochimilco/

References

Abson DJ, Fischer J, Leventon J, et al. (2017) Leverage points for sustainability transformation. *Ambio.* 46, pp. 30–39. https://doi.org/10.1007/s13280-016-0800-y
Alkire S, Deneulin S (2009) The human development and capability approach. In: Deneulin S, Sahani L (eds) *An Introduction to the Human Development and Capability Approach.* London: Earthscan, pp. 22–48.
Barnett J, Graham S, Mortreux C, et al. (2014) A local coastal adaptation pathway. *Nature Climate Change* 4: 1103–1108. https://doi.org/10.1038/nclimate2383

Charli-Joseph L, Siqueiros-Garcia JM, Eakin H, et al. (2018) Promoting agency for social-ecological transformation: A transformation-lab in the Xochimilco social-ecological system. *Ecology & Society* 23. https://doi.org/10.5751/ES-10214-230246

Eakin H, Siqueiros-García JM, Hernández-Aguilar B, et al. (2019) Mental models, meta-narratives, and solution pathways associated with socio-hydrological risk and response in Mexico City. *Frontiers in Sustainable Cities* 0: 4. https://doi.org/10.3389/FRSC.2019.00004

Hewson M (2010) Agency. In A. Mills, G. Durepos, & E. Wiebe (Eds.), *Encyclopedia of case study research* (pp. 12–16). London. SAGE Publications, Inc.

Lumosi CK, Pahl-Wostl C, Scholz G (2019) Can 'learning spaces' shape transboundary management processes? Evaluating emergent social learning processes in the Zambezi basin. *Environmental Science & Policy* 97: 67–77. https://doi.org/10.1016/j.envsci.2019.04.005

Lyon C, Parkins JR (2013) Toward a social theory of resilience: Social systems, cultural systems, and collective action in transitioning forest-based communities. *Rural Sociology* 78: 528–549. https://doi.org/10.1111/RUSO.12018

Maiese ML (2019) Transformative learning and the affordance of flexible habits of mind. *Learning, Design, and Technology* 1–22. https://doi.org/10.1007/978-3-319-17727-4_153-1

Manuel-Navarrete D and Pelling, M (2015). Subjectivity and the politics of transformation in response to development and environmental change. *Global Environmental Change*, 35, pp. 558–569.

Manuel-Navarrete D (2015) Double coupling: Modeling subjectivity and asymmetric organization in social-ecological systems. *Ecology & Society* 20. https://doi.org/10.5751/ES-07720-200326

Marshall F, Dolley J, Priya R (2018) Transdisciplinary research as transformative space making for sustainability: Enhancing pro-poor transformative agency in peri-urban. *Ecology & Society* 10249: 1–28. https://doi.org/10.5751/ES-10249-230308

Marshall F, Van Zwanenberg P, Eakin H, et al. (2021) Reframing sustainability challenges. *Transform Pathways to Sustainability* 187–205. https://doi.org/10.4324/9780429331930-15

Mezirow J (2000) Learning to think like an adult. *Learn as Transform Crit Perspect a Theory Progress* 5: 3–33. https://doi.org/10.1097/00001416-200407000-00021

O'Brien K (2012) Global environmental change II: From adaptation to deliberate transformation. *Progress in Human Geography* 36: 667–676. https://doi.org/10.1177/0309132511425767

O'Brien K, Sygna L (2013) Responding to climate change: The three spheres of transformation. *Proceedings of Transformation in a Changing Climate* 16–23.

Olsson P, Galaz V, Boonstra WJ (2014) Sustainability transformations: A resilience perspective. *Ecology & Society* 19. https://doi.org/10.5751/ES-06799-190401

Pahl-Wostl C (2006) *The Importance of Social Learning in Restoring the Multifunctionality of Rivers and Floodplains*. https://www.jstor.org/stable/26267781?seq=1#metadata_info_tab_contents. Accessed 31 December 2021

Pathways N (2021) *Transformative Pathways to Sustainability: Learning Across Disciplines, Cultures and Contexts*, 1st ed. Amsterdam: Routledge.

Patterson J, Schulz K, Vervoort J, et al. (2015) Transformations towards sustainability. In: *Emerging Approaches, Critical Reflections, and a Research Agenda. Earth System Governance Working Paper No 33. Lund Amsterdam Earth System Governance Project*. Abdingdon.

Pelling M (2014) Transformation: A renewed window on development responsibility for risk management. *Journal of Extreme Events* 1: 1402003. https://doi.org/10.1142/S2345737614020035

Pelling M, O'Brien K, Matyas D (2015) Adaptation and transformation. *Climate Change* 133: 113–127. https://doi.org/10.1007/s10584-014-1303-0

Pereira L, Olsson P, Charli-Joseph L, et al. (2021) Transdisciplinary methods and T-Labs as transformative spaces for innovation in social-ecological systems. *Transform Pathways to Sustainability* 53–64. https://doi.org/10.4324/9780429331930-6

Pesch U (2015) Tracing discursive space: Agency and change in sustainability transitions. *Technological Forecasting and Social Change* 90: 379–388. https://doi.org/10.1016/j.techfore.2014.05.009

Ramadier T (2004) Transdisciplinarity and its challenges: The case of urban studies. *Futures* 36: 423–439. https://doi.org/10.1016/J.FUTURES.2003.10.009

Redman CL (2014) Should sustainability and resilience be combined or remain distinct pursuits? *Ecology & Society* 19. https://doi.org/10.5751/ES-06390-190237

Scoones I, Stirling A, Abrol D, et al. (2020) Transformations to sustainability: Combining structural, systemic and enabling approaches. *Current Opinion in Environmental Sustainability* 42: 65–75. https://doi.org/10.1016/j.cosust.2019.12.004

Scott SM (2003) The social construction of transformation. *Journal of Transformative Education* 1: 264–284. https://doi.org/10.1177/1541344603001003006

Sen AK (1999) Introduction – development as freedom. *Devlopers as Free* 3–12.

Shah SH, Rodina L, Burt JM, et al. (2018) Unpacking social-ecological transformations: Conceptual, ethical and methodological insights. *Anthropology Review* 5: 250–265. https://doi.org/10.1177/2053019618817928

Siqueiros-García JM, Manuel-Navarrete D, Eakin H, et al. (2022) Sense of agency, affectivity and social-ecological degradation: An enactive and phenomenological approach. *Frontiers in Psychology* 13. https://doi.org/10.3389/fpsyg.2022.911092

Stirling A (2008) "Opening up" or "closing down": Analysis, participatoin and power in the social appraisal of technology. *Science, Technology, & Human Values* 33: 262–294. https://doi.org/10.1177/0162243907311265

Tschakert P, Das PJ, Shrestha Pradhan N, et al. (2016) Micropolitics in collective learning spaces for adaptive decision making. *Global Environmental Change* 40: 182–194. https://doi.org/10.1016/j.gloenvcha.2016.07.004

Westley F, Laban S (2012) *Social Innovation Lab Guide*. Waterloo, Canada: University of Waterloo.

Westley F, Olsson P, Folke C, et al. (2011) Tipping toward sustainability: Emerging pathways of transformation. *Ambio* 40: 762–780. https://doi.org/10.1007/s13280-011-0186-9

Westley F, Tj O, Schultz L, et al. (2013) A theory of transformative agency in linked social-ecological systems. *Ecology & Society* 18(3): 24026. Article 27. https://doi.org/10.5751/ES-05072-180327

Chapter 49

The 4D framework

A holistic approach to countering climate change misinformation

John Cook

1. The problem of climate misinformation

Misinformation about climate change damages society in multiple ways. It reduces climate literacy and support for climate action (Ranney and Clark, 2016), polarises the public (Cook et al., 2017), cancels out attempts to communicate accurate information (van der Linden et al., 2017), unduly influences how scientists report their results (Lewandowsky et al., 2015), and causes the public to self-censor rather than discuss climate change with family and friends (Geiger and Swim, 2016).

Climate misinformation campaigns have persisted since the 1990s. This decades-long poisoning of public discourse has been a major contributor to the high degree of public polarisation (Hornsey et al., 2018), which has intensified to the point that the strongest factor driving climate attitudes is political affiliation (Hornsey et al., 2016). Social media has exacerbated the problem with misinformation more likely to be shared online than accurate information (Qiu et al., 2017).

These negative impacts underscore the imperative of countering misinformation. However, misinformation is a complex, interconnected problem, with politics, culture, psychology, and technology interacting to create a ubiquitous, wicked problem. Piecemeal solutions are likely to be inadequate. For example, debunking interventions designed by communication scientists will have limited societal impact if they cannot be deployed at scale. Similarly, technical solutions can be ineffective or even counterproductive if they fail to adopt recommendations from psychological research. Solutions to climate misinformation need to consider the interconnected nature of misinformation.

2. The 4D framework: A holistic approach to countering climate misinformation

This chapter offers a holistic, interdisciplinary framework for addressing misinformation that incorporates four themes: detection, deconstruction, debunking, and deployment. Detection involves building a more comprehensive understanding of the misinformation landscape and developing the ability to automatically detect specific contrarian claims. Deconstruction adopts critical thinking methods to deconstruct and analyse misinformation, identifying reasoning fallacies. Debunking adapts the content gleaned when deconstructing misinformation into correction interventions, using communication recommendations from psychological research. Deployment applies the findings from the

DOI: 10.4324/9780429282348-54

first three themes in technical or public engagement projects, ideally in forms that can be scaled up for maximum societal impact.

2.1 Detecting climate misinformation

In order to counter climate misinformation, one must first understand it. This requires developing a landscape of the various claims found in climate misinformation. There have been a number of efforts to document different aspects of climate misinformation, such as Rahmstorf's early categorisation of climate science misinformation into trend, attribution, and impact scepticism (Rahmstorf, 2004). More recently, a comprehensive taxonomy of contrarian claims expanded earlier efforts to five main categories of climate contrarian claims: it's not real, it's not us, it's not bad, climate solutions won't work, and the experts/science are unreliable (Coan et al., 2021).

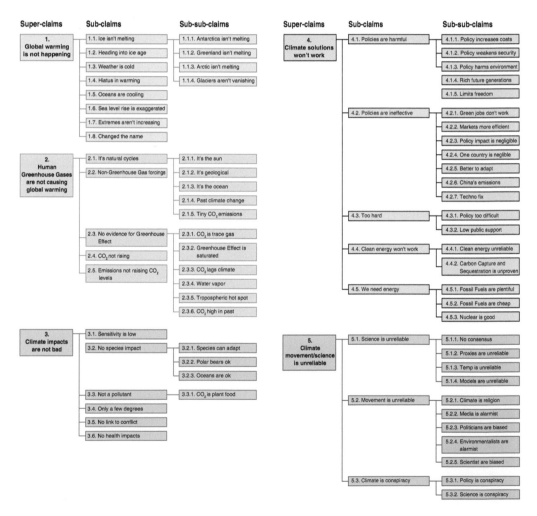

Figure 49.1 A taxonomy of climate contrarian claims

Source: Coan et al., 2021

There are various ways to identify prominent climate myths that require countering. Surveys and social listening methods can identify which misperceptions are most held among the general public. The most promising approach for misinformation detection comes from machine learning, which is ideally suited for big data problems. Unsupervised machine learning – where no human training is used – has identified different topics in climate misinformation, finding an emphasis on attacks against climate scientists' integrity (Boussalis and Coan, 2016). One limitation of this approach is that while it can identify general topics, it cannot identify specific contrarian claims.

Supervised machine learning – where human coding helps train machine learning models – shows more promise in identifying specific instances of climate misinformation. This approach has been used to detect the most prominent contrarian claims in blogs and conservative think-tank websites, finding that attacks on scientists are the most common form of climate misinformation and arguments against climate solutions are on the relative increase (Coan et al., 2021). This quantitative approach indicates that the goal of climate misinformation is to erode public trust in climate science and ultimately delay climate action.

2.2 Deconstructing misinforming arguments

Once contrarian claims have been classified, they need to be analysed to ascertain which claims are fallacious and identify any misleading rhetorical techniques or logical fallacies. Once fallacies have been determined, this information can be used to explain how the misinformation misleads. Cook et al. (2018) developed a step-by-step critical thinking-based methodology for systematically deconstructing and analysing misinformation. While their research focused on common myths about climate change, this approach can be applied to misinformation across any topic.

A framework describing different rhetorical techniques and logical fallacies is useful when identifying fallacies. Diethelm and McKee (2009) described five techniques of science denial: fake experts, logical fallacies, impossible expectations, cherry picking, and conspiracy theories (summarised with the acronym FLICC). This framework has subsequently been developed into a more comprehensive taxonomy (Cook, 2020).

2.3. Debunking (or prebunking) misinformation

Numerous psychological studies have established the difficulties in removing the influence of misinformation once people have come to believe them (Ecker et al., 2022). Correspondingly, many studies have developed a set of recommended best-practices for correcting misinformation effectively. Among the suggested techniques, two elements are especially key in correction messages. First, factual explanations should replace the gap left in people's mental models from the dislodged myth. Second, corrections should include an explanation of the technique or fallacy that the misinformation uses to mislead.

While debunking misinformation after people have received it is the most common approach to countering misinformation, an alternative approach is pre-emptive inoculation. This is based on inoculation theory, which applies the principles of vaccination to knowledge. While biological vaccination builds immunity by exposing people to a weak form of a disease, inoculation theory involves exposing people to a weakened form of misinformation, so they are less likely to be misled in subsequent encounters. Inoculation has been shown to effectively neutralise climate misinformation (van der Linden et al., 2017).

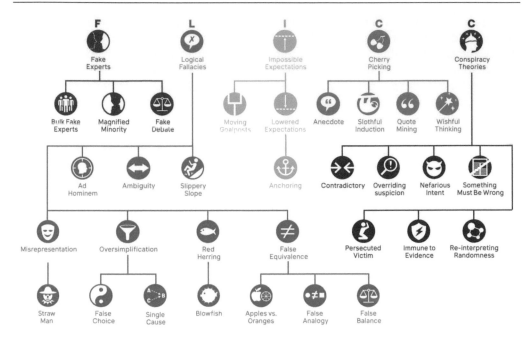

Figure 49.2 The FLICC taxonomy of denialist techniques and logical fallacies in climate misinformation

Source: (Cook, 2020).

The two main approaches to inoculation are fact-based – showing misinformation is wrong through factual explanations – and logic-based – explaining the techniques used to mislead. Both fact-based and logic-based approaches are effective. However, the logic-based approach is particularly attractive as it can convey resistance to a misleading technique across multiple topics (Cook et al., 2017).

2.4 Deploying solutions

Once misinformation has been detected, deconstructed, and debunked, the final step is to deploy corrective interventions, ideally in engaging ways that can scale up to maximise societal impact. There are many avenues and platforms through which to deploy corrections, including technical outlets such as social media, websites, online courses, browser extensions, and smartphone apps or more conventional outlets such as mainstream media and classrooms. It is important that technical solutions incorporate the findings of psychological research – an approach known as technocognition (Lewandowsky et al., 2017).

Classrooms offer the opportunity to apply the 4D framework in an educational context. Misconception-based learning involves teaching scientific concepts by examining misconceptions or misinformation and how they distort the science. This teaching approach results in stronger and longer-lasting learning gains relative to standard lessons. A technical application of misconception-based learning is the Massive Open Online Course *Making Sense of Climate Science Denial* which featured many video lectures inoculating students against fallacies found in climate misinformation (Swire-Thompson et al., 2021).

Gamification offers an engaging and effective technological solution to misinformation. An online game *Bad News*, which required players use misleading tactics to create fake news, was effective in teaching players to identify fake news articles (Roozenbeek and van der Linden, 2018). Similarly, the smartphone game *Cranky Uncle* inoculates players against the fallacies listed in the FLICC taxonomy and has been adopted widely as a classroom activity (Cook, 2022).

3. Conclusion

Climate misinformation is a multi-faceted societal problem, combining factors such as psychology, technology, culture, sociology, and a changing media landscape. Developing effective solutions requires a multi-disciplinary approach that synthesises research from computer science, critical thinking, climate science, and psychology – then couples the research with technological or educational solutions. While existing efforts have begun to incorporate such holistic approaches, more work is needed to develop these methods.

While techniques have been developed to detect and categorise misinformation, these techniques can be further developed to detect more specific contrarian claims as well as incorporate emerging forms of climate misinformation such as greenwashing. Existing research has deconstructed examples of climate misinformation but other forms of misinformation such as probabilistic-framed misinformation and character attacks require analysis. While there is already a great deal of research into debunking techniques, further research is required on exploring the relative efficacy of refutation approaches depending on factors such as social media platform, format, and topic.

The 4D framework offers a holistic, interdisciplinary framework for both research and practical applications that seek to counter the complex, ubiquitous problem of misinformation. Two examples highlighted in this chapter are classroom interventions and critical thinking games but there are countless other combinations. Through collaborations between researchers from different disciplines and communication practitioners (particularly builders of technological applications), innovative solutions can be developed at scale that engage and inoculate the public against misinformation.

4. References

Boussalis, C., & Coan, T. G. (2016). Text-mining the signals of climate change doubt. *Global Environmental Change, 36*, 89–100.

Coan, T. G., Boussalis, C., Cook, J., & Nanko, M. O. (2021). Computer-assisted detection and classification of misinformation about climate change. *Scientific Reports, 11*(22320).

Cook, J. (2020). Deconstructing climate science denial. In Holmes, D., & Richardson, L. M. (Eds.) *Edward Elgar Research Handbook in Communicating Climate Change.* Cheltenham: Edward Elgar.

Cook, J. (2022). Teaching about our climate crisis: Combining games and critical thinking to fight misinformation. *American Educator, 45*(4), 12.

Cook, J., Ellerton, P., & Kinkead, D. (2018). Deconstructing climate misinformation to identify reasoning errors. *Environmental Research Letters, 11*(2).

Cook, J., Lewandowsky, S., & Ecker, U. (2017). Neutralizing misinformation through inoculation: Exposing misleading argumentation techniques reduces their influence. *PLoS One, 12*(5), e0175799.

Diethelm, P., & McKee, M. (2009). Denialism: What is it and how should scientists respond? *The European Journal of Public Health*, *19*(1), 2–4.

Ecker, U. K. H., Lewandowsky, S., Cook, J., Schmid, P., Fazio, L. K., Nadia Brashier, N., Kendeou, P., Vraga, E. K., & Amazeen, M. A. (2022). The cognitive and social drivers of misinformation belief, persistence, and correction. *Nature Reviews Psychology*, *1*(1), 13–29.

Geiger, N., & Swim, J. K. (2016). Climate of silence: Pluralistic ignorance as a barrier to climate change discussion. *Journal of Environmental Psychology*, *47*, 79–90.

Hornsey, M. J., Harris, E. A., Bain, P. G., & Fielding, K. S. (2016). Meta-analyses of the determinants and outcomes of belief in climate change. *Nature Climate Change*, *6*(6), 622–626.

Hornsey, M. J., Harris, E. A., & Fielding, K. S. (2018). Relationships among conspiratorial beliefs, conservatism and climate scepticism across nations. *Nature Climate Change*, *1*.

Lewandowsky, S., Ecker, U. K., & Cook, J. (2017). Beyond misinformation: Understanding and coping with the "post-truth" era. *Journal of Applied Research in Memory and Cognition*, *6*(4), 353–369.

Lewandowsky, S., Oreskes, N., Risbey, J. S., Newell, B. R., & Smithson, M. (2015). Seepage: Climate change denial and its effect on the scientific community. *Global Environmental Change*, *33*, 1–13.

Qiu, X., Oliveira, F. M., Sahami Shirazi, D., Flammini, A., & Menczer, F. (2017). Limited individual attention and online virality of low-quality information. *Nature Human Behaviour*, *1*(7), 1–7.

Rahmstorf, S. (2004). The climate sceptics. *Weather Catastrophes and Climate Change*, 76–83.

Ranney, M. A., & Clark, D. (2016). Climate change conceptual change: Scientific information can transform attitudes. *Topics in Cognitive Science*, *8*(1), 49–75.

Roozenbeek, J., & van der Linden, S. (2018). The fake news game: Actively inoculating against the risk of misinformation. *Journal of Risk Research*, 1–11.

Schmid, P. and Betsch, C. (2019). Effective strategies for rebutting science denialism in public discussions. *Nature Human Behaviour, 3*(9), pp. 931–939.

Swire-Thompson, B., Cook, J., Butler, L. H., Sanderson, J. A., Stephan Lewandowsky, S., & Ecker, U. K. H. (2021). Evidence for a limited role of correction format when debunking misinformation. *Cognitive Research: Principles and Implications*, *6*(1), 1–15.

van der Linden, S., Leiserowitz, A., Rosenthal, S., & Maibach, E. (2017). Inoculating the public against misinformation about climate change. *Global Challenges*, *1*(2).

Further reading

The Conspiracy Theory Handbook. http://sks.to/conspiracy
The Debunking Handbook 2020. https://sks.to/db2020

Audio visual resource

YouTube Playlist: Critical Thinking About Climate. https://www.youtube.com/playlist?list=PL1xbdG-NAkB1fAyNAWErM4uEXyduldiew

Chapter 50

Decolonising human-nature relationships

Indigenous ontologies and development

Thomas Aneurin Smith

1. From local knowledges to indigenous ontologies

Indigenous ontologies can critique how development and environmental management are done. This chapter explores what Indigenous ontologies are, how they offer alternative ways to think about human-nature relations, whilst advancing a critical approach to concepts of ontology and Indigeneity.

Indigenous or local knowledges are one of the defining features of the participatory movement in environmental management. Initially local people's knowledges, for example of soils and vegetation, were viewed by development 'experts' as technical fixes to localised problems (Briggs, 2013). Local knowledges became tied to empowerment, and by the late 1990s, incorporating local or Indigenous knowledges into environmental management had become mainstream. Critiques emerged of this mainstreaming: it romanticised Indigenous peoples as 'in harmony' with nature and homogenised communities whilst ignoring grassroots politics (Smith, 2011).

2. Reframing human-nature relationships

A more meaningful engagement with Indigenous ontologies, worldviews, or ways of being, is now challenging conventional approaches to human-nature relationships in development. Rather than think of Indigenous understandings as a resource for experts to 'tap', regarding them as ontologies or radically different ways of understanding what exists in the world can decolonise environmental management and reframe how Indigenous groups engage with development. Modernism assumes there is one single reality, or 'nature', with different 'cultures' offering alternative interpretations of that reality. An alternative is that there are instead multiple ontologies, not just different interpretations but radically different worlds (Blaser, 2014). Ontology is not just a different word for culture; it recognises the radically different ways that the world is lived in and conceived of.

Taking multiple ontologies as radically different ways of understanding the world can be politically and methodologically powerful, challenging entrenched political formations. In the late 1990s and early 2000s a transnational movement of Andean Indigenous groups, notably across Bolivia, Peru and Ecuador, came to prominence. These groups brought to the national and international stage concepts such as *Pachamama* (Mother Earth) and *Vivir Bien* (living well) which bring together the rights of humans and non-humans (Zimmerer, 2015). These were widely popularised as alternative, holistic beliefs, promoting harmonious relationships with nature as opposed to Western development

DOI: 10.4324/9780429282348-55

and the exploitation of natural resources. These concepts became written into political constitutions, including of the Republic of Ecuador and the Plurinational State of Bolivia, renamed in 2009 to recognise the multiple Indigenous nations composing the state (Laing, 2015).

Beyond political recognition, how do Indigenous ontologies radically reframe human-nature relations, and approaches to environment and development? Harris (2017) provides a rich account of the Kallawaya, Indigenous people from the Apolobama Mountains, Bolivia, travelling healers on the High Andes. Central to their healing practices is the Kallawaya *Cosmovisión,* which regards the mountain and human body as analogous. Healing concerns the spiritual dissonance and disjuncture's between a person and their *ayllu,* which describes the whole community of beings that inhabit the world and their relationships: humans, plants, animals, mountains, rivers etc. The Kallawaya understand environmental change, the melting glaciers and shifting ecologies of the region, as a spiritual altering of community health. This understanding finds climate change felt in everyday life, disrupting social life as much as the environment, rather than responding to climate change through adaptation as a 'development fix'.

Examples of the embedded and spiritual relationships between Indigenous groups and local environments are found globally. In Tanzania, forest patches varying in size from small groves to large, forested areas, have been conserved for hundreds of years because of their spiritual or ancestral importance (Smith and Andindilile, 2017). Such 'sacred natural sites' have been recorded all over the world. Sacred forest groves are typically held as common property, with a series of taboos and norms that preserve forest sites. These sites have been found to support high species diversity, richness and endemism.

Conservationists have celebrated such ontologies for protecting forest habitats, but the mapping of Indigenous ontologies onto ecological management can insist on a modernist rationality which does not fully recognise ontological differences. Evidence from sacred natural sites in Tanzania (Smith, 2020) has shown there are complex relationships between local people, forests and the nonhumans that inhabit them. Forest sites at *Nyiha* villages in southwest Tanzania are protected by ancestral chief's graves, and permission to enter or harvest wood is granted by the existing chief. Trespassers are met by ancestral chiefs, manifest as dangerous snakes, who chase them away. Whilst these ontologies offer protection to forests, outsiders cannot adopt village rituals, because ancestors are particular to that village, which, given recent influxes of outsiders, and conversions to Christianity, make maintaining local rituals problematic, leading to increasing forest encroachment. These ontologies are also connected to social hierarchies and conflict: they rely on male-only lineages of chieftainship; traditional chiefs come into conflict with elected village governments and some rituals exclude women. These ontologies also rely on secrecy: the full consequences of rituals are not revealed to outsiders. These factors make the long-term status of Indigenous ontologies that act to protect sacred natural sites uncertain.

The conservation of specific species through local ontologies is also complicated. In Madagascar, certain animals are linked to *fadys,* localised ontologies that make some behaviours towards an animal taboo (Holmes et al., 2018). Many species of Lemur are believed to be the spirits of Malagasy ancestors, so killing them is taboo. *Fadys* have been celebrated by conservationists as leading to low levels of bushmeat hunting in Madagascar. However, other species are persecuted because of local ontologies: the aye-aye is often viewed as an evil presence, thus leading to its killing to ward off evil. *Fadys* are highly localised, and relationships with any species are not common throughout Madagascar. Simplistically aligning these ontologies with conservation disconnects them from their spiritual and political worlds.

Attempts to re-politicise Indigenous ontologies are well-documented in North and South America (Blaser, 2009; de la Cadena, 2010), through acts of resistance to conventional conservation management approaches which block Indigenous practices, such as hunting. There has been less attention to more ambiguous political relationships. Witchcraft, magic and the occult have been less prominent in contemporary understandings of people-environment relations, and neither have they featured much in studies of resistance to development. Murrey (2015) examines the role of witchcraft, or *la sorcellerie*, along the course of the Chad-Cameroon Oil Pipeline. The destruction of forests to make way for the pipeline resulted in accusations of witchcraft towards pipeline employees, whilst politicians and government officials were also suspected of witchcraft because of their invisible accumulation of oil wealth. Yet such accusations did not amount to a collective resistance, due to the invisible, taboo and highly personal nature of *la sorcellerie*. Whilst Indigenous ontologies have translated into political resistance, bringing together Indigenous groups and environmentalists, the ambivalence of some ontologies, such as witchcraft, towards political activism means they do not sit easily alongside environmental conservation.

3. Indigeneity: critical appraisals

The term Indigenous is used in different ways by different actors. Radcliffe (2017) argues that only particular features of being Indigenous are endorsed by states, including cultural distinctiveness, environmental knowledges and rootedness to place. This politics of recognition is evident in the USA, Canada and Australia under the 'modern contract' of inclusion, which rarely tolerates Indigenous 'ways of being' or demands for self-determination. The notion of being 'Indigenous' may be used cautiously. Being Indigenous in East African contexts can be associated with tribal identity, yet identification and categorisation of 'tribes' commonly dates to the colonial period, whilst post-colonial states have grappled with ethnic conflict associated with tribal identities. Even in these contexts, many groups, be they identify as 'Indigenous' or otherwise, still face erasure by the developmentalist state – and the missionary work of major religions – that straddles the colonial and post-colonial eras (Schneider, 2006).

In Bolivia, under the presidency of Evo Morales, the first self-proclaimed Indigenous president, conflict emerged around the politics of Indigeneity. Laing (2015) argues that the Bolivian state selectively positioned Indigenous human-environment relations as anti-Western, bringing concepts such as *Pachamama* and *Vivir Bien* to global climate activism, whilst simultaneously expanding the hydrocarbon and mining sectors at the expense of non-majority Indigenous groups. Equally, within non-majority Indigenous resistance movements there were diversities of opinion on the appropriate extent of resource exploitation in their territories. Indigeneity is therefore dynamic, linked to political mobilisation and sovereignty over land and resources, but there are competing mobilisations of Indigeneity.

Notions of ontology have also been scrutinised. For some, the idea of ontology represents another form of colonialism (Todd, 2016). From a decolonial perspective, using 'ontology' as a catch-all term for Indigenous ways of being is a continuation of the colonial violences perpetuated through development, where 'experts' impose external categories on Indigenous groups. How Indigenous ontologies have been used by the Western academy, by politicians and in the name of development, can erase the practices and embodied aspects of Indigenous ontologies (Hunt, 2014).

The colonial project made Indigenous peoples and their lands governable through making them visible, without their consent. Indigenous communities, rather than become increasing visible, may refuse to participate in the state, and knowledge production (Simpson, 2016). Yet there are also taboos about the analysis of the internal politics of Indigenous groups and the moral and ethical complexities of Indigenous ontologies (Ortner, 1995). It is perhaps for this reason that seemingly harmonious Indigenous human-nature relations are more readily adopted by environmental managers, whereas those more ambiguous, such as witchcraft, are ignored. Seeking solidarity with and between Indigenous groups is not straightforward for the decolonial project, given these complexities.

4. Conclusions

Indigenous ontologies sustainably manage environments, provide a holistic understanding of climate change and reframe human-nature relationships away from exploitation. Whilst Indigenous ontologies can accomplish these things in distinct places and times, the human-nature relations they denote are complex, embedded in social and political practices, and some, such as witchcraft, are ambiguous in character. Indigeneity does not have one meaning; neither does a turn to Indigenous ontologies entirely fulfil the possibilities of decolonising development. This does not mean Indigenous ontologies cannot be powerful ways to re-generate human-nature relations. Bawaka Country (2016: 470), demonstrates how writing and thinking as a co-production among Indigenous people, researchers and the country itself can generate indigenous-led theories, feel the grounded nature of environmental change, whilst recognising that 'there is no easy understanding'.

References

Blaser, M. (2009) The threat of the Yrmo: The political ontology of a sustainable hunting program, *American Anthropologist*, NS111(1), 10–20.

Blaser, M. (2014) Ontology and indigeneity: On the political ontology of heterogeneous assemblages, *Cultural Geographies*, 21(1), 49–58.

Briggs, J. (2013) Indigenous knowledge: A false dawn for development theory and practice? *Progress in Development Studies*, 13(3), 231–243.

Country, B. (2016) Co-becoming Bawaka: Towards a relational understanding of place/space, *Progress in Human Geography*, 40(4), 455–475.

De La Cadena, M. (2010) Indigenous cosmopolitics in the Andes: Conceptual reflections beyond "politics", *Cultural Anthropology*, 25(2), 334–370.

Harris, D. M. (2017) Mountain-bodies, experiential wisdom: The Kallawaya cosmovisión and climate change adaptation, *Third World Thematics*, 2(2–3), 376–390.

Holmes, G., Smith, T. A., & Ward, C. (2018) Fantastic beasts and why to conserve them: Animals, magic and biodiversity conservation, *Oryx*, 52(2), 231–239.

Hunt, S. (2014) Ontologies of indigeneity: The politics of embodying a concept, *Cultural Geographies*, 21(1), 27–32.

Laing, A. F. (2015) Resource sovereignties in Bolivia: Re-conceptualising the relationship between indigenous identities and the environment during the TIPNIS conflict, *Bulletin of Latin American Research*, 34(2), 149–166.

Murrey, A. (2015) Invisible power, visible dispossession: The witchcraft of a subterranean pipeline, *Political Geography*, 47, 64–76.

Ortner, S. B. (1995) Resistance and the problem of ethnographic refusal, *Comparative Studies in Society and History*, 37(1), 173–193.

Radcliffe, S. A. (2017) Geography and indigeneity I: Indigeneity, coloniality and knowledge, *Progress in Human Geography*, 41(2), 220–229.

Schneider, L. (2006) Colonial legacies and postcolonial authoritarianism in Tanzania: Connects and disconnects, *African Studies Review*, 49(1), 93–118.

Simpson, A. (2016) Consent's revenge, *Cultural Anthropology*, 31(3), 326–333.

Smith, T. A. (2011) Local knowledge in development (geography), *Geography Compass*, 5(8), 595–609.

Smith, T. A. (2020) Episodes of concealing: The invisibility of political ontologies in sacred forests, *Cultural Geographies*, 27(3), 333–350.

Smith, T. A., & Andindilile, W. (2017) Assemblages of forest conservation in Tanzania: Gradients between chiefs, snakes, spirits and witches, *Third World Thematics*, 2(2–3), 316–337.

Todd, Z. (2016) An indigenous feminist's take on the ontological turn: "Ontology" is just another word for colonialism, *Journal of Historical Sociology*, 29(1), 4–22.

Zimmerer, K. S. (2015) Environmental governance through "speaking like an indigenous state" and respatializing resources: Ethical livelihood concepts in Bolivia as versatility or verisimilitude? *Geoforum*, 64, 314–324.

Additional resources

Bawaka Collective, a research collective. https://bawakacollective.com

Carrithers, M., Candea, M., Sykes, K., Holbraad, M., & Venkatesan, S. (2010) Ontology is just another word for culture, *Critique of Anthropology*, 30(2), 152–200.

Howitt, R., & Suchet-Pearson, S. (2006) Rethinking the building blocks: Ontological pluralism and the idea of "management", *Geografiska Annaler: Series B, Human Geography*, 88(3), 323–335.

Nadasdy, P. (2007) The gift in the animal: The ontology of hunting and human–animal sociality, *American Ethnologist*, 34(1), 25–43.

"What kind of witchcraft is this?" Development, magic and spiritual ontologies, special issue of *Third World Thematics*. https://www.tandfonline.com/toc/rtwt20/2/2-3

Chapter 51

Water insecurity

Catherine Fallon Grasham

Introduction

Water is life

This simple yet profound statement can often be seen in large lettering at the entrances to water utilities in low- and middle-income countries (LMICs). According to the The Joint Monitoring Programme for Water Supply and Sanitation, (JMP by UNICEF and WHO), around one in three people in the world lack access to safely managed drinking water, the majority of which live in LMICs. Waterborne diseases are a leading contributor to infant mortality. In contrast to the way that access to clean water is commonly taken for granted in high-income countries (HICs),[1] in LMICs, the life-or-death matter of access to water is tangible.

Water insecurity goes beyond access to safe drinking water, perpetuating poverty through adversely affecting income-generating activities, food security, physical and mental health, hygiene and sanitation, as well as education. The concept enables approaches to water management that consider water as a co-produced resource, situated within broader physical and socio-political structures. Water insecurity is an important concept for the Sustainable Development Goals (SDGs) and the leave no one behind (LNOB) agenda, since it embeds values, capabilities, responsibilities, equity, and justice.

Historically, approaches to water management have focussed on ensuring adequate supply to overcome water scarcity with technological solutions. To this day, many practical approaches remain ensconced in this way of thinking. This has implications for the way that water is controlled and allocated, who decides how and who benefits from interventions to reduce water insecurity. In the context of unprecedented global population growth and the climate crisis, equity-based approaches to water security are more important than ever. This chapter begins by discussing the plurality of water security, followed by explanation of the intimate relationships between water insecurity, development and climate resilience. Finally, I conclude by offering one idea towards answering: why do technical-based solutions to water insecurity continue to persist even though we know that water insecurity is inherently relational?

The plurality of water insecurity

Water security has been the dominant paradigm in the water sector since the 1990s, following the 1994 Human Development Report. It emerged at a time when approaches

DOI: 10.4324/9780429282348-56

to water management were going beyond the physical availability of fresh surface and ground water, to include social factors (Cook and Bakker, 2012). Prior to this, technical approaches to water management understood water purely as a physical resource that could be objectively measured and managed. For example, the framework of Integrated Water Resources Management (IWRM) – that argued for basin-scale water allocation approaches predominantly with engineering solutions – was widely adopted by water ministries across the world. The concept of water insecurity is hotly debated and there are myriad definitions.

One of the most commonly cited definitions of water security is Grey and Sadoff's (2007: 545),

> The availability of an acceptable quantity and quality of water for health, livelihoods, ecosystems and production, coupled with an acceptable level of water-related risks to people, environments and economies.[2]

What this definition calls for in practice is understanding physical water systems – where and when water is available, its chemical composition, flood management, ensuring water access through droughts/dryspells and in drylands with chronic water scarcity, while avoiding water pollution and developing appropriate water treatment technologies.

Grey and Sadoff's definition has been criticised for underemphasising issues of diversity, equity, power and sustainability, meaning that it is often not the most suitable to operationalise in the context of development work. The definition is apolitical and does not allow much space for talking about water rights, gender or examining asymmetrical water governance arrangements. For example, the focus on an 'acceptable level' of risk has very little focus on capabilities and well-being. Moreover, the idea of an 'acceptable level of water-related risks' is not well-aligned with the leave no-one behind (LNOB) agenda since it does not ask the question of who defines acceptability, who characterises risk, who is experiencing the risk and what the implications of that are for development.

There are other, more recent, scholarly approaches to water insecurity see water resources as co-produced by biophysical and social processes, arguing that water security is relational – meaning that it is intimately shaped by social, cultural and political processes. For instance, Jepson et al. (2017: 2) 'envision water security as both grounded in the social relations of access to water as well as critical to a set of relations and functionings that advance human flourishing' (2017). Water security is not just about the socio-political relations that affect water's availability and quality but is predicated on the diverse needs and wants of people. Development studies needs an approach that contextualises how water can meet these needs and wants and how those fit into broader aspirations and notions of well-being, subjectively defined.

Deciding which definition of water security to adopt and operationalise depends on the method of measurement, the scale, and the boundary. For example, if we adopt Jepson et al.'s idea that water security is a relational concept, then it cannot be objectively measured, while Grey and Sadoff's definition lends itself well to quantifiable metrics, of which there are many, including the development of a plethora of water security indices (Octavianti and Staddon, 2021). Moreover, the scale and boundary of the enquiry, for example a sprawling urban metropolis, versus a small, remote village demand different approaches to understanding water insecurity. Therefore, it is important to be aware of the plurality of water security and to determine appropriately how to approach it.

Water and development

Water is intimately connected to development; is it fundamental for lives and livelihoods, as well as environmental sustainability. However, your first thought of water and development might take you to a handpump in a small village where women and girls are collecting water to take home. This is unsurprising given that there is such a strong focus on water, sanitation and hygiene (WASH) interventions in the humanitarian and development sectors. Though a critical part of the water insecurity problem, particularly pertinent to the LNOB agenda of the Sustainable Development Goals (SDGs), water for domestic use makes up less than 10 per cent of global freshwater demand. This image is only a small piece of the puzzle.

The complexity and depth of the concept of water security is captured, to some extent, in SDG6. There are eight water-related targets within SDG6, only two of which are directly related to WASH (6.1 and 6.2), while the other six cover alternative aspects of water security related to livelihoods and water governance, among others. The eight targets are being unevenly monitored; four of which are severely lacking in data (6.3 water quality, 6.5 transboundary water management, 6.4 water use efficiency and 6.6 water-related ecosystems). Moreover, where data is available, it is clear that many countries are unlikely to achieve the SDG6 targets by 2030, with spill-over impacts on other goals in relation to the environment, food security, health, energy, gender equality, and poverty (Nkiaka et al., 2021).

Access to sufficient water for lives and livelihoods is the dominant framing in water and development, while in practice there are different frameworks for addressing water insecurity in the development space. In addition to the WASH framework, there is also a strong scholarly focus on agricultural water management, and comparatively less engagement with water in non-agricultural livelihoods. This makes sense since the majority of the world's poorest are engaged in subsistence agriculture and most freshwater use worldwide is for agriculture (70 per cent). However, the current technical focus on adequate water supply does not go far enough to advance livelihood capabilities (Shah, 2021), particularly if equity-based distributional issues are not considered.

Water security and climate resilience

Climate resilience is not possible without water security. Climate resilience strengthens the ability of all to mitigate vulnerability to risks from – and adapt to changing patterns in – climate hazards and variability (Grasham et al., 2021). Moreover, most climate adaptation strategies are related to water, whether that be a nomadic pastoralist travelling to find water for their livestock or a large investment in an inter-basin water transfer infrastructure to supply a city in a water-scarce area. In the midst of the climate crisis, we already see rain patterns changing, with floods and droughts intensifying, causing unprecedented climate-driven migration. However, it is not just extreme weather events that have adverse impacts on lives and livelihoods; the effects of slow onset events – or smaller variations in rainfall – reverberate, entrenching poverty and inequality, often in ways that are not captured by the 'acceptable' levels of risk framing of water insecurity by Grey and Sadoff.

The global impacts of the climate crisis will be unevenly distributed: poorest countries will be hit hardest, and within communities, some socially vulnerable groups will be disproportionally affected. As governments worldwide seek to climate-proof their econo-mies, national policies are being re-oriented within a framing of 'climate resilience'. In the water sector, 'climate resilient water security' is an emerging paradigm, with donor fund-ing being mobilised for climate resilient WASH programmes across LMICs. But what does climate resilient water security mean in practice?

When approached from an external, technical position, it involves constructing dams to regulate flooding, water storage for cities or drilling wells in the desert. Such techni-cal approaches can result in maladaptation – when adaptation strategies inadvertently increase vulnerability to climate change. When approached from a bottom-up relational position, it requires understanding – to strengthen – community resilience and enforcing regulatory frameworks that strive for more equity-based approaches to water security. Climate resilient water security is a matter of life and death, not just a technical issue.

Conclusion

Water security is necessary for sustainable development and climate resilience. Water security is plural, but a relational approach, rather than a technical one, is more suitable for development studies that actively seek to understand and challenge socio-political structures governed by asymmetrical power relations. Despite this, the scholarly and practitioner foci remain on increasing water availability, while technical approaches con-tinue to outnumber efforts to transform the inequitable socio-political structures that (re)-produce water insecurity (Shah, 2021). Could this be because social transformation is more complex and lengthier? Drilling a borehole and installing a handpump (for exam-ple) is far more politically favourable than exposing (to dismantle) the political structures that keep water flowing to power.

Notes

1 This is not to say that people living in HICs are all water secure. HICs share many similar dynamics with LMICs vis-a-vis associations between socio-economic and political marginalisa-tion and water insecurity.
2 Other widely accepted definitions are those from GWP, UN-WATER, OECD and UNESCO.

References

Cook, C., & Bakker, K. (2012). Water security: Debating an emerging paradigm. *Global Environ-mental Change*, 22(1), 94–102. https://doi.org/10.1016/j.gloenvcha.2011.10.011
Grasham, C. F., Calow, R., Casey, V., Charles, K. J., de Wit, S., Dyer, E., Fullwood-Thomas, J., Hirons, M., Hope, R., Hoque, S. F., Jepson, W., Korzenevica, M., Murphy, R., Plastow, J., Ross, I., Ruiz-Apilánez, I., Schipper, E. L. F., Trevor, J., Walmsley, N., & Zaidi, H. (2021). Engaging with the politics of climate resilience towards clean water and sanitation for all. *NPJ Clean Water*, 4(1), 1–4.
Grey, D., & Sadoff, C. W. (2007). Sink or swim? Water security for growth and development. *Water Policy*, 9(6), 545–571.
Jepson, W., Budds, J., Eichelberger, L., Harris, L., Norman, E., O'Reilly, K., Pearson, A., Shah, S., Shinn, J., Staddon, C., & Stoler, J. (2017). Advancing human capabilities for water security: A relational approach. *Water Security*, 1, 46–52.
Nkiaka, E., Bryant, R. G., Okumah, M., & Gomo, F. F. (2021). Water security in sub-Saharan Africa: Understanding the status of sustainable development goal 6. *Wiley Interdisciplinary Reviews: Water*, 8(6), e1552.

Octavianti, T., & Staddon, C. (2021). A review of 80 assessment tools measuring water security. *Wiley Interdisciplinary Reviews: Water*, 8(3), e1516.

Shah, S. H. (2021). How is water security conceptualized and practiced for rural livelihoods in the global South? A systematic scoping review. *Water Policy*, 23(5), 1129–1152.

Additional Resources

Amorim-Maia, A. T., Anguelovski, I., Chu, E., & Connolly, J. (2022). Intersectional climate justice: A conceptual pathway for bridging adaptation planning, transformative action, and social equity. *Urban Climate*, 41, 101053.

Gerlak, A. K., House-Peters, L., Varady, R. G., Albrecht, T., Zúñiga-Terán, A., de Grenade, R. R., Cook, C., & Scott, C. A. (2018). Water security: A review of place-based research. *Environmental Science & Policy*, 82, 79–89. https://doi.org/10.1016/j.envsci.2018.01.009

Grasham, C. F., Charles, K. J., & Abdi, T. G. (2022). (Re-)orienting the concept of water risk to better understand inequities in water security. *Frontiers in Water*, 3. https://doi.org/10.3389/frwa.2021.799515

Marcal, J., Antizar-Ladislao, B., & Hofman, J. (2021). Addressing water security: An overview. *Sustainability*, 13(24), 13702.

Chapter 52

The blue economy

Kate Symons

Introduction

Major conservation and development institutions are paying more attention to ocean and marine environments and the four billion people who depend on directly on ocean resources (Cohen et al., 2019). The ocean's role in development is firmly on the agenda, with goal 14 of the Sustainable Development Goals, entitled 'Life Under Water', focusing on the need to preserve ocean environments alongside use ocean economies like fisheries, aquaculture and tourism to increase economic benefits to small island developing states and least developed countries. This incorporation of the ocean into sustainable development discourse and practice has taken place as threats to ocean environments increase, and scientists warn of an 'emerging mass extinction in the oceans' driven by climate change, industrial fishing, extraction and pollution. (Payne et al., 2016: 1). At the same time, understanding of the role of the ocean as providing vital functions such as climate stability, renewable energy and supporting public health and food security is also growing. The ocean's changing role in security, territorialisation, the impacts of sea level rise and through dynamics of connectivity and culture is also foregrounded in debates about geopolitics and sea level rise. Against this backdrop, ideas about blue growth have emerged as a way of protecting ocean resources while simultaneously exploiting them for economic gain. In this chapter, I will explore two key ways in which the ways in which the blue economy has been conceptualised; the first as a continuation of ideas about environmental sustainability and the second as a geopolitical project. Drawing on examples, I will discuss some of the environmental and political contradictions at the heart of sustainable blue growth, before concluding with some thoughts about where the blue economy might go next.

The blue economy as sustainable development

One way to see the blue economy is as part of a growing focus on the sustainability of the ocean. Borrowed from the green economy, blue economy models combine environmental protection of oceanic and coastal areas, economic growth based on sustainable marine resources and use of technology and methods of knowledge construction (such as natural capital accounting) which claim to develop a sustainable relationship between the ocean and society. WWF defines it as 'the use of the sea and its resources for sustainable economic development' (WWF, 2015: 2), while Keen et al. (2018: 334) define it as activities which 'balance sustainable economic benefits with long-term ocean health in a

DOI: 10.4324/9780429282348-57

manner which is consistent with sustainable development and its commitment to intra- and inter-generational equity'. Similarly, the World Bank states that '[t]he blue economy concept seeks to promote economic growth, social inclusion, and the preservation or improvement of livelihoods while at the same time ensuring environmental sustainability of the oceans and coastal areas' (World Bank, 2017: vi), suggesting that 'economic development and ocean health are compatible propositions' (World Bank, 2017, viii). These definitions clearly place the blue economy within sustainable development discourses, which argue that several competing social, economic and environmental benefits can be obtained with the right mix of governance, incentive and technology. This focus on livelihoods, human health and sustainable economic opportunities for poverty reduction is largely the language and definition reflected in SDG 14, too.

The blue economy can also be seen as a system of economic governance which formally recognises the possibility of ocean sustainability along with broader economic growth opportunities (Cohen et al., 2019). While this moves the emphasis away from smaller-scale sustainable activities towards growth and production, this is not 'business as usual'. An ecological economics lens, some critics argue, shifts the blue economy away from a conventional economic emphasis on growth and production towards a recognition of the dependence of economies on ecosystems (Keen et al., 2018). However, a paper by Voyer et al. (2018a) highlights different economic activities grouped under blue economy policies, including fossil fuel extraction and undersea mining, fisheries and aquaculture, biotechnology, tourism, shipping and renewable energy, along with valuation and protection of ecosystems. These activities involve multiple actors and scales and have different approaches to development, ranging from small-scale community managed fisheries and ecotourism projects which directly support livelihoods to large-scale state and private investment in ocean-based energy and extractives which use blue economy rhetoric to position the ocean as 'the new economic frontier' (Cohen et al., 2019: 2). These different forms of blue growth create winners and losers which cannot necessarily be mitigated by technocratic processes like marine spatial planning (ibid). The blue economy can therefore be defined as a flexible term which refers to several competing policy agendas and areas of economic activity, which can include larger-scale industrial and extractives development aimed at economic growth, rather than only focused on poverty reduction and sustainable livelihoods (Voyer et al., 2018a).

The idea of blue sustainability is critiqued by scholars who argue that blue economies are primarily intended to provide access to ocean resources and new opportunities for economic expansion (Brent et al., 2020; Childs and Hicks, 2019). This can operate in two ways. First, blue economy strategies create new opportunities for the expansion of economic activities, such as the physical expansion of new frontiers as well as the restructuring of regulation to enable new products and services. At the same time, aspects of blue economies focused on sustainability, such as renewable energy and conservation, address perceived crises of legitimacy to enable less sustainable economic activities to continue (Ekers and Prudham, 2015). For example, Mozambique's blue economy highlights the

> great potential for important mineral resources such as heavy minerals and gems, sands, potassium evaporites, manganese nodules, sulphides, among others . . . . that can contribute to Mozambique's economy, [along with] potential for onshore and offshore exploration of strategic and rare resources beyond oil and gas.
>
> (Government of Mozambique, 2021: 8–9)

while also discussing ocean-based ecotourism, media campaigns on ocean protection, development of new marine protected areas and the protection of mangrove forests, along with policy and finance mechanisms like carbon credits and biodiversity offsetting. Under this reading, the idea of blue growth is a way of expanding new economic opportunities, like those identified as being part of blue economies like deep sea mining, while at the same time creating ways to offset or mitigate the environmental cost through technologies, discourses and policy fixes aimed at sustainability. This speaks to the idea that green/blue economies are a direct 'denial of the negative environmental impacts of "business as usual" capitalism' (Büscher and Fletcher, 2015: 273). In short, a critical reading of the sustainability claims of blue economies is vital, in order to highlight the ecological contradictions inherent in ideas about 'blue growth'.

The politics of the blue economy

A different way to understand the blue economy is to examine how it is being used to enhance geopolitical territorialisation and national development in specific contexts (Keen et al., 2018; UNECA, 2016; World Bank, 2017). This approach sees the blue economy as a financing regime to provide an overarching strategy for investment into ocean-based projects, as well as a vehicle for political co-operation, so-called blue diplomacy. For example, in coastal African contexts, the blue economy is encouraging extractive projects in the ocean such as gas extraction, sea-bed mining and industrial fishing while enhancing maritime security and providing opportunities for investment into fisheries infrastructure and tourism (Bond, 2019; Childs and Hicks, 2019). In small island states, the agenda provides a way to incorporate the ocean into national development strategies, thereby extending the reach and connectedness of their territories (physically and metaphorically) through their exclusive economic zones of 200km out into the ocean from their land boundary (Kothari and Wilkinson, 2010).

Blue economy policies are increasingly providing political rationale for major coastal infrastructure and large-scale financial loans. For example, China's Belt and Road initiative is financing ports and other export facilities in Kenya, Ethiopia and Sri Lanka, attaching these locations to China through physical links, political partnership and ties of debt (Bond, 2019). Mozambique's emerging blue economy also highlights strategic political agendas behind blue economies. A major emphasis of Mozambique's strategy is enhancing financial, security and institutional co-operation between nations to 'illegal, unreported and unregulated fishing' (Government of Mozambique, 2021: 9), further highlighting how regional level power and bilateral relations are important goals in the blue economy agenda. Some scholars therefore consider the blue economy should be regarded as primarily a geopolitical project inherently linked to attempts to secure ocean space, capture trade routes and to deepen economic ties with political allies (Bond, 2019; Voyer et al., 2018b).

Such large-scale political agendas create winners and losers, so it is important to highlight political critiques which foreground issues of power, justice and agency and how blue economies can reinforce uneven power relations across social groups (Andriamahefazafy and Kull, 2019; Bennet et al., 2019, 2021; Bond, 2019; Kalina et al., 2019). This research questions the idea that the ocean can simultaneously be a source of economic growth while maintaining the protection of traditional livelihoods, cultures and natural environments by highlighting the effects on ocean-dependent communities (Okafor-Yarwood et al., 2020). Cohen et al. (2019) argue that seemingly apolitical marine spatial planning efforts under blue economy rationale 'hold strong parallels with other significant conversions of community-held resource into private goods . . . and risks a similar disenfranchisement of the maritime equivalent of peasant farmers' (page 4). Consequently, a small but growing body of research explores different ways to protect community rights and gender equity, as well as the needs of more-than-human life in ocean governance (Bavinck et al., 2018; Cohen et al., 2019; Keen et al., 2018, Radomska and Åsberg, 2021).

Where next for the blue economy?

Clearly, then, there are divergent ideologies and political agendas attached to the blue economy concept as well as differing examples of policy frameworks and projects that claim to be examples of blue growth. Before concluding, I would like to point to two areas of future research where oceans, and the blue economy are developing in prominence – ideas about the right to the ocean and marine justice, and the growing prominence of the ocean in addressing the climate crisis. These are two major topics in and of themselves, so what follows is a sketch of emerging research rather than an extensive discussion.

Turning first to the idea of rights-based and pluralistic approaches to the blue economy, this involves reframing the notion of the economy from top-down technocratic construct to consider the different ways in which multiple communities engage with and depend on ocean resources. As noted, over four billion people depend on ocean resources, often using artisanal and small-scale aquiculture and fishing to support livelihoods. Bennet et al. (2019) highlight multiple social injustices towards people who depend on these small-scale fisheries from blue growth schemes, including: environmental degradation, pollution and destruction of life-supporting ecosystems; loss of access to resources; greater inequality, especially marginalisation of women; human rights abuses such as forced relocation and exclusion from decision-making processes. Similarly, Okafor-Yarwood et al. (2020) contrast large top-down blue economy schemes in Africa which perpetuate the loss of the right to nature for local communities and tend to disregard environmental and social concerns against schemes which promote inclusive governance, transparency, localised decision-making and collaborative management based on communal principles (see also Cohen et al., 2019 who articulate a rights-based and participatory approach to fisheries management). Consequently, I envisage an important line of future research and practice is to develop blue economies which better represent and recognise the rights of communities to the ocean.

Turning now to the issue of the role of the ocean in climate change, it is notable the COP26, held in Glasgow in November 2021, heralded the ocean as playing a vital future role in the fight against climate change. Blue carbon is a distinct concept to the blue economy and refers to carbon that is stored in coastal marine plants and environments (Steven et al., 2019). These coastal ecosystems such as seagrasses, sediment, mangroves forests and marshes are very effective at storing carbon but are being destroyed at a rapid rate by development. Consequently, blue carbon is of great interest to countries who want to meet their net zero targets, as well as industries which want to increase or protect blue carbon in order to sell blue carbon credits (these are units of carbon which can be bought and sold internationally to help companies or nations meet their climate change pledges). The blue economy and blue carbon are therefore linked through these two dynamics – encouraging coastal planning which restores and protects blue carbon and then finding ways to attach economic value to units of carbon. It is suggested that these processes can work hand-in-hand if projects can find ways to protect coastal ecosystems through sustainable livelihood development and conservation efforts, while at the same time using the blue carbon for broader national development opportunities (Steven et al., 2019; Vanderklift et al., 2019). However, there are few examples of projects which would successfully meet these goals at time of writing.

To conclude, the past decade has seen a rapidly growing interest in the role of the oceans in human development. In the literature and in my own research in Mozambique, the blue economy has been associated with different political agendas and multiple activities, making it a difficult concept to define. Like sustainable development, it attempts to unify contradictory goals and activities under a 'win-win' agenda, suggesting that with the right mix of policy and technology, the rich ocean resource base can be exploited for economic growth without social or environmental cost. At the same time, the blue economy is also a geopolitical agenda to further national economic growth, along with regional diplomatic and security co-operation. The blue economy is clearly an evolving concept, with different examples of how it is being enacted. Future research focuses attention on the rights of resource-dependent ocean communities and attempts to protect blue carbon skinks through attaching economic value to blue carbon credits. What I have hoped to do in this chapter is point out the divergent practices and politics attached to this malleable and appealing concept.

Bibliographic references

Andriamahefazafy, M. and Kull, C.A., 2019. Materializing the blue economy: tuna fisheries and the theory of access in the Western Indian Ocean. *Journal of Political Ecology*, 26(1), pp. 403–424.

Bavinck, M., Jentoft, S. and Scholtens, J., 2018. Fisheries as social struggle: A reinvigorated social science research agenda. *Marine Policy*, 94, pp. 46–52.

Bennett, N.J., Cisneros-Montemayor, A.M., Blythe, J., Silver, J.J., Singh, G., Andrews, N., Calo, A., Christie, P., Franco, A.D., Finkbeiner, E.M., et al., 2019. Towards a sustainable and equitable blue economy. *Nature Sustainability*, pp. 1–3.

Bond, P., 2019. Blue Economy threats, contradictions and resistances seen from South Africa. *Journal of Political Ecology*, 26(1), pp. 341–362.

Brent, Z.W., Barbesgaard, M. and Pedersen, C., 2020. The blue fix: What's driving blue growth? *Sustainability Science*, 15(1), pp. 31–43.

Büscher, B. and Fletcher, R., 2015. Accumulation by conservation. *New Political Economy*, 20(2), pp. 273–298.

Childs, J.R. and Hicks, C.C., 2019. Securing the blue: political ecologies of the blue economy in Africa. *Journal of Political Ecology*, 26(1), pp. 323–340.

Cohen, P.J., Allison, E.H., Andrew, N.L., Cinner, J., Evans, L.S., Fabinyi, M., Garces, L.R., Hall, S.J., Hicks, C.C., Hughes, T.P. and Jentoft, S., 2019. Securing a just space for small-scale fisheries in the blue economy. *Frontiers in Marine Science*, 6, p. 171.

Ekers, M. and Prudham, S., 2015. Towards the socio-ecological fix. *Environment and Planning A*, 47(12), pp. 2438–2445.

Government of Mozambique, 2021. Growing Blue Conference, November 2021, Final Communique – Comunicado-Final-versao-final-23.11.2021.pdf (crescendoazul.gov.mz)

Kalina, M.R., Mbereko, A., Maharaj, B. and Botes, A., 2019. Subsistence marine fishing in a neoliberal city: a political ecology analysis of securitization and exclusion in Durban, South Africa. *Journal of Political Ecology*, 26(1), pp. 363–380.

Keen, M.R., Schwarz, A.M. and Wini-Simeon, L., 2018. Towards defining the blue economy: Practical lessons from Pacific ocean governance. *Marine Policy*, 88, pp. 333–341.

Kothari, U. and Wilkinson, R., 2010. Colonial imaginaries and postcolonial transformations: Exiles, bases, beaches. *Third World Quarterly*, 31(8), pp. 1395–1412.

Okafor-Yarwood, I., Kadagi, N.I., Miranda, N.A., Uku, J., Elegbede, I.O. and Adewumi, I.J., 2020. The blue economy–cultural livelihood–ecosystem conservation triangle: The African experience. *Frontiers in Marine Science*, 7, p. 586.

Payne, J.L., Bush, A.M., Heim, N.A., Knope, M.L. and McCauley, D.J., 2016. Ecological selectivity of the emerging mass extinction in the oceans. *Science*, 353(6305), pp. 1284–1286.

Radomska, M. and Åsberg, C., 2021. Fathoming postnatural oceans: Towards a low trophic theory in the practices of feminist posthumanities. *Environment and Planning E: Nature and Space*, 5(3), pp. 1428–1445.

Steven, A.D., Vanderklift, M.A. and Bohler-Muller, N., 2019. A new narrative for the blue economy and blue carbon. *Journal of the Indian Ocean Region*, 15(2), pp. 123–128. DOI: 10.1080/19480881.2019.1625215

UNECA, 2016. *Africa's Blue Economy: A Policy Handbook*. Addis Ababa: UN Economic Commission for Africa (UNECA).

Vanderklift, M.A., Gorman, D. and Steven, A.D., 2019. Blue carbon in the Indian Ocean: A review and research agenda. *Journal of the Indian Ocean Region*, 15(2), pp. 129–138.

Voyer, G.Q., Mcilgorm, A. and Azmi, K., 2018a. Shades of blue: What do competing interpretations of the blue economy mean for oceans governance? *Journal of Environmental Policy & Planning*, 20(5), pp. 595–616. DOI: 10.1080/1523908X.2018.147315

Voyer, M., Schofield, C., Azmi, K., Warner, R., McIlgorm, A. and Quirk, G., 2018b. Maritime security and the blue economy: Intersections and interdependencies in the Indian Ocean. *Journal of the Indian Ocean Region*, 14(1), pp. 28–48.

World Bank and United Nations Department of Economic and Social Affairs, 2017. *The Potential of the Blue Economy: Increasing Long-term Benefits of the Sustainable Use of Marine Resources for Small Island Developing States and Coastal Least Developed Countries*. Washington, DC: World Bank.

WWF, 2015. Principles for a Sustainable Blue Economy. *WWF Baltic Ecoregion Programme* [online] Available at:https://wwf.panda.org/wwf_news/?247477/Principles%2Dfor%2Da%2D Sustainable%2DBlue%2DEconomy

Chapter 53

Fisheries and development

Carole Sandrine White

1. Fisheries in the context of the Global South

Fishing is an important livelihood and source of food for many people in the Global South. Around 60 million people are engaged directly in wild capture fisheries and aquaculture (FAO, 2022; FAO/Duke/Worldfish, 2023). For wild capture fisheries, this number rises to over 120 million people when processing activities are included and to 600 million people when their dependents are included, with the majority (97 per cent) living in developing countries (World Bank, 2012; FAO/Duke/Worldfish, 2023). Of those 120 million fishers and fishworkers, 94 per cent work what can be defined as small-scale fisheries (SSF; FAO/Duke/Worldfish, 2023: Table 1), which produce two-thirds of catches destined for direct human consumption around the world (FAO, 2022). The roles of men and women in fisheries are extremeley gendered with the majority of men working in the catching sectory and women working in processing activities (Figure 53.2).

The role of fish in nutrition and food security in developing countries is increasingly recognised as significant (Hicks et al., 2019) particularly where fish is the main source of available protein and micronutrients in people's diets (FAO, 2022). At the same time, global fish exports from the Global South have also risen (now representing 50 per cent of global exports in 2020 which are often destined to the Global North; FAO, 2022). Globally, fish consumption and demand has increased at a rate that is twice as high as population growth since 1961, resulting in increasing pressures on fisheries resources. Although global fish production is still increasing overall, reaching 178 million tonnes in 2020; this was due to a sustained increase in the aquaculture sector particularly from Asia, which now supplies most of the world's fish (FAO, 2022, Figure 53.1). Meanwhile, catches from wild fisheries have declined over the last 30 years, from 86 million tonnes in 1996 to 78.8 million tonnes in 2020, with an increasing number of fisheries from developing countries assessed as overexploited, presenting particular risks for people in the Global South who depend on fisheries (FAO, 2022).

The role of fisheries in income generation and food security clearly have an important part to play in meeting the Sustainable Development Goals, not only for meeting SDG 14 'Life Under Water' but also through contributions to SDGs on hunger, gender equality and poverty. However, the relationship between fisheries and poverty alleviation is still debated which continues to influence how development programmes, investments and policies are designed and implemented around the world. This chapter initially explores key debates in how fisheries contribute to development. It then explores the main international policy agendas for small-scale marine fisheries in the Global South. Finally, it

DOI: 10.4324/9780429282348-58

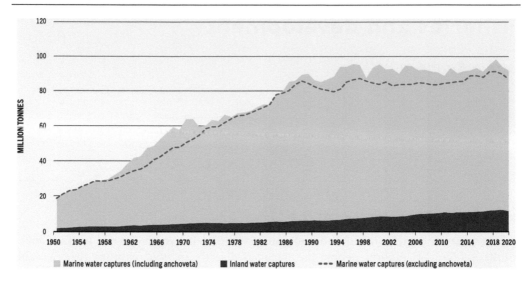

Figure 53.1 World capture fisheries and aquaculture production in 2020

Source: FAO, 2022. The State of World Fisheries and Aquaculture 2022. Towards Blue Transformation. Rome, FAO. https://doi.org/10.4060/cc0461en

Table 53.1 Regional and global estimates of small- and large-scale fisheries employment (part- and full-time) and subsistence activities in 2016, extrapolated from household-based surveys for 78 countries, by segment of the value chain

Region	Pre-harvest	Harvesting		Subsistence fishing		Post-harvest Inland and marine		Total
	inland/ marine	Inland	Marine	Inland	Marine	Processing	Trading	
Small-scale fisheries								
Africa	306090	2653300	1358476	3451521	1207386	1366190	3276366	13619328
Asia	1012241	1124832	9797426	3165015	14801713	5236511	18838478	92584866
Europe	156087	58259	170880	0	0	177591	415566	978383
Oceania	21034	111256	177088	688286	336430	77760	141650	1553504
Americas	230578	527170	1359168	207443	494203	634160	849072	4301794
Total	1726030	1459837	1286308	3599745	1683972	7492211	2352113	113037876
Large-scale fisheries								
Africa	20981	21 763	186067			296371	303388	828 569
Asia	242878	675343	1161775			846902	1796162	4723060
Europe	153731	7269	190245			275391	285738	912374
Oceania	1090	0	13798			15005	22236	52128
Americas	47112	11265	185718			280084	271844	796023
Total	465791	715640	1737603			1713752	2679368	7312154

Source: FAO/Duke University/Worldfish, 2023

SSF Livelihoods

491.7 million (in 2016)

Includes those directly engaged in SSF activities – either employed along the SSF value chain or engaged in subsistence activities – and their household dependents

Full- and part-time employment in SSF (for income and food)

60.2 million

or 89.2% of total employment in fisheries along the value chain including inland and marine

20.9 million women

39.3 million men

Pre-harvest

1.7 million

or 2.9% of total SSF employment

300 thousand women
(or 18% of total pre-harvest segment)

1.4 million men
(or 82% of total pre-harvest segment)

Harvesting

27.5 million

or 45.6% of total SSF employment

14.6 million in inland SSF
(or 53% of total harvesting segment)

12.9 million in marine SSF
(or 47% of total harvesting segment)

5.1 million women
(or 18.7% of total harvesting segment)

Post-harvest

31 million

or 51.5% of total SSF employment

15.5 million women
(or 50% of total post-harvest segment)

15.6 million men
(or 50% of total post-harvest segment)

Additional livelihoods dependent upon employment in SSF

192.8 million people dependent at least partly upon employment in SSF, of which 69.5 million people are fully dependent upon employment in SSF

Subsistence work (for food only)

52.8 million:

36.0 million in inland SSF
(or 68.1% of total subsistence)

16.8 million in marine SSF
(or 31.9% of total subsistence)

23.8 million women
(or 45.2% of total subsistence)

Additional livelihoods dependent upon subsistence activities in SSF

185.9 million people dependent at least partly upon subsistence activities in SSF

Figure 53.2 Infographic with overview of global estimates of small-scale fisheries (SSF) employment, engagement in subsistence activities and additional livelihoods dependent on small-scale fisheries in 2016, extrapolated from household-based surveys for 78 countries

Source: FAO/Duke University/Worldfish, 2023

outlines some of the emergent themes shaping the discourse on fisheries and development in policy and practice.

2. Debates in fisheries and development

2.1 Fishing as a last resort or a way of life?

Fisheries have been portrayed as having a significant role in poverty alleviation through providing livelihoods and food for fishers, their families and for the wider economy. However, despite this, many fishing communities around the world are also found to be poor (Allison and Ellis, 2001; Béné et al., 2010). Fishing has long been one of the few livelihoods open to the landless with limited capital, especially where access to marine resources is under an open access regime or is governed as a common property without entry restrictions. Because of this, fishing livelihoods have been described as a 'last resort' occupation, implying that people fish *because* they are poor and have no alternatives (Allison and Ellis, 2001; Béné et al., 2010). However, a common counternarrative is that fishers are poor *because* they fish. In other words, their livelihood has led to their ruin, due to their inability of individuals to exploit the fishery sustainably as an organised collective. Either way, both narratives seem to conclude that 'fishing rhymes with poverty' especially in the Global South where fishers are often considered 'trapped in poverty' due to debt and limited alternative livelihoods.

In this section, two models are described for understanding the complex relationship between fisheries and development and their implications for policy briefly explored. These are referred to here as a 'wealth' and 'well-being' models (see also Figure 53.2).

2.2 Wealth model

A popular model for understanding how fisheries develop was inspired by Hardin's 'Tragedy of the Commons' (1968) following a Malthusian model which frames the underlying problem for management, as one where 'too many [poor fishers] chase too few fish'. This views the outcome of exploiting common resources without state or market intervention as inevitable collective ruin since each fisher acts independently to protect their own interests. As fisheries become overexploited, fishers catch and earn less and less compared to the effort they invest – measured in terms of 'catch' or 'profit per unit of effort' (CPUE or PPUE).

Viewed as a common property problem, the solution put forward under a 'wealth model' is to grant fishers property rights, to limit their access to the resource and create a sense of personal responsibility for its management (Hardin, 1968). This leads to reducing the number of fishers and/or how much they can catch. The resulting greater catch per fisher and higher levels of income for fishing households with access rights is viewed as successful in reducing poverty. However, while this system reverses the issue of 'too many [poor] fishers chasing too few fish', it often privileges those that are already most effective and better off. Proponents of this model suggest that those excluded from accessing the fishery can diversify their income by taking up alternative livelihoods e.g development of aquaculture, tourism-based livelihoods or environmentally friendly occupations. However, former fishers may not easily take up other forms of livelihood, particularly where these alternative sectors are yet to be developed. They also may not wish to especially where fishing is more than a means of earning a living and is associated with a sense of identity (Pollnac et al., 2012).

The focus of this model on rent maximisation tends to encourage efficiency including investment in technological development, or increased competition via market-based instruments (e.g Individual Transferable Quotas, catch limits which may be traded or sold, export-orientated production). This succeeds in its aim of increasing profitability within the fishery. However, it results in altering the relational dynamics of a community and in access to resources being concentrated and owned by the wealthiest and most powerful individuals (Béné et al., 2010).

2.3 Well-being model

An alternative to the rent maximisation approach proposed in the welfare model by conventional fisheries economists, is a 'well-being model' which highlights the welfare function of fisheries (Allison and Ellis, 2001; Béné et al., 2010). As well as their potential for revenue generation, being a fisher is often also expressed as a way of life, associated with a strong sense of occupational or community identity (Pollnac et al., 2012). Adopting a 'livelihoods approach' for fisheries, Allison and Ellis (2001) considered the different resources different households draw on (natural, economic, social, cultural) which influence SSF are exploited and governed. This framing highlighted that fishers are resourceful and able to diversify their livelihood activities in response to an often changing and uncertain climate, rather than being helplessly trapped in poverty. More recently and building on an alternative understanding of poverty, based on capabilities rather than poverty defined principally as wealth or utility levels, a 'social well-being approach' has been applied to studying fisheries (Coulthard et al., 2011). Using a well-being framing – that understands 'living well' as occurring when human needs are met, where one can act meaningfully to pursue one's goals, and where one can enjoy a satisfactory quality of life – it illustrates how fisheries help people fulfil their goals and needs beyond income (Coulthard et al., 2011). By recognising the essential role that social relations play in people's fishing livelihoods in accessing and using resources, these approaches also highlight the social and cultural functions that fisheries contribute to by fostering a common sense of identity and belonging.

This model recognises that fisheries may also have other functions beyond revenue generation. For instance, fisheries provide a valuable source of essential nutrients for many poor communities (Hicks et al., 2019). They might also be a safety net for vulnerable people facing poverty or livelihood loss as a result of shocks such as hurricanes (Béné et al., 2010).

3. International policy and emerging issues for fisheries

3.1 The blue economy – are SSF being squeezed out?

In the past few decades, the development of the 'blue economy' has been part of national and international policy debates around the world and advocated by NGOs, the private, business leaders and donors alike (Cohen et al., 2019). In a similar vein to the 'wealth model' described above, the 'blue growth agenda' promises to generate revenues from previously untapped, undercapitalised natural resources. This involves both developing new economic sectors – perceived as more efficient in terms of the wealth they generate

	Wealth-based model (for example, World Bank and FAO, 2008)		Welfare model (for example, Jul Larsen, 2003)	
	Poverty reduction: Fishery contributes to lift people out of poverty		Poverty and vulnerability prevention: Fishery contributes to maintain a minimum standard of living	
Level	Contribution	Mechanisms	Contribution	Mechanisms
Household level/sector	• Generation of wealth	• Effective capture of fishery rent (capital accumulation) • High level of commercialisation • Access to effective market mechanisms • Fish as cash crop for investment and diversification	• Safety-net function (transient poverty)	• Reduces vulnerability and mitigates poverty effects • Food security through direct contribution (subsistence) but also fish as immediate source of cash income to cover basic needs (health, education, food)
Local level	• Engine for rural development	• Increased demand for goods and services • Rise in wages	• Labour buffer / safety valve for the poor (chronic, poverty)	• Alternative sources of income, food and/or employment.
National level	• Economic growth	• Trickle up to government through taxes and foreign-exchange earnings (regional or international trade)	• Social-redistributive system (welfare)	• Income and employment multipliers
Type of management system	Restricted access (exclusive management system)		Common pool / semi-open access (inclusive management system)	
Institutional context	Developed countries Developing countries		Developing countries Small-scale fisheries	

Source: Modified from Béné et al. (2007).

Figure 53.3 Comparison of the wealth-based and welfare models in small-scale fisheries

Source: Béné et al., 2010. (permission received by author from Christophe Béné to use this figure)

or in offering sustainable solutions compared to traditional activities such as SSF – and using coastal space differently for the purpose of marine conservation, coastal development and industrialisation. Aquaculture for global seafood export markets, offshore energy and mineral extraction prospecting are examples of the types of economic activities proposed. Conserving tropical coastal ecosystems (e.g., mangroves or coral reefs) is advocated due to their role as carbon sinks, as nurseries for fish species, coastal protection and their potential for tourism. However, important questions related to justice have been raised around the extent to which people can participate in decision-making processes linked to blue investments, the recognition of local people and what they value and how equitably distributed the benefits will be (Bennett et al., 2021). In particular, it is feared that people who are already marginalised (women, Indigenous and other disadvantaged groups) may be increasingly squeezed out and replaced by investments related to the blue economy (Cohen et al., 2019).

3.2 Future directions in international policy for small scale fisheries

Organisations such as the FAO or initiatives including the Too Big to Ignore network advocate that SSF must be safeguarded because of the economic, nutritional and employment benefits to many coastal communities that the 'well-being model' highlights. However, at the same time, threats to SSF around the world are being voiced amid pressures for development encouraged by a blue growth agenda. Both the Voluntary Guidelines for Securing Sustainable Small-Scale Fisheries in the Context of Food Security and Poverty Eradication (SSF Guidelines, FAO 2015) and the Voluntary Guidelines on Responsible Governance of Tenure of Land, Fisheries and Forests in the Context of National Food Security (VGGT, FAO 2012) outline the need to respect customary and informal tenure rights. Managing fisheries sustainably faces significant challenges, not least because fisheries statistics are often incomplete, particularly when it comes to small-scale, subsistence, recreational and inland fisheries, which are often excluded from national reporting. This not only means that it may be difficult to set limits necessary for achieving sustainable exploitation but also that the value of SSF may be underestimated, economically, socially and culturally. Working to address this challenge, the FAO, WorldFish and Duke University have collaborated on new report published in 2023 'Illuminating Hidden Harvests' which deepens past analysis on SSF and their socio-economic contributions, with a gender lens from the outset (Figure 53.2). Until now, models for understanding fisheries and development including the wealth and well-being models tended not to consider communities in a disaggregated manner. In particular women's contributions to functioning fisheries and their vulnerabilities have been absent from official sources of data on fisheries.

The 2030 Agenda for Sustainable Development (2030 Agenda) offers a vision for a fairer world in which no one is left behind. It includes goals for the sustainable development of fisheries and aquaculture in economic, social and environmental terms including food security and nutrition. In particular, Goal 14: 'Life under Water' sets out objectives for fisheries, including objective 14.b to 'provide access for small-scale artisanal fishers to marine resources and markets'. Achieving the 2030 vision, will depend on a recognition of the essential role of fisheries and aquaculture for development, not just in terms of wealth generation but their important welfare function for achieving social well-being in the context of climate change in the developing world.

References

Allison, E.H. and Ellis, F., 2001. The livelihoods approach and management of small-scale fisheries. *Marine Policy*, *25*(5), pp. 377–388.

Béné, C., Hersoug, B. and Allison, E.H., 2010. Not by rent alone: Analysing the pro-poor functions of small-scale fisheries in developing countries. *Development Policy Review*, *28*(3), pp. 325–358.

Bennett, N.J., Blythe, J., White, C.S. and Campero, C., 2021. Blue growth and blue justice: Ten risks and solutions for the ocean economy. *Marine Policy*, *125*, p. 104387.

Cohen, P., Allison, E.H., Andrew, N.L., Cinner, J.E., Evans, L.S., Fabinyi, M., Garces, L.R., Hall, S.J., Hicks, C.C., Hughes, T.P. and Jentoft, S., 2019. Securing a just space for small-scale fisheries in the blue economy. *Frontiers in Marine Science*, *6*, p. 171.

Coulthard, S., Johnson, D. and McGregor, J.A., 2011. Poverty, sustainability and human wellbeing: A social wellbeing approach to the global fisheries crisis. *Global Environmental Change*, *21*(2), pp. 453–463.

FAO, 2022. *The State of World Fisheries and Aquaculture (SOFIA) – Meeting the Sustainable Development Goals*. Rome, Italy: Food and Agriculture Organization.

FAO, Duke University & WorldFish, 2023. *Illuminating Hidden Harvests – The Contributions of Small-Scale Fisheries to Sustainable Development*. Rome: Food and Agriculture Organization. https://doi.org/10.4060/cc4576en

Hardin, G., 1968. The tragedy of the commons. *Science*, *162*, pp. 1243–1248.

Hicks, C., Cohen, P., Graham, N., Nash, K., Allison, E., D'Lima, C., Mills, D., Rosher, M., Thilsted, S., Thorne-Lyman, A. and MacNeil, M.A., 2019. Harnessing global fisheries to tackle micronutrient deficiencies. *Nature*, *574*, pp. 95–98.

Pollnac, R., Maarten, B. and Iris, M., 2012. Job satisfaction in fisheries compared. *Social Indicators Research*, *109*(1), pp. 119–133.

World Bank, 2012. *Hidden Harvest: The Global Contribution of Capture Fisheries*. Washington, DC: World Bank.

Further reading

Allison, E.H. and Horemans, B., 2006. Putting the principles of the sustainable livelihoods approach into fisheries development policy and practice. *Marine Policy*, *30*(6), pp. 757–766.

Béné, C., Arthur, R., Norbury, H., Allison, E.H., Beveridge, M., Bush, S., Campling, L., Leschen, W., Little, D., Squires, D. and Thilsted, S.H., 2016. Contribution of fisheries and aquaculture to food security and poverty reduction: Assessing the current evidence. *World Development*, *79*, pp. 177–196.

Béné, C. and Friend, R.M., 2011. Poverty in small-scale fisheries: Old issue, new analysis. *Progress in Development Studies*, *11*(2), pp. 119–144.

FAO, 2012. *The Voluntary Guidelines on Responsible Governance of Tenure of Land, Fisheries and Forests in the Context of National Food Security*. Rome: FAO. http://www.fao.org/3/a-i2801e.pdf

FAO, 2015. *Voluntary Guidelines for Securing Sustainable Small-Scale Fisheries in the Context of Food Security and Poverty Eradication*. Rome: FAO. http://www.fao.org/3/i4356en/I4356EN.pdf

Finkbeiner, E.M., Bennett, N.J., Frawley, T.H., Mason, J.G., Briscoe, D.K., Brooks, C.M., Ng, C.A., Ourens, R., Seto, K., Swanson, S., Urteaga, J. and Crowder, L.B., 2017. Reconstructing overfishing: Moving beyond Malthus for effective and equitable solutions. *Fish and Fisheries*, *18*(6), pp. 1180–1191.

Johnson, D.S., 2018. The values of small-scale fisheries. In *Social Wellbeing and the Values of Small-Scale Fisheries* (pp. 1–21). Cham: Springer.

Kleiber, D., Harris, L.M. and Vincent, A.C., 2015. Gender and small-scale fisheries: A case for counting women and beyond. *Fish and Fisheries*, *16*(4), pp. 547–562.

OECD, 2016. *The Ocean Economy in 2030*. Paris: OECD Publishing. https://www.oecd.org/environment/the-ocean-economy-in-2030-9789264251724-en.htm

Pauly, D. and Zeller, D., 2016. Catch reconstructions reveal that global marine fisheries catches are higher than reported and declining. *Nature Communications*, *7*, p. 10244.

UN Sustainable Development Goals Knowledge Platform. SDG 14. Conserve and sustainably use the oceans, seas and marine resources for sustainable development https://sustainabledevelopment.un.org/sdg14

Chapter 54

Famine

Stephen Devereux

Introduction

'Famine' can be defined as a widespread and protracted disruption in access to food, which will result in acute malnutrition and mass mortality unless alternative sources of food are available. An estimated 70 million people died in famines during the 20th century (Devereux, 2000). On the other hand, by the 1970s famine was eradicated from historically famine-prone Europe (Russia) and most of Asia (China, India, Bangladesh). Nonetheless, since the 'Band Aid' famine in Ethiopia in 1984, a number of famines have occurred in Africa (Malawi, Niger, Somalia, Sudan) and Asia (North Korea). The persistence of famines into the 21st century is paradoxical, given recent impressive advances in agricultural technology, communications, early warning systems and international humanitarianism.

Historical famine trajectories

Pre-20th-century famines were often triggered by natural disasters that destroyed the subsistence basis of agrarian communities, whose vulnerability was exacerbated by under-development – weak markets, undiversified livelihoods, no food aid system. Since the late 19th century, improvements in transport and communications integrated isolated communities into the wider economy, allowing governments and traders to respond promptly to food crises. Another factor that reduced famine vulnerability was the emergence of nation-states, but this also exposed previously autarchic communities to potent global forces. During the colonial period in India, catastrophic famines occurred that cost millions of lives, until the British introduced the 'Famine Codes' in the 1880s to prevent further famines and legitimise their rule (Davis, 2001). In parts of Africa, starvation was used as a way of crushing initial resistance to colonisation. The penetration of colonial capitalism into subsistence-oriented economies – the commodification of food crops and expansion of 'cash crops' – was blamed for causing the 1970s famine in the West African Sahel (Meillassoux, 1974).

After independence, historically famine-prone countries took one of two routes. Some, like India, made progress in reducing vulnerability factors, through strengthening political accountability for famine prevention and through improvements in food production associated with 'Green Revolution' biotechnology. There has been no major famine in south Asia since Bangladesh in 1974. By contrast, independence in Africa was associated with increased political instability and the emergence of 'war famines', the first occurring

DOI: 10.4324/9780429282348-59

in Biafra, Nigeria, in the 1960s. Subsequently, many African countries suffered food crises related to conflict and militarisation, including Angola, Mozambique, Liberia, Sierra Leone and Uganda. In the Horn of Africa, the lethal combination of wars and droughts contributed to the persistence of famine to the present (von Braun et al., 1998).

Theories of famine

Theoretical explanations for famine tend to reflect the disciplinary specialisation of their authors. This section reviews the main theories by discipline: demography (Malthusianism), economics ('entitlements'; market failure), politics (democracy, international relations) and conflict (war).

• Malthusianism

The most famous theory of famine was conceived in the 1790s by an English priest, Thomas Malthus, who argued in his *Essay on the Principle of Population* (Malthus, 1798) that population could not grow indefinitely in a world of limited natural resources. Eventually, the number of people needing food would exceed global production capacity, when famine would intervene to regulate population growth and balance the demand and supply of food. Although Malthusianism remains extremely influential, as a theory of famine it has many limitations (Dyson and Ó Gráda, 2002). Malthus developed his theory before the industrial revolution moved vulnerable people out of agriculture and into urban centres, before advances in transport and communications allowed food surpluses to be shipped around the world and before scientific advances in agricultural research dramatically increased crop yields.

Moreover, in no country has population growth ever been 'regulated' by famine. When 30 million Chinese died in the 'Great Leap Forward' famine around 1960, the population of China was 650 million; it now exceeds one billion. In poor countries, high fertility rates and 'dependency ratios' persist as a vulnerability factor, but it is economic growth that solves this problem, not food crises. In wealthy countries, a 'demographic transition' to zero population growth (when births equal deaths) has already occurred. Global population is projected to stabilise during the 21st century, at a total well within global food production capacity.

• Economics and Sen's 'entitlement approach'

Amartya Sen's *Poverty and Famines* was published in 1981 (Sen, 1981) and was immediately recognised as the most influential contribution to famine thinking since Malthus. Sen argued that a person's 'entitlement to food' derives from four sources – production, trade, labour and gifts – and that famine is not primarily determined by national food availability but by failures of access to food at the level of groups or individuals. Even when a drought causes crop failure, only some groups suffer 'entitlement failure' and face starvation – wealthy families and urban residents are rarely affected. Paradoxically, farmers who produce food are often most vulnerable to famine; this is because their 'entitlement' derives from a single unreliable source – rain-fed agriculture.

The entitlement approach complements economic explanations for famines that are based on poverty and market failures. When food production or market supplies of food fall, prices rise and, because food is essential for survival, people who cannot afford to buy the food they need face starvation. In Bangladesh in 1974, rapid food price rises caused by expectations of harvest failure made rice unaffordable for landless labourers, because wealthy people and traders hoarded rice and created an artificial scarcity that caused 1.5 million deaths (Ravallion, 1987).

• Politics: Famine as 'act of man'

Famines have always been 'political', but the role of political factors in either creating or failing to prevent famine is increasingly recognised. First, famines affect people who are politically and economically marginalised. Most famines occur in countries that have little global economic or geopolitical significance – Ethiopia, Sudan, Malawi, North Korea. Within countries that suffer famine, the entire population is never at risk – invariably, minority ethnic groups and poor people living in remote rural districts with negligible political influence are most vulnerable.

Second, famines are related to lack of democracy. Apart from his 'entitlement approach', Sen also made an invaluable contribution to the political analysis of famines. Sen argued that two features of democracies protect people against famine: a vigilant free press, so that emerging food crises are neither ignored nor covered up and free and fair elections to ensure state accountability, because a failing government can be dismissed by the electorate (Drèze and Sen, 1989). To illustrate this argument, Sen contrasted the experiences of India and China. Since achieving independence in 1947, India has effectively prevented famine, while China suffered the world's worst famine in history in 1958-1962. Sen attributes India's success largely to its democratic institutions of campaigning journalism and active opposition politics, both of which were absent in communist China, where lack of information meant the famine was not predicted, while lack of accountability allowed the state to escape unpunished.

Another famine theorist, Alex de Waal, adapted Rousseau's concept of the 'social contract' to argue that post-independence governments in India have upheld an 'anti-famine contract' with their population (de Waal, 1997). Conversely, the persistence of famine in other countries might be explained by the absence of such a 'contract'. If the state faces no pressure to prioritise the basic needs of its citizens, human rights abuses carry no political cost, and this explains why famines are more likely to occur in authoritarian regimes (Stalin's Soviet Union, Mao's China, Mengistu's Ethiopia, Kim Jong-Il's North Korea) or during wars, rather than in stable democracies. The famines in the Soviet Union were attributable to punitive government policies, such as forced collectivisation and grain seizures. The 1984 famine in Ethiopia was concealed by the military government, which was fighting a civil war against drought-affected Tigray at the time, until it was exposed by the foreign media. The 1990s famine in North Korea occurred under a repressive regime that does not respect basic human rights and is not responsive to international pressure.

A third political factor in contemporary famines is the role of aid donors. The emergence since World War II of an international humanitarian community that assumes some responsibility for protecting lives across the world has played a significant role in reducing famine deaths and has taken the pressure off governments for ensuring the food security of their citizens. When food aid is not delivered in time to prevent a food crisis, the national government blames the international donors, and the donors blame the government. Ultimately, no one is held accountable. Political tensions between aid donors and national governments have also played a role – food aid has even been used as a political weapon. During the 1974 famine in Bangladesh, the United States withheld food aid because Bangladesh was trading with Cuba in violation of U.S. trade sanctions. In 1984, the United States delayed sending food aid to Ethiopia in an attempt to undermine the Marxist Dergue regime, and in 2000 emergency food aid to Somali Region was delayed because the donors disapproved of Ethiopia's border war with Eritrea and feared that food would be diverted to feed Ethiopian soldiers. Failures of accountability by national governments and international donors have been identified as a fundamental cause of the persistence of famine in 21st-century Africa (Devereux, 2009).

• War and 'complex political emergencies'

Conflict has become a feature of many African famines, and in the 1990s the phrase 'complex political emergencies' was coined to characterise these famines. But the relationship between war and famine is not new. Famine has been used as a weapon of war at least since the Middle Ages, when cities were besieged until their inhabitants either surrendered or starved. During World War II, parts of the Netherlands were blockaded by the German army, food stores were seised and all imports were prohibited into cities like The Hague and Leiden, where an estimated 10,000 people died. More recently, the town of Juba was subjected to similar starvation tactics during a civil war in Sudan.

Famine conditions can also be created as an unintended consequence of conflict, because of its devastating multiple impacts on food production, marketing systems and relief interventions. First, conflict disrupts agricultural production – farmers are displaced, conscripted, disabled or killed; food crops and granaries are destroyed; livestock are raided or slaughtered. Second, war undermines food marketing – trade routes are disrupted, markets are bombed, traders stop operating because they fear for their safety. Third, relief interventions are undermined by logistical constraints and security risks – governments often ban humanitarian agencies from operating in conflict zones, food convoys are attacked and aid agencies withdraw their staff. People in conflict areas become cut off from any source of food, and starvation follows (de Waal, 1997; Duffield, 2001).

The 2011 famine in Somalia is the most recent example of a 'complex emergency' famine, where food production was first disrupted by drought, then trade and aid responses were undermined by conflict and insecurity. The combination of a weak national state, weak international donors that withdrew under pressure from the militant group Al-Shabaab and United States counter-terrorism laws that effectively criminalised aid flows into southern Somalia, and the obstructive practices of Al-Shabaab, all contributed to a lethal failure of humanitarian response, despite excellent early warning information. Unless accountability for allowing this famine to occur is enforced, it seems likely that similar avoidable tragedies will occur in future, in Somalia or elsewhere.

Future famines

Modern famines are less widespread and less severe than historical famines: fewer countries are vulnerable, and fewer people die. On the other hand, recent famines have been exacerbated by a number of new factors – including flawed processes of economic liberalisation and political democratisation, rising prevalence of HIV/AIDS, and problematic relationships between national governments and international donors. Modern famines are also more complex, being caused by multiple failures – of the weather, 'coping strategies', markets, local politics, national governments and the international community. The Malawi famine of 2002 is typical of these 'new famines' (Devereux, 2007): though triggered by erratic rains, food subsidies and parastatal marketing outlets had been abolished under structural adjustment reforms, food prices spiralled as traders failed to respond, high HIV-prevalence had undermined community coping capacity and donors intervened too late, trying to force the government to admit that it had corruptly sold the national strategic grain reserve.

All contemporary famines are fundamentally political. Decisions taken by governments, donors and humanitarian organisations either contribute to creating famine conditions or are responsible for failures to prevent famine. It follows that political will to prevent famine is essential, at every level. The 'right to food' is enshrined in the Universal Declaration of Human Rights of 1948, but little progress has been made in enforcing this right to date. Some writers have argued that famine should be criminalised as a crime against humanity in international law (Edkins, 2000). Until an 'anti-famine contract' is established at the global level, whether voluntarily or enforced under international law, future famines will continue to be tolerated, rather than eradicated.

Further Reading

de Waal, A. (1997), *Famine Crimes: Politics and the Disaster Relief Industry in Africa*, Oxford: James Currey.
Alex de Waal takes an explicitly political approach to explaining contemporary African famines, blaming African governments and failures of the international community.
Devereux, S. (1993), *Theories of Famine*, Hemel Hempstead: Harvester Wheatsheaf.
This book summarises the main theoretical explanations for famine causation, including climate shocks, Malthusianism, 'entitlement', market failure, government policy and war.
Devereux, S. (ed) (2007), *The New Famines: Why Famines Persist in an Era of Globalisation*, London: Routledge.
This edited collection critically reviews recent developments in famine theory and includes studies of recent famines in Ethiopia, Madagascar, Malawi, Sudan, Iraq and North Korea.
Sen, A. (1981), *Poverty and Famines: An Essay on Entitlement and Deprivation*, Oxford: Clarendon Press.
The most influential book on famine since Malthus' *Essay on the Principle of Population*, in which Sen introduces his 'entitlement approach' to the analysis of poverty and famines.

References

de Waal, A. (1997), *Famine Crimes: Politics and the Disaster Relief Industry in Africa*, Oxford: James Currey.
Davis, M. (2001), *Late Victorian Holocausts: El Nino Famines and the Making of the Third World*, London: Verso.
Devereux, S. (2000), 'Famine in the Twentieth Century', *IDS Working Paper*, 105. Brighton: Institute of Development Studies.
Devereux, S. (ed) (2007), *The New Famines: Why Famines Persist in an Era of Globalisation*, London: Routledge.
Devereux, S. (2009), 'Why Does Famine Persist in Africa?', *Food Security*, 1(1): 25–35.
Drèze, J. and Sen, A. (1989), *Hunger and Public Action*, Oxford: Oxford University Press.
Duffield, M. (2001), *Global Governance and the New Wars*, London: Zed Books.
Dyson, T. and Ó Gráda, C. (eds) (2002), *Famine Demography: Perspectives from the Past and Present*, Oxford: Oxford University Press.

Edkins, J. (2000), *Whose Hunger? Concepts of Famine, Practices of Aid*, Minneapolis: University of Minnesota Press.

Malthus, T. (1798, 1976 edition), *An Essay on the Principle of Population*, New York: W.W. Norton.

Meillassoux, C. (1974), 'Development or Exploitation: Is the Sahel Famine Good Business?', *Review of African Political Economy*, 1: 27-33.

Ravallion, M. (1987), *Markets and Famines*, Oxford: Oxford University Press.

Sen, A. (1981), *Poverty and Famines: An Essay on Entitlement and Deprivation*, Oxford: Clarendon Press.

von Braun, J., Teklu, T. and Webb, P. (1998), *Famine in Africa: Causes, Responses, and Prevention*, Baltimore: Johns Hopkins University Press.

Chapter 55

Renewable energy and development

Andrew Lawrence

The Sustainable Development Goals (SDG) progress report 2020 reiterates widespread recognition of energy as an enabler of sustainable development, while the IPCC Sixth Assessment Report of 2021 underscores the urgency of dramatically reducing greenhouse gas (GHG) emissions to prevent ecological catastrophe. Together, these mandates underscore the indispensable centrality of renewable energy (RE) for development worldwide. While RE endowments are universally accessible, their optimal development depends not only on effective and participatory design, distribution and consumption strategies, in order for the SDG goal 7.1 – Universal Access to Energy – to be achieved; it also depends on concerted transnational action to foster and finance this goal.

This chapter discusses major RE technologies, some salient social and environmental challenges and criticisms associated with their development, as well as strategies for overcoming these challenges. Its first section discusses key pro-developmental advantages of RE policy adaptation. It then highlights barriers to 100 per cent RE generation, distribution, and infrastructure development. The final section explores various strategies for overcoming these barriers, in terms of financing, production, design, and implementation.

RE production, distribution, and adaptation

Renewable energy (RE) commonly refers to both specific technologies harnessing widely (often universally) accessible energy sources (e.g., wind, solar, hydro, biogas, geothermal, and tidal), as well as the socio-technical systems in which these technologies are embedded. Notwithstanding their specific attributes, these technologies and systems share several characteristics. Per unit of energy, all produce more local employment and – from the development of their various inputs, to use and disposal – fewer greenhouse gas (GHG) emissions and other harmful pollutants than do fossil fuels (e.g., coal, oil, and natural gas) or other non-renewable sources (e.g., nuclear power; De Souza Mendonca et al., 2020). Their use also correlates more strongly with adjusted net savings, for high-income and lower-income countries alike – a key determinant of socioeconomic development outcomes (Güney, 2019). In addition, almost all such sources are abundantly accessible in every major world region (albeit with local and regional variations) and amenable to progressively increasing end-use electrification (Jacobson et al., 2017). Finally, in both their extraction as well as generation or end use of energy, all are typically more scalable and more rapidly deployable than their non-renewable counterparts, as well as increasingly less expensive, particularly once negative externalities from pollution are internalised.

DOI: 10.4324/9780429282348-60

For these reasons, they are also better equipped for responding to hazards associated with global warming (such as droughts and floods) and thus strengthening resiliency in the face of climate-related threats.

However, the aforementioned advantages over fossil fuel sources together have proven inadequate thus far to ensuring the adoption of RE at a pace and extent that ensures realisation of the COP26 goal of limiting warming to 1.5C. Neither do they guarantee that in the production of various elements of the RE value chain, conditions of work are less exploitative than is typical for the oil, coal, and natural gas industries, for example. Exploitative and unfree work conditions are well documented for many of the extractive mineral inputs needed for the solar, wind, and battery storage sectors, for example (Business and Human Rights Resource Centre, 2021), as well as for significant segments of RE manufacturing (Murphy and Elima, 2021).

Barriers to 100 per cent RE generation, distribution, and development

Achieving a transition of energy systems to complete electrification generated entirely by renewable sources entails several distinct – but interconnected – struggles. These include replacing the financing and subsidisation of fossil fuels for that of RE sources; extending the grid where feasible and where not, providing off-grid solutions to RE access; promoting demand-side energy savings and efficiency improvements; and redesigning markets and ownership models in order to maximise systemic security and access, as well as wages and work conditions, while keeping retail prices at a minimum. Supply intermittency, intrinsic to some RE sources such as wind and solar, requires qualitatively different design systems from those dominated by baseload generation such as from fossil fuels, since storage of excess generation attains greater prominence. These in turn require an integrated, trans-ministerial approach by governments (addressing the policy areas of e.g., generation, finance, transport, and agriculture simultaneously), and – undergirding this – high levels of information for, consultation of, and participation by, affected communities.

Financing remains grossly inadequate to the dual goals of limiting warming to 1.5C and attaining SDG7 worldwide. From 2009 to 2019, the RE proportion of global energy generation net capacity growth more than tripled, approaching two-thirds of net capacity additions with public and private sector investments of $2.7 trillion in new non-hydro RE power capacity. For the decade to 2030, however, only $1 trillion has been committed – less than an eighth of the amount needed to attain the COP target limiting warming to 1.5C and a tenth of what is spent annually on fossil fuels (FSFM, 2020: 11).

Strategies for overcoming RE barriers

Meaningful decarbonisation worldwide necessarily entails the continuous and steep reduction of fossil fuel extraction and use, particularly among the historically largest greenhouse gas (GHG) emitters, individually and nationally, as well as by corporation and sector. For both national and global climate and energy justice claims, therefore, *contextual framing* is a key element of common redistributive strategies. For example, Global South demands refer to the fact that cumulative US emissions amount to one fifth of the world's total, almost double (and on a per capita basis, six times) the level of the second largest cumulative emitter, China. The US, together with five other large historical emitters – Russia, Germany, the UK, Japan, and Canada – comprise only 15 per cent of the world's population but are responsible for 60 per cent of total emissions since 1850 (Evans, 2021). During the period of most rapid emissions growth, from 1990 to 2015, the richest one percent alone were responsible for fifteen percent of cumulative emissions – twice as much as the poorest half of the world's population. This injustice is compounded by those of the 2008 financial crisis, the Covid crisis, and the cumulative and accelerating crisis of global warming, each of which impose disproportionate effects upon the livelihoods and life chances of the world's poorest populations – those least responsible for causing these crises in the first place.

OECD-based energy justice campaigns in turn point to local inequalities, such as the fact that the per capita emissions rate of the bottom half of US households by income – while roughly equivalent to China's – is less than a fifth the average for the top ten percent of US households (Wagstyl et al., 2021). Similar national emissions patterns throughout the OECD, while not as extreme as global inequalities, also impose disproportionate burdens upon the poorest members of each country, who also suffer the localised effects of fossil fuel pollution and energy poverty disproportionately. For struggles in both the Global North and South, therefore, justice-based claims for energy access entail more equitable funding for and access to RE infrastructure and employment, equalisation of incomes, and compensation or reparation for past harms, which could be secured through (national and transnational) debt forgiveness, more favourable terms of access to credit, and the decommodification of public goods as a means of strengthening climate resilience.

Whereas the fossil fuel-based global energy economy is characterised by highly geographically concentrated sites of extraction, requiring extensive, long-distance, and highly endogenous transport infrastructure, its emerging RE counterpart is much more geographically dispersed, requiring (and facilitating) localised generation, storage, and use (Lawrence, 2021). The former's concentrated and uneven landscape of extraction and distribution encouraged labour and development strategies of controlling nodal 'chokepoints' to gain leverage and access rents. RE geographies, by contrast, typically entail dispersed struggles over land-use, localised generation, distribution, and consumption, down to the household level. This intrinsically 'disruptive' potential can be channelled towards more empowering, developmental ends, such as ownership and control of not only energy infrastructure, but other elements of autonomous, solidaristic economies as well, such as food and water systems, and manufacturing and service sectors. Localisation can also serve to further develop the trend, necessitated by the Covid pandemic, of rethinking and reworking the car-centric geographies of work and urban development bequeathed by the 20th century's 'auto-industrial age', as well as reinscribing the centrality of mass transit as a public good.

Maximising the benefits of RE technological diffusion necessitates, however, reversing the current trend of tariff escalation in the turbine, PV, and li-ion battery sectors in particular. Solidarity campaigns that combine the strengthening and extension of RE global value chains as well as the rights of all value chain workers, in conjunction with rapidly accelerated levels of RE financing in middle- and low-income countries, would be the best means of achieving the dual promise of universal energy access and net zero emissions worldwide (Goldthau et al., 2022).

References

Business & Human Rights Resource Centre (2021). *Transition Minerals Tracker.* https://trackers. business-humanrights.org/transition-minerals/

De Souza Mendonca, A. K., Barni, G. D. A. C., Moro, M. F., Bornia, A. C., Kupek, E., & Fernandes, L. (2020). Hierarchical modeling of the 50 largest economies to verify the impact of GDP, population and renewable energy generation in CO2 emissions. *Sustainable Production and Consumption*, 22, 58–67.

Evans, S. (2021). Analysis: Which countries are historically responsible for climate change? *Carbon Tracker*, 5 October, 2021.

FSFM (2020). *Frankfurt School of Finance Management. Global Trends in Renewable Energy Investment 2020.* Frankfurt, Germany: Frankfurt School of Finance Management. https://www. fs-unep-centre.org/wp-content/uploads/2020/06/GTR_2020.pdf

Goldthau, A., Hughes, L., & Nahm, J. (2022). *Assessing the Reshoring of Supply Chains in Low Carbon Technologies: The State of Affairs and Implications* (March 1, 2022). Available at SSRN: https://ssrn.com/abstract=406604

Güney, T. (2019). Renewable energy, non-renewable energy and sustainable development. *International Journal of Sustainable Development & World Ecology*, 26(5), 389–397. https://doi.org/1 0.1080/13504509.2019.1595214

Jacobson, M. Z., Delucchi, M. A., Bauer, Z. A. F., et al. (2017). 100% clean and renewable wind, water, and sunlight all-sector energy roadmaps for 139 countries of the world. *Joule*, 1, 108–121. https://doi.org/10.1016/j.joule.2017.07.005

Lawrence, A. (2021). Reconceptualizing contemporary energy markets. *Competition & Change*, 25(5), 631–650.

Murphy, L., & Elima, N. (2021). *In Broad Daylight: Uyghur Forced Labour and Global Solar Supply Chains. Report*, 1 May. Sheffield: Sheffield Hallam University.

Wagstyl, S., Bernard, S., & Bruce-Lockhart, C. (2021). Climate change is becoming less a battle of nations than rich vs poor. *Financial Times*, 21 May, 2021. https://www.ft.com/content/4788beae-9035-4449-b5cd-200dc7b6ea9d

Five additional resources (further reading/online resources/policy documents, etc.):

Bogdanov, D., Farfan, J., Sadovskaia, K., et al. (2019). Radical transformation pathway towards sustainable electricity via evolutionary steps. *Nature Communications*, 10(1), 1–16, 1077.

Lazard (2021). *Lazard's Levelized Cost of Energy Analysis – Version 15.0*, October 2021. https:// www.lazard.com/media/451905/lazards-levelized-cost-of-energy-version-150-vf.pdf

Papaefthymiou, G., & Dragoon, K. (2016). Towards 100% renewable energy systems: Uncapping power system flexibility. *Energy Policy*, 92, 69–82.

Pearse, R., & Bryant, G. (2021). Labour in transition: A value-theoretical approach to renewable energy labour. *Environment and Planning E: Nature and Space*, 5(4), 1872–1894.

Stavropoulos, S., & Burger, M. J. (2020). Modelling strategy and net employment effects of renewable energy and energy efficiency: A meta-regression. *Energy Policy*, 136, 111047.

Chapter 56

Climate adaptation

Rónán McDermott, Karsten Schulz, Lummina Horlings, and Lorenzo Squintani

Introduction

Climate change is significantly threatening the global development gains of recent decades, highlighting the manifold links between climate change adaptation and development. On the one hand, climate change increases the intensity and frequency of extreme weather events, including heatwaves, droughts, catastrophic wildfires, desertification, and sea-level rise. On the other hand, decisive adaptive actions on a large-scale will be required to avoid – or at least buffer – the negative effects of climate change on human societies that cannot be avoided anymore, for example mass migration, political instability and conflicts, food-supply and value-chain disruptions, infectious disease outbreaks and other health impacts, and biodiversity loss.

While the broad global and regional trajectories of climate change impacts have been linked with historical and a range of future emissions scenarios, the precise impacts in terms of timing, frequency and severity is less certain, particularly at more localised scales. This arises as a result of tipping points and uncertainty around feedbacks within the earth system as well as risk cascades emerging from instability within human systems, such as conflict and political instability (Kemp et al., 2022). Such inherent unpredictability and uncertainty concerning climate impacts demand that the field of climate adaptation adopts a risk management approach. A risk-informed approach to sustainable development accordingly marks the 2030 Agenda for Sustainable Development, the 2015 Paris Agreement, and the Sendai Framework for Disaster Risk Reduction 2015–2030.

Tackling the severe challenges of climate change mitigation (decarbonisation) and adaptation (building resilience) simultaneously requires human societies to transformatively address the root causes of vulnerability from the perspective of sustainable development (Fedele et al., 2019: 118). Many of the perennial topics of human development, including poverty reduction, gender equality, ecosystem conservation and food security are, therefore, directly linked to the cross-cutting issue of adaptation. The adaptation endeavour is thus inherently far-ranging and demands coordination between a wide breadth of disciplines and sets of expertise, from the natural sciences to development and governance-related research. With this complexity in mind, the chapter provides a broad overview of climate risk awareness and assessment, followed by an overview of how climate adaptation is implemented. The example of food systems will be used throughout to illustrate the key points made.

DOI: 10.4324/9780429282348-61

Risk awareness and assessment

Climate risk can be considered a function of hazard, exposure, and vulnerability (IPCC, 2014). Hazards concern the direct climate impacts themselves and ultimately constitute the stimuli that prompt adaptation. These include long-term trends in mean temperatures and precipitation and secondary effects such as sea level rise, variability of climate inter-seasonally to periods of decades, and variability that manifests as extreme events such as heatwaves and flooding (Pelling, 2010). A multi-hazards perspective is increasingly deployed to capture the intersection of two or more hazards, for example the increasing concurrence of heat and drought events that are causing crop production losses and tree mortality (Tilloy et al., 2019; IPCC, 2022). Exposure is a further key component of risk and can be reduced by avoiding settlement in hazardous areas such as floodplains or under certain circumstances through the relocation of individuals, households, or communities. Vulnerability is a commonly deployed concept with a plethora of often-competing definitions. At its core it refers to the 'propensity or predisposition to be adversely affected' (IPCC, 2022). Such propensity or predisposition is rooted in and dependent upon the development trajectories of any set of individuals or communities. This component of climate risk thereby incorporates consideration of wider non-climatic trends such as rapid urbanisation, rising inequalities, biodiversity loss, unequal trade relations and environmental degradation. As such, vulnerability is entrenched in social, economic, and political structures that are often distant in time and space from where it is manifested (Wisner et al., 2004). It is also socially differentiated; gender, age, class, caste, ability, and other social characteristics are important drivers of vulnerability depending on the context. The dynamic nature of vulnerability as well as the impacts of climate change, play out in spatially varied ways within and across households, communities, and countries as well as over time. The most vulnerable states and communities will continue to be the hardest hit in a warming world, exacerbating inequities (Kemp et al., 2022).

The combination of vulnerability, exposure, and hazard constitutes risk and the complexity of the interactions between the different components are increasingly understood under the concept of systemic risk (UNDRR, 2022). This concept considers the compounding and cascading potential of risk. As the IPCC (2022) notes, climate risks are becoming more complex and difficult to manage and are cascading across regions and sectors. Compounding risks are risks that arise due to the intersection or co-location of several sources of risk and cascading risks refer to the secondary and tertiary effects beyond the immediate impact of the climate-related hazard event. For example, food systems are vulnerable to cascading risks as exemplified by low water levels in the La Plata River restricting the export of grain and contributing to the drastic increasing of grain prices on the world market in 2021. This example also highlights the multi-scalar and systemic nature of risk; a climate impact in one locale can have significant impacts within the wider system.

Key characteristics of adaptation

While the significant reduction of greenhouse gas emissions, primarily by the more developed states, is urgently required in order to address climate risk, wide-ranging climate adaptation will also be required under every emissions scenario to complement such mitigation efforts, particularly for those most exposed and vulnerable to climate hazards.

The IPCC defines adaptation as follows:

In human systems, the process of adjustment to actual or expected climate and its effects, in order to moderate harm or exploit beneficial opportunities. In natural systems, the process of adjustment to actual climate and its effects; human intervention may facilitate adjustment to expected climate and its effects.

(IPCC, 2022: 5)

This definition has been widely used in academia as well as in the public and private sectors. However, criticisms acknowledge that a sole focus on actual or expected climate and its effects might preclude a deeper engagement with the root causes of human vulnerability and the promotion of resilience, for example through poverty reduction, which is also a central concern of (sustainable) development. Adaptation ought to thereby involve a risk management approach that addresses not only climate hazards but also the vulnerabilities and forms of exposure that contribute to risk (Wisner et al., 2004; Oliver-Smith et al., 2016).

Adaptation can be considered along axes of time, scale and cost. For example, an umbrella is a relatively low-cost solution to sheltering one or two individuals for the duration of a rain shower. The installation of irrigation systems on the other hand is relatively more expensive but constructed to protect swathes of agricultural land from drought over a timescale of up to several decades (Felli, 2021). Distinctions can be drawn between planned and autonomous forms of adaptation, as well as between anticipatory and reactive forms of adaptation. Reactive adaptation ought not to be confused with mere coping in the face of climate shocks or stressors. Coping may require strategies such as the sale of assets or the burdening of social networks that undermine the capacity to adapt to future shocks and stressors (IRIN, 2012). Planned and anticipatory forms of adaptation also have the capacity to ultimately undermine future adaptation pathways. Ostensibly successful adaptation in one locale can shift risk to other locales, for example the potential for aquifer-fed irrigation increases the likelihood of drought nearby. Thus, responses to climate change can in and of themselves contribute to risk alongside direct climate impacts, a process known as maladaptation (IPCC, 2022).

Reflecting the wider field of development, a distinction is often made between top-down approaches to climate adaptation and more bottom-up, community-based, locally led forms. The former is typified by state- or market-based solutions that hold out the promise of wide-ranging structural change, for example through the construction of large-scale grey infrastructure such as seawalls. However, in practice they can often further marginalise vulnerable groups. Bottom-up approaches on the other hand are rooted in communities beyond or even against the state. They tend to emphasise the role of local and indigenous knowledge in informing risk assessment and adaptation practices as well as the importance of deliberation and equitable social relationships in achieving their objectives (Huq and Reid, 2007; Leitheiser et al., 2021).

Shifting towards transformative adaptation

Adaptation can also, crucially, be considered in terms of the extent of system change envisaged. Given the speed of observed and projected changes to temperature and precipitation, incremental adaptation in many contexts may not be sufficient, prompting the need for more transformative forms of adaptation. Such latter forms of adaptation are distinguished by their restructuring, path-shifting, innovative, multi-scale, systemwide, and persistent nature (Fedele et al., 2019). Transformative adaptation can also confer wider co-benefits to development and the environment, for example by reducing vulnerability to a wider range of risks beyond those originating from climate-related hazards or by the reduction of greenhouse gas emissions (Schreuder and Horlings, 2022).

More incremental forms of adaptation within agriculture and food systems may involve for example the introduction of new irrigation practices while more transformative forms might involve broader fundamental reform of the governance of natural resources (Kates et al., 2012). Given the short timeframe for meaningful climate mitigation as well as the increasing risks to the conveyance of food from 'farm to fork', food systems in particular have the potential for multiple co-benefits between decisive climate action, adaptation, and sustainable development. Changes in food systems, for example, also have ripple effects for local communities, companies, and political decision-makers in the public sector, as they affect health systems, political stability, economic systems, and human development in equal measure.

A key enabling condition for transformative adaptation is a clear political agenda and an enabling and inclusive adaptive form of governance, which is oriented towards social justice and equality. This is likely to become more relevant as the rising costs of climate measures disproportionately affect those already vulnerable to climate risks. The empowerment and facilitation of communities and a shift of agency towards them, together with institutional learning to support processes of co-creation between public, private, and civic actors within different sectors and on multiple levels and scales, is urgently required (Holling and Meffe, 1996; Hügel and Davies, 2020; Schreuder and Horlings, 2022).

Conclusion

It is imperative that development pathways adequately take climate risk into account and move rapidly in a climate resilient direction. In other words, development ought to avoid exacerbating risks through continued emissions, and residual risks to development now and in the future ought to be adequately addressed in a transformative and systemic manner. Adaptation is ultimately embedded within the wider development endeavour, and additional opportunities for co-benefits arise from linking adaptation and sustainable development both conceptually and practically. Key enabling conditions in this regard include the forms of governance outlined earlier, support for civic collective action and bottom-up community responses, as well as financing. The latter involves honouring pledges for adaptation funding within the UN Framework Convention and Climate Change regime and ensuring that such funding is complementary to, rather than included within existing development aid commitments. An holistic, networked, and multi-level consideration of catastrophic climate scenarios and climate adaptation action is ultimately required to link the challenges of sustainable development and climate adaptation practically and to utilise the transformative potential of large-scale systemic shifts.

References

Fedele, G., Donatti, C. I., Harvey, C. A., Hannah, L., & Hole, D. G. (2019) Transformative adaptation to climate change for sustainable social-ecological systems. *Environmental Science and Policy*, 101, 116–125. https://doi.org/10.1016/j.envsci.2019.07.001

Felli, R. (2021) *The Great Adaptation: Climate, Capitalism and Catastrophe*. London: Verso Books.

Holling, C. S., & Meffe, G. K. (1996) Command and control and the pathology of natural resource management. *Conservation Biology*, 10, 328–337.

Hügel, S., & Davies, A. R. (2020). Public participation, engagement, and climate change adaptation: A review of the research literature. *Wiley Interdisciplinary Reviews: Climate Change*, 11(4), e645.

Huq, S., & Reid, H. (2007) Community-based adaptation: A vital approach to the threat climate change poses to the poor. London: International Institute for environment and development.

IPCC (2014) Climate change 2014: Impacts, adaptation, and vulnerability. *Summary for Policymakers*, 1–44.

IPCC (2022) Intergovernmental panel on climate change, climate change 2022: Impacts, adaptation, and vulnerability. In *Contribution of Working Group II to the Sixth Assessment Report of the Intergovernmental Panel on Climate Change*, H.-O. Pörtner et al., Eds. Cambridge: Cambridge University Press.

IRIN. (2012) *Coping versus adapting.* [online] Available at: https://www.thenewhumanitarian.org/news/2012/04/02/coping-versus-adapting

Kates, R. W., Travis, W. R., & Wilbanks, T. J. (2012) Transformational adaptation when incremental adaptations to climate change are insufficient. *Proceedings of the National Academy of Sciences USA*, 109, 7156–7161. https://doi.org/10.1073/pnas.1115521109

Kemp, L., Xu, C., Depledge, J., Ebi, K. L., Gibbins, G., Kohler, T. A., Rockström, J., Scheffer, M., Schellnhuber, H. J., Joachim, H., Steffen, W., & Lenton, T. M. (2022) Climate endgame: Exploring catastrophic climate change scenarios. *Proceedings of the National Academy of Sciences USA*, 119(34).

Leitheiser, S., Trell, E. M., Horlings, I., & Franklin, A. (2021) Toward the commoning of governance. *Environment and Planning C: Politics and Space*, 40(3), 744–762.

Oliver-Smith, A., Alcántara-Ayala, I., Burton, I., & Lavell, A. (2016) The social construction of disaster risk: Seeking root causes. *International Journal of Disaster Risk Reduction*, 22, 469–474.

Pelling, M. (2010) *Adaptation to Climate Change: From Resilience to Transformation* (1st ed.). London: Routledge.

Schreuder, W., & Horlings, L. G. (2022) Transforming places together: Transformative community strategies responding to climate change and sustainability challenges. *Climate Action*, 1(24). https://doi.org/10.1007/s44168-022-00024-3

Tilloy, A., Malamud, B. D., Winter, H., & Joly-Laugel, A. (2019) A review of quantification methodologies for multi-hazard interrelationships. *Earth-Science Reviews*, 196, 102881.

UNDRR (2022) *Global Assessment Report for Disaster Risk Reduction.* https://www.undrr.org/publication/sendai-framework-disaster-risk-reduction-2015-2030 retrieved 28th March 2021.

Wisner, B., Blaikie, P., Cannon, T., & Davis, I. (2004) *At Risk: Natural Hazards, People's Vulnerability, and Disasters.* London: Routledge.

Chapter 57

Global environmental justice

Adrian Martin

Origins, meanings, and aims

Environmental justice came into common usage in 1980s America following a series of protests against the location of toxic waste sites and polluting industries (Murdock, 2020). Environmental protest was not new, but claims that the location of pollution was driven by racism – and the central role of the black civil rights movement – certainly was. In 1982 the small and predominantly African-American community of Warren County, North Carolina, was selected for landfilling PCB-contaminated soils, leading to a now famous protest in which hundreds were arrested. Since that time, researchers have shown that Warren County was not an isolated case but part of a regular pattern of environmental injustice in which communities of colour suffered disproportionately from pollution because of institutionalised racism (Bullard, 1990).

Justice traditionally refers to the fair treatment of people or 'what we owe to each other'. Injustice occurs where people are treated differently – and unfairly – often due to arbitrary biological, cultural or geographical characteristics such as the colour of their skin, gender, beliefs or simply where they are born. Environmental justice examines and opposes the patterns of injustice that arise when such social differences come to predict an individual or group's level of exposure to environmental harms such as pollution and hazards or their ability to benefit from nature's goods, such as healthy food and green spaces. A natural hazard, such as hurricane Katrina in 2005, may have natural causes, but when people of colour suffer disproportionately because institutional racism has made them more vulnerable, it is a human-influenced disaster that can be framed in terms of environmental justice.

Environmental justice is therefore about the interface between major societal inequalities such as racism, patriarchy and coloniality and environmental problems such as industrial pollution, natural hazards, climate change and biodiversity loss. It is a framework of analysis that contributes to understanding how various forms of social discrimination – both singularly and in combination – undermine the chances that some groups of people have to live well, by burdening them with environmental harms and denying them benefits.

Environmental justice scholars also observe how this interface between social and environmental problems works the other way around. Just as environmental degradation intensifies and perpetuates social inequalities, so too social inequalities are drivers of unsustainable practices. One way this happens is where inequality leads to low

DOI: 10.4324/9780429282348-62

environmental standards. A tragic example of this is when the US-based Union Carbide Corporation operated far lower safety standards in its pesticide plant in Bhopal, India, than in its plant in West Virginia, leading to a devastating gas leak that killed many thousands and injured half a million others. Another way inequality undermines sustainability is by producing barriers to collective action to address the current environmental emergency. This is illustrated by asking questions such as: how can a highly unequal society develop the shared vision and cooperation to tackle climate change? How can the proven capacity of indigenous peoples to care for their territories be harnessed when they are marginalised within their nation-states? How can perverse incentives be transformed when large corporations continue to exercise such power over political and economic decision-making? An example of the latter is that the UK government effectively subsidises air travel by not taxing aviation fuel or tickets.

The aim of environmental justice analysis is to reveal and understand how power relations lead to the unfair treatment of some and in particular how social status determines exposure to environmental harms and access to environmental benefits. It is also to confront and change these structural conditions that undermine good quality of life for many.

The expanding circle

In his book *The Expanding Circle,* the philosopher Peter Singer (2011) notes the tendency for humans to use their reason to expand the scope of their moral duty, starting with care for close family and kin, eventually expanding to people all around the world – and even beyond – to other species and entities. The metaphor of an expanding circle applies quite nicely to the evolution of environmental justice scholarship and activism since the early 1980s (e.g., Schlosberg, 2013).

Expanding categories and subjects of justice

It has been argued that the primary purpose of critical social science is to shed light on the character of subordination that predominates in any society (Fraser, 1985). To achieve this it is necessary for scholars to employ appropriate social categories – ones that are fit for purpose for revealing patterns of dominance and denigration. For example, if patriarchy is a significant form of subordination, then 'gender' and gendered analysis must be part of the critical vocabulary. Whilst early environmental justice work focused mostly on the category of race, it has since expanded its concern to wealth, class, gender, coloniality, disability, age, indigeneity and other social categories. Recently, for example, concerns that the costs of climate change mitigation and adaptation are being pushed onto low-income groups (e.g., energy poverty, loss of traditional employment), on rural locations (land use changes such as livestock reduction that may conflict with traditional rural livelihoods) and on women (e.g., due to gendered activities of water collection). This expansion in social categories is also linked to intersectionality – the idea that social categories often intersect with each other, so one's experience of environmental injustice is not shaped only by singular categories, such as being a person of colour but through multiple intersecting categories, such as also being on a low income. As the poet Audre Lorde put it, 'There is no such thing as a single-issue struggle as we don't live single-issue lives'. Such an insight into intersectionality underlies calls for a wide-ranging, progressive alliance across movements for e.g., environmental justice, feminism, peace, indigenous rights and so on.

Until fairly recently, the categories of discrimination were restricted to current humans and mainly then within the boundaries of a nation-state. This has expanded to incorporate concern for future generations (inter-generational justice) and for the other-than-human world (ecological justice), including for example sentient animals and entities such as rivers, mountains and mother earth (rights of nature).

> The eyes of all future generations are upon you. And if you choose to fail us, I say – we will never forgive you.
> (Greta Thunberg, UN Climate Summit, 2019, New York)

> Environmental justice affirms the sacredness of Mother Earth, ecological unity and the interdependence of all species, and the right to be free from ecological destruction.
> (Principle 1, Environmental Justice Principles,
> National People of Color Environmental Leadership Summit, 1991)

Expanding the circle to other-than-human entities introduces new categories of discrimination that are of concern, including speciesism and anthropocentrism. In currently dominant decision-making, other-than-human nature (other species, ecosystems, rivers, mountains etc.) tends to only be valued for its use to humans, such that a river is worthy of protection only in so far as it yields fish and other benefits to humans. Such anthropocentrism seems an inadequate basis for ensuring the fair treatment of other-than-human nature, for example because animals are only valued for their contribution to human economies, not as being valuable in their own right (and not as holding rights). Furthermore, anthropocentrism currently dominates other worldviews and this dominance is itself a form of discrimination against people who hold other beliefs, including many indigenous peoples who have non-anthropocentric worldviews (IPBES, 2022).

Expanding dimensions and scales

Whilst early EJ concerns may have paid most attention to distributional justice, notably the distribution of environmental harms, it was soon apparent that such distributional outcomes are connected to decision-making processes and that exclusion and marginalisation in decision-making stems from status inequalities between different groups of people. For example, in June 2022, following more than a decade of harassment, the Maasai people of Loliondo, in the Ngorongoro district of Tanzania have been ordered to leave their ancestral land by a government and its military that wants to turn their territory into high-end hunting and tourism (Ndoinyo, 2021). It is a distributional environmental injustice in which the Maasai lose the very essence of living well – their land – to enable accumulation of wealth for a few. It is also a procedural injustice because the Maasai and their interests are not well represented in decision-making. But underlying all is what we call a failure of recognition – a contemporary coloniality that denigrates other cultures and fails to even perceive the Maasai as worthy of being treated well, in a similar way to how patriarchy positions women as not worthy of the same treatment as men. It is a deep-rooted sense of the inferiority of the Maasai worldview and way of life, including a failure to recognise the long stewardship of the earth that springs from an understanding of human and other-than-human nature as inseparably bound up in a holistic relationship with land. Indigenous environmental justice scholars note how such loss of territory to settler colonialism has always involved forms of oppression that straddle distribution, procedure and recognition (Whyte, 2017).

The attention to expanded categories and subjects goes hand in hand with expanding the dimensions of injustice. For example, ecofeminist thinkers highlight how attention to gender necessarily goes beyond concerns about distribution to consider matters of representation and status (e.g., Warren, 1999). Analysis of environmental justice struggles is now commonly framed by different dimensions of harm, most commonly including distribution, procedure and recognition (Schlosberg, 2013). The idea of recognition has itself expanded in the past decade, initially focusing on institutionalised forms of discrimination along categories of wealth and identity, to increasingly consider forms of coloniality in which the plurality of knowledge systems and worldviews (Kothari et al., 2019) are neither recognised or respected. Such 'epistemic' or 'cognitive' injustice is a particular form of mal-recognition involving the failure to ensure respect, let alone parity, for diverse forms of knowledge, worldviews and ways of knowing and valuing nature. It is a key example of recognition injustice and one that is central to emerging decolonial thinking in environment and development scholarship.

Not surprisingly, environmental justice analysis and protest has spread to many countries of the world (see EJ Atlas, https://ejatlas.org) whilst also evolving from national to global level analysis of both the causes and consequences of injustice (Sikor and Newell, 2014). *Global* environmental justice engages with world-wide drivers of injustice, exploring how political and economic systems consistently produce 'unequal ecological exchange' that allows some to accumulate wealth at the expense of others. It is well documented that the enrichment of some parts of the world has involved forms of burden-shifting to other parts of the world – including resource use (such as minerals and food production), exploitation of the commons (such as oceans and carbon sinks) and pollution (e.g., by relocating polluting and resource-intensive industries). For many activists, the history of ecologically unequal exchange has accrued large ecological debts that should be factored into calls for justice (Roberts and Parks, 2009). For example, historical responsibility is a key fairness principle for many climate justice activists but one that historically high emission states are reluctant to incorporate into their Nationally Determined Contributions (the key vehicle for achieving the Paris Climate Agreement targets) (Williges et al., 2022).

From problems to solutions

The range of environmental justice concerns also points to a range of solutions. Solutions to distributive issues are in some ways the low-hanging fruits. It is sometimes possible to address these without addressing system-wide issues such as power structures. Solutions to procedural and recognition injustices often necessitate more profound and system-wide societal changes to reform institutions (social conventions, norms and laws) and the power structures that they mirror. Taking the example of biodiversity conservation, this explains why the majority of efforts to make conservation more people-friendly have involved *distributional* interventions: the most common among these are benefit sharing

Table 57.1 Examples of different biodiversity and climate justice issues

	Nature of concern	Biodiversity loss example	Climate change example
Distribution: cost of impacts	Some groups suffer disproportionately from the effects of environmental change	Smallholder farmers who suffer from lack of pollinating insects due to agro-industrial pesticide use	Current youth and future generations suffer disproportionately from historical carbon emissions
Distribution: cost of transition	Some groups are having to pay greater costs of resolving environmental problems	Indigenous peoples and local communities threatened by the expansion of protected area conservation	Local communities facing unemployment and other costs associated with energy and industrial transitions
Procedure	Some groups of humans and other-than-humans are marginalised in decision-making	The absence of rights of nature means the interests of non-human species are not considered	Gender inequality marginalises women in decisions about how to adapt to climate change
Recognition	Many people and worldviews are not treated as having equal status	Worldviews that value nature in relational terms (e.g., mother earth or totemism) are not respected	Local knowledge about climate change impacts or adaptation options is not respected

schemes (e.g., where local communities receive a fraction of park entry fees), compensation schemes (e.g., payment for damage to crops caused by park wildlife), relocation schemes (e.g., provision of alternative land away from a protected area), payments for ecosystem services (e.g., rewarding local people for their contribution to protecting forests), and alternative livelihoods (e.g., providing sources of income such as tourism or beekeeping that do not involve having to enter a park to collect resources). Whilst such 'solutions' can be helpful in certain contexts, it is also well known that such economic mechanisms rarely if ever can compensate for injustices of *recognition* (Martin, 2017). For example, providing alternative livelihoods to people denied access to territories may help economically but cannot address the harms that are related to their cultural attachments to land and place. In a similar fashion, efforts that appear to address environmental injustice by focusing on recognition, for example, through greater attention to local knowledge, can also fall short and become exploitative, for example giving rise to forms of biopiracy, where local knowledge about the pharmaceutical properties of a plant are exploited.

An important lesson here is that whilst environmental justice analysis benefits from a global perspective, environmental justice action must normally be rooted in local context. Environmental justice action certainly requires a global dimension, including alliances of people within connected, international social movements, and through lobbying for global rights. For example, successful global lobbying resulted on 28 July 2022 in a UN General Assembly declaration that access to a clean and healthy environment is a universal human right. But the thousands of ongoing environmental justice struggles are mainly local, with numerous communities on the frontline of environmentalism (the 'earth defenders'), protecting their places, values and ways of living against extractivism and global scale accumulation. Above all else, *global* environmental justice scholarship requires attention to the knowledge, experience and actions of these groups and seeking ways in which our actions can advance solidarity.

References

Bullard, R.D., 1990. *Dumping in Dixie: Race, Class, and Environmental Quality.* Boulder, Colorado: Westview Press.

Environmental Justice Network, 1991. *Principles of Environmental Justice.* [online] EJNet.org. Available at: https://www.ejnet.org/ej/principles.html

Fraser, N., 1985. What's critical about critical theory? The case of Habermas and gender. *New German Critique, 35*, pp. 97–131.

Greta Thunberg, 2019. UN News. *Greta Thunberg tells world leaders 'you are failing us', as nations announce fresh climate action.* [online] Available at: https://news.un.org/en/story/2019/09/1047052

IPBES, 2022. *Summary for Policymakers of the Methodological Assessment of the Diverse Values and Valuation of Nature of the Intergovernmental Science-Policy Platform on Biodiversity and Ecosystem Services.* Assessment Report on Diverse Values and Valuation of Nature | UNEP – UN Environment Programme.

Kothari, A., Salleh, A., Escobar, A., Demaria, F. and Acosta, A., 2019. *Pluriverse: A Post-Development Dictionary.* New Delhi, India: Tullika Books.

Martin, A., 2017. *Just conservation: Biodiversity, wellbeing and sustainability.* Abingdon: Routledge.

Murdock, E., 2020. A history of environmental justice: Foundations, narratives, and perspectives. In Coolsaet, B. (Ed) *Environmental Justice*, Abingdon: Routledge.

Ndoinyo, Y., 2021. Human rights violations in the name of conservation: Case of Ngorongoro district. In *Environmental Defenders* (pp. 30–36). Abingdon: Routledge.

Roberts, J.T. and Parks, B.C., 2009. Ecologically unequal exchange, ecological debt, and climate justice: The history and implications of three related ideas for a new social movement. *International Journal of Comparative Sociology, 50*(3–4), pp. 385–409.

Schlosberg, D., 2013. Theorising environmental justice: The expanding sphere of a discourse. *Environmental Politics, 22*(1), pp. 37–55.

Sikor, T. and Newell, P., 2014. Globalizing environmental justice? *Geoforum, 54*, pp. 151–157.

Singer, P., 2011. *The Expanding Circle: Ethics, Evolution, and Moral Progress.* Oxford: Princeton University Press.

Warren, K.J., 1999. Environmental justice: Some ecofeminist worries about a distributive model. *Environmental Ethics, 21*(2), pp. 151–161.

Whyte, K., 2017. The Dakota access pipeline, environmental injustice, and US colonialism. *Red Ink: An International Journal of Indigenous Literature, Arts, & Humanities, 19*(1).

Williges, K., Meyer, L.H., Steininger, K.W. and Kirchengast, G., 2022. Fairness critically conditions the carbon budget allocation across countries. *Global Environmental Change, 74*, p. 102481.

Additional Resources

Coolsaet, B. Ed., 2020. *Environmental Justice: Key Issues.* Abingdon: Routledge.

Environmental Justice Atlas: EJAtlas | Mapping Environmental Justice Loliondo and Maasai evictions: Oloshol Maasai community land rights struggle in Loliondo, Tanzania – YouTube; URGENT ALERT – Heavy Police Presence in Loliondo as Threats of Mass Evictions of the Maasai Intensify | The Oakland Institute.

Part 6

The economics of development. Editorial introduction

Economics has long been a cornerstone discipline in development studies. It offers a powerful set of conceptual and analytical tools, from supply and demand analysis and market equilibrium theories to game theory and econometric models. These tools have been crucial for analysing and addressing key concerns in development studies, such as poverty alleviation, income and wealth inequalities, the sustainable use of resources, the dynamics of global trade, and the impact of aid and remittances.

In the context of development studies, the discipline has over the years expanded, broadening its scope beyond traditional emphases on economic growth and industrialisation to reflect broader understandings of development and the complex challenges that modern societies face, from climate change to social inequalities. Moreover, there are many different schools of thought in economics. The dominant school, namely, orthodox – or neoclassical – economics has occasionally been critiqued for its narrow emphasis on material wealth, side-lining well-being, environmental sustainability, and broader concerns with social justice and equity.

In the second world post-war era, the Bretton Woods institutions, specifically the World Bank, the International Monetary Fund (IMF), and the World Trade Organisation (WTO), have been amongst the most powerful actors in the economic development policy landscape and poverty eradication. The World Bank, with its focus on long-term economic development and poverty reduction and the IMF, with its emphasis on macroeconomic stability, have been significant sources of loans. The WTO, in particular, has played a pivotal role in shaping global trade rules, often with direct implications for development strategies, trade balances, and local economic autonomy.

However, their collective role in development has faced criticism for prioritising the interests of former colonial powers and wealthier countries of the Global North at the expense of developing nations and marginalised communities. Furthermore, their promotion of policies foregrounded in orthodox economics, which emphasises free markets, privatisation, and deregulation, has served the interests of some more than others. The IMF, for example, drew criticism for its structural adjustment programs in countries of the Global South in the late 1980s and early 1990s, which contributed to the collapse of social protection systems, undermined local industries, led to the privatisation of essential public services, and set terms of trade more favourable to industrialised countries of the Global North, whilst exacerbating poverty and inequality in the Global South.

It's essential to note that the landscape of economics, encompassing both academia and institutional relationships is not nor ever has been static. Approaches to economic

DOI: 10.4324/9780429282348-63

policy and practice, especially those associated with the Bretton Woods institutions, are changing. Not only in response to academic critiques but also those of activists and social movements such as the Green Movement, Fair Trade advocates, and Occupy, which have highlighted the socio-economic disparities and environmental concerns arising from global economic policies. Consequently, these institutions have arguably undertaken some degree of reform, incorporating hitherto marginal approaches such as participatory development, civil society, and grassroots engagement to move towards more balanced and inclusive development strategies.

These changes are also underscored by the discipline of economics increasing embrace of broader approaches, such as sustainable and inclusive growth models, circular economies, and community-driven development to address concerns like environmental degradation, rising inequalities, and social welfare. Core assumptions of orthodox economics, such as the notion of the utility maximising 'rationale' actor, have been challenged and supplemented by insights from behavioural and experimental economics to feminist and ecological economics. For instance, feminist economics has shed light on gender disparities in economic opportunities, while ecological economics has emphasised the interplay between economies and ecosystems. These diverse perspectives have not only enhanced our analytical tools but also provided a broader lens through which to view developmental issues.

The term 'economics' extends beyond an academic discipline to refer to a set of relationships, practices, ideologies, and values that dictate how resources are allocated and utilised. This broader definition has allowed economics to transcend its traditional boundaries, drawing from and contributing to disciplines across the social sciences. Such interdisciplinary collaborations have further enriched the field, paving the way for a more sophisticated understanding of development challenges. Furthermore, while economics remains central to development studies, its dominance has been attenuated by insights from other fields such as anthropology, geography, sociology, political science, and environmental studies. For example, anthropological studies have deepened our understanding of how cultural values and beliefs influence economic decisions. Similarly, environmental economics has introduced mechanisms for incorporating the costs of environmental degradation into economic planning and decision-making, such as carbon trading and green bonds. This multidisciplinary collaboration, together with the insights gained from a wider range of schools of thought within economics, has fostered a richer understanding of the various forces at play in economic systems and processes and proffered new tools for tackling global challenges, from climate crises to socio-economic disparities and beyond.

The chapters in this section exemplify this diversity of economics. However, they are unified by several recurring themes, including the role of institutions, the impact of globalisation, the relationship among growth, aid, and development, and the significance of natural resources and sustainability. Central to these discussions are questions around equity, social justice, and inclusivity, mirroring broader concerns in contemporary development studies.

Chapter 58

Growth and development

Augustin Kwasi Fosu

Growth vs. development

Economic growth is defined as changes over time in the output of a given country. It is usually measured by the (percent) changes in the gross domestic product (GDP). Traditionally, its growth relative to a country's population – per capita GDP growth – has been employed as a measure of (economic) development. However, development represents the *well-being* of the country's citizenry. Thus per capita GDP growth is an inadequate measure of development for several reasons. First, GDP is simply too gross to properly reflect development. It includes earnings by foreigners who own portions of the factors of production. This portion is likely to be particularly large in countries dependent on natural resources. Even if this source is netted out, GDP still includes investment generally, the bulk of which is not available for possible consumption at the household level. Second, there are other forms of income received directly by households as direct transfers, which improve households' well-being and are not part of GDP. These are in the form of remittances sent from individuals residing in other countries. Third, the citizenry may benefit from in-kind transfers, whether directly or in the form of subsidies by government or from others such as development partners.

Current development measures and performance

It is now generally agreed that the United Nations Development Program's (UNDP) Human Development Index (HDI) is a reasonable measure of (human) development. The HDI reflects achievements in three dimensions: standard of living (income), education, and health. This human development (HD) measure has also been modified to be gender- and environment-sensitive, with variants that reflect gender inequality or sensitivity to impacts on the environment. Various attributes reflecting Sen's 'capability' concept of freedom (Sen, 1999), such as civil and political rights, have also been incorporated.

HDI and its components have increased steadily for developing countries, particularly since 1990, the benchmark year for the Millennium Development Challenge. Using data for 95 developing countries, the mean annual growth rate for HDI 1991–2017 is 0.86 per cent. The educational component, EDI, has shown the strongest annual growth at 1.40 per cent, followed by the life expectancy index at 0.64 per cent and the income index at 0.50 per cent.

In terms of differences across regions, the strongest progress on HDI has been demonstrated by South Asia (SA) at 1.29 per cent annual rate over the same 1991–2017 period,

DOI: 10.4324/9780429282348-64

followed by sub-Saharan Africa (SSA) and East Asia and Pacific (EAP) at 1.06 per cent and 1.03 per cent, respectively. Next are Middle East and North Africa (MENA) and Latin America and the Caribbean (LAC) at respective growth rates of 0.75 per cent and 0.71 per cent. The lowest growth rate was exhibited by Eastern Europe and Central Asia (EECA) at 0.50 per cent.

With the lowest HDI levels, SSA's and SA's relative high performance on HDI is to be expected if these regions are to catch up with the rest of the developing world. However, even with these higher growth rates, SSA and SA still ranked at the bottom in HDI levels by 2017, at 0.553 and 0.659, respectively. These compare with: 0.743, 0.728, 0.697, and 0.689 for EECA, LAC, EAP, and MENA, respectively.

Progress on poverty is another indicator of development. Indeed, countries with high levels of poverty tend to exhibit low levels of HDI, and conversely. Where inequalities in development indicators are very large, however, the average values may not sufficiently reflect the conditions of the poor, requiring the need to focus on poverty per se, in order to complement the information content generated by the HDI-type measures.

The most recognised indicator of (income) poverty is the headcount ratio, which measures the proportion of the population considered to earn an income less than the standard required for basic needs (poverty line); there are also the additional measures for the spread and severity of poverty, which inform us about the extent of poverty. The poverty line may vary from country to country and over time. However, to simplify comparability across countries and over time, the line has been standardised, currently at the daily income/expenditure of US$1.90 at 2011 international standards. As an indicator of extreme poverty, the poverty rates at this standard are the yardstick for measuring Goal 1 of the Sustainable Development Goals (SDGs), as was the case with Goal 1 of the Millennium Development Goals (MDGs).

Indeed, the developing world as a whole has experienced tremendous progress on poverty since the early 1980s, with the extreme poverty headcount declining by an average annual rate of about 3 percent. The measures of the spread and severity of poverty fell even faster, at 4 per cent and 5 per cent, respectively. There were major differences across regions, however, with East Asia and the Pacific dominated by China exhibiting the fastest decline and SSA with the slowest progress, that is, among regions with high initial poverty levels (see Fosu, 2017).

These poverty measures, however, do not necessarily reflect deprivation in human development. Thus in 1997 the UNDP introduced the Human Poverty Index (HPI) for developing countries. This measure is intended to reflect deprivations in the three dimensions of human development: long and healthy life, knowledge, and a decent standard of living. For more developed countries, HPI is further modified to reflect social exclusion.

The Multidimensional Poverty Index (MPI), which is currently being popularised in the literature and in policy circles, is similar in concept to the HPI, with dimensions: standard of living, education, and health. However, the UNDP measures are simplified as simple averages, while the MPI uses cutoffs across the dimensions to determine the level of poverty.

In spite of the usual criticisms levelled against the aforementioned measures of development, they can project reasonable pictures of how well various countries are performing beyond mere income growth. They are useful in that they can at least signal countries that require special attention on the development front. The HD and poverty indices are complementary for gauging the nature of the development challenge. For example, a low

HDI value concurrently with a high per-capita income suggests that growth is not being efficiently translated into HD. Similarly, if both HDI and HPI or MPI are high, then the HD achievement is not being sufficiently shared with those at the bottom of the income distribution. Ideally, HDI should be high and HPI or MPI low, which might require minimising inequality in HD constituents.

Association between growth and development measures

But, how well does growth reflect development? *On average*, growth has been the main engine for poverty reduction globally (Dollar and Kraay, 2002; Fosu, 2017). Nonetheless, in certain countries, income distribution has been the dominant factor not only in its impact on growth but also in transforming growth into poverty reduction (Fosu, 2017).

Suri et al. (2011) observe a bidirectional relationship between growth and measures of the non-income HD constituents (education and health). They find that while HD is the end product of economic growth, improvements in HD allow economies to sustain economic growth.

There appears to be a strong positive correlation between per capita GDP growth and changes in HDI and its components across and within countries. Using a five-year panel involving a global sample of 95 developing countries over 1991–2017, I find significant positive correlations between per capita GDP growth, on the one hand and changes in HDI and its components, on the other.[1] In contrast, the correlations are quite weak for annual panel data, suggesting that a longer period is required for translating per capita GDP growth into HD.

Policies for development

What policies are considered pro-poor or pro-development? Employment generation is a particularly salient factor in channeling economic growth into poverty reduction and development generally, on a sustainable basis. Policies that augment the demand for labour are therefore essential for generating desirable social-impact outcomes for developing economies (Fosu, 2013). Thus, appropriate policies are those that increase employment in sectors with relatively large levels of labour productivity.

Reducing poverty would entail raising income levels for the less well-off. This could occur via own market activities, or transfers such as remittances from individuals, organisations and government. Government could additionally provide in-kind benefits, such as through education and health subsidies. Societies may also benefit from global public goods, including health improvements via technological advancement or through global efforts to fight diseases. Obviously, the reverse could also occur if rich countries' effluence led to adverse climatic change for instance, with negative consequences for poor countries.

The nature of the sociopolitical environment is a particularly important determinant of the effectiveness of transforming growth to development or poverty reduction. For example, political instability may not only lead to lower economic growth but also to reduced effectiveness with which growth is translated into HD (Fosu, 2002). Hence, managing sociopolitical instability, which tends to additionally adversely affect growth (Fosu, 2001), is crucial.

Health is critical. Impaired health exacerbates poverty and undermines development, whether directly or indirectly via lowering growth, with adverse implications for development. Both morbidity and mortality are important contributors to development and poverty woes emanating from diminished health. (Hamoudi and Sachs, 1999) Health standards, as exemplified by Goal 6 of the Millennium Development Goals (MDGs) and Goal 3 of the Sustainable Development Goals (SDGs), for instance, are essential for attaining poverty eradication and development success.

Education, especially literacy, is also crucial. Traditionally, education has been viewed as the requisite capability for obtaining decent employment, which entail relatively high wages and nonwage benefits, both of which serve to improve the standard of living for workers. Literacy is also a necessary capability for accessing the tremendous global database of knowledge on all aspects of HD. Yet, high levels of education cannot necessarily improve expected standard of living unless there is also the complementary investment in physical capital, in the form infrastructure, to augment the derived demand for labour. Unfortunately, in many developing countries, there is the tendency to subsidise higher education at the detriment of basic education and complementary physical capital financing, leading to increasingly significant excess labour supply and unemployment. Since labour market performance is likely to be the main process involved in generating a virtuous circle between EG and HD, a balanced policy approach that takes such complementarity seriously is called for.

Conclusion

It is often observed that economic growth (EG) is necessary – but not sufficient – for development. Yet, it is not entirely convincing that EG is even necessary, given its potentially adverse impacts on the non-income components of human development (HD). Yet, as observed, EG in the form of per capita GDP growth is positively associated, at least in the medium term, with changes in HD as measured by HDI or its components.

Income growth is found to have been the main driver behind poverty reduction. However, income distribution remains an important factor, not only in its ability to influence growth but also in its tendency to affect the transformation of EG into poverty reduction. Such importance seems to differ across countries, though. Thus understanding the idiosyncratic characteristics of countries in this regard is essential.

Also observed is the bidirectional nature of the EG and HD relationship. While EG is found to positively lead to HD as an end, the latter may in turn sustain EG.

Maintaining the previous virtuous circle between EG and HD would require a vibrant labour market, in order to ensure gainful employment across the income distribution. Governments in especially developing countries have a critical role to play in this regard. Solid basic education is critical and so is basic health. And, complementary infrastructures would foster the necessary demand for meaningful labour market participation. Finally, the global community must ensure that there is at least a level playing field for especially the low-income developing countries.

Note

1   The respective correlation coefficients involving logarithmic changes in HDI and its components of life expectancy, education and income are .46, .12, .08, and .81, all statistically significant at the .01 level, except that of education that is significant at the .05 level. The correlation coefficients involving algebraic changes are even larger: .61, .18, .21, and .83, respectively, all significant at the .01 level.

References

Dollar, D. and Kraay, A. (2002). Growth is Good for the Poor. *Journal of Economic Growth*, 7(3): 195–225.

Fosu, A. K. (2001). Political Instability and Economic Growth in Developing Economies: Some Specification Empirics. *Economics Letters*, 70(2): 289–294.

Fosu, A. K. (2002). Transforming Growth to Human Development in sub-Saharan Africa: The Role of Elite Political Instability. *Oxford Development Studies*, 30: 9–19.

Fosu, A. K. (2013). Achieving Development Success: Synthesis of Strategies and Lessons from the Developing World. In A. K. Fosu (Ed.), *Achieving Development Success: Strategies and Lessons from the Developing World*. Oxford: Oxford University Press: 1–22.

Fosu, A. K. (2017). Growth, Inequality, and Poverty Reduction in Developing Countries: Recent Global Evidence. *Research in Economics*, 71(2): 306–336.

Hamoudi, A. and Sachs, J. (1999). *Consequences of Health Status: A Review of the Evidence. CID Working Paper 030*. Cambridge: Harvard University. http://www.cid.harvard.edu/cidwp/030.htm

Sen, A. K. (1999). *Development as Freedom*. Oxford: Oxford University Press.

Suri, T., Boozer, M. A., Ranis, G. and Stewart, F. (2011). Paths to Success: The Relationship Between Human Development and Economic Growth. *World Development*, 39(4), 506–522.

Chapter 59

Aid and growth

Ines A. Ferreira

The question of whether aid helps countries grow is of central importance for both donors and recipient countries and one that has generated immense discussion for five decades. The debate was particularly intense within the development discourse in the late 1990s and early 2000s, dividing those who argued that aid had failed to achieve its goals and had been harmful instead and those who claimed that although partially correct, these arguments were exaggerated. Prominent on the former side of the debate were Peter Bauer, who argued against aid, joined some decades later by William Easterly and Dambisa Moyo, whereas Jeffrey Sachs was a persistent voice in the latter group, calling for a 'Big Push' in aid to lift people out of poverty. Within the academic literature, opinions were also divided between those who argued for a positive effect of aid, those who did not find evidence for the aid-growth relationship, and a third group claiming that aid helped to promote growth but only under certain conditions.

The difficulties in addressing this question are manifold and begin with the fact that the landscape of aid is multifaceted. Given that this is a broad topic, this chapter focuses on the impact of aid on growth at the macro level and provides an overview of the main insights provided by the empirical aid effectiveness studies. Within this literature, aid commonly refers to official development assistance (ODA), which designates flows of resources to recipient countries and territories and to multilateral agencies with the objective of promoting economic development and welfare, and which are provided on concessional terms (OECD-DAC, 2019). Due to the macroeconomic perspective of these studies, the effectiveness of aid is generally measured by its average contribution to economic growth across a sample of countries.

A closer look at the evidence

Aid 'works; it doesn't; it can, but that depends'. This excerpt of the title of McGillivray et al.'s (2006) paper on the disputes over the evidence on the aid-growth link summarises the three broad views identified by Radelet (2006):

- On average, aid has a positive effect on growth, but with diminishing returns.
- Aid has no effect on growth and may do more harm than good.
- Aid helps to accelerate growth but only under certain circumstances.

Following Hansen and Tarp's (2000) chronological overview of the aid effectiveness literature, the early studies of aid, spanning from the 1960s until the 1990s, found general

DOI: 10.4324/9780429282348-65

support for the view that aid has a positive effect on growth. According to the first generation of studies, an inflow of aid would correspond to an increase in total savings and investment and therefore to growth, based on a Harrod-Domar growth model. Overall, the results lent support to a positive impact of aid on savings, and therefore on growth, though the amount of the increase on savings was smaller than the amount of aid flow (Hansen and Tarp, 2000). Some studies in the second generation considered the link via investment and found support for a positive link between aid and investment, whereas others concluded that, as long as there is a positive relationship between savings and growth, the link between aid and growth will also be positive (Hansen and Tarp, 2000).

A new generation of studies emerged in the 1990s. Some explored a potential non-linear relationship between aid and growth and concluded that this positive effect had diminishing returns, which refers to the idea that the first million dollars of aid will have a larger marginal effect than the second million, and so on. A couple of authors tested whether this could be explained by considering a limited absorptive capacity of countries to take up large inflows of foreign capital or the problem of the Dutch disease effects, which results from a real exchange rate appreciation caused by aid inflows (McGillivray et al., 2006).

In contrast with this view, Boone's (1996) paper sided with the second of the views identified by Radelet (2006) with results suggesting that aid had no effect, neither on investment nor on growth. However, his claim encountered a lot of criticism, at the same time as Burnside and Dollar's (2000) seminal paper added another twist to the aid effectiveness debate. This marked the beginning of a strand of literature that continued throughout the 1990s and viewed the effect of aid as conditional on certain factors.

According to Burnside and Dollar (2000), aid had a positive effect on growth only in countries pursuing 'good' policies, where policy was measured by an index capturing inflation, trade openness and the budget deficit as percentage of gross domestic product (GDP). This result found support in other studies informed by the policy recommendations of international aid agencies, such as the World Bank, which echoed the view that allocating aid to 'good performers' would promote aid effectiveness in spurring growth. However, their study was also heavily criticised and triggered responses from different scholars who cast doubt on the robustness of their results by showing that they did not hold when changes were made to the data sample and the empirical specification used.

Their work was accompanied by other studies that considered the characteristics of recipient countries. They found that aid effectiveness was conditional on, among others, institutional quality, the political system and its stability, the level of social capital, as well as the geographic conditions of a country. A smaller group of authors also examined the effect of external and climatic factors, namely, trends in terms of trade, short-term export instability, and natural disasters on the aid-growth link.

Furthermore, some academics drew attention to the potential conditional effects emerging from the supply side of aid. In addition to the analysis of diminishing returns already mentioned, some examined the effects of donor motives (e.g., recipient need vs. self-interest). Others tested whether the effect of aid on growth was affected by aid volatility (i.e. the uncertainty surrounding aid flows), aid proliferation (which characterises situations where large numbers of donors and projects create pressure on the capacity of the recipient government to manage the flows of aid), or aid fragmentation (i.e. when there is a larger number of donors each contributing with a smaller share of aid).

In parallel, some authors challenged a few of the assumptions adopted, namely that aid had a unique contemporaneous effect on growth, and that different kinds of aid would have the same effect. Instead, they considered different criteria to disaggregate aid, such as the timing of its effects (e.g., short-impact and long-term impact), its purpose (e.g., social and economic), its modalities (e.g., project and financial), as well as the type of donor (e.g., multilateral and bilateral). Rajan and Subramanian (2008) proposed a comprehensive approach that included a new instrumentation strategy, accounted for different time horizons and periods as well as different sources, types of aid and expected timing for its effects, and considered different specifications and samples. Their results lent support to the more sceptical view, as they found little evidence for any relationship. However, their result was defied, for instance, by Clemens et al. (2012) who revisited their empirical strategy (alongside the results in other influential studies) by considering only the portion of aid that is expected to have an influence on growth within a few years. They found a positive, though modest, effect of aid in contrast with the null effects found before.

Rajan and Subramanian's (2008) work marked the beginning of a new generation of pessimistic studies, which were once again met by another set of work supporting a cautious, but optimistic view of a positive effect of aid (Addison et al., 2017). The latter seems to be the prevailing view in the most recent empirical studies, which provide stronger evidence that, on average and when considered over an extended time period, aid has had a positive effect on growth, than the view that aid is ineffective (Addison et al., 2017).

The challenges of establishing causality in aid effectiveness studies

This lack of consensus throughout five decades of aid-growth studies begs the question of why there is so much disagreement among academics. This chapter highlights three important reasons. The first was hinted at in the introduction and is related to the complexity of aid and the fact that it comes in different forms, through different channels, and with different purposes. As mentioned before, some attempts have been made to disaggregate aid in order to gain a better understanding of its effects on growth. Although this chapter focuses on the macro effects of aid on growth across countries, one should note that some studies have focused on particular countries or regions. The literature also abounds with studies using impact evaluation to assess different projects, as well as with work examining the meso effects of aid on sectoral outcomes, such as health or education.

The second reason highlighted here is related to the availability and quality of data. Since the beginning of the aid effectiveness literature, and especially since the 1990s, there has been a substantial increase in data available on aid flows as well as a big improvement of its reliability. The most common sources for aid data are the OECD-DAC dataset and, more recently, AidData.

Finally, one of the main reasons underlying the disagreement in the literature stems from the technical challenges of demonstrating that aid *causes* growth. The literature frequently refers to the problem of endogeneity of aid as one of the hardest problems to tackle in aid-growth regressions. On the one hand, this may be result from an omitted variable problem, i.e. not accounting for the effect of a third variable that may affect aid and growth. On the other hand, it is plausible to assume that aid inflows might be affected by economic growth, which can lead to identification problems. A significant part of the discussion within the studies emerging after Rajan and Subramanian (2008) is related to the quest for finding appropriate ways of dealing with this problem.

Conclusion

In short, 50 years of research on the impact of development aid have shown a struggle to reach a consensus on whether aid works or not in promoting economic growth. A significant part of the existing studies argues that this effect is conditional on certain factors, not only on the recipient side of aid, such as the macroeconomic policies or the level of institutional quality in the country, but also on the supply side, e.g., donor motives or aid volatility. Overall, despite the contestation of the empirical results, the growing evidence in recent years has lent more support to the optimistic view that aid is an effective and important instrument to economic development.

References

Addison, T., Morrissey, O. & Tarp, F. 2017. The Macroeconomics of Aid: Overview. *The Journal of Development Studies*, 53, 987–997.

Boone, P. 1996. Politics and the Effectiveness of Foreign Aid. *European Economic Review*, 40, 289–329.

Burnside, C. & Dollar, D. 2000. Aid, Policies and Growth. *The American Economic Review*, 90, 847–868.

Clemens, M. A., Radelet, S., Bhavnani, R. R. & Bazzi, S. 2012. Counting Chickens when They Hatch: Timing and the Effects of Aid on Growth. *Economic Journal*, 122, 590–617.

Hansen, H. & Tarp, F. 2000. Aid Effectiveness Disputed. *Journal of International Development*, 12, 375–398.

McGillivray, M., Feeny, S., Hermes, N. & Lensink, R. 2006. Controversies Over the Impact of Development Aid: It Works; It Doesn't; It Can, But that Depends . . . *Journal of International Development*, 18, 1031–1050.

OECD-DAC. 2019. *Official Development Assistance – Definition and Coverage* [Online]. http://www.oecd.org/dac/stats/officialdevelopmentassistancedefinitionandcoverage.htm   [Accessed 28/09/2019].

Radelet, S. 2006. *A Primer on Foreign Aid. CGD Working Paper Number 92.* https://www.cgdev.org/sites/default/files/8846_file_WP92.pdf [Accessed 30/09/2019].

Rajan, R. G. & Subramanian, A. 2008. Aid and Growth: What Does the Cross-Country Evidence Really Show? *Review of Economics and Statistics*, 90, 643–665.

Additional reading

Roodman, D. 2007. *Macro Aid Effectiveness Research: A Guide for the Perplexed. CGD Working Paper Number 134.* http://www.cgdev.org/sites/default/files/15003_file_Guide_Perplexed.pdf [Accessed 30/09/2019].

Temple, J. 2010. Aid and Conditionality. In: Rodrik, D. & Rosenzweig, M. (eds.) *Handbook of Development Economics*. Oxford: Elsevier.

Chapter 60

Foreign aid in a changing world

Stephen Brown

Foreign aid during the cold war

During the Cold War, there were two main donor groups, centred on the United States and the Soviet Union. The leading Western donors formed the Development Assistance Committee (DAC) within the Organisation for Economic Co-operation and Development (OECD), the main coordination body of the Western industrialised countries. The Soviet Bloc established its own version, the Council for Mutual Economic Assistance (CMEA, also known as COMECON). Each provided assistance to its allied states in the developing world. Many DAC members also provided assistance to officially non-aligned socialist countries. Much of Western foreign aid during this period served as a tool of donor country foreign policy, especially in the case of the United States (Morgenthau, 1962). Still, many donors also provided significant amounts for more altruistic purposes (Lumsdaine, 1993).

Most foreign aid was – and still is – provided directly from government to government, known as bilateral aid, but a significant amount was channelled through multilateral institutions such as the United Nations and the World Bank. The focus of the assistance varied according to trends in development thinking. For instance, an initial emphasis on large infrastructural projects, such as hydroelectric dams, in the 1960s and 1970s failed to produce the expected 'trickle-down' benefits for the poor. As a result, donors placed greater emphasis in the 1970s on meeting basic needs. In the 1980s, facing persistent poverty, donors focused on macroeconomic stability and liberalisation, but that approach also proved ineffective.

Early reconfigurations: aid in the 1990s

The collapse of the Soviet Bloc and the end of the Cold War presented the West with a tremendous opportunity to reallocate massive defence spending into more productive areas, including international development and promote democracy, good governance and free markets. However, the 'peace dividend' never materialised. Faced with severe fiscal deficits, Western governments, rather than reallocate military budgets funds to aid, actually cut their budgets in both areas. Knowing that their former rivals could not take advantage of the situation, they reduced aid spending without fear. As Figure 60.1 illustrates, total DAC aid disbursements rose steadily in the 1980s but fell after 1992.

The US's 'unipolar moment' (Krauthammer, 1990/91) also changed the nature of foreign aid. For instance, donors placed greater emphasis on democratisation and peacebuilding.

DOI: 10.4324/9780429282348-66

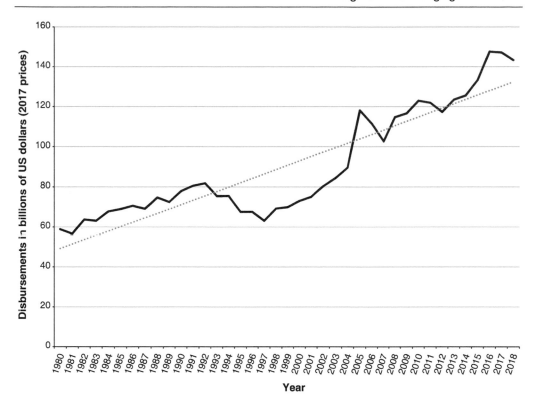

Figure 60.1 Total net DAC official development assistance, 1980–2018

Source: OECD Query Wizard for International Development Statistics, https://stats.oecd.org/qwids (accessed 5 August 2019)

However, those efforts soon proved disappointing as well. Though donors helped liberalise political systems to a certain degree in recipient countries, the results often fell short of liberal democracy. In many cases, donors' economic and security priorities meant that they did not translate their pro-democracy rhetoric into action (Brown, 2005). Though several civil wars, which had been fed by superpower rivalry, did end in the 1990s, such as Mozambique's, a number of new conflicts also emerged, including the Rwandan genocide and two deadly civil wars in Zaire/Democratic Republic of the Congo.

21st-century development cooperation

Changing context

Early in the new millennium, two events profoundly reshaped the context of foreign aid and justified increased expenditure. In 2000, the United Nations' member states unanimously adopted eight Millennium Development Goals (MDGs), which committed them to achieving ambitious poverty-reducing targets by 2015. This helped refocus foreign aid away from macroeconomic reforms towards measures with more concrete short- to medium-term effects on the quality of life of billions of people living in poverty, including by reducing child mortality and achieving universal primary education.

The following year, however, the MDGs were overshadowed by a new threat that replaced Communism in Western mindset: terrorism. The attacks of 11 September 2001 led many donors, notably the United States, to reorient their development assistance to sectors and countries that played a central part in the 'war on terror'. After US-led invasions, Iraq and Afghanistan became top aid recipients of foreign aid. Donors used aid – with limited success – to try to 'win hearts and minds' in conflict areas, rather than fight poverty per se, often administered by military actors instead of civilians. Increasingly, donors recast underdevelopment as a source of terrorism and a threat to Western countries, rather than a black mark on human conscience that needed to be reduced or eliminated for altruistic, ethical reasons.

Later in the decade, in 2007–2008, a third event sharply affected the foreign aid context: the global economic crisis. The latter hit OECD countries especially hard, leading many DAC members to cut their aid levels, although some protected their aid budgets. (Figure 60.1 shows a drop in total expenditure in 2010–2012.) The slow and uneven recovery not only threatened global aid flows but also discredited Western development models, especially when rapid-growing non-Western countries such as China remained relatively unaffected (Birdsall, 2012).

New actors and modalities

The stagnation of OECD aid after 2010 accelerated the rise in importance of non-DAC donors, already an important trend (Woods, 2008). Often referred to as 'emerging donors', the growth in aid from countries such as Brazil, China, India, Saudi Arabia, Turkey and Venezuela not only multiplied the number of donors but also provided assistance on different terms from 'traditional' ones.

Following the end of the Cold War, especially after the early 2000s, DAC donors increasingly recognised their own part in the often disappointing results of development assistance. They recognised that not only was *more* aid needed, but also – perhaps more importantly – *better* aid. For instance, they have progressively implemented the principle that foreign aid should not be 'tied' to the purchase of goods from the donor country, as this practice adds to the costs of aid without any commensurate benefit from the recipient's perspective (Jepma, 1991). Donors have also increasingly worked together to provide joint assistance to government programs, for instance in the education sector, rather than working in isolation and supporting uncoordinated projects, such as the construction of schools.

In 2005, DAC members agreed to five basic principles, laid out in the Paris Declaration on Aid Effectiveness: a developing country's lead role in designing and implementation its own strategies (known as ownership), donors' alignment with the latter, coordination among donors (harmonisation), mutual accountability and a focus on results. Despite donors' enthusiastic embrace of these principles, to date they have largely failed to put them into practice (OECD, 2011).

However, emerging donors, even if they endorsed the Paris principles as aid recipients, do not want to be bound by them when providing assistance of their own. In fact, they do not all share the basic concept of altruism that has to various degrees characterised the theory, if not the practice, of six decades of Western foreign aid. Whereas DAC members generally separate 'official development assistance' (defined as government assistance whose main purpose is improving economic or social well-being in developing countries) from other financial flows, Southern donors frame development cooperation (as distinct from aid) as being of mutual benefit. They emphasise the value of bundling of aid and non-aid instruments, including investment, loans and trade, often relying on tied aid.

As the MDGs expired in 2015, the United Nations promulgated their successors, the Sustainable Development Goals (SDGs). The latter comprised 17 bold and broad-ranging objectives to be met by 2030, including the complete elimination of poverty. Several trillion dollars a year of additional funding would be required to achieve them, which far surpassed the amount that donor countries would be willing to provide. In fact, as can be seen in Figure 60.1, aid levels dropped from US$147 billion in 2016 to $143 billion in 2018. A consensus emerged that the vast sums needed could only be mobilised by 'unlocking' the wealth of the private sector. To help fill this funding gap, donor countries and multilateral institutions increasingly supported 'blended finance' and other 'innovative' mechanisms to mobilise private funds. Early signs suggest that nowhere near the required level of funding will be met. Moreover, because the private sector's primary concern is profitability, many questions remain about the effectiveness of these modalities at lifting people out of poverty and achieving full gender equality and women's empowerment, among other goals.

Coordination problems and the future of the international aid architecture

An additional challenge to coordination in the field of development assistance is not only the growing number of actors but also of *types* of actors involved, many of which are reluctant to coordinate with other donors. Non-state development actors have become prominent – especially in the health sector – and often have a different approach to development. Like emerging donors, private foundations, such as the Bill and Melinda Gates Foundation, and 'vertical funds', including the Global Fund to Fight AIDS, Tuberculosis and Malaria, often operate outside traditional donor coordination mechanisms, as does the private sector. The lack of cooperation can lead to duplication of efforts, as well as contradictory approaches and activities.

In the past, DAC members set foreign aid norms. To date, they have driven discussions of aid effectiveness, but this dominance is waning, and it is not yet clear who or what will play a similar role for development cooperation in the future. The G20 and the UN Development Cooperation Forum are possible successors, as is the 'multi-stakeholder' Global Partnership for Effective Development Cooperation, but the international aid 'architecture' is still under construction – without a designated architect.

With the current focus on state-led development strategies, non-governmental organisations from both the Global North and the South tend to be marginalised in development policy discussions. They are increasingly vocal and organised transnationally, which may help them claim a larger role in the future.

The implications of all these changes for foreign aid remain unknown. A convergence might develop among the practices and principles of the various actors in the field. Some of these new trends could help reduce poverty and inequality, but the effects of others may not be so positive.

References

Birdsall, N. (2012). *The Global Financial Crisis: The Beginning of the End of the "Development" Agenda? CGD Policy Paper 003*. Washington, DC: Center for Global Development. http://www.cgdev.org/content/publications/detail/1426133. Consulted 6 May 2012.

Brown, S. (2005). Foreign Aid and Democracy Promotion: Lessons from Africa. *European Journal of Development Research*, 17(2), pp. 179–198.

Jepma, C.J. (1991). *The Tying of Aid*. Paris: OECD.

Krauthammer, C. (1990/91). The Unipolar Moment. *Foreign Affairs*, 70(1), pp. 23–33.

Lumsdaine, D.H. (1993). *Moral Vision in International Politics: The Foreign Aid Regime, 1949–89*. Princeton, NJ: Princeton University Press.

Morgenthau, H. (1962). A Political Theory of Foreign Aid. *American Political Science Review*, 56(2), pp. 301–309.

Organisation for Economic Co-operation and Development. (2011). *Aid Effectiveness 2005–10: Progress in Implementing the Paris Declaration*. Paris: OECD.

Woods, N. (2008). Whose Aid? Whose Influence? China, Emerging Donors and the Silent Revolution in Development Assistance. *International Affairs*, 84(6), pp. 1205–1221.

Additional resources

Brown, S. (2017). National Development Agencies and Bilateral Aid. In: P.A. Haslam, J. Schafer and P. Beaudet, eds. *Introduction to International Development: Approaches, Actors, and Issues*, 3rd ed. Oxford: Oxford University Press, pp. 141–160.

Davies, R. and J. Pickering. (2017). How should Development Co-Operation Evolve? Views from Developing Countries. *Development Policy Review*, 35(S1), pp. O10–O28.

Horner, R. and D. Hulme. (2019). From International to Global Development: New Geographies of 21st Century Development. *Development and Change*, 50(2), pp. 347–378.

Mawdsley, E. (2018). The "Southernisation" of Development? *Asia Pacific Viewpoint*, 59(2), pp. 173–185.

Prizzon, A., R. Greenhill and S. Mustapha. (2017). An "Age of Choice" for External Development Finance? Evidence from Country Case Studies. *Development Policy Review*, 35(S1), pp. O29–O45.

Chapter 61

Aid conditionality

Jonathan R. W. Temple

Aid donors sometimes impose conditions on grants, loans or technical assistance. The donors demand that, in return for aid, certain actions should be carried out by the recipient government. This practice has been important enough to acquire a name, conditionality. As a term, it sounds dry, technical and unexceptionable, but it has long been one of the most controversial aspects of international development policy. Its implementation has sometimes led to fury directed at the World Bank and IMF.

Some think conditionality represents aid at its worst, allowing powerful donors to impose their will on poorer countries. There are others who think conditionality rarely goes far enough and is not properly enforced. The introduction of structural adjustment lending in the 1980s intensified the debate. The World Bank and IMF made loan disbursement conditional on macroeconomic stabilisation and often on wider reforms. That approach had a mixed record, and conditionality has evolved significantly over time.

The case for conditionality

Attaching conditions to aid starts from a reasonable assumption, namely that recipient governments typically have multiple objectives, not all of which are aligned with those of donors. This should not be controversial. Even in long-standing democracies, politicians often have mixed motives and do not simply seek the national interest. The objectives of political leaders are likely to be similarly compromised in aid recipients, many of which are autocracies of various kinds, or countries where democracy is fragile and not long-established.

For now, assume (a little naively) that the donor is altruistic, while the recipient government has differing objectives. How should the donor proceed? The conflict of interest corresponds to the 'principal-agent' problem in the economic theory of incentives. The donor is the principal, setting the terms for transfers to the recipient government, the agent. Usually, the optimal policy for the principal is to make transfers conditional on the actions of the agent (if these are observed) or outcomes that are linked to those actions. In the case of aid, conditionality seeks to make aid more effective: the donor induces or 'buys' certain policy actions or reforms which, if the donor is altruistic and well-informed, should benefit the poor.

The arguments against conditionality

A case against conditionality can proceed in several directions. One argument is that, whatever the technical case for imposing conditions, it is morally illegitimate for donors

DOI: 10.4324/9780429282348-67

to use their economic strength to influence the choices of sovereign nations. This has more force than economists typically allow, but the importance of sovereignty needs to be carefully weighed. Since few recipient countries are stable, effective democracies, a distinction can be made between the sovereignty and interests of recipient governments and the interests of their populations. If conditions help to make a recipient government more effective or more responsive to the domestic population, it might be a mistake to give the loss of sovereignty too much weight.

Some reject the assumption that the donor is altruistic. In this view, donors seek major reforms, such as trade liberalisation or privatisation, primarily to serve their own economic interests. Conditionality is seen as a new form of economic imperialism. A full assessment is beyond the scope of this chapter, but many economists would be sceptical that reforms in recipient countries, of the type usually promoted, have major economic benefits for donors. When economists make a case for specific policies, such as trade liberalisation, these are usually based on projected benefits for the reforming country.

The use of 'projected' raises further objections, however. The donor is not necessarily well-informed, and there is room for controversy about major policy reforms and growth strategies. The logic of policy conditionality then risks hubris. The appropriate mix of policies will vary across countries and over time, and recipient governments will have their own development priorities.

These arguments are important, but the extent to which they apply varies. Policy conditionality has sometimes gone too far, imposing conditions that were questionable or intrusive. But some conditions are designed to forestall the damaging long-run consequences of sustained budget deficits and high inflation, in circumstances where recipient governments are failing to address these problems. Other conditions may be relatively innocuous, such as those relating to the internal organisation of the recipient state, in areas such as budgetary administration and financial planning.

Another set of objections is more practical. Some argue that traditional conditionality does not work: it fails to influence recipient governments. One explanation is that donors find it hard to commit to their conditions. In theory, if the conditions are not met, the aid should not be disbursed. In practice, that decision may be hard to sustain. A donor may want to disburse the aid in any case, perhaps because this would lead to the best remaining outcome for the country's poor. A recipient government that understands this has no incentive to reform, and conditionality becomes ineffective. This problem is compounded by 'budget pressure' effects, in which the officials of aid agencies face institutional pressures to disburse aid, regardless of whether conditions have been met. Sometimes, defensive lending is likely, where new loans are disbursed to help a country avoid default on past loans.

The modern view is that much depends on the political economy of the recipient country. Conditionality can strengthen the hand of a reforming government, but in other cases, political leaders may view a donor's threat to terminate aid flows as outweighed by other pressures and constraints. The recipient government may pay lip-service to the need for reform and even pursue it temporarily, but lasting change requires a domestic constituency that is willing to champion a given set of policy measures and priorities.

Country ownership

The political economy perspective leads naturally to the idea of 'country ownership'. In some interpretations, this is an action to be taken by recipient governments. Donors often want recipient governments to take ownership of a set of policy measures and commit to seeing them through to a successful conclusion. As Khan and Sharma (2003) put it, the country carrying out reforms has to be committed to their spirit as well as their letter.

But achieving that commitment may be a consequence of ownership rather than the thing itself. Ownership is not something that recipient governments can simply will into existence. Instead, it must emerge through a deliberative process. Recipient governments will be more committed to a policy agenda if they have helped to shape it. Hence, recipient governments must retain some autonomy, and the concepts of ownership and autonomy are linked. In response, a donor might add that even autonomous individuals and governments should be open to advice.

Since 2000, the major donors have used poverty reduction strategies to encourage ownership. Working with the World Bank and IMF, aid recipients draw up long-term plans to address poverty, linked to an overall development strategy. In its early years, this approach prompted the charge of 'ventriloquism': aid recipients are led to understand which policies are needed to gain donor support. If countries lack the capacity to formulate development strategies, or donors remain too prescriptive, the stated commitments to autonomy may be no more than window-dressing.

The evolution of conditionality

Conditionality has evolved over time, not least in the approach taken by the IMF. For obvious reasons, the IMF attaches macroeconomic policy conditions to its adjustment loans. More controversially, the conditions have sometimes included detailed structural reforms. This could be counter-productive. The attempt to impose unpopular reforms can undermine the perceived legitimacy of an overall programme, threatening its chances of success. Multiple conditions also increase commitment problems: should the donor disburse when the conditions have been only partially fulfilled? In some cases, the conditions became overly demanding. In the wake of criticism, the IMF sought a more streamlined approach, with fewer conditions. In principle, this should emphasise only conditions seen as critical to the IMF's core responsibilities and the repayment of its adjustment loans. But successful implementation remains difficult, not least when decisions are taken over the timing and extent of fiscal austerity.

The appropriate route is even less clear for long-term development lending, such as that undertaken by the World Bank, the regional development banks, sometimes the IMF and bilateral donors. As we have seen, traditional policy conditionality is often thought to have failed. The World Bank's 1998 report *Assessing Aid* advocated selective aid allocation: instead of imposing ineffective conditions, aid should be targeted at countries with good institutions and policies. The Millennium Challenge Account of the US has been one example of this approach.

Some economists argue that sustained pressure has led to long-term improvements in the policies of recipient governments. Newer instruments such as 'floating tranche' conditionality introduce greater flexibility in the timing of reforms and disbursement. The recognition of the importance of domestic politics has encouraged a shift towards 'process' or governance conditionality. In this approach, donors attach broad conditions to governance, public management and policy making. These can include simple, concrete changes that are likely to lead to better outcomes, such as presidential term limits. This approach seems more legitimate than attempts to micro-manage economic policy, especially when it strengthens the accountability of leaders to their domestic populations. A more radical idea is that conditionality should shift from input and process variables, to an emphasis on measured performance in areas such as education, healthcare or water and sanitation. The World Bank's Program-for-Results financing has grown steadily since its introduction in 2012. As of April 2019, there were 99 active programmes under this heading, adding up to $40 billion of World Bank financing.

Another new development is the emergence of non-traditional donors, including 'rising powers' such as China and India. As some large developing countries become richer and more active donors, the sources of aid will diversify, and south-south aid will become more common. This will have implications for traditional donors. For example, Hernandez (2017) finds that the World Bank imposes fewer loan conditions on African countries that are receiving significant aid from China.

Conclusion

Sometimes, the conditionality debate assumes that a good strategy could be applied universally. In contrast, Bourguignon and Sundberg (2007) describe a three-track model: countries with good governance and policies receive largely unconditional budget support, intermediate countries face something more like traditional conditionality but with more emphasis on governance and performance and fragile states are aided through humanitarian assistance and aid that bypasses the state, for example by funding NGOs. In practice, however, the tracks can become blurred: budget support has sometimes been withdrawn or modified in response to democratic regress or concerns about corruption (Molenaers et al., 2015).

The three-track model recognises that unconditional aid has problems. There is an interesting parallel in the revenues that some developing countries receive from natural resources. According to the conventional wisdom, those revenues have often led to diminished accountability and poor outcomes, at least in countries with weaknesses in governance. Given this observation and the technical case for imposing conditions, it seems likely that donors will continue to use conditionality but in ways that acknowledge at least some of the weaknesses of the past.

References

Bourguignon, F. and Sundberg, M. (2007). Aid effectiveness – opening the black box. *American Economic Review*, 97(2), 316–321.

Hernandez, D. (2017). Are "new" donors challenging World Bank conditionality? *World Development*, 96, 529–549.

Khan, M. S. and Sharma, S. (2003). IMF conditionality and country ownership of adjustment programs. *World Bank Research Observer*, 18(2), 227–248.

Molenaers, N., Gagiano, A., Smets, L. and Dellepiane, S. (2015). What determines the suspension of budget support? *World Development*, 75, 62–73.

Further Reading

Grant, R. W. (2012). Chapter 6 examines when IMF loan conditions are legitimate. In *Strings Attached*. Princeton University Press, Princeton.

Koeberle, S., Bedoya, H., Silarszky, P. and Verheyen, G. (eds.) (2005). *Conditionality Revisited: Concepts, Experiences and Lessons*, World Bank, Washington, DC.

Morrissey, O. (2016). Aid conditionality. In *The New Palgrave Dictionary of Economics*, online edition.

Temple, J. R. W. (2010). Aid and conditionality. In D. Rodrik and M. R. Rosenzweig (eds.) *Handbook of Development Economics*, Vol. 5, Amsterdam: North-Holland Publishing Company, The Netherlands, pp. 4415–4523. The above essay draws on the longer discussion in this literature review.

Chapter 62

Trade liberalisation and economic development in the developing countries

Kalim Siddiqui

Introduction

This chapter critically analyses the theoretical and empirical basis of trade liberalisation and finds that the arguments of many mainstream economists concerning the static and dynamic gains from free trade are based on weak theoretical grounds. It also discusses the impact of trade liberalisation on the industrial and agricultural sectors and shows how the performance of both sectors has a long-term impact on local industrialisation, food security, employment and the well-being of people in developing countries.

Free trade theory finds widespread support among the international financial institutions, namely the International Monetary Fund (IMF), World Bank, and World Trade Organisation (WTO; Siddiqui, 2016). This free trade approach deepens the process of uneven development and unequal exchange as seen, for instance, in the Trump administration's attempts to hinder China's economic development by means of disadvantageous trade agreements. At present, Chinese developmental policies challenge US global corporations such as Boeing and Microsoft because they require some control over the nature of the US investment by granting China a degree of technology transfer. Existing WTO-enforced intellectual property rights, from which US corporations benefit, provide, among many other things, exorbitant patent rights for medicines and grant Microsoft Windows an effective monopoly on operating systems. With genuine free trade consumers in the US and elsewhere could get cheaper medicines and have more choice in operating systems, but US corporations would not have the levels of profit guaranteed by current arrangements (Siddiqui, 2018a).

Theoretical background

Free trade theory emphasises that if protections are removed resources should flow from high-cost to low-cost products resulting in an increase in productivity. David Ricardo's theory of comparative advantage provides a foundation for understanding the nature of mutually advantageous international free trade and forms the basis of arguments generally used to defend a laissez-faire approach (Siddiqui, 2018b). Protection is seen as interference in the free play of beneficent market forces (Kruger, 1996). Ricardo's model assumes that all resources will be fully employed, but in reality we find that in developing countries mass unemployment and mass poverty have often existed alongside vast but under-exploited resources. During the British colonial period, for example, the imposition of free trade policy on India made it possible for the Lancashire cotton industries

DOI: 10.4324/9780429282348-68

to prosper while hand loom production in India was systematically undermined by the active intervention of the British authorities (Bagchi, 2000). Another notable example could be cited here: in 1699 with the Wool Act Britain banned the export of woollen cloth from the colonies to other countries. This proved to be a severe blow to the Irish wool industry.

Britain adopted 'free trade' policies in the 19th century when it possessed relatively more advanced technologies and industries compared with those of other European countries. These policies were extended to the colonies to further Britain's business and trade interests. From the mid-19th century, Africa and Latin American countries were also integrated into the world economy as suppliers of primary commodities, as envisaged by the 'comparative advantage' model.

At the same time that colonies were encouraged to specialise in the production and export of primary products rather than manufactured goods, Britain abolished import duties on raw materials produced in the North American colonies. Thus, Britain slowed – or completely prevented – modern industrial development in the colonies and in other territories in which it enjoyed pre-eminent influence. Bagchi observes:

> In the victory of private enterprise, the construction of a state fostering its growth played a critical role, and free trade as a policy did not gain ascendancy until Britain had already emerged as the most powerful nation in the world economically, militarily and politically . . . it had begun preaching the doctrine of free trade to others, even enforcing it with gunboats and soldiers, as in the case of opium war.
>
> (Bagchi, 2000: 403–404)

It is claimed that if all countries adopt free trade policies then, 'the world economy can achieve a more efficient allocation of resources and a higher level of material well-being than it can without trade' (McConnell and Brue, 2005: 696). In contrast to this, Bieler and Morton found:

> Trade liberalisation has often implied deindustrialisation and import dependence. An analysis of the consequences of trade liberalisation in Africa and Latin America during the 1980s and 1990s, for example, reveals widespread job losses, increasing unemployment and declining wages in both continents.
>
> (Bieler and Morton, 2014: 40)

In the late 1980s and 1990s, at the behest of the World Bank and the IMF and as a condition for their loans, most of the Latin American countries adopted Structural Adjustment Programmes or SAPs (i.e., neoliberal reforms), while China, which was not a member of the WTO, was able to maintain greater control over both trade and foreign capital investments. The Chinese government was able to encourage foreign investors to establish joint ventures with local companies that included agreements on technology transfer. In China rapid urbanisation and higher growth also resulted in a sharp increase in the scale of the domestic market. By 2010 China became a net importer of food, the largest importer of soy and accounted for half of the world's total imports.

WTO and trade liberalisation

The WTO aims to liberalise world agricultural markets by eliminating subsidies to inefficient producers, tariffs and the practice of holding food stocks by governments

(WTO, 2013). The result was supposed to be a rise in agricultural commodity prices through a de-regulated market that would benefit farmers. At the same time increased competition was supposed to generate greater efficiency and thus bring down prices to the benefit of consumers. However, such assumptions ignored the fact that agricultural trade is in fact characterised by large economic, social and political inequalities.

Due to climate limitations developed countries cannot produce coffee, cocoa, bananas or tea but want these commodities for their food processing industries and they want to acquire these from deregulated markets. In the cotton and sugar markets, however, distortions exist because of subsidies given to producers in both the United States and the European Union and therefore these products are protected from liberalisation. With the signing of the WTO's international treaty *Agreement on Agriculture* (AoA) in 1995, developing countries were granted little access to new markets in the developed countries but were required to accept significantly more imports. This depressed local investment and production, ultimately exacerbated food deficits and undermined food security in developing countries.

The agricultural sector plays an important role not only in maintaining a healthy rural environment and ecology but also in the economic development of a country. It makes a significant contribution to per capita income and employment, especially in the developing countries. Neoliberal policy reforms in agriculture alter the situation in this sector and restructure the economic fabric of the society. Food security and self-sufficiency are important contributing factors to the stability and economic growth of regional and international economies. Accordingly we should examine the impact of trade liberalisation (i.e., free trade) on the agricultural sector and food security issues in the developing countries.

The WTO has become as an important international multilateral institution, not only by bringing liberalisation of trade in agriculture, manufacturing and services but also through its dispute settlement mechanism. In particular, the WTO's negotiations at Doha in 2001 resulted in policies made largely to protect the interests of agro-business corporations based in the West, while offering few benefits to farmers in the developing countries (Stiglitz and Charlton, 2006).

The WTO wants to introduce the idea that agriculture and food production should be treated as any other form of production and be subjected to the rules of competition in deregulated and open markets similar to those in the industrial sector. The supporters of this approach claim that if such policies are follwed in the developing countries they will increase output under competitive conditions and achieve levels of surplus and prosperity similar to those enjoyed by Europe and North America even though those regions do not, in fact, apply such policies in their domestic markets. The developing countries as a group would be wise therefore to defend their interests and seek reform of the WTO in order to protect their agriculture, manufacturing and service sectors and their interests in general.

The proponents of free trade choose to forget that in the 18th and 19th centuries the transition of European and North American agriculture towards greater use of technology and capitalist large-scale production took place at the same time their industrial sectors were expanding and their surplus populations were migrating to the Americas, Australia and South Africa. These developments resulted in the largest land-grabbing and resource-extraction exercises in human history, during which indigenous populations were eliminated or enslaved and their land and natural resources expropriated. Because developing countries have no such possibilities, adoption of the WTO's agriculture reform policy inevitably leads to greater poverty and to ecological destruction exacerbated by climate change.

Free trade in agriculture undermines food sovereignty and adversely affects the possibilities for autonomous development and food self-sufficiency in the developing countries. For example, the WTO's 1994 Agreement on Trade-Related Investment Measures (TRIMS) do not allow the use of local content specification to increase linkages between foreign investors and local manufacturers or restrictions on the outflows of capital by investors. Other WTO policies such as the Agreement on Trade-Related Aspects of Intellectual Property Rights (TRIPS) further allow privatisation and concentration of knowledge in the hands of global corporations.

Unlike manufacturing, agricultural production cannot take place throughout the whole year and therefore prices cannot be lower during the harvest season than during the rest of the year. In the developing countries subsidies were aimed at reducing production costs by providing inputs lower than market prices, but in the developing countries farmers are often forced to sell their products soon after harvest due to difficulty with storage and the need for money to repay debts, which is known as *stressed sale*. To ensure prices governments buy agricultural commodities at prices higher than markets to protect farmers from market fluctuations. Additionally, during shortages, governments release agricultural products from storage to stabilise prices in the market. Under WTO rules, such food stock holdings are prohibited and farmers in the developing countries are left entirely at the mercy of the market (Siddiqui, 2015).

The farmers in North America and the European Union operate highly mechanised capital-intensive agriculture and productivity ranges between 10,000 and 20,000 quintals of cereals per farmer per year. In the developing countries, especially in Africa and Asia, farming is far less mechanised and capital intensive and productivity ranges from just 100 to 500 quintals per farmer per annum (Siddiqui, 2018b).

In 1991with the adoption of neoliberal reforms in India the government reduced its investment in irrigation and extension services in agriculture and for the last two decades the crisis in rural communities has deepened (Siddiqui, 2016). In addition to cuts in government spending and greater emphasis on market forces farmers have had to suffer the demonetisation and cash crisis of 2016. This was done soon after monsoon harvest and due to lack of banknotes farmers were unable to sell their products or buy inputs to sow winter crops. The agrarian crisis has been reflected in the increasing number of farmers' suicides and forced migration to the cities. Over the same period the availability of institutional finance to farmers has been reduced, meaning that the cost of borrowing has risen and also global agricultural commodity prices have declined, particularly since 2017. As a result profitability in the agriculture sector has decreased. India is the largest producer of wheat and second largest producer of rice. In 2017, the production of wheat in India was nearly 96.6 million tonnes and consumption was about same. However, India still exported nearly 3 million tonnes from government stocks. In the same year, rice production was 105 million tonnes and consumption was 103 tonnes, but India exported 11 million tonnes of rice from government stocks. The balance between global food prices, food security, and the living standards of the poor is thus extremely precarious under neoliberal policies, even in an economy as large as that of India.

References

Bagchi, A.K. 2000. "The Past and the Future of the Development State", *Journal of World-System Research*, 6(2): 398–442.

Bieler, A. and A.D. Morton. 2014. "Uneven and Combined Development and Unequal Exchange: The Second Wind of Neoliberal 'Free Trade?'", *Globalizations*, 11(1): 35–45.

Kruger, A. 1996. *The Political Economy of Trade Protection*. Boston: National Bureau of Economic Research.

McConnell, C.R. and S.L. Brue. 2005. *Economics: Principles and Policies*, 16th ed. Boston: McGraw Hill Irwin.

Siddiqui, K. 2015. "Trade Liberalisation and Economic Development: A Critical Review", *International Journal of Political Economy*, 44(3): 228–247. Taylor & Francis. ISSN 0891–1916. Print/1558–0970. https://doi.org/10.1080/08911916.2015.1095050.

Siddiqui, K. 2016. "International Trade, WTO and Economic Development", *World Review of Political Economy*, 7(4): 424–450, Winter. Pluto Journals.

Siddiqui, K. 2018a. "U.S. – China Trade War: The Reasons Behind and Its Impact on the Global Economy", *The World Financial Review*, 62–68, November/December. ISSN 1756–3763. http://www.worldfinancialreview.com/?p=36411.

Siddiqui, K. 2018b. "David Ricardo's Comparative Advantage and Developing Countries: Myth and Reality", *International Critical Thought*, 8(3): 1–28, September. Taylor & Francis Group. https://doi.org/10.1080/21598282.2018.1506264.

Stiglitz, J. and A. Charlton. 2006. *Fair Trade for All*. Oxford: Oxford University Press.

WTO (World Trade Organisation). 2013. *The Case for Open Tarde*. https://www.wto.org/english/thewto_e/whatis_e/fact3_e.htm. (Accessed on 10 December 2018)

Further Reading

Ghose, A.K. 2004. "Global Inequality and International Trade", *Cambridge Journal of Economics*, 28: 229–252.

Reinert, E. 2007. *How Rich Country Got Rich and Why Poor Countries Stay Poor*. London: Constable & Robinson.

Rodrik, D. 2004. *How to Make Trade Regime Work for Development*. Cambridge, MA: Harvard University Press.

Sen, S. 2005. "International Trade Theory and Policy: What is Left of Free Trade Paradigm?", *Development and Change*, 36(6): 1011–1029.

Siddiqui, K. 2016. "Will the Growth of the BRICs Cause a Shift in the Global Balance of Economic Power in the 21st Century?", *International Journal of Political Economy*, 45(4): 315–338. Routledge Taylor & Francis. https://doi.org/10.1080/08911916.2016.1270084.

Chapter 63

The knowledge based economy and digital divisions of labour

Mark Graham, Sanna Ojanperä, and Martin Dittus

> The new international economy creates a variable geometry of production and consumption, labor and capital, management and information.
>
> (Castells, 1989: 348 – in Downey, 2008)

Information is the raw material for much of the work that goes on in the contemporary global economy, and there are few people and places that remain entirely disconnected from international and global economic processes. Information – and ultimately knowledge – is the carrier for the myriad signals needed for such markets to constantly be enacted, performed and understood.

As such, it is important to understand who produces and reproduces, who has access and who and where are represented by information in our contemporary knowledge economy. This chapter discusses inequalities in traditional knowledge and information geographies, before moving to examine the Internet-era potentials for new and more inclusionary patterns. It concludes that rather than democratising platforms of knowledge sharing, the Internet seems to be enabling a digital division of labour in which the visibility, voice and power of the world's economic cores are reinforced rather than diminished.

Information geographies

> Information is not knowledge, Knowledge is not wisdom, Wisdom is not truth.
>
> – Frank Zappa, 1979

As Frank Zappa points out, it is important to distinguish among information, knowledge and other signals, representations and understandings. While this chapter is not the right venue for a detailed discussion and problematisation of the differences between such terms, it is important to clarify what is meant by *information* and *knowledge*. *Information* is generally used to refer to codified descriptions that can answer questions such as 'who', 'what', 'where', and 'why'. *Knowledge*, in contrast, usually refers to the structuring, process, organising or internalisation of information.

Traditionally, information and knowledge about the world have been highly geographically constrained. The transmission of information required either the movement of people or media capable of communicating that knowledge. Historical maps offer perhaps the best illustration of the geographic limitations to knowledge transmission. The

DOI: 10.4324/9780429282348-69

13th-century *Carta Pisana* (the world's oldest navigational chart), for instance, which was produced somewhere on the Italian peninsula,[1] depicts relatively accurate information about the Mediterranean, less accurate information about the fringes of Europe (Great Britain is represented as a strange rectangular shape) and no information about any parts of the world that are farther afield.

The example of the *Carta Pisana* starkly illustrates the constraints placed on knowledge by distance. 13th-century transportation and communication technologies (e.g., ships and books) allowed some of the constraints of distance to be overcome by the map's Italian cartographers. But, in the 13th century those technologies were not effective enough to allow detailed knowledge about the Americas, East Asia and much of the world to be represented on the map.

These highly uneven geographies of information matter. They shape what is known and what can be known, which in turn influences the myriad ways in which knowledge is produced, reproduced, enacted and re-enacted. Importantly, it is not just artifacts from the Middle Ages that display such uneven patterns. Almost all mediums of information (e.g., book publishing, newspaper publications and patents) in the early 21st century are still characterised by huge geographic inequalities: with the Global North producing, consuming and controlling much of the world's codified knowledge and the Global South largely left out of these processes.

Figure 63.1 starkly illustrates some of these patterns by visualising the locations in which academic journals are published. The maps use data from all 11,600 journals included in the Web of Knowledge Journal Citation Reports (JCR) database and shade of each country corresponds to the number of journals published from within it. The JCR database is an especially crucial metric not only because its owners claim that it offers a 'systematic, objective means to critically evaluate the world's leading journals' (Clarivate, 2019), but also because it forms an important part of the ways that academics, departments and universities are evaluated (i.e. non-JCR publications are generally considered to be less valuable than those in the JCR database).

The map reveals a staggering amount of inequality in the geography of the production of academic knowledge. The United States and the United Kingdom publish more indexed journals than the rest of the world combined: 54 per cent of the natural science and 74 per cent of the social science journals. Western Europe, in particular Germany and the Netherlands, also scores relatively well. The majority of the rest of the world then hardly features in these rankings. One of the starkest contrasts is that the Netherlands (where several major publishers admittedly hail from) produces over 20 times as many natural science journals and over ten times as many social science journals than the entire continent of Africa.

These geographies of information reveal how knowledge and economic power are closely intertwined and undoubtedly both reflect and reproduce positionalities of centrality and marginality in the global knowledge economy. Despite the entrenchment of much of the world's codified knowledge in the Global North, many people are pointing to the potential for significant changes in such patterns. The Internet and other information and communications technologies (ICTs) provide and enable possibilities for fundamentally different communications media, methods, platforms and practices. In other words, while movements and control of information were previously constrained by the significant limitations of communication and transportation technologies, such constraints

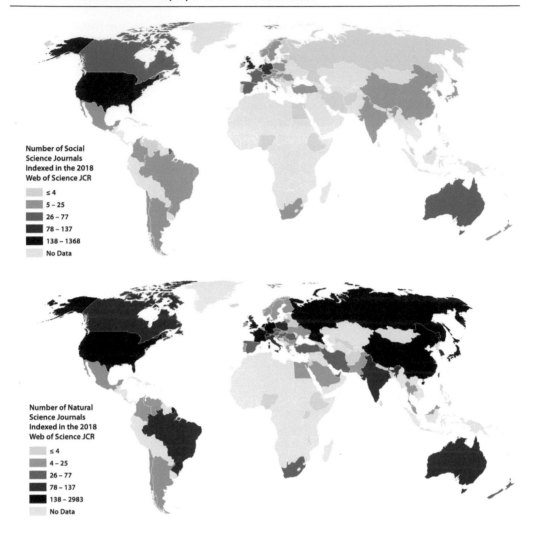

Figure 63.1 The location of academic knowledge

Source: Graham, M, Ojanperä, S. and Dittus, M. (2019)

rarely apply in the Internet-age. Movements of information are almost instantaneous and can be transmitted across the world for minimal costs. As such, there are very real potentials for the geographic and temporal frictions that traditionally constrained and limited the movements of information to be overcome.

The potentials of ICTs and reconfigured information economies

Access to ICTs is by no longer confined to an elite few. At the end of 2018, there were more mobile-cellular subscriptions than people on the planet (International Telecommunication Union, 2018). This means that most people on our planet now have some form of access to telecommunications services, and indeed, most mobile devices are now in use in the Global South. In 2018, the world crossed the halfway line with 51.2 per cent

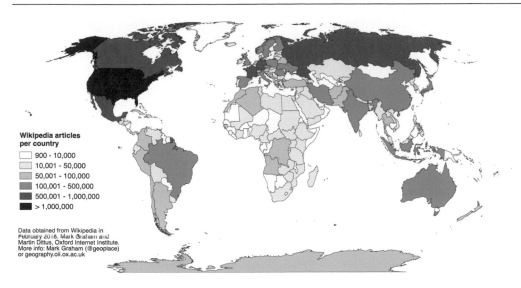

Wikipedia articles per country

- [ ] 900 - 10,000
- 10,001 - 50,000
- 50,001 - 100,000
- 100,001 - 500,000
- 500,001 - 1,000,000
- > 1,000,000

Data obtained from Wikipedia in February 2018. Mark Graham and Martin Dittus, Oxford Internet Institute. More info: Mark Graham (@geoplace) or geography.oii.ox.ac.uk

Figure 63.2 Map of all geotagged Wikipedia articles across Wikipedia's 300 languages

Source: Graham, Ojanperä, Dittus (2019)

or 3.9 billion people using the Internet around the world at the end of the year (International Telecommunication Union, 2018). In other words, over half of all human beings have some form of online access.

Concomitant with this broadening of access to communication technologies has been a fairly widespread belief that now, for the first time in human history, many of the geographic frictions that traditionally contributed to concentrations of information can be overcome. The central idea in those visions is that the Internet is able to bring into being an ethereal alternate dimension with two key characteristics (Graham, 2015). First, a 'space' that is infinite and everywhere (because everyone with an Internet connection can enter) and, second, one that is simultaneously fixed in a distinct (albeit non-physical) location that allows all willing participants to arrive into – and interact in – the same virtual space (Graham, 2011). It is thus important to closely examine the difference that the Internet has made in bringing about potentially new information geographies.

Unfortunately, what most contemporary mappings of information demonstrate is that the Internet has failed to enable a more distributed geography of codified information creation and use. Figure 63.2, for instance, maps contributions to Wikipedia, which is one of the world's largest online platforms of user-generated content. Despite the fact the platform is potentially available and open to most[2] of the 3.9 billion people on Earth with an Internet connection, that millions of people have contributed across the platform's 300 languages and that millions of places around the world have been described, we still see an incredibly concentrated geography of codified knowledge. For example, Africa has only 15 per cent the number of articles as Europe and yet almost twice the population. More articles have been written about Antarctica than most countries in Africa, and many in Latin America and Asia. It is not just Wikipedia that displays such skewed patterns of online information geographies. Many other platforms, repositories of content and online databases exhibit similar spatial cores and peripheries of knowledge (Graham and Zook, 2011; Ojanperä et al., 2017; Ojanperä et al., 2019).

While earlier information sources (like the *Carta Pisana*) had more apparent lacunae and absences and local origins, online platforms can be more duplicitous in their appeals to be neutral, objective and comprehensive. Despite the many ways of understanding Internet geographies (Zook, 2007), there remains a widespread assumption that the Internet is a neutral space facilitating many-to-many relationships and allowing access to what Wikipedia's founder refers to as 'the sum of all human knowledge' and Google's founders describe (Google Inc., 2004) as their 'unbiased and objective' results. However, the Internet has only enabled amplifications of earlier unequal patterns of information geographies.

Digital informational divides

The dense clouds of information – or 'repositories of experience' (Grabher, 2002) about some parts of the world are likely self-reinforcing because of the ways that exchanges of both codified and tacit knowledge are significantly facilitated by spatial proximity (Benner, 2003). These initial uneven geographies of information were brought into being by the spatial fixes of physical telecommunication networks, rates of literacy, patterns of access to existing knowledge, capital and other resources necessary to produce and publish and a range of other social, economic and political patterns, practices and processes. And despite the changing ways in which space is produced through spatial fixes, and changing geographies of literacy, knowledge and access, dense clusters of information persist in many places because of the self-perpetuating nature of knowledge transfer discussed earlier.

The stickiness of information cores and peripheries, even in an age of supposed friction-free communications, is concerning because of Harley's (1989) observation that spatial configurations of information both have power and reproduce power. Because of its uneven geographies, the power/knowledge nexus is thus inherently inclusionary and empowering for some people and places and inherently exclusionary and disempowering for others. Knowledge clusters that are reinforced by repeated rounds of spatial fixes thus result in – and reinforce – a landscape of uneven geographic development (Downey, 2008). While the earlier 'New International Division of Labour' heralded a movement of production from the Global North to the Global South (Dicken, 2010), we now seem to be witnessing a new digital division of labour in which much of the world's knowledge work is produced in the global cores.

Ultimately, despite a rapid growth in education and Internet access for much of the world, a significant proportion of the people on our planet are still entirely disconnected from global platforms of knowledge sharing. Even amongst those 3.9 billion that are now online, a significant proportion of those that are connected are still left out of global networks, debates and conversations. Digital divisions cannot be simply bridged through connections and open online platforms, and much more work needs to be done to overcome inequalities in visibility, voice and power in an increasingly networked world. In other words, while connectivity is clearly a pre-requisite for participation in 21st-century platforms of knowledge sharing and participation, connectivity and access are by no means a determinant of knowledge access, creation and sharing.

Notes

1 The precise origins of the map are unclear.
2 Wikipedia is sporadically censored in some countries: most notably China.

Bibliography

Benner, C. (2003) Learning Communities in a Learning Region: The Soft Infrastructure of Cross-Firm Learning Networks in Silicon Valley. *Environment and Planning A*, 35, 1809–1830.

Clarivate (2019) *Journal Citation Reports*. https://clarivate.com/products/journal-citation-reports/

Downey, G. (2008) Human Geography and Information Studies. *Annual Review of Information Science and Technology*, 41(1), 683–727.

Google Inc. (2004) *Amendment No. 9 to Form S-1 Registration Statement*. https://www.sec.gov/Archives/edgar/data/1288776/000119312504142742/ds1a.htm#toc59330_1

Grabher, G. (2002) Fragile Sector, Robust Practice: Project Ecologies in New Media. *Environment and Planning A*, 34, 1911–1926.

Graham, M. (2011) Time Machines and Virtual Portals. *Progress in Development Studies*, 11(3), 211–227.

Graham, M. (2015) Contradictory Connectivity: Spatial Imaginaries and Techno-Mediated Positionalities in Kenya's Outsourcing Sector. *Environment and Planning A*, 47, 867–883.

Graham, M., Ojanperä, S., & Dittus, M. (2019) Data Shadows and Digital Divisions of Labor. In Graham, M., & Dutton, W. H. (eds). *Society and the Internet: How Networks of Information and Communication are Changing our Lives*. Oxford: Oxford University Press.

Graham, M., & Zook, M. (2011) Visualizing Global Cyberscapes: Mapping User-Generated Placemarks. *Journal of Urban Technology*, 18(1), 115–132.

Harley, J. B. (1989) Deconstructing the Map. *Cartographica*, 26, 1–20.

International Telecommunication Union (2018). *Measuring the Information Society Report 2018*. International Telecommunication Union. https://www.itu.int/en/ITU-D/Statistics/Pages/publications/misr2018.aspx (Accessed April 28, 2019).

Ojanperä, S., Graham, M., Straumann, R., De Sabbata, S., & Zook, M. (2017) The Geography of Engagement in the Knowledge Economy: Regional Patterns of Content Creation. *Information Technologies in International Development*, 13, 33–51.

Zook, M. (2007). The Geographies of the Internet. *Annual Review of Information Science and Technology*, 40(1), 53–78.

Further Reading

Castells, M. (1989) *The Informational City: Information Technology, Economic Restructuring, and the Urban-Regional Process*. New York: Blackwell.

Dicken, P. (2010) *Global Shift*. London: Sage.

Graham, M. (2018) The Virtual Palimpsest of the Global City Network. In Ren, X., & Keil, R (eds). *The Globalizing Cities Reader*. Abingdon: Routledge, 198–204.

Graham, M., De Sabbata, S., & Zook, M. (2015) Towards a Study of Information Geographies: (Im)mutable Augmentations and a Mapping of the Geographies of Information. *Geo: Geography and Environment*, 2(1), 88–105. DOI: 10.1002/geo2.8

Ojanperä, S., Graham, M., & Zook, M. (2019) The Digital Knowledge Economy Index: Mapping Content Production. *The Journal of Development Studies*, 55(12), pp. 2626–2643.

Chapter 64

New institutional economics and development

Philipp Lepenies

New Institutional Economics (NIE) is an expansion of neo-classical economic theory. Its merits stem from the fact that it has identified efficient institutions to be a prerequisite for development. Yet, as neo-classical economics is more and more rejected as a useful basis for social analysis, the prominent role of NIE in development will probably diminish over time. However, NIE still strongly influences development policy. Thus, it is useful to understand the theoretical context of NIE as well as its shortcomings.

1. What is NIE?

NIE attempted to incorporate a theory of institutions into economics. It was a deliberate attempt to make neo-classical economic theory more 'realistic'. Neo-classical economic theory assumed that information flows freely between the actors in competitive markets and that, as a result, institutions do not matter. In contrast, NIE postulated that information is distributed asymmetrically (*asymmetrical information*) and that market transactions come at a cost (i.e. the cost of gathering information or *transaction costs*). Consequently, institutions had to be formed to reduce these costs.

NIE retained the neo-classical assumptions that individuals seek to maximise their utility from scarce resources subject to budget constraints and that collective outcomes rest on the choices made by rational individuals (i.e. *methodological individualism*). However, it discards the concept of *instrumental rationality*, which implies that the choices made by each individual are completely foreseeable. With all information readily available to everyone (*perfect information*), there is no uncertainty in human actions. Institutions become unnecessary and efficient markets characterise economies.

The necessity for a modification of neo-classical theory arose from the fact that so-called *social dilemmas* could not be explained by it. Social dilemmas are situations in which the choices made by rational individuals yield outcomes that are socially irrational. This is obvious in the case of *market failures* that can be caused by *negative externalities* (i.e. a cost arising from an activity which does not accrue to the person or organisation carrying on the activity, e.g., pollution) and *public goods* (i.e. goods that are open to all, free of charge and thus not usually supplied by the market), but also applies to cases of *asymmetrical or imperfect information* (i.e. information is not fully available to everyone). Imperfect information might cause *moral hazards* (i.e. the danger that one of two parties of a contract knowingly alters her behaviour in order to maximise her utility at the other parties expense), *adverse selection* (e.g., imperfect health insurance contracts attract those who have high health risks) and/or *principal-agent dilemmas* (i.e., the problem of how a

DOI: 10.4324/9780429282348-70

'principal' can motivate an 'agent' to act for the principal's benefit rather than following his or her self-interest).

NIE's core argument is that institutions provide the mechanisms whereby rational individuals can transcend social dilemmas and economise on transaction costs (Bates, 1995: 29). Institutions are thus 'the rules of the game of society . . . the humanly devised constraints that structure human interaction. They are composed of formal rules (statute law, common law, regulations), informal constraints (conventions, norms of behaviour and self-imposed rules of conduct), and the enforcement characteristics of both' (North, 1995: 23).

The term 'New Institutional Economics' was coined in the 1970s by Oliver Williamson to distinguish it from an earlier attempt to incorporate institutions into economic theory at the beginning of the 20th century, the so-called (Old) Institutional Economics whose main authors were Thorstein Veblen and John R. Commons.

NIE came into being in 1937, when Ronald Coase explained the existence of firms. Ironically, neo-classical theory could not explain why firms existed and why market transactions were not carried out solely by individuals as methodological individualism suggests. Coase departed from the Walrasian notion of market transactions being made costless on the spot by an invisible and omniscient auctioneer – i.e. the idea of perfect information. Instead he assumed that 'the main reason why it is profitable to establish a firm would seem to be that there is a cost of using the price mechanism' (Coase, 1937: 390). Transactions thus involve the cost of discovering what the relevant prices are.

With time, the idea of transaction costs, probably the single-most important concept of NIE, was developed further. Some authors distinguish different transaction costs in accordance with the three big areas of analysis within NIE, i.e the market, the firm and the state. Thus, there are:

- Market transaction costs which are those described by Coase.
- Management transaction costs within a firm that come as a result of administrative procedures, strategic planning as well as supervision of the work force.
- Political transaction costs, which are the costs of establishing, enforcing and utilising a political system.

Others identify transaction costs according to the process of transacting, i.e. information and search costs, costs of negotiating contracts and the costs of enforcing them. Yet, all transaction costs 'have in common that they represent resources lost due to lack of information' (Dahlman, 1979: 148).

It is important to point out that the NIE is not a homogeneous school of thought. Rather, it consists of a variety of theoretical writings by a large number of different authors. NIE includes research on transaction costs, political economy, contract theory, property rights, hierarchy and organisations, public choice and development.

2. NIE and development

Since the 1990s, NIE has had a tremendous impact on development policy and theory. This can be demonstrated by tracing out the obvious influence that NIE has had on the World Bank, by presenting Douglass C. North's NIE-inspired theory of development and by describing the relevance of NIE for development practitioners.

2.1 NIE and the world bank

Since free markets alone cannot be relied upon to ensure development, NIE emphasises the necessity for development policy to design favourable growth-inducing institutional settings. This was reflected in the new role ascribed to the State in the World Bank's World Development Report (WDR) *'The State in a Changing World'* (1997). Therein, the State, after having been viewed as an obstacle to the functioning of competitive markets in the years before, was suddenly identified as an important facilitator of favourable institutional arrangements.

The appointment of Joseph Stiglitz, a major theorist of the NIE, as chief economist of the World Bank, also reflected the influence of the NIE on the Bank's policy. In 1986, Stiglitz stated that the assumptions of neo-classical economics were 'clearly irrelevant' for the analysis of developing countries. Instead, he showed that asymmetrical information prevailed in most markets (1986: 257). During his spell as chief economist of the World Bank, the bank began to define itself as a 'knowledge bank' whose responsibility was to gather and disseminate information transparently on a global scale. NIE dominated the WDR 1998/1999 *Knowledge for Development*, which highlighted the general importance of overcoming asymmetrical information in development. The influence of NIE was also obvious in the WDR 1999/2000 *Entering the 21st Century*, which summarised the lessons learnt from the last 50 years of global development policy. One lesson plainly read: 'Institutions matter' (World Bank, 1999: 1).

2.2. Explaining institutional change and underdevelopment

A major branch of NIE is concerned with the analysis of institutional change and underdevelopment. Its most prominent author is Douglass C. North who added a historical perspective to neo-classical economics. Historically, societies had to learn how to solve the problem of scarcity.

> The key . . . is the kind of learning that organisations acquired to survive. If the institutional framework made the highest pay-off for organisation's piracy, then organisational success and survival dictated that learning would take the form of being better pirates. If on the other hand productivity-raising activities had the highest pay-off, then the economy would grow.
>
> (North, 1995: 21)

Thus, developmental outcomes in the world differ according to how people learned to cope with scarce resources.

For North, the Western capitalist system has been flexible enough to adapt itself to the institutional necessities induced by the higher division of labour, minute specialisation, impersonal exchange and world-wide interdependence. However, in a country with inefficient institutions, only a process of internal re-contracting can change the institutional setting. As long as those holding the bargaining power have an incentive to defend the status quo and inefficiencies are perceived to be rewarding, the situation will not improve (*path dependence*). This is a major deviation from the neo-classical notion of long-term equilibrium – and a more pessimistic one where underdevelopment becomes plausible.

2.3. The still existing practical relevance of NIE for development

The practical relevance of NIE is twofold. *First*, concepts such as 'asymmetrical information', 'transaction costs', 'adverse selection', 'moral hazard' and 'principal-agent dilemmas' are currently widely used *tools for socio-economic analysis. Second*, they also *serve as the basis for individual project design.* 'Institution-building' itself has become the raison d'être of many development projects in the last decade.

Parting from the definition of 'institutions', any attempt to establish 'rules of the game' and their enforcement characteristics (be it new laws, regulations or governance structures etc.) can consequently be seen as an application of NIE.

A prominent case where concepts of NIE are used is that of the analysis of financial services for the poor. The banking sector usually does not offer financial services to the informal sector because information is asymmetrically distributed between the potential borrower and lender. The lender does not have sufficient information on the borrower whom he does not know personally, who usually does not keep written accounts or business plans and who can not offer physical collateral. Thus, the lender cannot calculate the risk of default. As a result, credit to the informal sector is rationed since lenders are reluctant to give out credit. If financial services to the poor are to be provided, these problems have to be addressed with adequate institutional design.

In the absence of physical collateral, for instance, group-based lending could be an institutional design option to overcome the problems of asymmetrical information. By introducing peer-monitoring as a control mechanism of the borrowers and by linking future payments to group members to the repayment performance of the entire group during monitored weekly meetings, the risks posed by the lack of information described above are minimised.

However, just as NIE is a heterogeneous theory, there is also no such thing as a clear-cut NIE-approach to development. Few development practitioners or theorists who make use of NIE concepts would define themselves as being 'of the NIE'. Yet, the fact that elements of the NIE are used so widely and that the importance of institution-building has been generally acknowledged is arguably the strongest sign of how much NIE has already become commonplace in development.

3. Critique and conclusion

NIE has 'challenged the dominant role ascribed to the market . . .[by highlighting that] neither state nor market is invariably the best way in which to organise the provision of goods and services' and that efficient institutions are the key to successful development (Harris et al., 1995: 1). It is, without a doubt, the strongest merit of NIE to have put the issue of institutions on the development agenda.

However, NIE is not without limitations. As noted before, NIE is an attempt to change neo-classical economics 'from within'. This alone is praiseworthy. Yet, as it maintains the basic assumption that individuals rationally pursue the maximisation of their utility at all times, little or no room is given to any behaviour which might not be guided by the individual's rationally calculated quest for utility-maximisation. Hence, NIE is, so to speak, still not realistic enough as it maintains a simplistic and incomplete model of human behaviour.

The major flaw of the NIE, be it in development or elsewhere, is that many concepts of the NIE are hard to measure, sometimes even hard to define as 'a clear cut definition of transaction costs does not exist' (Eggertsson, 1990: 14). From this derives the difficulty in measuring exactly what transaction costs are. The same applies to the notion of asymmetrical information or the simple question, what 'information' means – especially when one takes into account that 'information' might mean different things to different people. Research which utilises concepts of NIE might thus bring forth insights for a special case. Nevertheless, it is often not comparable with other findings.

Laudably, NIE's historical analysis of development and institutional change rejects the simple idea of market-driven institutional progress. All the same, the attempt to explain persistent underdevelopment by analysing if and how institutions have used resources efficiently in the past is not as straightforward as it seems. A major problem arises out of the way in which history is interpreted. Different interpretations of the past might give rise to various interpretations of the present, especially of the reasons for underdevelopment. Therefore, a historical interpretation might not be shared by everyone. Any historical analysis is just one possible point of view – among many others.

In North's approach, the reasons for persistent underdevelopment are by assumption endogenous. As long as the bargaining power rests with those forces of society that have in interest in perpetuating inefficient institutions, no efficient institutions can emerge. Yet, the role that external factors (e.g., international political or economic power structures) can play in the explication of underdevelopment is not particularly highlighted, a severe omission given the global economic and political interdependencies. More problematic, still, is his notion of path-dependence. With this idea, it seems that countries are trapped in their inefficiencies. This view is overly pessimistic and eclipses the possibility of active development or development cooperation.

Notwithstanding, NIE has rightly identified institution-building as a necessary developmental activity. Nevertheless, identifying a problem through NIE-inspired analysis does not automatically lead to infallibly designed institutions. As it is now generally accepted by the economics profession that humans do not behave as modelled in neo-classical economics, extreme caution is advised when attempting to create 'rules of the game' assuming an economically rational and utility-focused behaviour of the target group. NIE carries the risk that institutions are designed and planned by development experts from scratch – based on the inner logic of the NIE and less on an in-depth analysis of the complex social, political and cultural contexts and the subjective aspirations of a target population. If, however, one is conscious of the limitations of neo-classical economics and gives due weight to thorough social analysis, NIE can still provide important ideas.

Guide to Further Reading

Furobotn, E. G. and Richter R., 1998, *Institutions and Economic Theory. The Contribution of the the New Institutional Economics*, Ann Arbor: Michigan University Press. A very thorough and detailed analysis of practically all aspects of NIE.

References

Bates, R. H., 1995, 'Social Dilemmas and Rational Individuals: An Assessment of the New Institutionalism', in Harris, J., Hunter, J. and Lewis, C. M. (eds.), *The New Institutional Economics and Third World Development*, London: Routledge, 27–48.

Coase R., 1937, 'The Nature of the Firm', *Economica*, Vol. 4, 386–405.

Dahlman, C., 1979, 'The Problem of Externality', *Journal of Law and Economics*, Vol. 22, 141–162.

Eggertsson, T., 1990, *Economic Behavior and Institutions*, Cambridge, UK: Cambridge University Press.

Harris, J., Hunter, J. and Lewis C. M., 1995, 'Introduction: Development and Significance of NIE', in Harris, J., Hunter, J. and Lewis, C. M. (eds.), *The New Institutional Economics and Third World Development*, London: Routledge, 1–13.

North, D. C., 1995, 'The New Institutional Economics and Third World Development', in Harris, J., Hunter, J. and Lewis, C. M. (eds.), *The New Institutional Economics and Third World Development*, London: Routledge, 17–26.

Stiglitz, J. E., 1986, 'The New Development Economics', *World Development*, Vol. 14, 357–265.

World Bank, 1997, *World Development Report 1997: The State in a Changing World*, Oxford: Oxford University Press.

World Bank, 1999, *World Development Report 1999/2000: Entering the 21st Century*, Oxford: Oxford University Press.

Chapter 65

Development and consumption

Cecile Jackson

The anthropological interest in consumption, the meaningful end use of products, has rather passed interdisciplinary development studies by. Ben Fine is right that consumption should not be analysed in isolation from production; he argues that Marxist political economy does attend to use values and consumption (2002: 60). He reminds us of importance of the materiality of goods and the social relations of production and exchange and that separating the economy (production and exchange) from culture (consumption) does not make sense. I agree. But in interdisciplinary development studies culture and consumption are relatively neglected terms, and my purpose here is a re-balancing. Here I argue that a focus on consumption is analytically necessary to understand the core concerns of development; well-being and inequality. Anthropologists have argued that consumption is a lot more than individual satisfaction of needs and wants but is profoundly social, cultural, political and oriented to life-projects of the kind of person one wants to be; it is learned as well as creative, it involves performances and props, expresses agency and achieves social change. Anthropologists of consumption, (Douglas and Isherwood, 1979; Bourdieu, 1986; Appadurai, 1986; Miller, 1995; McCracken, 1988) have argued against the privileging of relations of production as the primary optic on social relations, inequality, social reproduction, change and resistance. Goods constitute and communicate identities, concretise intangible and ephemeral social relations and patrol inequalities, and consumption processes are the means of everyday agency and resistance. The properties of objects are also important to their social possibilities and potential as a means of change; since goods are not fixed in either their manner of use or their affordances, they differ in the scope they offer for creativity of use. Things offer many kinds of usage and meaning, way beyond the expectations of their producers, and they may be deployed to resistant purposes or conversely can patrol or extend relations of inequality.

Well-being, identities, and consumption practices

In his capabilities and functioning model, Amartya Sen (1985) famously argued that what matters is not goods themselves but what they allow and enable, the well-being that individuals can obtain with goods, and this view is compatible with Douglas and Isherwood's expanded – and more social – account of capabilities as the life-projects that consumption of goods enable (Douglas and Isherwood, Douglas and Ney, 1998). Consumption processes are important for how capabilities are converted to functionings, to well-being. Anthropological conceptions of well-being are profoundly social and relational, and objects play an important part in constituting identities and social

DOI: 10.4324/9780429282348-71

relations, i.e. things are 'devices for reproducing relations between persons' (Appadurai, 1986: 25).[1]

As the technology of literacy, letter writing, enables new experiences of the self, so today mobile phones are shaping how people see themselves and their social relations. Women's women use of phones in India is a source of conjugal friction: men and mothers-in-law control phones, wives have to relinquish personal phones on marriage and are not allowed unsupervised access. If a husband eventually does allow his wife a phone it must be a new one, not brought from her natal home, and husbands check their wives' phone content, to limit communication with natal families and friends (Doron, 2012). At the same time phones are dramatically changing the nature of marriage; restricting phone access of unmarried girls is difficult, unarranged romantic relationships can flourish and even betrothed partners can talk extensively and intimately and build companionate bonds without breaking taboos on physical contact. Phones offer women new freedoms from patriarchal control of their socialities. Thus gender relations may be as much changed by the consumption of mobile phones as by changing gender divisions of labour or production processes.

Identities are constituted and expressed through consumption practices. Membership of social categories requires having and using particular goods which signal membership to others (clothing, hair styles, cultural products), not for their utility but for what they say about you. Goods speak. The goods commonly used in development practice to define the poor all signal identities and have meaning beyond practical value. Important as they are, food is more than calories and micronutrients, housing is more than shelter and security and clothing is more than protection from the elements. Ethnic identities are powerfully constituted through food consumption, ingredients, recipes, meals and styles of eating and houses are homes in which space and décor express changing gender identities. For the *sapeurs* of Congo, unemployed, unmarried youth make their self-hood through consuming extravagantly costly designer clothing and find a kind of kinship with others in their ranked hierarchies of clubs (Friedmann, 2004) as do voguing black drag queens in Harlem (Lawrence, 2013). Consumption is a terrain for agency, choosing goods to perform and communicate your self-chosen identities to others.

Consumption does more than constitute and signal social categories; it is embedded in our subjectivities, in our inner experience of self-hood, in a personal, biographical and affective sense. Our stuff is meaningful to us in quite profound ways, as Belk argues in his concept of the extended self. We consider our things to be part of us, the loss of possessions, e.g., by theft, induces grief unrelated to the value of objects, and we invest ourselves in objects, especially over time and with the accretion of memory (1988).

The investments we make in goods changes them too and, recursively, their meaning in our lives. This is not just an individual but also a collective phenomenon whereby goods acquire character from the collective and cumulative effects of the identities of their consumers. For example, goods can become feminised when mostly consumed by women, such as cosmetics (Burke, 1996) or cloth (Buggenhagen, 2012) or classed when consumed by the aspirant middle class. Branding of goods aims to benefit from the way that character is conferred on goods by consumers, in order to induce object attachments and manipulate demand. So too with the consumption of space; kitchens are feminised by virtue of the constant presence of women inhabiting them, and boardrooms are masculinised by virtue of the repeated presence and performative use by men. The identities and desires of consumers actively create the character of goods.

So when the alienation of workers from products of their labour in mass production of manufactured goods strips identity from goods, culture works to reinstate it – collectively and individually – and thus to res-ocialise goods through consumption. When commodities become possessions they acquire personal qualities and meanings, and the longer they are possessed the more memories accrete, which attaches people to their stuff, however ordinary. My bedside drawer contains one small, very worn, ballet shoe once used by my now adult daughter. It proves quite difficult for goods to avoid being re-socialised through consumption, since it is not a passive but a meaning-making process by consumers.

The way in which commodities, goods acquired through market exchanges and with monetary value, are re-socialised in the ways that consumers attribute personal meaning to goods through consumption offers opportunities for agentic and creative consumption which refigures social relations, including those of inequality, which I turn to next.

Inequality and resistance

Since levels of absolute poverty in the developing world are declining, incomes increasing and inequalities are on the rise everywhere, the question of status signalling and social mobility through consumption acquires intensified relevance as a terrain of agency. Inequality has clearly been patrolled and reproduced in many settings by strong norms, if not laws, of appropriate consumption. In India low caste Nadar women used not to be allowed to cover their chests or to carry water on their hips in the upper caste style, two storey houses were forbidden to low castes and the use of umbrellas by the dark skinned continues to be mocked (Osella and Osella, 1999). Social mobility via consumption has become a real progressive possibility

> [f]or a community struggling to throw off the stigma of a production-based caste role, as toddy-tapper and agricultural manual labourers, an increasing focus on consumption as a source of identity is advantageous.
>
> (Osella and Osella, 1999: 1003)

new clothes are considerably more accessible for male social mobility than land acquisition.

Membership of the huge new middle classes in India are now defined more by consumption than occupation, income or education (Donner, 2011), as James also finds for South Africa (2017). But social mobility through consumption requires skilled performances and 'doing' consumption correctly is an important skill, so that consumption does not incite envy (Gell, 1986; Dean, 2013), is socially approved and achieves desired effects.

Commodification[2] in development studies is generally treated as a driver of class and gender inequality (Bernstein, 1977). But can market exchange offer scope to women to challenge gender orders? Social boundaries of inequality are often maintained by controlling the means of consumption of key goods. Historically, gerontocratic patriarchies in Zimbabwe were based on control by older men of cattle required for bridewealth. Intergenerational relations between men changed when wage labour became common and the monetarisation of bridewealth allowed younger men to finance bridewealth and women greater freedom (Schmidt, 1992). During the same period the discovery by businesses of Zimbabwean women as consumers allowed them to re-invent themselves as responsible household money managers (Burke, 1996). More recently mass manufactured cloth in west Africa may have displaced local weaving, but it also enabled cloth consumption by women and the networks of exchange which fundamental to women's power in urban Senegal (Buggenhagen, 2012: 19). Commodified consumption has created freedoms and opportunities for poor rural women, but has consumption expectation unsettled male identities and gender power balances?

Consumption also lies at the heart of the crisis is male gender identities in the south since having dependents, i.e. responsibilities for enabling the consumption of others, notably wives and children, in the provider identities of men, is central to masculinities yet increasingly difficult for men to achieve (Jackson, 2015). In post-agrarian scenarios social inclusion for men requires a capacity to enable the consumption of others, and in South Africa, for example, the scarcity of employment impinges much more acutely on men than women who have never been so singularly reliant on labour-based identities. The identity of dependent wife is an honourable one, and inclusion via motherhood and kin-based identities a backstop against the vagaries of labour – and other – markets. Women have been able to sustain – and extend – meaningful matrifocal social worlds (Ferguson, 2013).

Consumption performances are also how men express gender conflict where they are experiencing perceived displacement by the ascendancy of women they can find their belonging in the homosocial exclusion of women. In Arusha, Tanzania, 'men measure their masculinity relative to their attachments to women, casting praise and blame on those who demonstrate greater or lesser disdain for women' (Weiss, 2009: 87), and masculinity is called into question if a man puts a woman above his male friends. In the struggles of men against the perceived ascendancy of women, consumption can be used to produce homosocial spaces, like the barbershops celebrating misogynist hip-hop culture and excluding women. Women are seen as antithetical to social life in barbershop interactions, and much of the barbershop talk denigrates women, yet women dominate the informal sector in which these barbershops exist. Barbershops are physical spaces where the erasure of women from urban sociality is attempted, in circumstances which are evidently increasingly matrifocal, with many young men (poorer and better off) living under the authority of a mother or mother's sister and related through matrilateral ties. 'Arusha is a city where women have long been active in exactly the kinds of commercial activities in which these young men are now struggling' (Weiss, 2009: 84).

Thus consumption is more relevant than ever to social relations of inequality; it has allowed goods to become the basis of networks of solidarity and assertions of women's worth, it is a primary means of class based social mobility, it helps explain the differential social impact of unemployment and in settings where the balance of gendered power is shifting somewhat towards women, consumption allows men to accommodate and resist.

Notes

1 This idea is shared with Marx's 'commodity fetishism', i.e. how under capitalism commodities appear as items of value detached from the social relations of production. Value appears intrinsic to the object itself.
2 Commodification is the older term, but commoditisation has come to be used interchangeably with it, and broadly I take both to mean the 'penetration into reproduction of commodity relations' (Friedman, 1980: 158).

References

Appadurai A (1986) *The Social Life of Things: Commodities in Cultural Perspective*. Cambridge: CUP.

Belk R (1988) Possessions and the extended self. *Journal of Consumer Research* 15: 139–167.

Bernstein H (1977) Notes on capital and peasantry. *Review of African Political Economy* 4: 60–73.

Bourdieu P (1986) *Distinction: A Social Critique of the Judgement of Taste*. London: Routledge.

Buggenhagen B (2012) *Muslim Families in Global Senegal: Money Takes Care of Shame*. Bloomington and Indianapolis: Indiana University Press.

Burke T (1996) *Lifebouy Men, Lux Women: Commodification, Consumption and Cleanliness in Modern*. Durham, North Carolina, Zimbabwe: Duke University Press.

Dean M (2013) From evil eye anxiety t the desirability of envy: Status, consumption and the politics of visibility in urban south India. *Contributions to Indian Sociology* 47(2): 185–216.

Donner H (ed) (2011) *Being Middle-Class in India: A Way of Life*. New York: Routledge.

Doron A (2012) Mobile persons: Cellphones, gender and the self in North India. *The Asia Pacific Journal of Anthropology* 13(5): 414–433.

Douglas M and Isherwood B (1979) *The World of Goods: Towards an Anthropology of Consumption*. London: Basic Books.

Douglas M and Ney S (1998) *Missing Persons: A Critique of the Personhood in the Social Sciences*. Berkeley: University of California Press.

Ferguson J (2013) Declarations of dependence: Labour, personhood, and welfare in southern Africa. *Journal of the Royal Anthropological Institute* 19: 223–242.

Fine B (2002) *The World of Consumption: The Material and Cultural Revisited*. London: Routledge.

Friedman J (ed) (2004) *Consumption and Identity*. Harwood Academic Publishers.

Friedmann H (1980) Household production and the national economy: Concepts for the analysis of Agrarian formations. *The Journal of Peasant Studies* 7(2): 158–184.

Gell A (1986) Newcomers to the world of goods. In Appadurai A (ed) *The Social Life of Things; Commodities in Cultural Perspective*. Cambridge: CUP, pp. 110–140.

Jackson C (2015) Modernity and matrifocality: The feminization of kinship? *Development and Change* 46(1): 1–24.

James W (2017) Not marrying in South Africa: Consumption, aspiration and the new middle class. *Anthropology Southern Africa* 40(1): 1–14.

Lawrence T (2013) Listen and you will hear all the houses that walked there before. *A History of Drag Ball and the Culture of Voguing*. http://www.timlawrence.info/articles2/2013/7/16/listen-and-you-will-hear-all-the-houses-that-walked-there-before-a-history-of-drag-balls-houses-and-the-culture-of-voguing

McCracken G (1988) *Culture and Consumption: New Approaches to the Symbolic Character of Consumer Goods and Activities*. Bloomington: Indiana University Press.

Miller D (1995) Consumption and commodities. *Annual Review of Anthropology* 24: 141–161.

Osella C and Osella F (1999) From transience to immanence: Consumption, life-cycle and social mobility in Kerala. *South India Modern Asian Studies* 33(4): 989–1020.

Schmidt E (1992) *Peasants, Traders and Wives: Shona Women in the History of Zimbabwe (1870–1939)*. New Hampshire: Heinemann Educational Publishers.

Sen A (1985) *Commodities and Capabilities*. OUP India.

Weiss B (2009) *Street Dreams and Hip-Hop Barbershops: Global Fantasy in Urban Tanzania*. Delhi: Bloomington and Indianapolis: Indiana University Press.

Chapter 66

Rethinking 'work' from the cities of the south

William Monteith

The brave new world of work

> Across much of the world people lacking access both to land and to waged employment form an increasingly prominent part of our social and political reality. Equally important, those occupying such ill-defined and precarious social locations are both pioneering new modes of livelihood and making new kinds of political demands.
>
> (Ferguson, 2015: 23)

Sustainable Development Goal 8 aims to promote 'full and productive employment and decent work for all'. However, its progress has been hampered by a lack of consensus on what constitutes 'decent work' and whether full employment is a realistic – or even desirable – objective in contexts where the majority of the population subsist outside of formal employment. At the time of writing, more than 60 per cent of the world's workers are located in the informal economy and 44 per cent are self-employed (ILO, 2019). Historical forms of exploitation have been accompanied by more recent waves of casualisation, leaving a growing proportion of the workforce flexible, poor, and devoid of the protections associated with the standard employment relationship.

The decline – or historical absence – of formal wage employment in many regions of the world poses a number of challenges to the ways in which researchers, governments, and citizens think about 'work' (Denning, 2010). How do people make a living outside of wage employment? What types of activities do they pursue? How do these activities disturb conventional roles (e.g., employed/unemployed) and entitlements (e.g., welfare)? And what possibilities do they provide for imagining alternative economies and societies? James Ferguson (2015) argues that there are opportunities, as well as challenges, in wage-scarce societies. By expanding our understandings of work beyond the 'proper job', it is possible to imagine alternative visions of how welfare and subsistence might be constituted, based, for example, on a new politics of distribution.

This chapter provides an outline of some of the dominant ways in which work in the Global South has been categorised and understood. I argue that the ideal of the formal wage worker has remained influential since the colonial era, with implications for the ways in which researchers and policy makers conceptualise the experiences of workers today. The chapter ends by outlining a more recent body of ethnographic literature that seeks to expand our understandings of 'work' through an engagement with the lived experiences of people subsisting outside of wage employment in the cities of the South.

DOI: 10.4324/9780429282348-72

Work in Global South: a short history

The concepts of 'work' and 'labour' were developed in Europe and exported via trade and colonialism. In the colonial era, 'the labour question' referred to how European governments could ensure a consistent labour supply in the colonies, while 'the work question' examined the individual activities to be carried out (Cooper, 2017). The answer to both questions was often violence and dispossession. Indeed, the enduring history of wage labour in much of the Global South is one of a form of violence and technique of governance – as a method of disciplining bodies in order to advance the interests of the colonial economy (Mbembe, 2001). For example, in Uganda, the British colonial authorities attempted to mobilise indigenous labour through the enforcement of land reforms and poll taxes, compelling people to move from their rural homesteads and engage in forms of work outside of local realms of productivity. By dividing the day between the employer's time and one's own time, capitalist wage labour made it possible to think of the 'social' realm as being separate to the 'economic' realm; of 'work' as being separate to 'life' (Millar, 2018). Yet this way of thinking about work and economy – common among theorists in the Global North – is of limited value for understanding the diverse range of livelihood activities present in the Global South.

The category of the wage worker remained influential in the early postcolonial era as newly independent states pursued strategies of economic nationalisation, giving rise to a series of trade union movements in South Asia and sub-Saharan Africa (Cooper, 2007). Large numbers of people across the Global South left rural livelihoods in search of employment opportunities in the city. However, the total number of wage workers did not increase as expected. Instead of being swept up in an industrial revolution that would pull them into the wage workforce, urban migrants more often ended up in informal settlements where they sought to eke out a living through a complex range of livelihood activities (Ferguson, 2015). As a result, researchers and policy makers went in search of alternative categories and concepts in order to better understand the emerging landscapes of work.

Work and the informal economy

In the early 1970s, the anthropologist Keith Hart (1973) coined the concept of the 'informal sector' (now broadened to the 'informal economy') to describe the livelihood activities of urban migrants in Ghana. The sector comprised all forms of work outside of the boundaries of state regulation, including 'legitimate' activities, such as petty commerce, personal services, and home-based production, and 'illegitimate' activities, such as prostitution, pickpocketing, and scavenging (ibid.). At the time of Hart's intervention, much of the development industry – including the World Bank and ILO – understood the informal sector as a 'backward' or 'traditional' realm of the economy in need of modernisation. The dominant policy prescription was thus to modernise and formalise the sector through a shift to more technologically advanced and capital-intensive forms of production capable of generating formal wage employment. Yet informal and alternative forms of work continued to proliferate, particularly in the aftermath of economic crises. For example, in Latin America, the share of non-agricultural workforce in informal employment rose from 43 per cent to 51 per cent between 1990–2002, at which point it was even higher in Asia (71 per cent) and sub-Saharan Africa (72 per cent) (Chant, 2014).

Informal work also began to emerge within global production networks as the production of goods and services was subcontracted to informal firms and industrial outworkers. Today, women dominate the smaller-scale and poorer paid categories of informal work, while men dominate the comparatively better paid informal workforces of larger firms (Kinyanjui, 2014; ILO/WIEGO, 2019).

Dualist conceptions of the informal sector, which imagine a marginal or traditional set of activities separate from the formal capitalist sector of the economy, have come under substantial critique since the time of Hart's writing. An alternative, structuralist perspective instead conceives of 'informalisation' as a process through which employers in the formal sector, acting with the support of the state, seek to reduce wage costs and enhance flexibility by making use of unprotected workers (Meagher, 1995: 259). Missing in both accounts is an acknowledgement of the historical origins of many of the types of work now described under the banner of the informal economy. Indeed, the concept of informality has been critiqued for rendering deviant sectors of economic activity that have long constituted the basis of the 'real' or 'popular' economy in many Southern contexts (Gago, 2017; MacGaffey, 1991; Simone and Pieterse, 2017). While the concept emphasises the evolving relationship between workers and the state, it dampens the role of other relationships and institutions which continue to exert a considerable influence on the experiences of workers in the cities of the South.

Entrepreneurship and the gig economy

The stubborn endurance of work outside of state regulation forced the development industry to revise their approach to work in the late 1970s. As accelerating rates of urbanisation increased concerns about urban poverty and unemployment, the ILO began to see the informal economy as a potential solution. The World Bank, ushering in a new era of neoliberal structural reform, reimagined the informal economy as a space of entrepreneurial innovation away from the interventions of the state. In a departure from the dualist and structuralist perspectives (outlined earlier), a Voluntarist school began to conceptualise the informal economy as a collection of entrepreneurs deliberately seeking to avoid the cumbersome regulations of the state (Chen, 2012). Rather than precarious survivalists in need of state protection and regulation, workers in cities across the Global South were recast as self-employed 'entrepreneurs' in need of capital investment. Market-based approaches to development began to proliferate, supporting the growth of the microfinance industry. A substantial scholarship has since critiqued such approaches for undermining the role of the state and emboldening private sector interventions that indebt small-scale producers and traders (Roy, 2010).

Today an overwhelming 72 per cent of all workers in the Global South are self-employed, of which approximately 2 per cent are employers (predominantly men), 50 per cent are own account workers (men and women), and 20 per cent contributing family members (predominantly women; ILO/WIEGO, 2019). An increasing number of self-employed men and women now generate an income through digital labour platforms in the so-called gig economy. Recent studies in Southeast Asia and sub-Saharan Africa have shown that such platforms provide some tangible benefits for workers, including autonomy and flexibility. However, they are characterised by forms of algorithmic control which result in low pay, social isolation and exhaustion (Wood et al., 2019).

Where the concept of informality foregrounds the relationship between workers and the state, prioritising rights-based approaches to development, entrepreneurship instead emphasises the relationship between workers and the capitalist market, prioritising resource-based approaches (Monteith and Giesbert, 2017). The influence of both approaches is visible in SDG 8, through the simultaneous emphasis on 'decent work' and 'productive employment'. Civil society organisations including WIEGO (Women in Informal Employment: Globalising and Organising) have worked to ensure that the ILO's Decent Work agenda takes into account the experiences and aspirations of female workers in the cities of the South. Nevertheless, the agenda is limited by a reliance on universal indicators, which tend to idealise the formal wage worker – regulated, pensioned and unionised – based on the historical experiences of wage workers in the Global North (Monteith and Giesbert, 2017). It thus idolises a form of work is the historical *exception* rather than the norm in much of the Global South. There remains a need then for approaches which de-centre Northern theories of wage labour in order to examine the generative potential of the diverse forms of livelihood found in other regions of the world.

Rethinking work: ethnographic approaches

A growing body of ethnographic scholarship draws attention to the various forms of subsistence, sociality, and 'hustle' that are characteristic of urban economies in the Global South (see e.g., Ferguson, 2015; Kinyanjui, 2019; Millar, 2018; Thieme, 2017). By eschewing the paradigms of informality and entrepreneurship, this scholarship emphasises the significance of a broader range of socioeconomic relations to people's working lives, including interdependency, reciprocity, and redistribution. As Kathleen Millar (2018: 4) argues in her work with waste pickers in Rio de Janeiro, Brazil:

> Persistent notions of informal labor as a product of scarcity or a last resort leave little room to ask why this work is *taken up* by those who pursue it, how it emerges from and fashions particular social and political relations, and how it expresses different visions of what life is for.

Millar argues that in order in order to understand activities such as waste picking in their proper context, we must undo the separation of 'work' and 'life' created capitalist conceptualisations of wage labour (see previous mention). By conceptualising waste picking as a 'form of living' that includes *but is not limited to* concerns with income, sustenance and livelihood, she shows that many waste pickers choose to return to the dump rather than seek out other available forms of employment owing to the various forms of sociality, support, and political association that it provides. In other words, while waste picking in Rio involves various harms and hazards, there is a 'decency' to it that is not captured by universal measures such as the Decent Work agenda.

Similarly, Tatiana Thieme (2017) broadens conventional understandings of work through the conceptual frame of the 'hustle' in Nairobi, Kenya. She argues that young people deploy the hustle in ways that 'combine the urgency of everyday economic survival with aspirational urban identities anchored in making the everyday struggle meaningful and culturally significant' (2017: 13). In this way, Nairobi youth also reject the separation of 'work' from 'life' by embedding their economic activities in a series of social relations and cultural references that can render it 'decent' in spite of poor regulation and material returns.

These contributions reflect Ferguson's (2015) emphasis on expanding our understandings of work beyond the 'proper job'. However, he encourages us to go one step further by imagining post-work interventions based on these relations and references. For example, in the context of Southern Africa, he proposes a 'new politics of distribution' in which redistributive payments are seen as a fundamental part of social organisation, rather than the interruption of a 'normal' situation (e.g., the wage-earning household). This proposal draws parallels with broader debates on the 'future of work' and universal basic income in the Global North. In order to realise this new politics, researchers and policy makers will be required to break the spell of 'work-related melancholia' (Barchiesi, 2011: 246) by better understanding the ways in which attitudes towards work, income and redistribution are underpinned by moral ideas of fairness and deservingness (see Fouksman, 2020).

In contemplating the decline of wage employment, researchers and policy makers in the Global North have much to learn from regions of the South (Breman and van der Linden, 2014). However, this learning process will require the suspension of dominant theories and categories – including those of modernisation and informality – in order to understand the diversity of ways in which people make a living in the world and the possibilities these activities provide for imagining futures in which work, identity, and security might be woven together differently.

References

Barchiesi, F. (2011) *Precarious Liberation: Workers, the State, and Contested Social Citizenship in Postapartheid South Africa*. Scottsville: University of KwaZulu-Natal Press.

Breman, J. and van der Linden, M. (2014) Informalizing the Economy: The Return of the Social Question at a Global Level. *Development and Change* 45(5): 920–940.

Chant, S. (2014) The Informal Economy in Cities of the South. In Desai, V. and Potter, R. (eds) *The Companion to Development Studies*, 3rd Ed. Abingdon: Routledge, 200–207.

Cooper, F. (2007) From Enslavement to Precarity? The Labour Question in African History. In Adebanwi, W. (ed) *The Political Economy of Everyday Life in Africa: Beyond the Margins*. Rochester: Boydell & Brewer, 135–156.

Cooper, F. (2017) From enslavement to precarity? The labour question in African history. *The political economy of everyday life in Africa: Beyond the margins*, pp. 135–156.

Denning, M. (2010) Wageless life. *New left review*, 66(6), pp. 79–97.

Ferguson, J. (2015) *Give a Man a Fish: Reflections on the New Politics of Entitlement*. New York: Duke University Press.

Fouksman, E. (2020) The Moral Economy of Work: Demanding Jobs and Deserving Money in South Africa. *Economy and Society* 49(2): 287–311.

Gago, V. (2017) *Neoliberalism from Below: Popular Pragmatics and Baroque Economies*. Durham: Duke University Press.

Hart, K. (1973) Informal income opportunities and urban employment in Ghana. *The journal of modern African studies*, 11(1), pp. 61–89.

ILO/WIEGO (2019) *Women and Men in the Informal Economy: A Statistical Picture*. Geneva: ILO. https://www.ilo.org/global/topics/employment-promotion/informal-economy/publications/WCMS_711798/lang-en/index.htm, accessed 10/06/2019

Kinyanjui, M.N. (2014) *Women and the informal economy in urban Africa: From the margins to the centre*. Bloomsbury Publishing.

Kinyanjui, M.N. (2019) *African Markets and the Utu-Buntu Business Model: A Perspective in Economic Informality in Nairobi*. Cape Town: African Minds.

MacGaffey, J. (1991) *The Real Economy of Zaire: The Contribution of Smuggling and Other Unofficial Activities to National Wealth*. London: James Currey.

Mbembé, A. (2001) *On the Postcolony*. Berkeley: University of California Press.

Meagher, K. (1995) Crisis, Informalization and the Urban Informal Sector in Sub-Saharan Africa. *Development and Change* 26(2): 259–284.

Millar, K. (2018) *Reclaiming the Discarded: Life and Labor on Rio's Garbage Dump*. New York: Duke University Press.

Monteith, W. and Giesbert, L. (2017) When the Stomach is Full, We Look for Respect: Perceptions of 'Good Work' in the Urban Informal Sectors of Three Developing Countries. *Work, Employment and Society* 31(5): 816–833.

Roy, A. (2010) *Poverty Capital: Microfinance and the Making of Development*. London: Routledge.

Simone, A. and Pieterse, E. (2017) *New Urban Worlds: Inhabiting Dissonant Times*. Cambridge: Polity.

Thieme, T.A. (2017) The Hustle Economy. *Progress in Human Geography* 42(4): 529–548.

Wood, A.J., Graham, M., Lehdonvirta, V. and Hjorth, I. (2019) Good gig, bad gig: autonomy and algorithmic control in the global gig economy. *Work, employment and society*, 33(1), pp. 56–75.

Additional resources

Ferguson, J. and Li, T. (2018) *Beyond the 'Proper Job': Political-Economic Analysis after the Century of Labouring Man, PLAAS Working Paper 51*. Cape Town: UWC.

Lindell, I. (2010) *Africa's Informal Workers: Collective Agency, Alliances and Transnational Organizing in Urban Africa*. London and New York: Zed Books.

Monteith, W. and Camfield, L. (2019) Business as Marriage, Marriage as Business: Female Entrepreneurship in Kampala, Uganda. *Geoforum* 101(May): 111–121.

WIEGO Publication Series (Online). http://www.wiego.org/wiego/wiego-publication-series

Chapter 67

Rural livelihoods in a context of the global land rush

Annelies Zoomers and Kei Otsuki

Introduction

According to a widely accepted definition, a livelihood 'comprises the capabilities, assets (including both material and social resources) and activities required for a means of living' (Chambers and Conway, 1991: 6). It encompasses income, both cash and in kind, as well as social institutions (kin, family, village), gender relations, and property rights required to support and to sustain a given standard of living (Bebbington, 1999: 2022). During the 1990s, research on livelihoods became central to rural development practice and thinking (Scoones, 2009). The research fed into an establishment of the so-called sustainable livelihood frameworks that conceptualised people's assets as different types of capitals with which people coped with the social and environmental changes, shocks, and uncertainties (Carney, 2002).

In this chapter, we argue that we need to re-engage with the livelihood research in order to understand new changes faced by rural people and places. Since the late 2000s, together with the global appearance of the food, energy, and climate crisis, along with the booming amount of large scale land investments, the new changes have emerged as the land rush, which often generated negative local impacts. In order to find ways to optimise the developmental outcomes of land investments, it is important to deepen our understanding of the full range of livelihood changes that are taking place 'under the radar'. We contend that existing approaches to assessing the impacts of investments on local development usually stop at the level of 'local communities', focusing on whether the people affected have benefited in terms of employment and income generation or whether the inflow of money has resulted in the loss of natural resources or displacement. A more holistic and agency-oriented assessment of people's livelihoods is necessary as a starting point, not only to clarify how people became vulnerable in the first place to such an extent that they are negatively impacted by investments but also to understand people's capabilities to cope with the vulnerability and to generate an alternative agenda both on their own and in collaboration with investors (Zoomers and Otsuki, 2017).

In the following section, we will discuss how the livelihood research helps to better grasp what is happening on the ground and how to improve the situation of individuals within communities in the current context of the global land rush.

DOI: 10.4324/9780429282348-73

Re-engaging with the livelihood research

During the 1990s, a number of scholars engaged in research on livelihoods have shown that 'the poor' cannot be seen as a homogeneous group and that livelihood strategies are extremely diverse and dynamic (Bebbington, 1999; Chambers and Conway, 1991; De Haan and Zoomers, 2003, 2005; Scoones, 2009). Based on this acknowledgement of diversity, a number of so-called Sustainable Livelihood (SL) frameworks have been developed and adapted by donor agencies, NGOs, and research organisations.

While the DFID's SL framework is one of the most known frameworks (Carney, 2002), the same general principles apply to all: they start from the assumption that people are agents who actively shape their own futures. Instead of focusing on what poor people lack (e.g., land, as is often the case in discussions about the land rush), the attention is focused on what people are entitled to (Bebbington, 1999; Sen, 1981). People's livelihoods largely depend on their access to the different types of capital assets (which form the basis for their livelihoods): human capital (skills, education), social capital (networks), financial capital (money), natural capital (land, water, minerals), and physical capital (houses, livestock, machinery, irrigation infrastructure). Cultural capital is sometimes added to emphasise the historical process of livelihoods, or the physical or financial capital is replaced by produced capital (De Haan and Zoomers, 2003, 2005). Political capital is emphasised in order to address the issue of power (Scoones, 2009).

The livelihood research has helped to show how people cope with stress and changes by substituting one capital for the other and making flexible combinations. For example, if a person does not possess land to cultivate (natural capital), she will try to purchase a parcel (financial capital) or enter into sharecropping relations through her network of social relations (social capital). Many poor households are in fact characterised as having 'constructed an increasingly diverse portfolio of activities and assets in order to survive and to improve their standard of living' (Ellis, 2000: 15). Multitasking and multilocal livelihoods are often used to compensate for insufficient income or deal with temporary crisis situations. It is recognised as the most common way to escape poverty, cope with insecurity, or reduce risk.

The term 'sustainable' in this context refers both to such a people-centred characteristic of a livelihood that endures the various shocks and uncertainties likely to be encountered in the environment and to avoid contributing to long-term depletion of natural resources (Chambers 1987, in Scoones, 2009). According to Chambers and Conway (1991: 6), 'a livelihood is sustainable when it can cope with and recover from stresses and shocks and maintain or enhance its capabilities and assets both now and in the future, while not undermining the natural resource base'. The ways in which people respond – and their resilience – depend on their total portfolio of livelihood assets in combination with their priorities and aspirations, as well as their 'bundle of powers' (Ribot and Peluso, 2003). People who lose their land (and do not have power to gain, control, or maintain access to resources) still have various options: they could rely on their social networks (and ask family or friends for financial support) or get access to land elsewhere based on sharecropping relations; others (who have physical capital, such as a lorry or a tractor) might prefer to earn their incomes by offering services or opening shops. Other people might decide to move away or to migrate as part of their livelihood diversification strategy (Ellis, 2000): sending remittances will help significantly increase the financial capital of those back home. Whether people suffer (or benefit) from resources depends on their access to power and capability to derive benefits from things and ongoing development process (Ribot and Peluso, 2003: 153).

Adopting the SL framework and taking into account that people's capability to benefit from resources will depend on the bundle of powers helps us to 'unravel the fuzzy relation between globalisation and local development from an actor point of view' (De Haan and Zoomers, 2003). The framework will contribute to a better understanding of how people deal with global challenges, their active roles in exploring opportunities and their coping strategies when confronted with externally driven change. Given this knowledge about sustainable livelihoods organised by people's capabilities, and returning to the question of how to optimise the developmental impacts of land investments, it is clear that in current discussions (and policy goals, such as inclusive development and leaving no-one behind), the agency of people and the diversity of livelihoods are largely overlooked. We should accept that people have different priorities and different livelihoods, as there are many different ways to optimise the developmental outcomes of land investments, which go beyond the current set of policy measures aimed at providing people with land titling or offering financial compensation.

Global land rush and the livelihood framework

Since the start of the global land rush in around 2007, landscapes in the Global South have indeed undergone enormous transformations that have had important implications for people's livelihoods (Borras and Franco, 2010; Carmody, 2016; Cotula et al., 2009; Deininger and Byerlee, 2011; Kaag and Zoomers, 2014). Even though it is difficult to make an accurate estimate of the total area involved, there is no doubt that large-scale land investments have been involving millions of hectares globally, and the figure is still on the rise, especially when taking into account large-scale investments in food and biofuels (e.g., soy, sugarcane, and oil palm), mining, infrastructure, and urban land development. Not only foreign investors but also domestic investors and local elites are involved in such land deals, often facilitated by governments and donor-organisations.

From the very beginning, scholars have been debating the consequences of this global land rush, making an assessment of its consequences for local communities. In spite of being promoted for stimulating countries' economic growth by introducing technology, creating employment, and contributing to solutions for the energy and food crises, a large number of case studies illustrate the negative impacts on local communities in the form of land grabbing and dispossession as well as the destruction of the natural environment and commons (Borras and Franco, 2010; Cotula et al., 2009; Kaag and Zoomers, 2014). Over the last decade, international organisations, governments, NGOs, and businesses have been seeking ways to minimise the harmful effects of large-scale land investments while optimising the positive developmental impact. According to the first Sustainable Development Goal, inclusive development should 'leave nobody behind'.

Using the livelihood framework in this context of the global land rush means that people are not only victims of new dynamics coming from above ('land grabbing'). The impacts of land investments are not only felt in the material aspects of life from the perspective of specific, locally bound situations because people are capable of diversifying or at least using the multidimensional livelihoods to cope with and shape the dynamics themselves. Local transformations in relation to the ongoing landscape transformations can only be understood through the holistic nature of livelihoods, which must be secured not only through responsible land governance but also through comprehensive social and economic policies that guarantee the poor's access to basic services and sources of livelihoods under changing varieties of conditions.

Conclusion

Re-engaging with the livelihood framework will highlight possible ways to fill the gaps in the current framing of the impacts of large-scale land investments. It will help us to gain a better understanding of the diversity that exists within a local community and also enables us to descriptively (rather than prescriptively) grasp how people live their lives and what resources they are entitled to (Sen, 1981). By describing what types of capital a person has access to, we come to understand the existing endowments and capabilities. Furthermore, as a livelihood is essentially described and measured within a household unit, it will open the black box of intra-household decisions, engendering an increased awareness of diverging positions within a household. Rather than being harmonious entities pursuing an optimal balance, individual household members often pursue individual ways to improve their situation by diversifying their income source or by moving to a new location. Indeed, in many cases, traditional solidarity-based principles of pooling incomes, consumption, and labour power within households have been weakened considerably (De Haan and Zoomers, 2003).

With the livelihood framework, research on the global land rush can move in the direction of a genuinely people-centred bottom-up approach: supporting local groups in their capabilities to build up their own portfolios of plans that reflect their own needs and priorities. Inclusive development should then mean as the process by which investment agendas come to be included in people's agendas to improve their livelihoods, rather than the current process by which people are included in business plans or consulted to facilitate the business operating on their land. In other words, local benefits depend on the capabilities of local groups to develop their own plans and to attract investors who are willing to bring in necessary investments in line with local needs and priorities.

References and further reading

Bebbington, A. (1999) Capitals and capabilities: A framework for analysing peasant viability, rural livelihoods and poverty. *World Development* 27(12): 2021–2044.

Borras S. and J.C. Franco (2010) From threat to opportunity? Problems with the idea of a 'code of conduct' for land grabbing. *Yale Human Rights and Development Journal* 13(1): 507–523.

Carmody, P. (2016) *New Scramble for Africa*. 2nd edition. Cambridge: Polity Press.

Carney, D. (2002) *Sustainable Livelihoods Approaches: Progress, and Possibilities for Change*. London: DFID.

Chambers, R. and G. Conway (1991) *Sustainable Rural Livelihoods: Practical Concepts for the 21st Century. IDS Discussion Paper 296*. Brighton: Institute of Development Studies.

Cotula, L., S. Vermeulen, R. Leonard and J. Keeley (2009) *Land Grab or Development Opportunity? Agricultural Investment and International Land Deals in Africa*. London/Rome: IIED, FAO, IFAD.

De Haan, L. and A. Zoomers (2003) Development geography at the crossroads of livelihood and globalisation. *Journal of Economic and Social Geography (TESG)* 94(3): 350–362.

De Haan, L. and A. Zoomers (2005) Exploring the frontier of livelihood research. *Development and Change* 36(1): 27–47.

Deininger, K. and D. Byerlee (2011) *Rising Global Interest in Farmland. Can it Yield Sustainable and Equitable Benefits?* Washington, DC: The World Bank.

Ellis, F. (2000) *Rural Livelihoods and Diversity in Developing Countries.* Oxford: Oxford University Press.

Kaag, M. and A. Zoomers (eds.) (2014) *The Global Land Grab. Beyond the Hype.* London: Zed Books.

Ribot, J.C. and N.L. Peluso (2003) A theory of access. *Rural Sociology* 68: 153–181.

Scoones, I. (2009) Livelihoods perspectives and rural development. *The Journal of Peasant Studies* 36(1): 171–196.

Sen, A. (1981). *Poverty and Famines: An Essay on Entitlement and Deprivation.* Oxford: Clarendonk Press.

Zoomers, A., M. Leung and G. van Westen (2016) Local development in the context of global migration and the global land rush: The need for a conceptual update. *Geography Compass* 10(2): 56–66.

Zoomers, A. and K. Otsuki (2017) Addressing the impacts of large scale land investments: Re-engaging with livelihood research. *Geoforum* 83: 164–171.

Chapter 68

Migration and transnationalism

Katie Willis

Introduction

International migration has become one of the key characteristics of the increasingly interconnected world of the early 21st century (de Haas et al., 2019). While many of these migrants are fleeing persecution or natural disasters, economic migration is highly significant. Between 1970 and 2020, the number of international migrants more than doubled to 281 million. While Europe and Asia had the largest share of the world's international migrants in 2020, hosting about 86 million migrants each, international migrants made up the largest percentage of the population in Oceania (22 per cent) and Northern America (16 per cent; McAuliffe and Triandafyllidou, 2021: 24).

Because of improved communication technology and transport, links across national borders are easier than they were in the past. This means that many migrants now live what have been termed 'transnational lives'. According to Basch et al. (1994: 6) 'transnationalism' in relation to migration is 'the process by which transmigrants, through their daily activities, forge and sustain multi-stranded social, economic, and political relations that link together their societies of origin and settlement, and through which they create transnational social fields that cross national borders'. This ability to live both 'here' and 'there' has implications for the construction of migrant identities, but the focus of this chapter will be on the role of transnational migration in development practices in the Global South.

Brain drain

While emigration may be a positive strategy for individuals and their families, for communities and countries in the Global South such migration has often been represented as a loss of resources; a so-called brain drain. This is particularly because more educated, skilled and dynamic individuals are over-represented within economic migrant flows. Within agricultural communities, this outmigration (to both national and international destinations) can leave insufficient labour to maintain agricultural production.

The medical sector has received particular attention in relation to the brain drain and international migration (Connell, 2010). Training doctors, nurses and other medical personnel is very expensive for governments but is viewed as a wise investment as these workers will contribute to a country's social and human development once they have been trained. However, as migration becomes more affordable and logistically possible, medical workers my choose to migrate overseas for better salaries and working

DOI: 10.4324/9780429282348-74

conditions. Jobs are available due to shortages in hospitals in the Global North or the unmet demand for care workers for the elderly, infirm and children. The impacts of the migration of medical personnel are particularly acute in sub-Saharan Africa, where formal healthcare resources are already limited due to insufficient funding for staff, infrastructure and medicines.

Remittances

Remittances are money and goods sent back 'home' by migrants. This process has become much easier due to developments in the banking system and international money transfer services. Remittances may also be returned through less formal means, such as being sent with friends or relatives on a trip 'home'. The transnational connections, both virtual and real, help facilitate the flows of remittances.

Remittances can clearly be used by family members for personal consumption, such as home improvements and property, health or education expenditure. Such expenditure is often characterised as being 'economically non-productive', particularly money is spent on imported goods. However, in some cases, money received from abroad may be invested in small businesses, so contributing to local level economic development. Improvements in housing and expenditure on health and education may also have wider development benefits. While the remittance expenditure benefits the recipients, it may lead to increasing inequalities within communities (De Haas, 2006).

As the level of international migration and flows of money have increased, remittances have attracted greater attention from governments and development policymakers. While it is impossible to provide an exact figure for the level of remittances due to informal and illegal transfers, the World Bank estimates that global remittances reached US$766.85 billion in 2022, with US$588 billion of this going to low and middle-income countries (World Bank, 2024). To put this into perspective, in 2022 overseas development assistance (ODA) from the countries of the Development Assistance Committee (DAC) of the Organization for Economic Cooperation and Development (OECD) was US$213 billion (OECD, 2024). This means that remittances far outweighed the amount of bilateral aid from the world's richest countries. India and Mexico are the recipients of the largest amount of remittances, but as a proportion of gross domestic product (GDP), smaller countries are reliant on remittances (see Table 68.1). Civil unrest and economic downturns can have significant effects on migrant job opportunities, and therefore remittances.

Given the size of these money flows, it is not surprising that governments in certain key labour exporting countries have sought to encourage migrants to remit. This may be in the form of individual remitting or through community organisations investing in projects. International migrants from the same village or urban district often end up migrating to the same location overseas. Hometown associations (HTAs) or equivalent may be set up to provide support for migrants overseas, but they may also raise money for projects 'at home'. These can be crucial for local infrastructure developments such as schools or health centres, but it is important not to over-romanticise these organisations and the capacity for exploitation or enhanced inequality which may be involved (Page and Mercer, 2012).

Flows 'back home' through transnational networks may also be in the form of intangible 'social remittances' (Levitt, 2001) incorporating ideas and practices regarding politics, environment and gender relations, for example. These social remittances are transferred

Table 68.1 Top ten countries by total remittance flows and per cent of GDP, 2021

		US$ millions			% of GDP
1	India	89,375	1	Tonga	46.2
2	Mexico	54,130	2	Tajikistan	32.7
3	The Philippines	36,685	3	Kyrgyz Republic	30.2
4	France	32,832	4	Samoa	29.4
5	Egypt	31,487	5	Lebanon	27.5
6	Pakistan	31,312	6	The Gambia	26.8
7	China	22,479	7	El Salvador	25.4
8	Germany	20,783	8=	Jamaica	25.3
9	Nigeria	18,956	8=	Honduras	25.3
10	Ukraine	18,060	9=	Bermuda	23.2

Source: (World Bank, 2024)

not only through individuals but also collectively through HTAs, as Levitt and Lamba-Nieves (2011) demonstrate through their work with Dominican migrants in the US.

Return migration

International migration can contribute to development processes of the 'home' country through more than just remittances. While the brain drain involves a loss of talent and human resources, skills and experience gained during migration can provide very positive inputs if migrants return. The influx of social, cultural and human capital, as well as financial capital, can be an important trigger to economic development. This process has been represented as 'brain gain'.

Of course, not all return migrants contribute in this way. For some, return may represent the end of their working lives, while in other cases, the reality of coming 'home' fails to live up to their expectations and further mobility follows as they cannot settle. Return migration may also be associated with the transference of potentially destructive practices, such as the development of US-linked criminal gangs in Mexico and Central America. However, there are many cases of positive outcomes for both individuals and communities as return migrants invest in new businesses, provide an impetus in local politics or add to the pool of skilled labour (Conway and Potter, 2006). In some cases, governments have been so keen to encourage certain kinds of return migration that they have offered tax incentives.

As with many forms of migration, it is impossible to identify return migration definitively, as a migrant may move away again in the future. In certain parts of the world, transnational connections have developed around businesses, thus creating migration circuits. These businesses often started by providing goods, such as food, for immigrant communities who were missing their home comforts (Bagwell, 2014).

The state and transnational migration

Increasing international mobility and the intensification of transnational exchanges does not mean the end of the nation-state; in fact without the continued presence of the nation-state the concept of 'transnationalism' would be redundant. However, processes of transnationalism and globalisation have required national governments to change their development strategies and relationships between states and citizens. As outlined earlier, international migration provides opportunities for development capital and expertise. This has meant governments paying increased attention to citizens living elsewhere and encouraging continued links. This is a break from past practices where governments focused almost completely on citizens living within the national boundaries.

Given the financial possibilities provided by remittances, governments are increasingly facilitating temporary labour outmigration through formal government-approved schemes. Such activities are not new, for example, thousands of Mexicans migrated to the US as part of the Bracero Program 1942–1964, but the current scale of government involvement in such schemes is unprecedented in countries of the Global South. The government of the Philippines, for example, regulates the activities and fees of agencies recruiting workers through the Philippine Overseas Employment Administration (POEA). There is also a strong political rhetoric of the overseas Philippine workers being heroes and heroines of the nation, so encouraging these migrants to feel a continued attachment to the country.

Such attachment is also encouraged through the increased relaxation of dual nationality regulations. While some countries, such as India and China, do not allow dual citizenship, many others are allowing their nationals to take on passports of another country. There is also increased flexibility in relation to voting in elections and standing for political office. All of these legal changes reflect national government attempts to seize the opportunities presented by transnational migration (McKenzie and Yang, 2015).

Conclusions

Migration is often viewed as a response to development failure, as individuals seek to find a better life for themselves and their families elsewhere. Increasing international migration flows and greater possibilities for transnational economic, social and political practices have meant that emigration does not mean the severing of ties with home that it often did in the past. Rather, international migration is increasingly being seen as a positive contributor to development in both the 'sending' and 'receiving' countries. This perspective has meant national governments, international development organisations and donor agencies have begun to develop policies to maximise the benefits accruing from international migration. Of course, international mobility can involve great hardships for the people involved and those left behind, and it can exacerbate existing inequalities at a local, national and global scale, but there is also considerable potential for migration to contribute to improvements in the standard of living and quality of life.

References

Bagwell, S. (2014) 'Transnational entrepreneurship amongst Vietnamese businesses in London', *Journal of Ethnic and Migration Studies*, 41, 329–349.

Basch, L.G., Glick Schiller, N. and Blanc-Szanton, C. (1994) *Nations Unbound: Transnational Projects, Post-Colonial Predicaments, and Deterritorialized Nation-States*, Longhorne, PA: Gordon and Breach.

Connell, J. (2010) *Migration and the Globalisation of Health Care*, Cheltenham: Edward Elgar.

Conway, D. and Potter, R.B. (2006) 'Caribbean transnational return migrants as agents of change', *Geography Compass*, 1, 25–50.

De Haas, H. (2006) 'Migration, remittances and regional development in Southern Morocco', *Geoforum*, 37, 565–580.

De Haas, H., Castles, S. and Miller, M. (2019) *The Age of Migration: International Population Movements in the Modern World*, Sixth Edition, London: Red Globe Press.

Levitt, P. (2001) *The Transnational Villagers*, Berkeley: University of California Press.

Levitt, P. and Lamba-Nieves, D. (2011) 'Social remittances revisited', *Journal of Ethnic and Migration Studies*, 37 (1), 1–22.

McAuliffe, M. and Triandafyllidou, A. (eds) (2021) *World Migration Report, 2022*, Geneva: International Organization for Migration.

McKenzie, D. and Yang, D. (2015) 'Evidence on policies to increase the development impacts of international migration', *The World Bank Research Observer*, 30 (2), 155–192.

Organization for Economic Cooperation and Development (OECD) (2024) 'OECD international development statistics'. www.oecd.org, accessed 8 February 2024

Page, B. and Mercer, C. (2012) 'Why do people do stuff? Reconceptualising remittance behaviour in diaspora-development research and policy', *Progress in Development Studies*, 12 (1), 1–18.

World Bank. (2024) 'World Bank open data', https://data.worldbank.org/, accessed 8 February 2024

Additional resources

De Haan, H. (2010) 'Migration and development: A theoretical perspective', *International Migration Review*, 44 (1), 1–38.

www.migrationinformation.org Migration Information Source run by the Migration Policy Institute, a Washington DC based think tank.

http://migratingoutofpoverty.dfid.gov.uk Migrating Out of Poverty research programme consortium funded by the UK Department for International Development (DFID).

United Nations Development Programme (UNDP) (2009) *Human Development Report 2009: Overcoming Barriers: Human Mobility and Development*, Basingstoke: Palgrave Macmillan.

Chapter 69

The measurement of poverty

Francesco Burchi and Howard White

Introduction

Poverty eradication has been at the heart of development cooperation and government interventions for decades. In order to alleviate poverty, it is necessary to understand what poverty is and how poverty can be measured. This chapter is concerned with these issues. The next section outlines key concepts which underpin the various poverty measurements discussed in the subsequent section. Finally, the relationship between income and multidimensional poverty is empirically investigated.

The concept(s) of poverty

In everyday usage the term 'poverty' is synonymous with a shortage of income. However, a considerable number of scholars and most international organisations now recognise the multi-dimensionality of poverty. This recognition is captured in the formulation of the Target 1.2 of the 2030 Agenda to 'reduce . . . poverty in all its dimensions'. In addition to material consumption, dimensions, such as health, education, decent employment, participation in social life, environmental quality, and political freedom all matter. Deprivation with respect to any one of these can be called poverty.

Supporters of the income-based approach to poverty argue that income is what really matters. Other arguments supporting this view are i) income and other measures of well-being such as health and education are highly correlated; ii) that it is easier for governments to influence income than, say, spiritual well-being; and, iii) the fungibility of income, which can be used to purchase different commodities depending on household needs, thus making government interventions less paternalistic.

But there are strong arguments in defence of multi-dimensionality. First, people usually value income instrumentally, as means to achieve better nutrition, health, or education, rather than as an end in itself. Poor people themselves often rank other dimensions as being more important than income. For example, based on a participatory assessment, Jodha (1988) showed with Indian data that the welfare of the poor had risen by measures they considered important – such as wearing shoes and separate accommodation for people and livestock – whereas surveys showed their income to have fallen. Therefore, as most notably argued by Sen (1985), we should focus directly on the dimensions people value. Second, the correlation between income poverty and deprivations in other dimensions is not as strong as claimed, and there are still few studies that adequately

DOI: 10.4324/9780429282348-75

compare income poverty with a composite measure of multidimensional poverty across many countries.

The definition of poverty has important policy implications. When poverty is defined solely in terms of income, then it is unsurprising that economic growth is found to be the most effective way to reduce it. But if health or education are valued then the development strategy is likely to put more emphasis on social policy. Ranis et al. (2002) showed that countries attempting growth-led strategies fail to achieve either human development or sustained growth, whereas countries investing in human capital move to a high growth path. It is, therefore, important to broaden the focus of poverty-reducing efforts to dimensions beyond monetary ones.

Two further conceptual issues are worth mentioning: the distinction between absolute and relative poverty and that between temporary and permanent – or chronic – poverty. A person is usually defined *absolutely* poor when he or she cannot have access to the basic needs; a person, instead, is *relatively* poor if his or her living standards are below societal standards, such as those of the median citizen of that society. As relative poverty has to date mostly been used in high-income countries, the following sections will largely refer to absolute poverty. However, relative poverty is of growing importance in low-income countries, being embodied in the SDGs, and a new set of global poverty estimates (Ravallion and Chen, 2019).

The measurement of income poverty

A low level of GNP per capita is sometimes used to label a country as poor. However, the GNP is not a good measure of poverty primarily because it does not take into account the distribution of income in the population.

An adequate measurement of income-poverty relies on household surveys. The first step consists in setting a poverty line, i.e. a level of income necessary for the satisfaction of basic needs. Then, a poverty measure has to be identified. The most common one is the headcount ratio, i.e. the percentage of the population falling below the poverty line (see, for example, Ravallion, 1992). However, this measure takes no account of how far people are from the poverty line, so that a rise in the income of the poor which leaves them below the poverty line appears to have no effect. Hence another measure, the poverty gap, is often used, which is the product of the headcount and the average distance of the poor from the poverty line (expressed as a percentage of the poverty line). Finally, if these distances are squared – and therefore greater weight is given to the poorest individuals – we obtain the squared poverty gap, which accounts also for inequality among the poor. These three measures – the headcount, the poverty gap and the squared poverty gap – are known collectively as the Foster et al. (1984) poverty measures and labelled P0, P1, and P2 respectively.

The income-poverty measures have several limitations. The first one concerns all these indices: income data are collected at the household level. It is therefore impossible to get figures on the number of women or children living in poverty (despite the tendency of some international organisations to report such figures), but only on the percentage of women and children living in households whose income is below the poverty line. Hypotheses about intra-household allocation of resources need to be made. Assuming equal distribution among all household members, as done by the World Bank for the international figures on poverty, is problematic as the consumption requirements of different individuals varies: children consume less than adults and, more controversially, women may need to consume less than men. This problem is catered for by the use of an adult equivalents scale which expresses the consumption needs of women and children as a fraction of those of an adult male. A related problem concerns the existence of economies of scale in household consumption – that is, two can live together more cheaply than they could apart as there are shared expenses (living space, utilities and many household items). Failure to take account of these economies will overstate poverty in large households (White and Masset, 2002).

A number of problems, then, emerges from setting the poverty lines. In applying the poverty line, consumption (expenditure) is commonly used rather than income. First, survey respondents usually have a far clearer idea of their expenditure than their income. Second, when income is uneven households will smooth consumption (i.e. even it out over time) so that at any point in time current consumption is likely to be a more accurate measure of welfare than current income. However, surveys may miss important forms of consumption for the poor such as wild foods and festivals.

Probably even more relevant is the fact that prices vary across time and space. Allowance must be made for these price differences in order for the poverty line to be comparable. There are even greater difficulties in comparing countries, partly since market exchange rates do not reflect differences in purchasing power. To alleviate this problem, purchasing power parity (PPP) exchange rates are usually adopted. However, the PPPs are not uniformly available and do not focus adequately on the goods and services consumed by the poor (Reddy and Pogge, 2010).

This brief discussion shows that the measurement of income-poverty is a difficult and problematic exercise: it is therefore necessary to consider non-monetary indicators of poverty, which are discussed in the next section.

The measurement of multidimensional poverty

For some decades several social indicators have been employed to capture living conditions in different countries. Indicators, such as child mortality, illiteracy, and unemployment, have been used as measures of poverty in the health, education and employment dimensions, respectively. This section focuses in particular on the *composite* indicators that have been proposed since the early 1990s. These measures can be divided into two categories: i) those that aggregate first across individuals/households and then across dimensions, ii) those that aggregate first across dimensions and then across individuals/households.

Measures that aggregate first across individuals and then across poverty dimensions

The most notable indicator in this category is UNDP's Human Development Index (HDI), a composite indicator of life expectancy, educational attainment and GDP per capita. While one could argue that a country is multidimensionally poor if it has a low score in the HDI, the HDI is not a good measure of poverty as it does not account for the distribution of these dimensions within the society. A better example is another indicator proposed by UNDP, the Human Poverty Index (HPI) as it focuses on deprivations. Specifically, the HPI is calculated as the average of the percentage of the population not expected to live to 40, the percentage who are illiterate and what is called the 'deprivation in living standards' (the average of those without access to water and healthcare, and the percentage of under-fives who are underweight).

The pros of these indicators are that they could both provide a single number as well the possibility to focus on each dimension separately, and they can rely on data from different sources. Critiques, instead, concern the real value added of combining these dimensions into an overall index – instead of using a 'dashboard approach' – and the impossibility of identifying who is multidimensionally poor as national-level data are used.

Measures that aggregate first across poverty dimensions and then across individuals

To address some of the above-highlighted limits, new measures of multidimensional poverty that rely on one single household survey have been recently proposed. From one survey it is possible to gather information on achievements of a household/individual in different dimensions of poverty, and therefore the resulting measures can focus on the interaction of deprivations in different dimensions, something not possible in the HPI. In order to measure poverty this way it is necessary to choose: i) the dimensions, ii) the indicators for each dimension; iii) the dimensional poverty line, to define who is poor in each dimension, iv) the identification criterion, to define who is multidimensionally poor; v) the weights (for each indicator); vi) the poverty measure. Most of the debate has concentrated on the last point, i.e. the identification of an aggregation function to combine deprivations in different dimensions into one single poverty index (or a family of indices).

The best known measure is the Multidimensional Poverty Index (MPI), which combines three dimensions: education, health, and standard of living (Alkire and Santos, 2015). The first two are measured each by two different indicators, while the last one by six indicators. Equal weights are assigned to the three dimensions and to the indicators within each dimension. The aggregation is based on the Alkire-Foster method (Alkire and Foster, 2011). This implies including a second poverty line: in the case of the MPI, each household is classified as multidimensionally poor if the number of weighted deprivations is equal or higher than 0.333. Like for income, the multi-dimensional headcount ratio is the share of the population that is multidimensionally poor. In a second step, it is necessary to compute the average deprivation share among the poor obtained by dividing the total number of (weighted) deprivations of the multi-dimensionally poor by the maximum number of deprivations that the poor could experience. The final MPI uses the adjusted headcount ratio, calculated as the product of the headcount ratio by the average deprivation share among the poor.

Some advantages of this type of measures concern the limited problem of cross-country comparability – as, for example, there is no need to use PPPs – and the easier collection of data on educational achievements, health, nutrition etc. as compared to data on income/consumption. Moreover, a measure like the MPI is decomposable by dimension – to detect the relative contribution of each dimension to poverty – and by group.

On the other hand, given the requirement of relying on one single survey, the choice of dimensions/indicators is seriously constrained by data availability. Other critiques are specific to the MPI. The first one is that, as the income poverty measures, the MPI is computed at household level, and therefore does not allow to detect intra-household differences. The other critiques concern the Alkire-Foster method, namely the arbitrariity of the second cut-off (0.333 is not theoretically grounded) and the impossibility to account for inequality among the poor. Other multidimensional poverty measures overcome this last set of limitations. For instance, the Global Correlation Sensitive Poverty Index (G-CSPI), which encompasses education, decent work, and health, is an individual measure (for 15–65 years old people) and accounts for inequality among the poor (Burchi et al., 2021; Burchi et al., 2022).

Comparing income poverty to multidimensional poverty in the world

To what extent is income poverty correlated with multidimensional poverty? Figure 1a plots the relationship between the MPI and the equivalent index of income poverty, the poverty gap – based on the US$2.15/day PPP international line for extreme poverty – across 74 low- and middle-income countries. The figure uses the most recent available data point for each country, allowing for a maximum of three years of discrepancy in the surveys used for the two indices. The graph shows that there is a positive but not very strong correlation (Person's coefficient = 0.53), and highlights the presence of many outliers.

This comparison, however, is problematic as income and multidimensional poverty are calculated on two different types of surveys, in most of the cases conducted in different years. The database constructed by Burchi et al. (2021) allows analysing the previous relationship using data from exactly the same year for both income and multidimensional poverty (through the G-CSPI) and in the majority of the cases also from the same survey. At the same time it is important to highlight that we are comparing a household-level index of income poverty with an index of multidimensional poverty focusing only on individuals between 15 and 65 of age. Based on their data (period 2005–2018) and updated statistics on extreme income poverty, Figure 1b plots the squared income poverty gap against the G-CSPI – both distribution-sensitive indices -across 91 countries. The correlation is higher than the one found earlier but still very far from the unity (0.67), and the relationship appears non-linear. There are several outliers: in particular, Zambia, Madagascar, and the Democratic Republic of Congo in relative terms have a much larger income poverty, while the opposite occurs in countries, such as The Gambia, Thailand, Guinea, and Burkina Faso.

This analysis shows that despite the fact that the two types of measures are closely related, income deprivations do not capture adequately deprivations in other dimensions. If poverty is deemed to be a multidimensional phenomenon, it is better to examine directly deprivations in such dimensions – either through a composite index or a dashboard of dimensional indicators.

Data sources

The main data source for income poverty is the World Bank. Data on different measures of income poverty are available from: https://pip.worldbank.org/poverty-calculator (accessed 15/9/2023).

Data on the MPI, instead, are provided by the UNDP, Human Development Report Office. Data are available from: https://hdr.undp.org/content/2023-global-multidimensional-poverty-index-mpi#/indicies/MPI (accessed 14/09/2023).

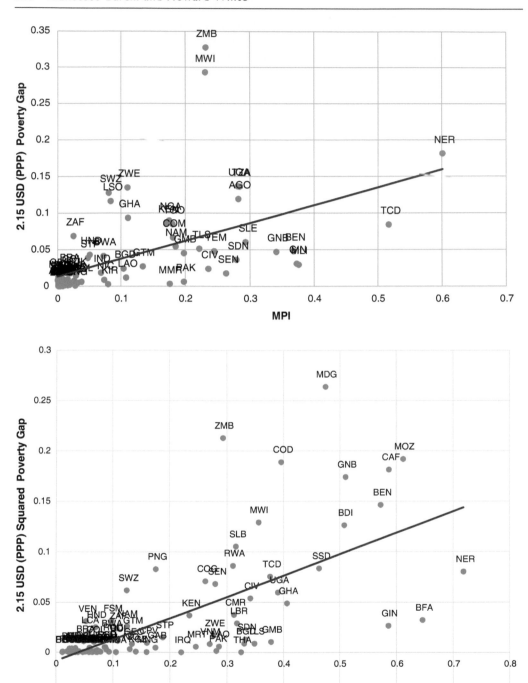

Figure 69.1a and b Scatter plots of the relationship between income and multidimensional poverty

References

Alkire, S., and Foster, J. E. (2011). Counting and multidimensional poverty measurement. *Journal of Public Economics* 95: 476–487.

Alkire, S., Roche, J. M., Ballon, P., Foster, J., Santos, M. E. and Seth, S. (2015). *Multidimensional poverty measurement and analysis*. New York. Oxford University Press, USA.

Burchi, F., Espinoza-Delgado, J., Rippin, N., and Montenegro, C. E. (2021). An individual-based index of multidimensional poverty for low- and middle income countries. *Journal of Human Development and Capabilities* 22(4): 682–705.

Burchi, F., Malerba, D., Montenegro, C. E., and Rippin, N. (2022). Assessing trends in multidimensional poverty during the MDGs. *Review of Income and Wealth* 68(2): S317–S346.

Foster, J., Greer, J., and Thorbecke, E. (1984). A class of decomposable poverty measures. *Econometrica* 52(3): S761–S766.

Jodha, N. S. (1988). Poverty debate in India: A minority view. *Economic and Political Weekly* 22(45–47): 2421–2428.

Ranis, G. and Stewart, F. (2000). Economic growth and human development. *World Development* 28(2): pp. 197–219.

Ravallion, M. (1992). *Poverty Comparisons: A Guide to Concepts and Methods, LSLS Working Paper 88*. Washington, DC: World Bank.

Ravallion, M., and Chen, S. (2019). Global poverty measurement when relative income matters. *Journal of Public Economics*, 177. https://doi.org/10.1016/j.jpubeco.2019.07.005.

Reddy, S. G., and Pogge, T. (2010). How not to count the poor. In J. Stiglitz, S. Anand, and P. Segal (Eds.), *Debates in the Measurement of Poverty*. Oxford: Oxford University Press.

Sen, A. (1985). *Capabilities and Commodities*. Amsterdam and New York: North-Holland.

White, H., and Masset, E. (2002) *Child Poverty in Vietnam: Using Adult Equivalence Scales to Estimate Income-Poverty for Different Age Groups, Working Paper 6*. Oxford: Young Lives.

Chapter 70

Behavioural economics and development economics

Bereket Kebede

Introduction

Neo-classical economics dominated microeconomics from around the end of the 19th century and it was formalised and standardised particularly after World War II following important theoretical contributions. Even though the neo-classical framework expanded into many non-traditional areas, the core element of its theoretical framework is based on the assumption of *homo economicus* where economic agents are characterised by rationality and stable and self-centred preferences. Rationality implies that economic agents are highly logical with great cognitive/computational capacity to solve complicated optimisation problems; the role of psychological factors such as emotions, inattention and impatience is ignored. Economic agents are also assumed to have stable preferences (utility functions) over time. In addition, the preferences of economic agents are self-centred, maximising their own interest, hence altruism, moral responsibility, spite and similar preferences are kept in the background.

Work challenging or extending the basic framework of neoclassical microeconomics was always a fruitful avenue of research. From these endeavours, the approach that integrated insights from psychology into microeconomics – behavioural economics as we know it now – emerged as the most influential research programme by the end of the 20th century. The prospect theory of Kahneman and Tversky (1979) is one of the influential theories that brought about this sea change.

Behavioural economics challenged key assumptions of neoclassical microeconomics by taking a psychological perspective. First, it emphasised limitations in the cognitive/computational capacities of economic agents relaxing the assumption of rationality. While broadly rational, economic agents have limited cognitive capacity constraining their ability to solve complicated problems (*bounded rationality*); individuals also differ in their cognitive capacities.

Second, behavioural economics questions the assumption of stable preferences. Stable utility functions imply that people have a clear idea of the satisfaction they get from consumption of different goods and this is stable over time. Behavioural economics argued for *reference dependent* rather than stable preferences; people usually compare the consequences of their decision to a reference, usually the status quo, not to the satisfaction over the whole range of consumption as stable preferences imply. Related to this, behavioural economics introduced the important concept of *loss aversion*; a loss significantly upsets individuals more than an equal amount of gain make them happy (e.g., a salary reduction

DOI: 10.4324/9780429282348-76

of £100 upsets an individual much more than an equal salary increase will make him/her happy).

Third, while recognising the importance of self-centred preferences, behavioural economics explicitly incorporated *other-regarding (social) preferences*. Individuals care about others without any private benefit in return (altruism, empathy). People relate to each other without a clear one-to-one exchange (reciprocity). Individuals may not like the success of others (envy) or, even worse, they may enjoy hurting others (spite). Deriving all these from self-centred preferences is rather difficult.

Behavioural economics rapidly expanded during the last few decades of the 20th century and it growth continues into the current period. One can convincingly argue that behavioural economics is no more a peripheral school of thought but part of the mainstream. The critique of behavioural economics has broadened the scope of neo-classical economics rather than being a fundamental rejection.

Behavioural economics heavily relies on the methodology of *experimental games*. In experimental games, people are asked to make incentivised decisions under a laboratory condition. First, the decisions in experimental games are incentivised meaning the participants earn money depending on the decisions they make. Incentives are provide as a means of revealing true preferences because decisions bear actual costs/benefits. For example, the responses of participants in a survey may not reveal their true preferences because of many reasons (e.g., they like to put themselves in good light, etc.). Second, the games are made in a lab. These labs may have networked computer terminals through which participants make interdependent decisions; while most labs are in universities of richer countries, some are in developing countries (e.g., the Busara Center for Behavioral Economics in Nairobi). In most development research, either portable electronic devices or pencil/paper in more natural settings (*lab-in-the-field*) are used. The games are played in a setting devoid of social contexts and mostly anonymously to avoid the confounding effects of social relationships unobservable to the researchers. Suppose we want to test how altruistic individuals are; dictator games are usually used for the purpose. In a dictator game the first person ('the dictator') is given money and then asked to give any amount of the money to a second anonymous person (the recipient) taking the remaining money for him/herself. The amount of money given to the recipient is a good measure of the 'dictator's' altruism. In real life, interpreting the action of an individual as altruistic after observing him/her giving money to another is problematic because it may be explained by other factors the researcher does not observe; e.g., if the recipient has power over the 'dictator', the money may be given due to fear rather than altruism. The wide use of experimental games in behavioural economics does not mean they are not used in other contexts (e.g., to test a neo-classical theory).

Behavioural development economics

The deviations from the standard assumptions of neoclassical economics highlighted by behavioural economics are largely universal; this is demonstrated by experiments conducted in rich and developing countries. The different circumstances and institutional set-ups likely make these behavioural deviations more significant in developing countries (Kremer et al., 2019). The consequences of decision biases are more severe in developing countries due to low standards of living. For instance, *present-biased preferences* (assigning more weight to payoffs that are in the nearer future when considering trade-offs between two future times) are one of the likely reasons for the widely observed low demand for actuarially fair weather insurance and low level of investment on preventive medicine (Kremer et al., 2019). The institutional set-up in developing countries is another reason. For example, the proliferation of small firms in developing countries likely increases the significance of behavioural factors as more formal modern management system is lacking (even though the latter is not completely immune from behavioural biases). Behavioural economics affects how development problems are diagnosed, how solutions are designed and how the scope of the problem is determined (Datta and Mullainathan, 2014).

Behavioural development economics on the one hand looks at how behavioural factors affect welfare outcomes; for example, asking whether high risk aversion and high impatience are reasons for poor people to stay poor is part of this research agenda (Cardenas and Carpenter, 2008). On the other hand, whether conditions of the poor may further exacerbate behavioural limitations is another research area ('psychology of poverty'). In this strand of literature, the cognitive capacity of the poor is compromised by their worry about scarcity, financial or otherwise (Kremer et al., 2019).

Experimental games have emerged as the main empirical workhorse of behavioural economics. Development researchers are now increasingly using experimental games mainly in the form of lab-in-the-field experiments. The use of experimental games has contributed towards addressing some questions that were not tackled in standard development economics. The last few paragraphs will provide examples of three widely used games: risk and time preferences and public goods games. Each of these games have many variants hence the following description focuses only on their basic features. In risk preference games, players are asked to buy a lottery from the money given to them by the researchers; the higher the amount they use for the lottery, the lower their risk-aversion. In time preference games, players are given the chance to get a smaller amount of money now or a larger amount later – the more they are willing to wait for the higher amount in the future, the lower their impatience. In public goods games, individuals are asked to contribute to a common pool of money that will be increased by the researchers and then equally distributed among the individual players; purely selfish individuals will not contribute any (free-riding). Public goods games capture the social dilemma between private and collective interests.

As discussed before, one aspect of reference dependent preferences is loss aversion. A study in the rural areas of Vietnam (Tanaka et al., 2010) shows that even though people are not necessarily afraid of uncertainty (income variations) they are loss-averse. The risk behaviour of people may be affected by social preferences. i.e., by comparing themselves with others, as predicted by behavioural economics. A risk preference game played with farmers in Ethiopia indicate that players seek to 'keep up with winners', they risk more when other participants have higher winnings affirming the importance of reference dependent and social preferences (Fafchamps et al., 2015).

Related research has shown the existence of present bias and its implication on development initiatives. Tanaka et al. (2010) find that people in rural Vietnam are present-biased irrespective of their income levels and locations of villages. There is a growing literature that looks at the implication of present-biased preferences on development initiatives. For example, Bauer et al. (2012) find that higher levels of present-biased preferences are robustly correlated to higher levels of micro-credit use by women in India.

Neoclassical economics predicts that no one will contribute in a public goods game. Similar to the results in Western university laboratories, public goods games played in developing countries also show that only a minority free-ride and there are significant differences in levels of cooperation in different societies (Cardenas and Carpenter, 2008). Recent work by Clist and Verschoor (2017) finds that the language in which a public goods game is played affects contributions because each language is likely related to latent social norms. The public goods game is also adapted to examine intra-household efficiency both in monogamous (Kebede et al., 2013) and polygynous (Barr et al., 2019) households; contrary to the assumption of many intra-household models, both types of marriages fail to attain efficiency.

Conclusion

This chapter highlights the main contributions of behavioural economics and how it has increasingly influenced development economics. With the current expansion of the literature, behavioural development economics will likely stay and play a more important role in the policy debate in developing countries as it has done in rich countries.

References

Barr, A., Dekker, M., Janssens, W., Kebede, B. & Kramer, B. 2019. Cooperation in Polygynous Households. *American Economic Journal: Applied Economics*, 11, 266–283.

Bauer, M., Chytilová, J. & Morduch, J. 2012. Behavioral Foundations of Microcredit: Experimental and Survey Evidence from Rural India. *American Economic Review*, 102, 1118–1139.

Cardenas, J. C. & Carpenter, J. 2008. Behavioural Development Economics: Lesson from Field Labs in the Developing World. *Journal of Development Studies*, 44, 311–338.

Clist, P. & Verschoor, A. 2017. Multilingualism and Public Goods Provision: An Experiment in Two Languages in Uganda. *Journal of Development Economics*, 129, 47–57.

Datta, S. & Mullainathan, S. 2014. Behavioral Design: A New Approach to Development Policy. *Review of Income and Wealth*, 60, 7–35.

Fafchamps, M., Kebede, B. & Zizzo, D. J. 2015. Keep up with the Winners: Experimental Evidence on Risk Taking, Asset Integration, and Peer Effects. *European Economic Review*, 79, 59–79.

Kahneman, D. & Tversky, A. 1979. Prospect Theory: An Analysis of Decision under Risk. *Econometrica*, 47, 263–292.

Kebede, B., Tarazona, M., Munro, A. & Verschoor, A. 2013. Intra-Household Efficiency: An Experimental Study from Ethiopia. *Journal of African Economies*, 23, 105–150.

Kremer, M., Rao, G. & Schilbach, F. 2019. Behavioral Development Economics. In: Bernheim, B. D., Dellavigna, S. & Laibson, D. (eds.) *Handbook of Behavioral Economics – Foundations and Applications 2: Volume 2 (Handbooks in Economics)*. Amsterdam, Netherlands: North Holland.

Tanaka, T., Camerer, C. F. & Nguyen, Q. 2010. Risk and Time Preferences: Linking Experimental and Household Survey Data from Vietnam. *American Economic Review*, 100, 557–571.

Further readings

Dhami, S. 2016. *The Foundations of Behavioral Economic Analysis*. Oxford, UK: Oxford University Press.

Kebede, B. & Zizzo, D. J. 2015. Social Preferences and Agricultural Innovation: An Experimental Case Study from Ethiopia. *World Development*, 67, 267–280.

Mullainathan, S. & Shafir, E. 2013. *Scarcity: The Cost of Not Having Enough*. London: Penguin Books.

Thaler, R. H. & Sunstein, C. R. 2009. *Nudge: Improving Decisions about Health, Wealth and Happiness*. London: Penguin.

Wilkinson, N. & Klaes, M. 2012. *An Introduction to Behavioral Economics*. Basingstoke, Hampshire and New York: Palgrave Macmillan.

Chapter 71

Financialisation and development

Ben Fine

From advanced country buzzword. . .

As indicated by its addition to this Companion, financialisation is a concept that is new to scholarship, in general let alone for development. It has experienced a meteoric rise across the social sciences over the last two decades preceding in origins, if accelerating in use after, the Global Financial Crisis, GFC, of 2007/8. It signifies a belated acknowledgement of the increasing weight of finance in the world economy over the past four decades, with the ratio of global financial assets to GDP having risen threefold. Do we really need so much more finance to smooth the wheels of commerce? And it can hardly serve as such as an invisible and neutral veil as posited in neoclassical economics where only the supposedly real economy of preferences, technologies, and resources have any long-term binding effects.

The most commonly referenced and otherwise implicit definition of financialisation is provided by Epstein (2005: 3) as 'the increasing role of financial motives, markets, actors and institutions in the operation of the domestic and international economies'. This has proven a more or less open invitation to perceive financialisation as any context in which money, markets, and/or monetary calculation and ethos, have some sort of presence. And the invitation has been taken up extensively and rapidly across many different topics and disciplines to such an extent that financialisation has become a buzzword, used in so many different ways that it inevitably loses precision.

Like other, if by no means all, buzzwords that have made their way to development, this has been the consequence of extrapolation from prior application to developed country contexts. Indeed, buzzwords tend to go through a number of stages, the first of which is application to existing discourses by introducing the new term (and possibly only nominally), followed by complaints of absence from other applications with demand for – and supply of – remedial action in the offing. In case of financialisation, the initial research was understandably driven heavily by the experience of advanced, especially the US and UK, economies.

This has allowed for a host of studies that might be dubbed finnX, in which the relationship between financialisation and some other variable X has been examined. Inevitably, X came to incorporate other countries and topics, ultimately developing countries and development. As a result, there is even emerging a cottage industry of literature surveys, special issues and handbooks, for finanancialisation as a whole, as well as for particular aspects including development (Bonizzi 2016 and Mertens et al. (eds) 2020). There have also been attempts to categorise contributions across common themes such as

DOI: 10.4324/9780429282348-77

rise of stakeholder capitalism at one extreme and impact on everyday life at the other. But such categorisations fail to accommodate the ever-expanding range of finnX and even induce a minority sport, with some justification, of denouncing and rejecting finnbuzz for being too amorphous and incoherent and not necessarily historically unprecedented (Christophers and Fine 2020).

. . . to drag on development

Where finnbuzz differs from most other buzzwords, especially within development, is that it does not derive from, neither has it been adopted by, the World Bank. Indeed, across the social sciences, what stands out is that the one discipline in which financialisation has failed to appear at all, at least at the time of writing, is in mainstream economics. Here, there is a sharp contrast with heterodox economics or political economy that has taken a leading role. In short, the financialisation literature is generally critical of finance and its role in capitalism so that finnX studies are generally dedicated to showing a negative relationship between finn and X where X is deemed as something good (for development) whether it be investment, employment, equality, growth, provision of social and economic infrastructure, technical change, poverty alleviation, economic and social volatility, and so on.

The reason why finnbuzz is absent from mainstream economics is because the latter's methodology is based upon the optimising, possibly more rounded, behavioural individual and so failing to consider the systemic role of finance. By the same token, traditional variables in use across the social sciences, and critical development studies – such as power and conflict – and associated relational structures and dynamics, are commonplace in study of (critical) development studies, financialisation, and the marriage of the two. Finnculture, of everyday life and beyond, is also of significance, but beyond the radar of mainstream accounts other than to educate the individual for the more rational use of credit (Bayliss et al. (eds) 2018).

However, mainstream accounts of finance have not stood still, if not moving to introduce the notion of financialisation. The standard thrust of scholarship, ideology, and policymaking has been that financial development and freedom of financial markets are good for (national) development (as opposed to what is otherwise deemed to be financial repression). This has, as is to be expected, taken a bit of a hit in the wake of the GFC and, unsurprisingly, marginal concessions have been made along the lines that there might be a case for capital controls and for limits on the ratio of finance to GDP in order to moderate excessive speculation.

For the latter so-called threshold analysis, the relationship between development (and especially growth) and finance is perceived to be an upside-down U in shape. But empirical analyses, for and against, have proceeded by disaggregating the different sorts and measures of finance involved, as well as the different putatively causal consequences, not least through the introduction of ever more arbitrarily chosen (so large is the set of options) conditioning variables. This sort of pursuit of finnX studies is similar to many others in the field of development where extremely complex relations are attempted to be reduced to a single one-way dependency (as with poverty on growth, trade on openness, and so on; Itaman 2017).

Unsurprisingly, the longstanding debate over the relationship between finance and development has been rejuvenated by the GFC and the disproportionate growth of finance more generally. But the critics of the (new) mainstream approaches are increasingly couching their contributions in terms of financialisation – whether it be around influence on macroeconomic volatility and policy, levels of investment, capital flight, distributional pressures, and so on. Although there is a stream in the financialisation literature on financial exploitation/appropriation (and the implications of household indebtedness for everyday lives; Lapavitsas 2013 and Fine 2014 for a critique and alternative) on its own terms, the finndevelopment literature is more advanced in light of critical appraisal of the negative effects of efforts at financial inclusion that have led to high indebtedness for consumption for the poor with handsome rewards for the managers of finnclusion, itself reflecting a step along the way from micro-finance, Bateman et al. (eds) (2018).

From methodological nationalism . . .

Otherwise, the finndevelopment literature, as indicated, lies in the shadow of its developed country origins albeit with one major exception where it is even able to offer a lead against the normal grain. Most of the financialisation literature is bound by, even committed to, methodological nationalism. What is the nature of this or that country's financial system, how has this led it to be financialised, and with what effects? Other than as one amongst other sources and applications of finance, this takes limited account of the global and systemic nature of financialisation. This is also characteristic of much of the finndevelopment literature but it also is much more aware of – and sensitive to – the power of global finance.

The simplest form that this takes is to situate critique of structural adjustment and stabilisation programmes – and corresponding IFI-promoted discourses – in relation to the imperatives of financialisation. In addition, quite apart from the dominance of international financial companies over domestic finance and trading (as well as the role of ratings agencies), some emphasis has been placed on the hierarchies of currencies and their functionings within the global financial system, with the US dollar as pinnacle. This, in turn, has given rise to notions of (semi-)peripheral and subordinate financialisation as characterising developing and emerging economies with strong connotations of what are now the less favoured scholarly (if popular) approaches of dependency and world systems theory – does one nation (part of the world) financially exploit another through finance as well as through other mechanisms?

. . . To variegated incidence and effects

To some extent, such oversimplified, structured postures are addressed by acknowledging the interactions of the global financial and production systems (as with global production networks), Powell (2018). But my own preference, one that is gathering momentum at the time of writing, is to refer to the variegated nature of financialisation and its effects. Take, for example, the major thrust of finnX research, that the short-term pursuit of quick returns, through securitisation of potential revenue streams, has depressing effects on X where X is somehow attached to development. But this is simply not true of (global) energy and food systems which are marked by excessive production (like the financial system itself). In this light, it is appropriate to examine how globally variegated financialisation is inserted within and unevenly distributed globally across the economic and social reproduction of different nations rather than seeing them as uniformly assigned to one or other side of a structure of subordination (although this is a major part of outcomes albeit with exceptions – how does China fit, for example).

Such grand postures can be made more specific through the following analytical structure. First is to recognise that the global economy is currently dominated by globally organised production networks and corporations. Second, over the past four decades, capitalism has been marked by neoliberalisation. Third, such neoliberalisation has been fundamentally based upon – but is not reducible to – financialisation. As acknowledged in the finnlit, the role of finance in what might be termed economic and social reproduction and restructuring is extensive and increasing. It has been variously analytically captured by related processes and concepts such as privatisation, commercialisation, deregulation, and austerity. Fourth, like many buzzwords before it, financialisation has gathered momentum without adding much novelty by way of theory, thereby allowing it to be applied willy-nilly. It is important, though, to distinguish financialisation from its effects. This can be done by sticking theoretically to a narrow definition of financialisation in which it is confined to the creation and deployment of financial assets alone. Its effects can then be separately discerned by locating them in relation to the corresponding economic and social activities to which they are attached, including the broader presence and influence of the ethos attached to financial interests such as commercialisation.

Such an approach allows for variegation to be addressed at macro, micro, meso, institutional, national, and global levels, as well as for social and cultural implications. Thus, for example, whilst mortgage finance has been appropriately prominent in the study of developed country finance-country financialisation of housing, this is inappropriate for developing countries. In this way, differentiation in the role and impact of financialisation has been broached through the system of provision approach that both acknowledges the increasing role of financialisation in economic and social development but in ways that are differentiated within and across both countries and provisioning by sector, Bayliss and Fine (2020).

Overall, financialisation is seen to have had both negative and distorting effects on developmental prospects both directly and through its influence on policies and cultures. Putting the genie of finance back in the bottle will be no easy task, and little attempt has yet been made. But it is the first step to be taken before more progressive and effective policies can be adopted to allow appropriate levels and forms of provisioning for economic and social development.

References

Bateman, M., S. Blankenburg and R. Kozul-Wright (2018) *The Rise and Fall of Global Microcredit: Development, Debt and Disillusion*, London: Routledge.

Bayliss, K. and B. Fine (2020) *A Guide to the Systems of Provision Approach: Who Gets What, How and Why*, London: Palgrave, forthcoming.

Bayliss, K., B. Fine and M. Robertson (eds) (2018) *Material Cultures of Financialisation*, London: Routledge.

Bonizzi, B. (2016) "The Changing Impact of Finance on Development", FESSUD Working Paper, no 124, http://fessud.eu/wp-content/uploads/2015/03/The-changing-impact-of-finance-on-development-working-paper-124.pdf

Christophers, B. and B. Fine (2020) "The Value of Financialization and the Financialization of Value", in D. Mertens, P. Mader and N. van der Zwan (eds) *International Handbook of Financialization*, London: Routledge, forthcoming.

Epstein, G. (2005) "Introduction: Financialization and the World Economy", in G. Epstein (ed) *Financialization and the World Economy*, Cheltenham: Edward Elgar, pp. 3–16.

Fine, B. (2014) "Financialisation from a Marxist Perspective", *International Journal of Political Economy*, vol 42, no 4, pp. 47–66.

Itaman, R. (2017) *A Critical Analysis of the Revised Finance-Growth Nexus: A Case for the Exclusion of Value Added of Financial Services from GDP*, PhD thesis, School of Oriental and African Studies, University of London.

Lapavitsas, C. (2013) *Profiting without Producing: How Finance Exploits Us All*, London: Verso.

Mertens, D., P. Mader and N. van der Zwan (eds) (2020) *International Handbook of Financialization*, London: Routledge, forthcoming.

Powell, J. (2018) *Towards a Marxist Theory of Financialised Capitalism*. University of Greenwich Business School, no GPERC62, https://gala.gre.ac.uk/id/eprint/20331/

Further reading

Bonizzi, B., A. Kaltenbrunner and J. Powell (2019) "Subordinate Financialization in Emerging Capitalist Economies", Greenwich Papers in Political Economy, no 23044, https://gala.gre.ac.uk/id/eprint/23044/13/23044%20POWELL%20Subordinate%20financialization%20GPERC%20WP%20BKP%20v2%2020050319.pdf

Christophers, B. (2013) *Banking across Boundaries: Placing Finance in Capitalism*, Chichester: Wiley-Blackwell.

Financialisation, Economy, Society and Sustainable Development, project funded by the European Union under the 7th Research Framework programme (theme SSH) Grant Agreement number 266800, http://fessud.eu/

Kaltenbrunner, A. and J. Panceira (2018) "Subordinated Financial Integration and Financialisation in Emerging Capitalist Economies: The Brazilian Experience", *New Political Economy*, vol 23, no 3, pp. 290–313.

Tori, D. and O. Onaran (2017) "Financialisation and Physical Investment: A Global Race to the Bottom in Accumulation?", Post Keynesian Economics Study Group Working Paper, no 1707, http://www.postkeynesian.net/downloads/working-papers/PKWP1707.pdf

Part 7

Conflict, violence, and peace. Editorial introduction

For many, the end of the Cold War (1945–1991) ushered in hopes for a new global order characterised by international cooperation and shared prosperity. However, by the dawn of the 21st century this vision had been shaken. The September 11, 2001 terrorist attacks prompted the War on Terror, leading to prolonged military engagements by the US and UK from Afghanistan to Iraq, straining global partnerships and intensifying geopolitical rivalries. Wars in the former Yugoslavia, Yemen, Syria, and Ukraine exemplified a world grappling with increasing fragmentation, tension, and mistrust, highlighting the fragility of international agreements and the shifting dynamics of global power in an increasingly multipolar world. Amidst these geopolitical shifts, economic challenges emerged, and the effects of the climate crisis intensified.

The climate crises and environmental degradation continue to generate insecurity and instability, fostering competition over dwindling resources. Economic crises – and some of the policies ostensibly intended to ease them – have further exacerbated inequalities and tensions, both between and within countries. Connected to these economic and environmental pressures, the world has seen a deepening division in ideologies and approaches. These divisions can be seen in a rise in populist nationalism, jeopardising cross-border solidarity and cooperation. The digital age has brought its own set of challenges and opportunities. The Arab Spring highlighted the power of technology and social media in mobilising mass protests. These digital platforms, alongside the emergence of artificial intelligence, have also been co-opted for surveillance, control, and disinformation by both states and non-state actors.

Present day conflict cannot be divorced from historical contexts. Colonial relationships, structures, and policies, often designed to extract resources and suppress local identities, have left indelible marks on contemporary conflicts, for example, in laying the foundations for ethnic tensions, internal divisions, and border conflicts. Moreover, colonialism has shaped gendered identities, practices, and relationships and the ways they play out to foster or enable violence. Furthermore, the ideological battles of the Cold War era continue to influence global politics and development strategies, with superpowers vying for influence through aid, interventions, and proxy wars.

As the world grapples with these 21st-century challenges, a vision of global peace and interconnectedness seems to have receded, giving way to fragmentation, tension, and a re-evaluation of the global order. While large-scale geopolitical events shape the global landscape, it's essential to recognise that the causes and consequences of conflict, violence, and prospects for peace are profoundly shaped by local, global, as well as personal

DOI: 10.4324/9780429282348-78

contexts. Personal interests, perspectives, and experiences manifest at multiple scales, from civil wars to violence within households, such that conflict, violence, and peace cannot be apprehended in solely geopolitical terms but as deeply personal experiences, informing and informed by everyday practices and relationships. Beyond the overt manifestations of conflict and violence, it's important to recognise the presence of 'structural violence'. This form of violence is often invisible, embedded in social structures, normalised through institutions, and reproduced in everyday social practice, norms, and traditions. It manifests in systemic ways, such as racism, sexism, ableism, and classism, which perpetuate disparities in power, wealth, and life opportunities. In contrast, the idea of 'positive peace' goes beyond the mere absence of direct violence. It emphasises the presence of social justice and equality. Positive peace recognises that true peace involves addressing the root causes of conflicts, including structural violence.

The consequences of conflict and violence are far-reaching. Beyond the harm to done to individual bodies, violence and conflict hinder economic growth, disrupt essential services and support systems, destroy infrastructure, deter investment, and displace populations, leading to a vicious cycle further fuelling conflict. When development processes exclude groups, distribute gains unequally, or privilege some over others, grievances can emerge and ignite violence.

Conflict, violence, and peace have taken centre stage as both challenges and context in development studies. This section of the Companion examines complex and connected themes of conflict, violence, and peace in the field. While the chapters in this section of the Companion examine a diverse range of topics related to conflict, violence, and peace, they share several key themes. First, they address the causes and impacts of conflict and violence, as well as the opportunities and challenges encountered in promoting peace and fostering security. Many of these discussions are closely tied to an analysis of the state's role in development processes, encompassing topics such as state capacity, state-society relations, security sector reform, and policing. A further theme is the intersection of conflict and violence with broader development issues like inequality, poverty, and livelihoods. The contemporary landscape is marked by geopolitical transformations, economic complexities, and technological advancements, which have contributed to a shifting global order. The global political context of multipolarity and fragmentation might seem to lend itself to conflict and violence rather than peace. Yet it is essential to recognise that these phenomena are rooted in personal interests and experiences. This understanding is crucial to address the underlying causes of conflict and violence and work towards a future characterised by social justice and equality.

Chapter 72

Fragile states

Tom Goodfellow

The concept of a 'fragile state' is rooted in the end of the Cold War and has been evolving constantly since; it is now used in such differing ways by a growing range of development agencies and academics that it is increasingly difficult to pin down. Nevertheless, the term reflects important ongoing efforts in recent decades to understand the relationship between patterns of violent conflict and the nature of the state in less developed countries. Today even market-oriented donor agencies such as the World Bank acknowledge that effective states fundamentally matter for development. The civil conflicts ravaging many parts of the globe not only cripple the capacities of states but according to some analyses are caused by the weakness or fragility of state structures and institutions. State fragility is therefore likely to remain a relevant concept, but clearer specification and more rigorous analysis are necessary to maximise its utility for development researchers and practitioners.

Emergence of the 'failed states' paradigm

The idea of 'state fragility' was preceded by that of 'state failure', which was essentially born after the fall of the Berlin Wall. During the Cold War, superpower rivalry allowed little scope for states to collapse completely. Civil conflicts were rife, as were failures on the part of many governments to deliver basic services to their citisens, but the United States and Soviet Union propped up client regimes and intervened militarily in countries where spiralling conflict threatened their geopolitical interests. With the demise of the Soviet empire in the early 1990s, however, states imploded places as diverse as Yugoslavia to Somalia. The proliferation of extremely violent civil conflicts around this time, often characterised as 'new wars' and linked to processes of globalisation and international criminal networks (Kaldor, 1999), represented state failure in a very fundamental sense (Helman and Ratner, 1992). At its most basic, state failure can thus be seen as the loss of the monopoly of violence by the central government within a given territory, reflecting a classically Weberian conception of the state (Ignatieff, 2002).

However, notwithstanding the tumultuous events in Eastern Europe and parts of Africa in the 1990s, it was September 11, 2001 that really projected the idea of failed states centre stage. Anarchy in developing countries was increasingly perceived as fostering international terrorism, so the capacity of far-flung states to govern their own territories became viewed as an urgent concern for developed countries in the Global North (Beall et al., 2006). Rotberg aptly captured this mood when he noted that after the 2001 attacks the problem of failed states acquired 'an immediacy and an importance that transcends

DOI: 10.4324/9780429282348-79

its previous humanitarian dimension' (Rotberg, 2002: 127). At the same time, research in the early 2000s claimed that civil wars were in part caused by 'weak' statehood (Fearon and Laitin, 2003), further bolstering the idea that building the capacities of states was critical for global security.

This renewed concern with state capacity was welcome and a necessary rebalancing after the late 20th century's overenthusiastic heralding of the 'retreat of the state' (Strange, 1996). Nevertheless, the 'failed states' discourse was highly problematic. The term was often used ideologically in reference to states deemed undesirable by the world's remaining superpower, the United States, even though cases such as Iran (or indeed pre-2003 Iraq) were hardly 'failed states' even if their governing regimes were considered abhorrent. The concept of 'failed states' thus came to have intrinsically normative undertones, begging the question of how one should define its opposite, a 'successful' state. Some authors propose broad lists of core functions to 'fix' failed states that go far beyond the concern with conflict from which the idea arose and which effectively stigmatise almost all developing countries as 'failing' (Ghani and Lockhart, 2008).

Let them fail? Echoes of European history

A related problem with the idea of state failure is that it raises the question of when the international community should intervene to bolster statehood or 'save' populations from their failing state. This is a pertinent question in view of the history of nation-states in more developed parts of the world: Tilly, for example, has argued that the process of state-building in medieval Europe was essentially a by-product of war and the need to raise taxes to fund it (Tilly, 1992). This idea that historically 'war made the state and the state made war' has been influential in contemporary debates on state failure, implying that it may be through the intrinsically violent efforts of 'coercion-wielders' to establish dominance over territory that capable states are consolidated. Some scholars effectively argue in favour of 'giving war a chance': if certain de jure states cannot provide security within a given territory then the international community should 'let them fail', enabling capable de facto states to emerge rather than protecting pseudo-states that exist in name only (Herbst, 2004).

To some degree the international community has indeed been 'allowing' failure in places such as Somalia and the Democratic Republic of Congo (DRC). However, even these virtually stateless places are treated as sovereign states in international law – a crucial difference from the period of European history discussed by Tilly. No matter how much a state today may be 'failing', its borders are reified by the international system, which some scholars argue undermines processes that could lead to successful statehood (Englebert, 2009). Yet also unlike in medieval Europe, in our globalised world of 24-hour media coverage the idea of completely abandoning nations to their fate is simply unacceptable to most people, even if it might produce more capable states in the long run. We therefore face a situation in which states with very little capacity do exist, but letting them collapse and vanish from the map is unpalatable, while simply labelling them 'failed' is unhelpful. The category of 'fragile states' represents an acknowledgement of this inescapable part of the contemporary international landscape.

From 'failure' to 'fragility'

The state fragility concept represents a more nuanced approach the enduring problems faced by states in conflict-affected parts of the world, recognising degrees of state weakness and not stigmatising states as 'failed' if they underperform. As the 2000s wore on, donor organisations including the World Bank, OECD, and the UK Department for International Development adopted the language of fragility. In many respects this discursive shift is a positive move, resulting in more aid going to the countries that need it most (Putzel, 2010) as well as shedding some ideological baggage of 'state failure' and refocusing more keenly on the specific issue of vulnerability to conflict (Naudé et al., 2011). The propensity to conflict is critical to defining fragile states, yet specifying exactly what makes states vulnerable in this way has proved problematic. Focusing solely on the monopoly of legitimate violence within a given territory is insufficient and renders the concept of fragility somewhat tautological.

The key problem in operationalising state fragility is therefore defining which capacities states have to be deficient in to be considered fragile and how deficient they have to be. The definitions used by many development agencies are extremely broad. A typical example is that used by the OECD, which defines fragile states as countries 'where there is a lack of political commitment and insufficient capacity to develop and implement pro-poor policies'. As Putzel notes, this definition fails to distinguish between states that are particularly prone to violent conflict and those characterised by poverty and underdevelopment generally. In other words, such definitions render fragile states equivalent to *all* developing countries, making the concept somewhat redundant (Putzel, 2010).

To clarify the nature of fragility it is helpful to consider what exactly constitutes the opposite of a fragile state. Just as the idea of 'failed states' intrinsically implied a normative ideal of 'successful' ones, it is difficult to conceive of fragility without defining its counterpart. Some commentators conceptualise the opposite of a fragile state as a 'developmental state' (Naudé et al., 2011: 5), but again this is problematic, not least because of the association of this term with the East Asian 'tigers', suggesting that any less successful states (i.e. all developing countries) are fragile. In fact, it is critically important to distinguish between underdeveloped countries that are particularly prone to conflict and those that are not. For example while Tanzania is still very poor and would hardly qualify as a 'developmental state', it is clearly not fragile in the way that the neighbouring DRC is. Having experienced no large-scale internal conflict, it can be thought of as *resilient*. Indeed, the logical opposite of a fragile state is a not a developmental one but, as Putzel (2010) notes, a resilient state.

Understanding exactly what makes states resilient to conflict (rather than 'developmental', which may depend on different factors), is therefore the central problem of fragility. The work of the Crisis States Research Centre at the London School of Economics provides some insights in this regard, focusing on four fundamental characteristics of fragility. As well as the monopoly of the legitimate use of force, they highlight the degree to which i) states possess bureaucratic capacity, as represented by a monopoly over taxation; ii) institutions (or rules) of the state 'trump' those of non-state organisations (such as rebel groups or traditional authorities); and iii) states attain territorial 'reach' across within their borders (Putzel, 2010). Ultimately, the extent to which a state achieves these characteristics of resilience depends on the nature of the political settlement within a given territory, particularly in terms of the 'bargains' struck among elites (Di John and Putzel, 2009). Resilience is thus a profoundly political matter, about far more than enhancing a state's technical capacities.

Conclusion

Fragile states are a historically-specific feature of the contemporary global order, reflecting the proliferation of de jure states since decolonisation alongside the limited willingness and ability of the rich world to intervene and bolster their resilience. Despite the very real changes wrought by globalisation, the significance of the state is clearly not in retreat; without resilient and effective states people in many parts of the developing world will continue to suffer from chronic insecurity, with state fragility and violent conflict perpetuating one another in a vicious cycle. Breaking this cycle means not only attempting to end violence but also supporting politically contentious processes of state-building. This is difficult terrain for donors, who are wary of engaging explicitly in politics, but it may ultimately be the only substitute for the processes that built states historically in developed parts of the world, which often involved letting violent conflict itself play out.

References

Beall, Jo, Thomas Goodfellow, and James Putzel. 2006. Introductory Article: On the Discourse of Terrorism, Security and Development. *Journal of International Development* 18 (1): 51–67.

Di John, Jonathan, and James Putzel. 2009. *Political Settlements: Issues Paper*. Birmingham: Governance and Social Development Resource Centre.

Englebert, Pierre. 2009. *Africa: Unity, Sovereignty, and Sorrow*. Boulder, CO: Lynne Rienner.

Fearon, James D., and David D. Laitin. 2003. Ethnicity, Insurgency, and Civil War. *The American Political Science Review* 97 (1): 75–90.

Ghani, Ashraf, and Clare Lockhart. 2008. *Fixing Failed States: A Framework for Rebuilding a Fractured World*. Oxford: Oxford University Press.

Helman, G.B. and Ratner, S.R. 1992. Saving failed states. *Foreign policy*, (89), pp. 3–20.

Herbst, Jeffrey. 2004. Let them Fail: State Failure in Theory and Practice. In *When States Fail: Causes and Consequences*, ed. Robert Rotberg. Princeton, NJ: Princeton University Press, 302–318.

Ignatieff, Michael. 2002. Intervention and State Failure. *Dissent*, Winter.

Kaldor, Mary. 1999. *New and Old Wars: Organized Violence in a Global Era*. Stanford: Stanford University Press.

Naudé, Wim, Amelia U. Santos-Paulino, and Mark McGillivray. 2011. *Fragile States: Causes, Costs, and Responses*. Oxford: Oxford University Press.

Putzel, James. 2010. Why Development Actors Need a Better Definition of State Fragility. *Policy Directions*, Crisis States Research Centre (September).

Rotberg, Robert I. 2002. Failed States in a World of Terror. *Foreign Affairs*, July/August.

Strange, Susan. 1996. *The Retreat of the State: The Diffusion of Power in the World Economy*. Cambridge: Cambridge University Press.

Tilly, Charles. 1992. *Coercion, Capital and European States, AD 990–1992*. Oxford: Blackwell Publishing.

Further reading

Di John, Jonathan. 2010. The Concept, Causes and Consequences of Failed States: A Critical Review of the Literature and Agenda for Research with Specific Reference to Sub-Saharan Africa. *European Journal of Development Research* 22 (10–30). This article provides an overview of some of the main themes in the academic literature on failed and fragile states.

Gutiérrez, Francisco, D. Buitrago, A. González, and C. Lozano. 2011. *Measuring Poor State Performance: Problems, Perspectives and Paths Ahead*. London: London School of Economics and Political Science. A thorough analysis of the problems involved in measuring aspects of state fragility and critique of existing fragile states indices. Available online at: http://www.dfid.gov.uk/r4d/Output/186126/Default.aspx

Naudé, Wim, Amelia U. Santos-Paulino, and Mark McGillivray. 2011. *Fragile States: Causes, Costs, and Responses*. Oxford: Oxford University Press. An edited collection with a range of perspectives encompassing some of the latest thinking on state fragility.

Putzel, James. 2010. Why Development Actors Need a Better Definition of State Fragility. *Policy Directions*, Crisis States Research Centre (September). A short policy paper that provides a succinct overview of some of the main findings of the Crisis States Research Centre. Available online at: http://eprints.lse.ac.uk/41300/1/StateFragilityPD.pdf

Rotberg, Robert I. 2004. *When States Fail: Causes and Consequences*. Princeton: Princeton University Press. An edited collection containing many important perspectives on the idea of state failure, including the previously mentioned chapter from Jeffrey Herbst.

Useful websites:

- **Crisis States Research Network homepage:** http://www2.lse.ac.uk/internationalDevelopment/research/crisisStates/Home.aspx. Many publications from this ten-year project funded by UK DFID are available for download here.
- **Fund for Peace homepage:** http://www.fundforpeace.org/global/. This organisation publishes an interactive 'failed states index' every year, which can be arranged in accordance with the different criteria considered constitutive of 'failure'.
- **World Bank's 2011 report on Conflict, Security and Development:** http://wdr2011.worldbank.org/fulltext. The report is available for download and contains a wide range of statistics and perspectives on conflict and fragility in the 21st century.

Chapter 73

Resource wars

Emma Gilberthorpe and Elissaios Papyrakis

Introduction

The term 'resource wars' refers to the violent armed conflicts that emerge over the pursuit and possession of commercially valuable precious mineral resources such as oil, gas, gold, precious metals and gemstones, and commercially valuable natural resources such as timber, land, water, and plant/animal products. The term came into use in the late 1980s in response to the geopolitical conflicts that arose over the control of Middle Eastern oil and African minerals towards the end of the Cold War (see Klare, 2001; LeBillon, 2017). Whilst these and similar conflicts are entangled in the relentless scramble for resources that triggered epochs of European expansion and colonisation, the term 'resource wars' specifically describes the political and economic processes of the post-colonial, neoliberal era where localised political, geographic, and historical factors contribute to the nature of the conflict.

'Resource wars' are associated with a number of interlinked global and local processes. Globally, these include population growth, economic expansion, technological advances, increased demand for raw materials, dwindling resources, and growing concern for ethicalsourcing and green alternatives. Locally, they include the remoteness and inaccessibility of areas where valuable natural resources are located, the vulnerable status of the people who live there, long-standing ethnic animosities, and deeply embedded historical grievances (Berman et al., 2017; Klare et al., 2011). Synthesised, they generate political and economic instability leading to weakened governance and limited capacity within government sectors, as well as critical levels of human rights abuses (including widespread sexual violence, extreme poverty, and child labour), environmental devastation, destruction of infrastructure, cultural annihilation, and mass human displacement.

These impacts are far more widespread and more destructive in resource wars than in other types of violent conflict. Resource wars also tend to be far more intensive and enduring than other types of conflicts (Klare, 2001; LeBillon, 2017) with those countries inflicted by *resource wars* being widely identified as characteristic of the *resource curse* (Auty, 1993). *Resource wars* typically occur when there are disputes over i) *territory* (in particular when states dispute terrestrial or marine borders around the location of valuable resources), ii) *ownership* (when property rights are weakly specified and enforced), and iii) *control* (when there is contestation over the allocation of resource rents). Resource wars tend to be fought within nations (civil wars) or between border states. When the sovereignty of raw materials is threatened, groups often employ extremist military force to protect their resource-based interests. Militia groups recruit and build

DOI: 10.4324/9780429282348-80

armies of typically male youth, often through force and tactics of intimidation and coercion, seeking financial support from the fruits of extractive industry. Some countries suffering from 'resource wars' include the Democratic Republic of Congo, Angola, Nigeria, Sierra Leone, South Sudan, West Papua, Bolivia, Columbia, and Myanmar.

Resource wars: a review of key concepts and theories

The 'resource curse' theory

Low-income countries with an abundance of resource wealth appear to be amongst the most conflict-ridden countries in the world. This phenomenon has been explained in terms of the 'resource curse'. The 'resource curse' emerged as a theory in the early 1990s to explain the failure of countries with an abundance of resource wealth to reap the benefits it produces (Auty, 1993). The term soon came to represent the general affliction suffered by resource-rich nations where weak institutions and capacity led to the mismanagement of resource revenues by elites fostering cultures of greed, corruption, and exploitation at the expense of social, economic, and political stability.

The 'greed versus grievance' hypothesis

Several studies explore the causes of violent conflict in different parts of the world by adopting the so-called 'greed versus grievance' hypothesis (Collier and Hoeffler, 2004). *Greed* typically refers to the financial incentives associated with joining a rebellion: these anticipated rewards come in the form of increased access to valuable assets and power within a given state. In a cost-benefit context, combatants weigh the value of these expected benefits (taking into consideration also the probability of victory and distribution of benefits across one's group) against any foreseeable costs (as in the form of ammunition, loss of revenues from other productive activities, and the probability of retribution if unsuccessful). *Grievance*, on the other hand, captures motives associated with identity; certain groups (along the lines of ethnicity, religion, or social class) may decide to engage in warfare if they face unjust and discriminatory behaviour by other groups and the state. There is nowadays ample empirical evidence in favour of the greed hypothesis in the context of mineral-rich nations; valuable natural resources (as in the case of oil and diamonds) are particularly linked to the increased likelihood and duration of conflict. This is because profits from extractive activity can potentially both finance prolonged conflict, as well as provide lucrative assets to the group that ultimately wins the war. Precious mineral resources are also more likely to lead to *resource wars*, in countries where the population is ethnically or religiously fragmented (Berman et al., 2017). This is especially so when the population is *polarised* across groups of similar sizes (as in this case the population dynamics do not favour any particular group that could easily suppress any rebellion from other competing groups). In other words, conflict is much more likely to arise, last longer and intensify when both greed and grievance motives are present.

Type of natural resources and appropriability

Certain types of natural resources are more prone to induce conflict amongst competing groups. The so-called *point resources* (e.g., oil, diamonds, precious metals) tend to be more geographically concentrated and hence easily lootable and appropriable by certain interest groups. On the other hand, *diffuse* resources (as in the case of agriculture, fisheries, and forestry) are generally more dispersed and this hinders the control of any accrued income from the corresponding resource sectors by certain groups (Vahabi, 2018). In addition, point resources are less characterised by socio-economic *backward and forward linkages* to the rest of the economy, and largely operate as *enclaves* that generate little domestic employment and income in other sectors. This explains to a large extent why point resources are more conducive to rent-seeking behaviour and resource wars (Ross, 2004). This is also the case for narcotics (e.g., cannabis and coca) which, although not necessarily geographically concentrated, are controlled by few small groups due to their illegality.

Conflict resources (also known as conflict minerals or conflict commodities)

The term 'conflict resources' refers to resources that have been sourced from conflict zones, and whose trade fuels that conflict. Because the trade of valuable resources sustains the conflict, it is a key factor in the longevity of *resource wars*. The resources that commonly fuel *resource wars* are usually the most appropriable (oil, tin, gold, tungsten, tantalum, diamonds, emeralds, rubies, and sapphires), and most geographically contained (narcotics and, to a lesser extent, timber). Because *conflict resources* both fund and fuel *resource wars*, efforts are in place to stop the trade of conflict resources. More stringent regulations have been put in place over the last decade to encourage companies to responsibly source minerals and other resources and eliminate those identified as coming from conflict zones.

The role of the state

Whilst poor governance and lack of capacity have been identified as key variables in the perpetuation of *resource wars*, several studies verify the important and multifaceted role the state can play in preventing them. Some governments are better than others in establishing institutions conducive to a strong rule of law, enforceable contracts, and transparent administrations. Governments in developing countries often fail to adequately protect the property rights of legitimate resource owners; when this happens, interest groups will try to appropriate resources that originally belonged to other competing beneficiaries. Furthermore, we also know that state-controlled natural resources (in comparison to privately owned ones) tend to be associated with a lower risk of conflict – this might either be because rulers have a particular interest in protecting their own resources or because they establish patronage networks (through resource-financed privileges) to prevent rent-seeking and contesting activity (Wegenast, 2016). Many governments in resource-rich countries also face severe budget constraints (especially in periods when resource prices drop substantially and abruptly); as a consequence, they have limited resources to allocate towards preventive efforts (e.g., for additional investment in policing and law enforcement). Furthermore, grievances are largely driven by a sense of injustice towards specific groups; the onset risk of resource wars decreases significantly when governments distribute resource revenues more equitably across groups and allocate a large share of them as welfare expenditure. As a matter of fact, *vertical inequalities* (expressed as differences in income or asset ownership inequalities among individuals or households) do not seem to have an effect on the risk of conflict, while *horizontal inequalities* (that is, the same differences across groups) incite it.

The impact of 'resource wars'

Scholars have compiled a vast amount of evidence to show the widespread and chronic nature of *resource wars* and its impacts. Data show that *resource wars* not only compound social, economic and political stability but also generate some of the worst violations of human rights on the planet. This type of conflict is also responsible for devastating levels of environmental destruction, the flagrant destruction of infrastructure, and the mass displacement of people.

References

Auty, R. (1993). *Sustaining Development in Mineral Economies: The Resource Curse Thesis*. London: Routledge.

Berman, N., Couttenier, M., Rohner, D. and Thoenig, M. (2017). This Mine is Mine! How Minerals Fuel Conflicts in Africa. *American Economic Review* 107 (6): 1564–1610.

Collier, P. and Hoeffler, A. (2004). Greed and Grievance in Civil War. *Oxford Economic Papers* 56 (4): 563–595.

Klare, M.T. (2001). *Resource Wars: The New Landscape of Global Conflict*. New York: Henry Holt.

Klare, M.T., Levy, B.S. and Sidel, V.W. (2011). The Public Health Implications of Resource Wars. *American Journal of Public Health* 101 (9): 1615–1619.

Le Billon, P. (2017). *The geopolitics of resource wars*. Abdingdon. Routledge.

Ross, M. (2004). What Do we Know about Natural Resources and Civil War? *Journal of Peace Research* 41 (3): 337–356.

Vahabi, M. (2018). The Resource Curse Literature as Seen through the Appropriability Lens: A Critical Survey. *Public Choice* 175 (3–4): 393–428.

Wegenast, T. (2016). Oil, Natural Gas, and Intrastate Conflict: Does Ownership Matter? *International Interactions* 42 (1): 31–55.

Additional sources

Gedicks, A. (1994). *New Resource Wars*. Quebec: Black Rose Books.

Gilberthrope, E. and Papyrakis, E. (2015). The Extractive Industries and Development: The Resource Curse at the Micro, Meso and Macro Levels. *Extractive Industries and Society* 2 (2): 381–390.

Karl, T. (2007). *The Paradox of Plenty: Oil Booms and Petro-States*. California: University of California Press.

Le Billon, P. (2017). *The Geopolitics of Resource Wars: Resource Dependence, Governance and Violence*. Abingdon, UK: Routledge.

Chapter 74

Gender and conflict

Erika Forsberg

Gender and conflict

This chapter summarizes two fields of research of relevance to gender in peace and conflict. First, it reviews some of the aspects related to the gendered nature of participation and suffering in armed conflict. Second, it reviews extant research on the relationship observed between levels of gender inequality and the likelihood of political violence. The chapter ends by highlighting some remaining knowledge gaps and avenues for future research.

The gendered nature of peace and conflict

In issues related to peace, conflict, and security one pattern stands out as particularly striking: its gendered nature. While there is large variation across rebel groups and state armies, the overwhelming majority of combatants now and throughout historical record are men; less than 1 per cent of warriors throughout history are estimated to have been women. This male overrepresentation is larger than can be explained by men's superior warrior qualities based on inherent biological differences (e.g., physical strength), indicating that social structures and norms are crucial determinants (Goldstein, 2001). Men are also in majority as perpetrators of other types of violence, such as in riots, domestic violence, and other violent forms of criminality. Women and men also suffer from the consequences of violence and disruption in different ways. While systematic sex-disaggregated data is often missing or unreliable, existing records indicate that men are overrepresented in terms of fatalities resulting from wartime violence – and this is the case also for civilian fatalities (Bjarnegård et al., 2015). Women, on the other hand, are more often subjected to non-lethal forms of violence. In electoral violence, women are increasingly facing intimidation while men more often face physical assault. Women also suffer disproportionally from the long-term consequences of war, such as deteriorating health and decreased life expectancy (Plümper and Neumayer, 2006) and increased levels of maternal mortality post-war (Urdal and Che, 2013).

Lastly, women and men are also affected in different ways by post-conflict peacebuilding, making it paramount for successful peacebuilding to adopt a gender perspective. For example, prevalent peacebuilding and transitional justice mechanisms, such as setting up truth commissions or designing and implementing disarmament, demobilization and reintegration (DDR) programs, often carry different meaning and have different

DOI: 10.4324/9780429282348-81

consequences for men and women. Take DDR as an example: in most armed conflicts, most combatants are men. Providing an opportunity for these men to return to civil life and access regular employment opportunities is crucial to avoid relapse into criminal activities. For women, however, the return of male ex-combatants to society may generate insecurity, vicarious trauma, and increased levels of domestic violence. These diverse – and often gendered consequences – need to be taken into account when designing such programs.

Although the gendered pattern is striking, there are also important nuances. We therefore need to qualify the simplistic stereotype of the male combatant/perpetrator and the female victim. First, while indeed more men than women participate in wartime violence, many rebel groups and regular armies have a significant number of women among their ranks, such as FARC in Colombia or CPN-M in Nepal (Wood and Thomas, 2017). Additionally, women often play a supportive role in sustaining armed conflict and do not necessarily hold more peaceful attitudes. Existing research in fact finds no significant gender differences in terms of attitudes towards using conflict violence (Tessler et al., 2002). As developed further later, gender norms – rather than biological sex – may matter more. In other words, individuals (regardless of sex) that hold gender equal norms are less prone to support violence.

Second, related to the first point, while more men than women participate in armed conflict, it is important to note the very small number of men that do participate. In fact, a large majority of men never participate in armed conflicts and would not consider using violence unless as a last resort. The relevant question to ask here is why the minority that does use violence is predominantly male (Bjarnegård et al., 2015). Usually male participation in organised violence is simply taken for granted and not explained in research. Existing studies indicate that this minority often holds gender norms where the male role is characterised by aggression, honor, and not backing down (Bjarnegård et al., 2017). In addition, there are societal norms that create expectations on both men (as soldiers) and women (as care-takers). These norms about combat being a male activity may explain that female combatants tend to be underestimated or overlooked, and further prescribe a gendered hierarchy, with male dominance and female passivity 'normalized'. However, importantly, these (patriarchal) norms, at least to some extent, in turn may legitimize and normalize the use of violence.

Third, while the general patterns suggest that women and men suffer from conflict in different ways (with more men being killed and more women suffering from some forms of non-lethal violence), this varies considerably across contexts. The differences may therefore be smaller than we intuitively think. For instance, while women overall are more often subjected to conflict-related sexual violence than are men, there is evidence suggesting that such violence when directed towards men has been overlooked or not categorised as sexual (for instance sexual torture categorised as just 'torture' when directed towards men), presumably because sexual violence may be associated with even more shame and stigma for men (Leiby, 2009). Another example that challenge the stereotype that equates women (and children) with civilian victimhood is that existing data find no gender difference in refugee displacement.

In addition to these gendered patterns in participation and suffering, the level of *gender inequality* has been found to correlate with the risk of armed conflict. This is explored in the following section.

Gender inequality and conflict

While research on conflict and security has previously largely neglected gender inequality as a predictor of armed conflict, focusing instead on natural resources, poverty, regime stability etc., a growing research field now demonstrates an empirical link between gender inequality and political violence. The correlation is found at several levels – including when studying individuals, organisations, subnational units, and countries – and is considered empirically robust.

Starting with the broadest level – countries – a growing body of quantitative research has followed in the path of Mary Caprioli's groundbreaking study (Caprioli, 2000) in studying country-level measures of gender inequality and their correlation with political violence. The results all point in the same direction: countries that display high levels of gender inequality are more likely to be associated with both intrastate and interstate armed conflict, and they have worse human rights records. Conversely, peace established under more gender equal conditions is more likely to hold, and higher levels of gender equality are linked to countries experiencing the onset of non-violent campaigns rather than violent uprisings.

The relationship found at the country-level is also reproduced when examining smaller units than countries. For instance, examining districts in India (i.e. the second-order administrative unit) it has been found that districts categorised by gender inequality are more likely to experience political violence, controlling for a range of alternative explanations (Forsberg and Olsson, 2020). Gizelis (2011) studies local level data from the UN peacekeeping operations (i.e. individual peace-building events) deployed in Liberia and Sierra Leone and finds that areas with relatively better levels of gender equality had more successful peacebuilding outcomes. The relationship is also found at the level of political organisations. In a study of ethnopolitical organisations in the Middle East, Asal et al. (2013) find that organisations espousing a gender inclusive ideology are more likely to pursue a non-violent strategy.

Last, the link between gender inequality and violence is also reproduced at the individual level. As mentioned previously, using political violence is primarily a male activity. However, this is not necessarily the result of women being innately more peaceful, at least not regarding attitudes to using violence. Research findings are fairly consistent in showing that individuals holding more gender equal norms are less supportive of violence and less likely to use violence themselves. This relationship holds also when controlling for biological sex. In other words, a man holding gender equal norms is likely to hold more peaceful attitudes than a women holding strong patriarchal norms. Gender inequality is a societal structure that forms women and men alike.

What, then, explains that we observe these correlations across different levels of analysis? Several explanations have been suggested. First, societal norms promoting masculinized political culture may lower the threshold for using violence. A traditional male role can be said to be an ideal where boys and young men learn that they must show harshness and be prepared to use violence to gain respect. As part of this, it is considered legitimate, even expected, to dominate over women. The treatment of women in highly unequal societies may also serve as a blueprint for how grievances by other 'out-groups' are perceived. Hence, the unequal gender norm in this case can be seen as a broader form of intolerance legitimizing also the oppression and domination over other groups, such as ethnic minorities or political opposition groups – which in turn increase the risk of violent conflict. In contrast to the traditional male role there is also a more modern male role. Here perceptions of male and female tend to evolve in such a way that masculinity is not as strongly associated with warfare, violence, and the right to privilege, and women are not as defined by motherhood and subordination. The result is a society that is both more peaceful and more equal, and these effects are due to the fact that both men and women have changed in their perceptions of what is male and female.

Second, certain gendered demographical patterns – such as a surplus of unemployed young men – may strengthen the capacity to mobilize for conflict. A key resource for mobilizing and conducting violent conflict is access to soldiers and the primary target group to mobilize is young men. High levels of gender inequality can affect the demographic balance. For example, a society characterised by gender inequality in the form of a persistent and strong son preference may, as a consequence, end up with a large male surplus. This is particularly evident in some Asian countries. A long-term consequence of large cohorts of excess men, is that many young men will not find a partner or employment. This group has little to lose from joining an armed group (or crime, riots, etc.) and hence constitutes a key target group to mobilize for participation in organised violence (Hudson and den Boer, 2002).

Remaining lacunas and ways forward

Given the breadth and robustness of the findings related to gender inequality and armed conflict, it is surprising that this explanation is customarily dismissed in current research. The most frequently cited scholarly works that statistically investigate the correlates of civil war, as well as articles that review extant findings, make no mention of gender inequality. This is a puzzling omission, especially in light of the striking role gender appears to play in the mobilization and conduct of political violence of all kinds (Bjarnegård et al., 2015).

That said, while the empirical patterns are strong we still have limited understanding of which theoretical mechanisms explain the patterns. This is of course not unique for gender but is also the case for other correlates of political violence, e.g., income levels. The previous discussion summarized two of the mechanisms that have been suggested. Although not comprehensive, these two mechanisms illustrate that gender inequality can be linked to armed conflict through quite different pathways that require different types of empirical measures. One of the mechanisms – about a male surplus resulting from a son preference – is also highly contextual and not relevant outside the context of a few countries. In moving forward, the first area that needs to be addressed concerns identifying the most relevant indicators of gender inequality and finding valid and reliable data. Most researchers have relied on the same set of indicators (women in parliament, fertility rates, access to education, women's participation in the labour force) to assess what they clearly believe to be different types of mechanisms. As a result, one and the same indicator has been invoked in support of multiple competing hypotheses. Second, in addition to a better match between theory and measurement, we also need a better match between theories and the unit of analysis used to assess them empirically. Most studies have hitherto relied on overly aggregated country-level data, despite the fact that most manifestations of gender inequality vary significantly within countries. Also, explanations focusing on norms can only be assessed by crude proxies if using measures presumed to be manifestations of such norms, such as observed gender gaps in education.

For a better assessment of the role of norms, more data on individuals' attitudes (e.g., survey-based data) is necessary. A third way forward is to develop research designs to address concerns for reversed causality and omitted variable bias.

To conclude, while the field still needs to develop in order to move our understanding further of how gender – and gender inequality – shapes the risk of conflict and patterns related to participation and suffering, the evidence clearly support its relevance.

References

Asal, Victor, Richard Legault, Ora Szekely, and Jonathan Wilkenfeld. (2013) Gender Ideologies and Forms of Contentious Mobilization in the Middle East. *Journal of Peace Research* 50(3): 305–318.

Bjarnegård, Elin, Karen Brounéus, and Erik Melander. (2017) Honor and Political Violence: Micro-Level Findings from a Survey in Thailand. *Journal of Peace Research* 54(6): 748–761.

Bjarnegård, Elin, Erik Melander, Gabrielle Bardall, Karen Brounéus, Erika Forsberg, Karin Johansson, Angela Muvumba Sellström, and Louise Olsson. (2015) *Gender, Peace, and Armed Conflict*. SIPRI Yearbook. Oxford. OUP.

Caprioli, Mary. (2000) Gendered Conflict. *Journal of Peace Research* 37(1): 53–68.

Forsberg, Erika, and Louise Olsson. (2020) Examining gender inequality and armed conflict at the subnational level. *Unpublished manuscript*.

Gizelis, Theodora-Ismene. (2011) A Country of their Own: Women and Peacebuilding. *Conflict Management and Peace Science* 28(5): 522–542.

Goldstein, Joshua S. (2001) *War and Gender*. Cambridge: Cambridge University Press.

Hudson, Valerie M., and Andrea den Boer. (2002) A Surplus of Men, A Deficit of Peace: Security and Sex Ratios in Asia's Largest States. *International Security* 26(4): 5–38.

Leiby, Michele L. (2009) Wartime Sexual Violence in Guatemala and Peru. *International Studies Quarterly* 53(2): 445–468.

Plümper, Thomas, and Eric Neumayer. (2006) The Unequal Burden of War: The Effect of Armed Conflict on the Gender Gap in Life Expectancy. *International Organization* 60(3): 723–754.

Tessler, Mark, Jodi Nachtwey, and Audra Grant. (2002) Further Tests of the Women and Peace Hypothesis: Evidence from Cross-National Survey Research in the Middle East. *International Studies Quarterly* 43(3): 519–531.

Urdal, Henrik, and Chi Primus Che. (2013) War and Gender Inequalities in Health: The Impact of Armed Conflict on Fertility and Maternal Mortality. *International Interactions* 39(4): 489–510.

Wood, Reed M., and Jakana L. Thomas. (2017) Women on the Frontline. *Journal of Peace Research* 54(1): 31–46.

Chapter 75

Violence against women and girls

Cathy McIlwaine

Violence against women and girls (VAWG) has become recognised as a crucial element within of international development research and policy domains. Although the UN Declaration on the Elimination of Violence Against Women dates back to 1993, the widespread recognition of the phenomenon has been more recent. Although Sustainable Development Goal (SDG) 5 on achieving gender equality and empowerment for all women explicitly includes the elimination of violence against women and girls and of all harmful practices, it was largely neglected in the Millennium Development Goals. Yet in the last two decades and as a result of lobbying by the women's movement at local, national and international levels, VAWG is now firmly acknowledged as a major human and women's rights violation, a public health concern, an international peace and security issue and an impediment to development interventions. Yet the incidence of VAWG remains extremely high globally with approximately 35 per cent of women having experienced physical and/or sexual violence at some point in their lives (WHO, 2013). This varies according to a wide range of intersectional factors such as race, ethnicity, class, gender identity, sexuality, immigration status, disability and so on (Imkaan, 2019). The incidence is also likely to be higher given widespread under-reporting and stigma not to mention the challenges created by the COVID-19 pandemic which has led to exponential increases in incidence as women have been confined to their domestic spaces (UN Women, 2020).

Defining gender-based violence and violence against women and girls

While definitions of violence are highly contested, it usually involves the use of physical force that causes hurt to others in order to impose a wish or desire with the primary motivating factor, either conscious or unconscious, being gaining and maintaining of power. Gender-based violence is distinguished where the motive relates to the gender of the victim/survivor which in the case of VAWG is women and girls. However, although 'violence against women and girls' and 'gender-based violence' are often conflated, it is important to emphasise that the latter includes violence against men, boys, transgender, other sexual minorities and those with gender-nonconforming identities. From a feminist perspective, the term gender-based violence has become de-politicised on grounds that it diverts attention from the reality that women and girls suffer disproportionately from violence as a result of male privilege and female oppression, as well as incorrectly implying that all gender groups suffer equally (COFEM, 2017).

DOI: 10.4324/9780429282348-82

Definitions of gender-based violence and violence against women and girls are therefore contested. The benchmark is the 1993 UN Declaration of the Elimination of Violence against Women (DEVAW) in Article 1:

> Any act of gender-based violence that results in, or is likely to result in, physical, sexual or psychological harm or suffering to women, including threats of such acts, coercion or arbitrary deprivations of liberty, whether occurring in public or in private life.

Article 2 continues that it may occur in the 'family, community, perpetrated or condoned by the State, wherever it occurs' and may refer to assault, sexual abuse, rape, female genital mutilation and other 'traditional' practices, as well as sexual harassment, trafficking in women and forced prostitution (UN, 1993). These were articulated further in the Beijing Platform for Action with UN Women identifying 'Ending Violence Against Women' and 'Peace and Security' as two of their nine priority areas that are fundamental to women's equality. While the DEVAW definition uses the term 'gender-based violence' it also explicitly identifies violence against women with more recent definitions also including 'girls'. The latter is to ensure that types of violence are included such as child, early and forced marriage and female genital mutilation and cutting (FGM/C). Also important within the DEVAW definitions is the acknowledgement that VAWG is not only domestic violence but occurs in private and public spheres of life and is perpetrated by intimate partners, non-intimate partners, strangers and the state (Hughes et al., 2016).

VAWG is therefore hugely diverse, taking multiple direct and indirect forms in different places and cultures. This diversity is often conceived across a 'continuum' where different types of violence overlap and reinforce one another (McIlwaine and Evans, 2020). Some more specific types include acid violence which is prevalent in Bangladesh, Pakistan and India, involving men throwing acid on women's faces as a form of attack or female genital mutilation/cutting that is concentrated in African and Middle Eastern societies (UN Women, 2015). The term 'femicide', which refers to the killing of women and girls by men because they are female, is usually associated with Latin American countries to highlight and capture exceptionally high rates of violent murders of women and girls. While this can entail multiple forms such as 'intimate femicide' where a woman is killed by a current or former intimate partner, the term 'feminicide' has also emerged to emphasise the active involvement of the state in the murder of women through impunity, direct perpetration of violence and/or a failure to prevent or respond to the killings (Menjívar and Walsh, 2017). Violence against women and girls is also intersectional, with some women more likely to experience it as a result of interrelated identities such as race, ethnicity, class, occupation, disability, sexuality and so on. For instance, the phenomenon of homophobic 'corrective rape' has been associated with black lesbian women in South Africa where the assault is argued to impose heterosexuality or gender stereotypes (UN Women, 2015), but is ultimately an outcome of the sexist and racist operation of colonialism, apartheid and contemporary inequalities.

A range of types of VAWG occur during armed conflict linked with how masculinity plays out through widespread sexual assault and rape of women as a military tool or tactic (Cockburn, 2013). As part of the broader the women, peace and security agenda, these have been enshrined in the landmark UN Security Council Resolution 1325 in 2000 (and the subsequent nine related resolutions 1820, 1888, 1889, 1960, 2106, 2122, 2242, 2467 and 2493). Focused on enhancing women's decision-making and empowerment, the resolutions focus on participation, prevention, protection and relief and recovery. Although women can also be perpetrators of violence during armed conflict as combatants and men and boys can be victims of sexual and especially homophobic violence, women and girls remain the most affected as victims/survivors. Similar processes also operate in relation to climate-driven disasters where upheavals and dislocations can result in increased incidence of gender-based violence against women and girls (Bradshaw, 2015).

Trends in violence against women and girls

Under-reporting of VAWG is widespread mainly due to enduring taboos, as well as sensitivities around data collection. Despite this, reported levels are extremely high throughout the world. As many as 54 per cent of women experienced physical or sexual assault by an intimate partner in Bangladesh over their lifetime with 59 per cent of women aged 20 to 24 years first married or in union before age 18 (Table 75.1). In Egypt where figures appear to be lower, the proportion of women and girls aged between 15 and 49 years who have undergone FGM/C is 87 per cent. Within countries, the incidence varies enormously according to identities and context variations such as rural/urban residence with the former associated within higher levels of intimate partner violence with non-intimate partner violence more prevalent in cities (McIlwaine, 2013).

Causes of violence against women and girls

While 'traditional' explanations of violence against women were based on 'natural' biological factors among men and/or psychological reasons related to 'impaired masculinity', these have been discredited. Instead, the multiple causes of VAWG are acknowledged as rooted in male domination and women's subordination that play out through unequal

Table 75.1 Proportion of adult women experiencing physical and/or sexual violence by intimate partner (lifetime and last 12 months)

Country	Lifetime physical and/or sexual intimate partner violence	Physical and/or sexual intimate partner violence in last 12 months	Child marriage*
Afghanistan (2017/18)	51%	46%	35%
Bangladesh (2016)	54%	27%	59%
Brazil (2018)	17%	3%	26%
Democratic Republic of Congo (2014)	51%	37%	37%
Egypt (2014–2018)	26%	14%	17%
India (2015–2016)	29%	22%	27%
Peru (2017)	31%	11%	21.5%
Uganda (2016)	50%	30%	40%

Source: compiled from UN Women Global Database on Violence Against Women (https://evaw-global-database.unwomen.org/en/countries) (accessed 31/10/2021)

* Refers to the percentage of women aged 20 to 24 years first married or in union before age 18

distributions of power and resources between women and men and discrimination (Hughes et al., 2016). These are reinforced through social norms and beliefs, practices and systemic structures, which are interrelated across different levels: from individual, to relationships, to family, to community, society and the state. Gender inequalities and discrimination are influenced by policy and legal frameworks, war and colonisation, as well as structural distinctions linked with age, income, ethnicity, sexuality and location. While VAWG has different expressions in different cultures, it is not cultural traditions that explain the violence but structural inequalities and discrimination. Gender inequalities can shape social norms that can lead to the normalisation of gender-based violence and thus the greater likelihood of perpetration. These intersect with other risk factors or triggers that can make VAWG more likely to occur. These can be very specific and include male alcohol or drug abuse, low educational levels, high rates of unemployment, infidelity, as well as economic crisis, war, disasters and so on. Invariably, these risks are bolstered by weak legal sanctions coupled with state fragility and widespread tolerance as well as the more active role of the state as potential perpetrator, actively and passively (McIlwaine, 2013; UN Women, 2015).

Conclusion

VAWG is one of the most damaging phenomena affecting women's well-being in the world today; it not only kills women but it undermines their physical and mental health and that of their children and wider communities. There are direct costs for healthcare and judicial services and indirect effects on educational and economic productivity as women are unable to work, go to school, or participate in politics. Women's freedoms are fundamentally curtailed through fear and stigma even if they do not experience gender-based violence directly. Violence against women and girls is therefore not an isolated individual issue, but one which affects communities, countries and international development efforts. While there are many initiatives across the world to reduce and prevent VAWG among civil society organisations, national government and international agencies not to mention a groundswell of feminist movements calling out such violence (including #MeToo, #Niunameos, ##NotOneMore; Ellsberg et al., 2015), much remains to be done, not least in our para-COVID world.

Guide to further reading

Gender and Development (2013) 21:3 Special issue on Conflict and Gender.
Gender and Development (2016) 24:2 Special issue on Violence against Women and Girls (VAWG).
Heise, L. L. & Kotsadam, A. (2015) Cross-National and Multilevel Correlates of Partner Violence: An Analysis of Data from Population-Based Survey, *The Lancet Global Health*, 3: 6, e332–e340.
Imkaan (2019) *The Value of Intersectionality in Understanding Violence against Women and Girls*, Imkaan and UN Women: London. https://www2.unwomen.org/-/media/field%20office%20eca/attachments/publications/2019/10/the%20value%20of%20intersectionality%20in%20understanding%20violence%20against%20women%20and%20girls.pdf?la=en&vs=3339
UN Office on Drugs and Crime (UNODC) (2018) *Global Study on Homicide: Gender-related Killing of Women and Girls*, UNODC: Vienna.
World Health Organisation (WHO) (2013) *Global and Regional Estimates of Violence against Women: Prevalence and Health Effects of Intimate Partner Violence and Non-Partner Sexual Violence*, WHO: Geneva.

References

Bradshaw, S. (2015) Gendered Rights in the Post-2015 Development and Disasters Agendas, *IDS Bulletin*, 46: 4, 59–65.

Cockburn, C. (2013) War and Security, Women and Gender: An Overview of the Issues, *Gender and Development*, 21: 3, 433–452.

COFEM (2017) Reframing language of 'gender-based violence' away from feminist underpinnings, *Coalition of Feminists for Social Change*. http://raisingvoices.org/wp-content/uploads/2013/03/Paper-2-COFEM.final_.sept2017.pdf

Ellsberg, M., Arango, D. J. & Morton, M. et al. (2015) Prevention of Violence Against Women and Girls: What does the Evidence Say? *The Lancet*, 385: 9977, 1555–1566.

Hughes, C., Marrs, C. & Sweetman, C. (2016) Introduction to Gender, Development and VAWG, *Gender & Development*, 24: 2, 157–169.

Imkaan (2019) *The Value of Intersectionality in Understanding Violence against Women and Girls*, Imkaan and UN Women: London. https://www2.unwomen.org/-/media/f ield%20off ice%20eca/attachments/publications/2019/10/the%20value%20of%20intersectionality%20in%20understanding%20violence%20against%20women%20and%20girls.pdf?la=en&vs=3339

McIlwaine, C. (2013) Urbanisation and Gender-Based Violence: Exploring the Paradoxes in the Global South, *Environment and Urbanization*, 25: 1, 65–79.

McIlwaine, C. & Evans, Y. (2020) Urban Violence Against Women and Girls (VAWG) in Transnational Perspective: Reflections from Brazilian Women in London, *International Development Planning Review*, 42: 1, 93–112.

Menjívar, C. & Walsh, S. D. (2017) The Architecture of Feminicide: The State, Inequalities, and Everyday Gender Violence in Honduras, *Latin American Research Review*, 52: 2, 221–240.

UN Women (2015) *A Framework to Underpin Action to Prevent Violence Against Women*, UN Women: New York.

UN Women (2020) *Covid-19 and Ending Violence Against Women and Girls*, UN Women: New York. https://www.unwomen.org/-/media/headquarters/attachments/sections/library/publications/2020/issue-brief-covid-19-and-ending-violence-against-women-and-girls-en.pdf?la=en&vs=5006

United Nations (UN) (1993) *Declaration on the Elimination of Violence Against Women (A/RES/48/104)*. https://www.un.org/en/genocideprevention/documents/atrocity-crimes/Doc.21_declaration%20elimination%20vaw.pdf

Chapter 76

Global human exploitation

Trafficking, forced labour, and modern slavery

Louise Waite

Introduction

The problem of human exploitation is as old as work itself. Yet despite development through the ages and more recent technological advances in the modern era, extreme exploitation remains doggedly persistent in contemporary global economies. Whether its stories of homeless men being forced into construction work in Britain, child slaves in cocoa farms in West Africa, bonded labourers on brick sites in South Asia or female domestic servitude in the Middle East, the media circulates many accounts of appalling losses of freedom for people across the globe. This daunting scale of extreme human exploitation in the 21st century is the backdrop for this chapter. Yet it does not set out to measure or map human exploitation or elucidate a typology or offer any putative solutions to enduring experiences of exploitation. Rather, the chapter critically reviews the shifting sands of engagement within political, policy and practice spheres regarding human exploitation. Of particular concern here is to explore complexities and discuss problematic tropes and representations. Despite common circulation of the word exploitation in the pantheon of social injustices and human rights abuses, the term remains semantically slippery and challenging to define in legal instrument. This chapter therefore first critically appraises the predominant lenses through which contemporary human exploitation is usually approached; trafficking, forced labour and modern slavery. The chapter then explores anti-modern slavery activities across the globe, before ending with a short conclusion.

Approaching trafficking, forced labour and modern slavery

The task of understanding trafficking, forced labour and modern slavery is complex and contingent. It is important to note that these terms are not mutually exclusive. They are more appropriately thought of as overlapping categories. The term modern slavery is often offered as an umbrella category that *incorporates* forced labour but also other categories such as forced marriage. People *may* be trafficked *or not* into their experiences of exploitation (which could be forced labour or related categories of domestic servitude, sexual exploitation or criminal exploitation). This leaves an understanding of all trafficked persons as modern slaves but not all modern slaves as trafficked. The section that follows discusses trafficking, forced labour and modern slavery in turn, before closing by exploring the contested drivers of these related phenomena.

DOI: 10.4324/9780429282348-83

Trafficking

Public concern about human trafficking (post trans-Atlantic slave trade) dates back to at least the late 19th century 'white slave trade' between Europe and North America and the paternalistic cause of prostitution abolition. A number of international instruments were consequently adopted in the first three decades of the 20th century that attempted to prevent human trafficking. The issue of human trafficking didn't subsequently receive much attention until the 1990s following the Cold War. The end of bipolar rivalry triggered the opening of borders and a growing recognition of transnational threats. At this point sex trafficking was, and some argue still is, the most explored form of human trafficking. The 1990s were also a time when the key tenets of the enduring anti-trafficking discourse emerged: sex work and anti-trafficking as ripe for moralising interventions and restrictive immigration policies in an era of migration securitisation.

In 2000 the United Nations developed a *Protocol to Prevent, Suppress and Punish Trafficking in Persons* (also known as the Palermo Protocol). It was the first to define 'trafficking in persons', and the criminalisation of trafficking was its pivotal plank. Human trafficking is now widely understood to refer to the movement of a person from one place to another (or the recruitment and harbouring of a person) for the purposes of exploitation. Although the Protocol is broadly welcomed, there is concern around interpretive difficulties in establishing thresholds in levels of consent, coercion and depths of exploitation. More recent anti-trafficking discourse in the second decade of the 21st century has shifted towards recognising diversity within trafficking, for example, trafficking for labour exploitation. Attempts to measure numbers of trafficked persons vary widely; largely due to difficulties in estimating the 'hidden figure' of trafficking in persons; that which takes place but is not detected. Estimates of the actual (detected and undetected) number of people trafficked annually across international borders range from around 700,000 to 2 million, with some estimates going as high as 4 million; internal trafficking estimations tend to fall between 2 and 4 million per year (Farr, 2013).

Forced labour

Forced labour, like trafficking, has received many decades of attention, much of it from the International Labour Organisation (ILO). Forced labour is usually defined as a situation in which a person is required to undertake work or services involuntarily under a menace of penalty (as established in the ILO's 1930 (No.29) Forced Labour Convention). The inter-war period saw the ILO spear-head international cooperative efforts to improve working conditions, albeit within a colonial framework. The Palermo Protocol signalled an opening of conceptual conversations between the terms forced labour and trafficking, as forced labour was included as one of the potential outcomes for trafficked persons in the Protocol. From the 1990s onwards trafficking was drawing more media and political attention. This made it expedient for the ILO to intervene more energetically in the growing anti-trafficking movement by suggesting forced labour could be the umbrella concept to cover all forms of trafficking.

In terms of measurements, the ILO produced minimum global estimates in 2005, 2012 and 2016. Although not directly comparable due to methodological differences, it was reported 12.3 million persons were in forced labour in 2005 (with at least 2.4 million of these estimated to be trafficked) compared to 20.9 million in 2012. The latest 2016 report has been produced with Walk Free Foundation (and heavily critiqued; Gallagher, 2017) and estimates 40.3 million in modern slavery, with a subset of 24.9 million in forced labour and 15.4 million in a new category of forced marriage. It is unknown how many people in these categories are trafficked. Such global estimates can be critiqued, but it is less controversial to assert that forced labour resides in *both* mature economies and less developed ones (Lewis et al., 2014).

Modern slavery

Common understandings of slavery until the 21st century related to its historical (trans-Atlantic slavery) not modern manifestations. Yet over the last decade the term *modern slavery* has become more prominent in the lexicon of global human exploitation. Modern slavery is generally conceived to be an umbrella term, encompassing human trafficking and forced labour (and slavery and servitude; aggravated forms of forced labour). Although not defined in any international instrument, most commentators begin their conceptual musings by finding shortcomings in the League of Nations' 1926 Slavery Convention definition which considered slavery to encompass powers of *ownership*. In a modern era legal ownership rights are very rarely asserted by slaveholders, so a set of 2012 Bellagio–Harvard Guidelines state that for contemporary or modern slavery, the exercise of ownership should be now understood as *possession*.

There have been a growing number of concerns, however, about the term modern slavery and its deployment in contemporary discourse (Craig et al., 2019). Some argue historical parallels risk appropriating the suffering of black people, obscuring our understanding of ongoing legacies of racial chattel slavery (Beutin, 2017). Others have pointed out that incremental adoption of the term modern slavery (eclipsing the term forced labour) leads to 'exploitation creep' with the side-effect that states can use this terminology to heighten moral condemnation and reduce their efforts to the most exceptional forms of exploitation (Chuang, 2014). There is also concern that growing politico-elite interest in modern slavery fails to acknowledge the role of neoliberal states in *creating* the conditions for modern slavery to thrive (Waite and Lewis, 2019).

The described terminological conflation across the terms trafficking, forced labour and modern slavery is of relevance to the provenance, politics, and construction of estimates in particular categories. The concern with resulting 'truths' about numbers of exploited/unfree persons is that certain forms of knowledge become valorised. Notwithstanding differences across estimates, what appears clear over the last two decades is that the figures show an upward trend in global exploitation. But we should ask whether there is a *real* increase in numbers or whether tools of detection have changed such that we are now *counting* more trafficked persons, forced labourers and modern slaves. Critical scholars in this field (e.g., Broome and Quirk, 2015) also urge awareness of how apparently neutral and technocratic assessments are able to partially conceal normative agendas and ideological projects (for example, conservative moral agendas around prostitution abolitionism that bleed into anti-trafficking responses).

Drivers of extreme exploitation

Understanding the *causes* of human trafficking, forced labour and modern slavery are socio-temporally contextual and often idiosyncratic. Such complexities do not play well in policy and legislative spheres. Much simpler is to understand extreme exploitation primarily through a criminal justice lens, with a binary construction of victims as passive agents and perpetrators as criminal offenders. Critical researchers instead suggest more nuanced appreciations of the variegated power relations and structural conditions that generate severe exploitation. This realm covers multi-scalar experiences that might contribute to an emerging situation of unfreedom such as: gender and other hierarchical social relations, local livelihood constraints/poverty, conflict and violence, the global demand for cheap labour generated from processes of neoliberal globalisation that systematically reduce worker protections and pervasive political economies embodying conservative moral agendas and heightened immigration and border security (Lewis et al., 2014).

Anti-modern slavery responses

The anti-modern slavery field is vast and positioned across multi-governance institutions, states, policy bodies, non-governmental organisations, community and faith-based groups. This broad interest is reflective of human exploitation simultaneously being a problem *of* development and *for* development. In terms of the D/development landscape, the need to undertake 'effective measures to eradicate forced labour, end modern slavery and human trafficking' is Sustainable Development Goal 8.7 and the promotion of freely chosen employment is part of the ILO Decent Work Agenda, 2001. Although multi-governance institutional responses to human exploitation are important, there is a lack of state action following ratification and getting laws onto statute books, and an arguable skewed focus of national legislation (e.g., the UKs 2015 Modern Slavery Act criticised as being more for 'heroic saviours' than victims, Craig et al., 2019). This has led to gaps in statutory provision in many countries – particularly around victim support. These gaps are increasingly filled by NGOs and civil society groups. There is a vibrant scene here, with organisations focusing on everything from awareness raising to direct support for those exiting situations of exploitation. Such organisations have undoubtedly raised the public profile of extreme exploitation in the last two decades; plus many are engaged in vital empowerment oriented work to help individuals re-build their lives following exploitation experiences.

Such work is lauded in certain spheres. In other more critical spheres there is a degree of concern regarding the 'rescue discourse' that seams through some responses (McGrath and Watson, 2018) and often appears to coalesce around so-called neo-abolitionist stances with their troubling neo-colonial, patriarchal and racist resonances (O'Connell Davidson, 2015). Rescues appear altruistic and humanitarian, yet may simultaneously appropriate the suffering of Other racialised bodies and neglect to provide post-rescue support services. The neo-abolitionists' discourse of globalised prostitution, for example, has led to a tendency to depict the enslaved person (often poor female) in need of rescue, and the rescuers shrouded in a righteous robe of 21st-century abolitionism. This industry arguably sits on a 'compassionate conservatism' platform whereby assistance for the exceptional few rests on the related impulse to be punitive against the many (O'Connell Davidson, 2015) and a tacit tolerance of everyday forms of labour abuse. It remains concerning that the vast majority of governments across the world are officially committed to the anti-slavery cause, yet, as noted by Quirk and LeBaron (2015: n.p.), 'this rhetoric conflicts with official support for legal regimes and policy responses that promote forms of systemic abuse, vulnerability, discrimination and exploitation'.

Conclusion

This chapter has provided a critical overview of contemporary human exploitation. Rather than seeing a relatively simple picture of a perpetrator/employer exploiting an individual in a binary relationship, susceptibility to exploitation should be understood as produced within complex intersections of structural, social and material relations. An understanding of points of vulnerability is useful here, as this allows us to see multiple and clustering vulnerabilities of unfree persons linked to factors both prior to exploitation (e.g., previous experiences of exploitation, poverty, conflict) and after arrival in the space of exploitation (e.g., environments of deterrence or hostility to new arrivals and the vulnerable, deregulated labour markets, precarious labour). There is much work to be done. Arguably we still lack knowledge about the complexities of extreme exploitation. Critical perspectives must remain central and can play a vital role in envisioning alternative ways of responding to unfreedom and exploitation that build on broader attempts to improve workers' rights and conditions.

References

Beutin, L.P. (2017). Black suffering for/from anti-trafficking advocacy. *Anti-Trafficking Review* (9), 14–30.

Broome, A. & Quirk, J. (2015). Governing the world at a distance: The practice of global benchmarking. *Review of International Studies* 41(5), 819–841.

Chuang, J. (2014). Exploitation creep and the unmaking of human trafficking law. *The American Journal of International Law* 108, 609–649.

Craig, G., Balch, A., Lewis, H. & Waite, L. (2019). *The Modern Slavery Agenda: Politics, Policy and Practice in the UK*. Bristol: The Policy Press.

Farr, K. (2013). Human trafficking. In Gold, S. & Nawyn, S.J. (Eds) *The Routledge International Handbook of Migration Studies*. London and New York: Routledge.

Gallagher, A. (2017). What's wrong with the global slavery index? *Anti-Trafficking Review* (8), 90–112.

Lewis, H., Dwyer, P., Hodkinson, S. & Waite, L. (2014). Hyper-precarious lives? Migrants, work and forced labour in the Global North. *Progress in Human Geography* 39(5), 580–600.

McGrath, S. & Watson, S. (2018). Anti-slavery as development: A global politics of rescue. *Geoforum* 93, 22–31.

O'Connell Davidson, J. (2015). *Modern Slavery: The Margins of Freedom*. London: Palgrave Macmillan.

Quirk, J. & LeBaron, G. (2015). The use and abuse of history: Slavery and its contemporary legacies. *Open Democracy*, 28 March 2015. Accessed 10 June 2019. https://www.opendemocracy.net/en/beyond-trafficking-and-slavery/use-and-abuse-of-history-slavery-and-its-contemporary-leg/

Waite, L. & Lewis, H. (2019). Migrant illegality, slavery and exploitative work. In Craig, G., Balch, A., Lewis, H. & Waite, L. (Eds) *The Modern Slavery Agenda: Politics, Policy and Practice in the UK*. Bristol: The Policy Press.

Additional resources

Alliance 8.7. http://www.alliance87.org

Anti-Slavery International. www.antislavery.org

Beyond Trafficking and Slavery. https://www.opendemocracy.net/beyondslavery

ILO. http://www.ilo.org/global/topics/forced-labour/lang--en/index.htm

Kempadoo, K. (2015). The modern-day white (wo)man's burden: Trends in anti-trafficking and anti-slavery campaigns. *Journal of Human Trafficking* 1, 8–20.

Kotiswaran, P. (2019) Trafficking: A Development Approach (October 8, 2018). Faculty of Laws University College London Law Research Paper No. 4/2019. Accessed 11 June 2019. SSRN. https://ssrn.com/abstract=3349103

Chapter 77

Cities, crime, and development

Paula Meth

Introduction

Crime is a dominant social issue in many cities across the Global South, exacerbating key development challenges including poverty and inequality, and undermining residents' well-being. It is tied to the realities of cities in complex ways, variously shaped by globalisation, rapid urbanisation, poverty, welfare regimes and the effectiveness of urban governance and policing. Crime within cities is a development issue, tied to urban inequalities and injustices, often historically formed. Historical processes have long term impacts on city residents, undermining livelihoods, residence options and justice. These include for example forms of colonisation; authoritarian political strategies and economic practices including preferential tax concessions or the destruction of informal economies. Contexts of economic, political and social marginalisation generate crime. However, the crimes of the poor are not the only – or most significant – urban crimes and causes of crime are more complex. Elite and state crimes are also significant, although their relationship to development studies is less clearly understood. This chapter explores the experiences of – and trends in – crime across cities in the Global South and considers ways of defining or explaining crime. It discusses the intersections between cities and crime and concludes with a consideration of crime as a development concern.

Experiences of crime and crime trends

Much crime research focuses on analyses of crime statistics. This work is valuable because it illustrates the scale and transitions or consistencies in crime over time and place. However it often cannot reveal more qualitative experiences of crime or explain the causes and consequences thereof. Crime can exacerbate poverty and inequality, occurring through the loss of earnings, livelihood opportunities and resources. Theft, corruption, homicide and assault can extend individual or household's poverty. Noting high crime rates in particular cities (see Table 77.1), managing crime through the adoption of different strategies is part of everyday life for many residents. These include how residents travel through the city and community's actions to provide protection. Crime also impacts institutions and organisations within cities, including business. Goldberg et al. (2014) reveal how 75 per cent of firms in Brazil spend money on security, in addition to loss of income through theft, fraud and higher insurance costs. Crime impacts employees. Businesses can close or relocate. It can be 'good for business' however, if providing security, for example. Certain businesses fuel crime within cities through illegal practices.

DOI: 10.4324/9780429282348-84

Table 77.1 Homicide rates per 100,000 of the population in most populous city by country (alongside the national Gini Index measuring the extent of income distribution)

Country	City	2008	2016	National Gini Index*	National Gini Index*
Kenya	Nairobi	4.0	3.7	47.7 (2005)	40.8 (2015)
El Salvador	San Salvador	133.7	193 (2015)	48.5 (2005)	38 (2017)
Lesotho	Maseru	59.7	61.9**	52.5 (2003)	54.2 (2010)
Jamaica	Kingston	29	54.3	45.5 (2004)	45.5 (2004)
Venezuela	Caracas	127.0	122**	46.9 (2006)	46.9 (2006)
Mongolia	Ulan Bator	11.5	6.3	36.5 (2008)	32.3 (2016)
Philippines	Quezon City	2.0	5.7***	43.0 (2009)	40.1 (2015)
India	Mumbai	1.3	0.9***	33.4 (2005)	35.7 (2011)

Source: adapted from UNOCD, 2019.

* data in these columns are from World Bank, 2012, 2019
** this is the latest figure which is from 2009
*** this is the latest figure which is from 2013

However, cities are not necessarily the most significant sites of crime within a nation. Between 2003 and 2016 there was a 34 per cent decrease in crime in cities in the Americas, contrasting national declines of around 16 per cent (UNOCD, 2019b). Statistics however obscure the lived realities of urban crime and experiences of fear within cities. A growing sensitivity of researchers to the emotional geographies of residents across cities of the Global South points to fear (and coping with crime) as a key source of anxiety (see Auyero et al., 2015 in further reading). Analysing crime trends across cities is difficult because the reliability and comparability of statistics is limited. Variations relating to crime classification, recording procedures, criminal justice systems and policing practices all reduce comparability. Processes of recording crime within countries is politically determined and allegations of strategic under-reporting are not uncommon. The United Nations Office on Crime and Drugs (UNOCD) provides some insight into urban comparison through listing particular crimes per most populous city, and these can be considered in relation to the Gini index for those countries (see Table 77.1).

Table 77.1 reveals the extreme variety of homicide rates across cities, as well as tentative (but not conclusive) evidence of the links between inequality and crime. India's low prevalence raises important questions about the necessary relationship between poverty and the incidence of more violent crimes, such as homicide, and the rise in San Salvador's homicide rates is not reflected in apparent improvements in levels of national inequality

Ways of defining and theorising crime

Defining what is – and what is not – a crime is a complex socio-political issue, as the notion of crime is contested. Whether particular practices are criminalised (defined and treated as criminal) or not tells us much about power relations within a society (examples of these issues will be given later in relation to informality and also gender violence). How we understand crime's relationship to a city or country's historical context as well as wider socio-economic and political practices occurring therein produces different analyses and theorisations of crime. Carrington et al (2019) argue for a 'southern criminology' which makes use of the insights and theories from the Global South to understand crime in those and other contexts. They argue that most literature on crime globally is dominated by Western or northern research.

A further factor shaping understandings of crime is the legislative variations in definitions of crime, which vary over time and place. The UNOCD listed in 2012 the following categories of crime under which their statistics were organised to include homicide, assaults, sexual violence, robbery, kidnapping, theft, motor vehicle theft, burglary and drug-related crimes (UNOCD, 2012). By 2019 this list had been broadened to include financing of terrorism, smuggling of migrants, corruption and sexual exploitation, reflecting changing global concerns over types of criminal activities, as well as shifts in the politicisation of particular crimes. Evidently, crimes such as fraud and police violence are not included, yet these remain significant in many cities in the Global South.

The poor within cities are frequently criminalised and suffer as victims of crime. This is a significant development concern, as it is also starkly gendered and racially determined. Men, particularly young men, suffer from violent crime disproportionately (UNOCD, 2019b) with 81 per cent of homicide victims globally being male (UNOCD, 2019b: 15). However, women suffer extensively from particular crimes, including sexual violence and intimate partner violence. Alves' work in São Paulo, Brazil explores questions of race in relation to police violence (and responses in return by criminal gangs) noting that young black men in particular are viewed as 'enemies of the state and as permanent threats to civil society' (2016: 64). Alves notes that 20 per cent of the murders in São Paulo are carried out by the police (calling this 'racial terror') and of these, 66 per cent are young, black men (2016: 62). This trend of state-directed violence towards young black men within cities is evident in South Africa too.

More broadly, the urban poor are often criminalised through their everyday practices of survival (housing, employment and access to justice). Those living informally (in many cities this accounts for over 80 per cent of the housing form) may technically be deemed criminal because their building structures do not meet required planning legislation, they may not have ownership or rights to their plot and they may access services such as electricity illegally. In these cases, the failure of cities to provide adequate housing for the growing urban poor enforces their 'criminal' practices. In many cities authorities turn a blind eye to such housing, but there are now multiple cases of mass evictions of slum housing which use the illegality of such structures to enforce the law (see Mumbai, India and Abuja, Nigeria for instance). A similar argument can be made in relation to informal economies across cities in the Global South, where traders frequently suffer the consequences of police raids, imprisonment and eviction (see Omoegun et al.'s (2019) analysis of traders practices following eviction in Nigeria and Resnick's (2019) analysis of crackdowns on traders in Accra, Dakar and Lusaka between 2000 and 2016). The criminalisation of the poor's informal practices contrasts with the treatment of the middle-classes, who often adopt illegal practices, which are condoned. In Delhi, some state institutions overlook or facilitate middle-class water users' wasteful or unpaid for water consumption practices in planned areas, blaming the poor for wastage. (Truelove, 2018).

A final example of the difficulties of defining crime relate to gender-based violence against women. Domestic violence and rape are not fully recognised as crimes in all states across the world. Additionally, even when recognised, multiple forces silence such crimes, meaning women suffer without legal protection or healthcare support. Middle Eastern countries' religious, societal, cultural and patriarchal structures means gender based violence is often viewed as a private issue despite legal mechanisms existing to counter such crimes. In Saudi Arabia, general intimate partner violence (IPV) rates vary between 34.3 and 57.7 per cent, yet Alhalal et al. (2019) note that 80.5 per cent of Saudi women view IPV as a private matter, reflecting the society's collectivist and patriarchal nature.

Cities and crime

Globally, crime levels are higher in cities than rural areas, but this is not always the case (e.g., homicide rates in the Americas). Criminal practices can crosscut the rural-urban divide (such as the trade in drugs) or can be experienced through online platforms. Nonetheless, crime is often associated with city life and commonly used to label cities with purportedly high crime levels. The urban phenomenon of crime is shaped by population size and density, whereby the concentration of populations accelerates the occurrence of crime therein. Cities also provide a market for criminals through hosting a large population of potential victims often in close proximity. Poorly managed rapid urbanisation and population growth mean cities often lack governance structures necessary to cope with crime. Policing services across most of the Global South are often ineffective and stretched, and criminal justice institutions (courts and prisons) are over-burdened. This has led to varying responses to crime beyond 'legal' policing measures, including gangs and/or vigilante groups and privatised security. Visibly dealing with crime effectively is a key political issue for city governments. Crime is a core election issue. As a result, crime management can be the focus of multiple governance structures and institutions including partnership organisations working between the police and communities, local political parties, local authority structures etc. In South Africa, reducing crime is embedded in the aims of other governance interventions, including state housing provision, with policy linking housing provision to crime reduction. In reality, persisting urban poverty and inequality means while new housing can reduce some crimes, it also produces new criminal practices because competition over assets is heightened (Meth et al., 2018). Crime prevention can dominate the local governance agendas absorbing resources. Despite their relative ineffectiveness, police forces are key urban actors across cities of the Global South, who play important roles as mediators and protectors. Styles of policing shift over time but more militarised combative policing is common in contexts where crime is high and the politics of crime are significant. Policing within the cities of South Africa and Brazil are again good examples here (see Lamb's (2018) work on the South African police's 'war on crime'). Finally, the physical designs of cities increasingly aim at managing crime. The rise of gated communities and the securitisation of urban spaces are evidence of this.

To conclude, crime within cities is a core development issue. It impacts on the social, political, physical, economic and emotional lives of residents and responses to crime can also be a developmental concern. Crime is also directly affected by changes in wider development indicators, particularly employment and governance.

References

Alhalal, E., Ta'an, W., & Alhalal, H. (2019). Intimate partner violence in Saudi Arabia: A systematic review. *Trauma, Violence, & Abuse*, 22(3). https://doi.org/10.1177/1524838019867156

Alves, J. A. (2016). 'Blood in reasoning': State violence, contested territories and black criminal agency in urban Brazil. *Journal of Latin American Studies*, 48(1), 61–87. https://doi.org.sheffield.idm.oclc.org/10.1017/S0022216X15000838

Auyero, J., Bourgois, P., & Schepher-Hughes, N. (Eds) (2015). *Violence at the Urban Margins.* Oxford: Oxford University Press.

Carrington, K., Dixon, B., Fonseca, D., Rodríguez Goyes, D., Liu, J., & Zysman, D. (2019). Criminologies of the Global South: Critical reflections. *Critical Criminology*, 27, 163–189. https://doi.org/10.1007/s10612-019-09450-y

Goldberg, M., Kim, K. W., & Ariano, M. (2014). *How Firms Cope with Crime and Violence: Experiences from around the World.* Washington DC: World Bank. https://openknowledge.worldbank.org/bitstream/handle/10986/16539/9781464801013.pdf?sequence=1&isAllowed=y [site accessed 4 September 2019].

Lamb, G. (2018). Police militarisation and the 'war on crime' in South Africa. *Journal of Southern African Studies*, 44(5), 933–949. https://doi.org/10.1080/03057070.2018.1503831

Meth, P., Buthelezi, S. and Rajasekhar, S. (2019). Gendered il/legalities of housing formalisation in India and South Africa. *Environment and Planning A: Economy and Space*, 51(5), pp. 1068–1088.

Omoegun, A. O., Mackie, P., & Brown, A. (2019). The aftermath of eviction in the Nigerian informal economy. *International Development Planning Review*, 41(1), 107–128. https://doi.org.sheffield.idm.oclc.org/10.3828/idpr.2018.30

Resnick, D. (2019). The politics of crackdowns on Africa's informal vendors. *Comparative Politics*, 52(1), October 2019, pp. 21–51(31). https://doi.org/10.5129/001041519X15615651139961

Truelove, Y. (2018). Negotiating states of water: Producing illegibility, bureaucratic arbitrariness, and distributive injustices in Delhi. *Environment and Planning D: Society and Space*, 36(5), 949–967. https://doi.org/10.1177/0263775818759967

UNOCD (2019). *United Nations Office on Drugs and Crime.* dataUNODC. https://dataunodc.un.org/

UNOCD (2012). *United Nations Office on Crime and Drugs.* www.unocd.org [site accessed 18 April 2012].

UNOCD (2019b). *Global Study on Homicide: Executive Summary.* Vienna: United Nations Office on Drugs and Crime.

World Bank (2012). *Gini Index.* The World Bank. www.data.worldbank.org [site accessed 19 April 2012].

World Bank (2019). *Gini Index.* The World Bank. www.data.worldbank.org [site accessed 4 September 2019].

Further reading

Auyero, J., Bourgois, P., & Schepher-Hughes, N. (Eds) (2015). *Violence at the Urban Margins.* Oxford: Oxford University Press. Focusing on various urban contexts across the Americas (the US, Nicaragua, Colombia etc.), this book draws on the work of sociologists and anthropologists to examine the particularities of violence in multiple urban contexts. It considers how violence is experienced, its causes, its political and criminal use and its impact on differentiated urban residents.

Salahub, J. E., Gottsbacher, M., De Boer, J., & Zaaroura, M. D. (2019). *Reducing Urban Violence in the Global South: Towards Safe and Inclusive Cities, Routledge Studies in Cities and Development.* Abingdon: Routledge. This book explores a range of context-specific techniques and approaches in cities across the Global South attempting to respond to or reduce urban violence and crime. It does this by contextualising the realities of cities in the Global South, with a particular focus on poverty and inequality. It offers a critical reflection of these varied interventions.

Audio-visual source

Muggah, R. (2017). *TED Global Presentation Demonstrates why Cities should have Greater Political Power*, 10 November 2017. https://www.ted.com/talks/robert_muggah_the_biggest_risks_facing_cities_and_some_solutions [site accessed 4 September 2019].

Chapter 78

Policing and development

Charlotte Cross

Introduction

This chapter reviews efforts to improve policing, referring to practices aimed at maintaining order, as part of development initiatives. It considers Security Sector Reform (SSR), community policing and attempts to engage with policing providers beyond the state police. In outlining key critical perspectives on such endeavours, the chapter emphasises the parallels between interventions aimed at improving policing and those in other sectors of international development policy and practice, including the limitations of attempts to impose standardised templates on diverse contexts and the politics of development at different scales. In addition to being the subject of international development interventions, policing is important as it can influence how development is defined and pursued, and who benefits from it and who does not. Thus, the chapter concludes by considering the policing *of* development and how it is negotiated and resisted.

Reforming policing for development

Policing has received increasing attention within international development since the 1990s, reflecting the growing influence of the so-called security-development nexus, referring to the assumed interdependence of development and security. The most prominent manifestation of attempts to address security and development in tandem through development interventions is SSR, involving reform and restructuring of security institutions including the military, judiciary, and police. Like many other international development initiatives, this typically entails provision of training and 'capacity building', sharing of technical expertise, and the transfer of financial and other resources from the Global North to the Global South.

Perceptions and experiences of policing – and the extent to which it contributes towards protecting the physical security and livelihoods of individuals and groups and maintaining positive social relationships – can be seen as an important influence on assessments of well-being and thus a worthy focus for development policy. Policing also shapes access to justice and informs the interpretation of rights and freedoms to which individuals and groups are entitled and the extent to which they are realised.

In many contexts in which such initiatives have been implemented there are further compelling reasons for efforts to improve policing. The enduring influence of histories of colonialism has informed cultures and practices of policing oriented to protecting regimes rather than populations. Policing is heavily implicated in maintaining authoritarian

DOI: 10.4324/9780429282348-85

political structures, through the policing of protests and elections, the regulation of civic space and the relationship between policing and political violence. Thus, development interventions are often premised upon promoting 'democratic' policing, as part of support for the rule of law and pursuit of wider governance objectives (e.g., Bayley, 2001). Police reform has received particular attention as part of 'peacebuilding' and 'state-building' in post-conflict contexts, where police may be heavily implicated in past and current violence and police legitimacy is often severely compromised.

Increased spending on supporting policing and related security initiatives can also be understood, however, as part of a broader securitisation of development and of aid-recipient regions and states, whereby a perceived lack of development is interpreted as a security risk, affecting both the countries concerned and 'developed' nations. While described as provision of development aid, such endeavours thus reflect specific national security interests linked to preventing threats to donor countries, such as counter-terrorism or disruption of drug-trafficking routes.

Much like development interventions in other sectors, externally sponsored police reform has also been criticised for entailing the imposition of standardised models with little awareness of social or political context, predicated on the assumption that a form of universalised 'transnational' policing practice based on Western liberal values is both possible and desirable (Hills, 2009). Many such reforms do not succeed on their own terms, generating only superficial or short-term change and may have unintended consequences as they are reshaped in their implementation by a range of interests and in accordance with local histories and cultures of policing. Furthermore, SSR and other donor-led initiatives have often failed to engage sufficiently with those actors most likely to have an impact on citizens' safety in their daily lives, for example, neglecting the state police in favour of the military and marginalising the wide range of institutions beyond the state police through which 'everyday policing' (Buur and Jensen, 2004) takes place, as discussed further later.

Policing for communities?

One of the principal means by which it is hoped policing can be made more democratic is through the promotion of 'community policing'. Despite – or perhaps because of – the difficulty in defining with any precision what this amorphous and malleable term means, it has become a virtually ubiquitous component of police reform programmes world-wide. Broadly, it conveys a sense of policing that in some way reflects what communities want. What this means in practice and what might constitute 'successful' implementation is, however, sufficiently ambiguous to enable community policing to be supported by diverse actors with different and sometimes competing interests and objectives (Brogden and Nijhar, 2005).

Originating in the UK and North America in the 1980s, community policing has since been promoted by bilateral donors, international non-governmental organisations and private consultancy firms around the world. In addition to improving relationships between police and communities and supporting general crime prevention, community policing has also been proposed as a solution to a wide range of specific challenges including road safety, violent extremism, gender-based violence, security provision in refugee camps and the illegal wildlife trade.

As observed of donor-supported police reform more broadly, community policing led by development organisations often entails attempts to implant standardised models with limited consideration of local context (Brogden and Nijhar, 2005). A rich body of literature thus explores how community policing is in practice reshaped to reflect particular historical, social and political contexts and diverse understandings and expectations of policing and 'the community', often diverging from the 'democratising' ideal promoted by international actors. Understanding what community policing means in different contexts is further complicated by the range of relationships and practices that might be described in this way. Initiatives sponsored by external donors or the state police often exist alongside – and in relationship to – other forms of policing carried out by and within communities, which may entail limited or no state police involvement. Indeed, how to engage with a plurality of policing providers has emerged as a significant policy challenge for development organisations, discussed further in the following section.

Policing beyond the state

As noted earlier, some of the limitations of SSR – and related 'state-building' and 'peace-building' initiatives – have been attributed to an undue focus on state security providers. Such critiques have informed greater recognition of the more complex policing landscape in many contexts, in which 'everyday policing' is performed by a range of actors engaged in the production and defence of moral communities, for whom what constitutes 'crime' may diverge from the dictates of state law (Buur and Jensen, 2004). The importance of vigilantes, private security providers and, indeed, the extent to which state police often do not act in accordance with the formal laws they are supposed to obey and are implicated in wider social and political relations complicate attempts to draw neat distinctions between state and non-state policing. Some forms of everyday policing can usefully be characterised as 'twilight institutions' (Lund, 2006), which exercise public authority but are not the state, operating 'between state and society, between public and private' (ibid. 686). Such institutions may derive their authority from diverse sources, including: claims to represent 'the community'; association with traditional or religious authorities; perceived effectiveness in tackling crime and other problems or a connection to a particular political movement or cause. However, the extent to which the exercise of authority to maintain a particular form of order is legitimate and accepted may change over time.

The idea that such institutions should be recognised within police reform strategies had become influential by 2011 when the World Development Report argued for a 'best fit' approach to security provision, involving 'combinations of state, private sector, faith-based, traditional and community structures for service delivery' (World Bank, 2011: 106). Although enthusiasm for alternatives to state policing marks a shift, in rhetoric at least, from state-centric models of police reform within international development assistance, mobilisation or co-optation of policing providers other than the state police to meet particular ends is hardly a new phenomenon. Governments of countries including Ethiopia, Mexico, Nigeria, Peru and Tanzania have at various points offered some form of recognition to policing providers that might otherwise have posed a threat to state authority.

Both pragmatic and normative reasons are advanced for incorporating providers other than the state police into policing provision. In contexts of limited resources, particularly following conflict, it might be more feasible to support existing providers rather than take on the expensive commitment of rebuilding state security agencies. Alternative providers are also often assumed to have greater local legitimacy, to be rooted in local 'tradition' and culture and to reflect local norms and priorities (Bagayoko et al., 2016). This mirrors broader trends in development scholarship and practice, whereby critiques of the failures of 'good governance', entailing the imposition of models based on idealised versions of European and North American institutions, have led to calls to work with 'the grain' of existing institutional arrangements and to support for 'practical hybrids' referring to 'modern state structures that have been adapted to, or infused with, contemporary cultural preferences' (Booth and Cammack, 2013: 112).

While recognition of the complexity of policing provision in many contexts and the need to problematise assumptions based on European and North American experience is important, significant questions remain regarding the likely 'success' of attempts to instrumentalise the plurality and hybridity of policing for 'development'. These challenges are not unfamiliar within development studies and the discussion here is informed by the extensive literature on participatory or 'community-based' development.

First, community policing and promotion of hybrid policing models risks romanticising and homogenising communities, occluding hierarchies and social divisions within neighbourhoods. Much like other development initiatives, people's participation in policing may sometimes be better understood as part of personal political or economic strategies than representing collective objectives. Given that policing often entails the capacity and authority to exercise violence, such dilemmas are particularly challenging. Use of terms such as 'community' obscures the fact that policing ultimately involves 'suppressing the activities and practices of one group on behalf of another' (Brogden and Nijhar, 2005: 234), and indeed policing can contribute to how communities are formed and maintained, shaping who is included and who is not. Such processes are often contentiously linked to party political competition as well as broader negotiations over social relations including those of gender and generation. While often used to connote more democratic and inclusive policing, for those who are marginalised within particular groups, rather than improving policing, empowering certain actors to police their communities might amount to 'decentralized repression' (Ruteere and Pommerolle, 2002).

Second, while 'local' forms of policing are often assumed to be more legitimate and sustainable due to greater local ownership and compatibility with local norms, it should not be taken for granted that what is locally available necessarily represents the kind of service people actually want. Providing neighbourhood policing is often costly and dangerous, and thus rather than community-led policing, people may aspire to have more 'modern' and 'state-like' services and crucially greater financial and practical support from the government, despite negative experiences of state policing. Thus, there is a danger that co-opting alternative security providers really comes to resemble a form of 'bargain-basement governance' (Meagher, 2012: 1078), meeting the objectives of donors and governments rather than those who must rely on the services in question.

Conclusion

This chapter concludes by arguing for the importance of further research that goes beyond considering policing as a subject of development initiatives to explore the policing *of* development, investigating how policing, carried out by a range of actors, shapes whose ideas of positive change are permitted and realised and who benefits from 'progress' and who does not. Negotiations over what kind of order should be upheld and how shape development processes and their implications. Investigating policing thus offers a productive lens through which to analyse impacts of and responses to broader processes of change, such as those linked to urbanisation or technological innovation, as well as key development challenges such as the climate emergency or the COVID-19 pandemic. Furthermore, debates about how policing *should* be – and the contestation of its repressive and exclusive manifestations – also offer an important arena in which to explore diverse imaginations and articulations of desirable futures.

References

Bagayoko, N., Hutchful, E. and Luckham, R. (2016) 'Hybrid security governance in Africa: rethinking the foundations of security, justice and legitimate public authority', *Conflict, Security and Development* 16, 1: 1–32.

Bayley, D. H. (2001) *Democratizing the police abroad: what to do and how to do it.* Washington: National Institute of Justice.

Booth, D. and Cammack, D. (2013) *Governance for development in Africa: solving collective action problems.* London: Zed Books.

Brogden, M. and Nijhar, P. (2005) *Community policing: national and international models and approaches.* Cullompten: Willan.

Buur, L. and Jensen, S. (2004) 'Introduction: vigilantism and the policing of everyday life in South Africa', *African Studies* 63, 2: 139–152.

Hills, A. (2009) 'The possibility of transnational policing', *Policing and Society* 19, 3: 300–317.

Lund, C. (2006) 'Twilight institutions: public authority and local politics in Africa', *Development and Change* 37, 4: 685–705.

Meagher, K. (2012) 'The strength of weak states? Non-state security forces and hybrid governance in Africa', *Development and Change* 43, 5: 1073–1101.

Ruteere, M. and Pommerolle, M. (2002) 'Democratizing security or decentralizing repression? The ambiguities of community policing in Kenya', *African Affairs* 102, 409: 587–604.

World Bank (2011) *World development report 2011: conflict, security and development.* Washington, DC: World Bank.

Further resources

Cross, C. (2021) 'Dissent as cybercrime: social media, security and development in Tanzania', *Journal of Eastern African Studies* 15, 3.

Hönke, J. and Müller, M. (eds) (2016) *The global making of policing.* Abingdon and New York: Routledge.

Medie, P. A. (2020) *Global norms and local action: the campaigns to end violence against women in Africa.* Oxford: Oxford University Press.

Pratten, D. and Sen, A. (eds) (2007) *Global vigilantes: perspectives on justice and violence.* London: Hurst.

Waseem, Z. (2021) 'Policing COVID-19 through procedural informality in Pakistan', *Policing and Society*, 31, 5: 583–600. https://doi.org/10.1080/10439463.2021.1887869

Peace-building partnerships and human security

Timothy M. Shaw and Abigail Kabandula

It is easy to fault UN peacekeeping missions around the world. 'Blue helmets' are frequently wasteful, toothless and lack a clear line of command. . . . Nonetheless, the eight UN missions in Africa have made the difference between a new descent back into civil war and a slow but hopeful climb towards stability.

(Economist, 2012: 15)

Introduction

After two decades, human security as concept and practice still seeks to privilege personal economic and social rather than 'national' strategic concerns, including ecological, educational, food, habitat and health priorities (UNDP, 1994: 22–40). But, despite its relatively recent post-Cold War definition and advocacy, human or individual security along with development has become increasingly more problematic in the 21st century. In part, this reflects the interrelated impacts of a decade of the BRICs/BRICS (Brazil, Russia, India, China and South Africa) and a series of 'global' economic crises around the turn of the second decade. In turn, debates about its definitions and elusiveness have proliferated and intensified as indicated in the second section, including the articulation of the notion of 'citizen security' from the Global South (UNDP, 2012).

Any post-bipolar 'peace dividend' was shattered after a short decade by the 'global' shocks of 9/11 in the US in late 2001 then 7/7 in the UK in mid-2005. Subsequent 'wars on terrorism' and interventions in Syria, Iraq and Afghanistan have complicated previously rather simplistic or idealistic notions of peace-keeping roles and partnerships – R2P – for post-conflict reconstruction as articulated in ICISS (2001).

This chapter for the fourth edition of this Companion has been rewritten because the interrelated worlds of human security (as well as peace-building) and development are quite different at the beginning of the third decade of the 21st century. These shifts were reflected in revisionist UN (2003, 2004) and donor deliberations yet they remain resilient as indicated in a new overview of fragile states with a focus on a trio of cases: Afghanistan, DRC and Haiti (Brock, et al., 2012). But they remain as important as ever for 'development studies' and for state and non-state policies as suggested ten years ago (Shaw, 2008) especially now that the UNECA (2011) advocates a developmental state policy for Africa. Further, as indicated in the second section, there is growing anxiety within International Non-Governmental Organisations (INGOs) and related circles about the tendency towards the conflagration of development and security post-9/11 (World Bank Group, 2018; SIDA, 2016; CIDCA, 2018; Gates Foundation, 2020).

DOI: 10.4324/9780429282348-86

This contribution has two interrelated themes. The first is that human security, however defined, remains as relevant as ever to global development even if it was momentarily displaced by an apparent return to the hegemony of 'national security' after 9/11. And the second is that peace-building is likewise vital but it has now become much more dangerous or 'robust' than at the end of the 20th century. In short, development and security are more intertwined and inseparable than ever, leading to controversial notions like the securitisation of development as well as the privatisation of security as indicated later, even the militarisation of refugee communities and camps (Muggah, 2006).

Human security before and after 9/11 and 7/7

Human security was articulated in the post-bipolar world as an antidote to established notions of national security and balance of power: the privileging of individual rather than collective security against the threat and practice of violence as the principal referent of security. Symptomatic of contemporary policy development, like parallel discourses on human development and human rights, human security has been defined by international agencies and think-tanks. Such 'public diplomacy' reflects growing 'contracting-out' by national regimes to combinations of non-governmental organisations, multinational corporations and international institutions: the bases of 'new multilateralisms' of global mixed actor coalitions around 'new' security issues like land-mines and blood diamonds, small arms and child soldiers. The apex of the first period of human security deliberations was the late-1990s Ottawa Process around land-mines advanced by the 1400-member International Campaign to Ban Landmines (ICBL) and spearheaded by Lloyd Axworthy as Canadian foreign minister.

As conceived and advocated in the mid-1990s by the UNDP (1994: 22–40), human security includes interrelated community, economic, environmental, food, health, personal and political securities. Human security puts people and other aspects that support their existence at the centre of security and so challenges the traditional notion of security that privileges the state and its constituencies. Because of its focus on issues that affect the people, human security issues are also known as non-traditional security (NTS) threats (Peoples and Vaughan-Williams, 2015). This notion of security expands the security referent object (who/what is secured) to people, animals, environment and development. Further, the notion also expands the actors that play a role in providing security, as well as causing insecurity. Traditionally, security was the state's domain. However, with the expansion of security into novel problems, non-state actors, both armed and unarmed such as INGOs, NGOs, insurgents, terrorist groups and private security companies play important roles in providing security as well as causing insecurity. Therefore, human security not only highlights the complexity of security in the 21st century but also the significance of continuous transformation of security and development agencies and agendas.

Further, most human security concerns including infectious diseases, intra-state wars ('new wars'), transnational crime, migration and proliferation of small arms and now drones are distinct from traditional inter-state security threat because they are characterised by a higher probability and intensity with ordinary people as the main casualties (Krahmann, 2003). Moreover, most human security threats are transnational in nature, so challenging the authority and reach of individual states, thus elevating multilateralism in security.

Nonetheless, by the start of the third decade of the new century, uneven 'globalisation' had served to proliferate security issues especially around the latest generation of 'new' states which has led towards +/–200 today, South Sudan being the 193rd member of the

UN. Any lingering idealism was shattered by the shocks of 9/11 and the US unilateralist response: its declaration of a 'war on terrorism'. In new and fragile states, globalisation has intensified and complicated NTS threats and measures to promote security (Kabandula and Shaw, 2018). Information technology with globalisation has rendered time and distance obsolete. The development of information technologies has connected the world in ways never imagined. The internet and cell phone technologies have connected millions across the globe. However, the impact of new information communication technologies (ICTs) on security in fragile states is two-fold, positive and negative. ICTs have increased the visibility of NTS threats including human rights violations allowing for a multiplicity of actors – individuals, NGOs, and the private sector – to respond. These actors can mobilise for human security over several interconnected communication networks like email, YouTube, Facebook, Twitter and WhatsApp. In this sense, globalisation has a positive effect in addressing NTS threats and promoting human security.

Nevertheless, globalisation can also exacerbate NTS threats and strain measures to address insecurity. The effects of globalisation are felt at local and community levels as new social forces change traditional ways of life, cultures and language, plus erode traditional livelihoods and economies. The erosion, in some cases, has caused identity crises arising from uncertainty and anxiety in daily life as people's political, economic and psychological activities are dislocated. Religious extremism manifesting as terrorism and perverse violence are some ways of coping and self-affirmation (McRae, 2001). These insecurities are transported to the global level through global interaction and interconnection aided by technology. Insecurity can travel through the spread of extremist ideologies and long-distance projection of violence with new technology such as drones (Stiglitz and Kaldor, 2013). Further, information technologies can prolong and intensify 'new wars', encourage state failure and dwarf development. Information technologies such as the internet and offshore banking institutions enable non-state actors to acquire resources to continue war (Axworthy, 2001) and make it difficult to protect people. Non-state actors with illegal control of local resources such as gold and diamonds can trade on the international black markets and use the money to pay for guns. They are also able to keep and access their funds in international offshore banks where they are protected from state regulation. The wars in Somalia (2006 to present) and Nigeria (2009 to present) are cases in point.

Symbolically, the development of a human security doctrine and elaboration of the related 'responsibility to protect' (R2P) in those communities and countries where it was threatened was in process when 9/11 diverted attention. So, the December 2001 report of the blue-ribbon Canadian-supported International Commission on Intervention and State Sovereignty (ICISS, 2001) was overshadowed by the new preoccupation with international terrorism. Thus, while the notion of human security is now almost three decades old, its definition and realisation are more problematic than ever (MacLean et al., 2018). The ICISS (2001: xi) intended to extend the notion of international law for the new millennium from state sovereignty to the protection of people:

> Where a population is suffering serious harm, as a result of internal war, insurgency, repression or state failure, and the state in question is unwilling or unable to halt or avert it, the principle of on-intervention yields to the international responsibility to protect.

However, as the war on terrorism has dragged on and its costs – human, financial, regional etc. – have become ever more apparent, so analysts have begun to rediscover human rather than other varieties, such as national or regional even global security. Thus, the 2003 report of the UN Commission on Human Security and subsequent reports on 'new' security threats and responses around the millennium summit in 2004–2005 served to both rehabilitate and refine the concept. The post-bipolar and 9/11 pre-summit panel report on 'A More Secure World; our shared responsibility' recognised that:

> The threats are from non-State actors as well as States, and to human security as well as State security. The central challenge for the twenty-first century is to fashion a new and broader understanding . . . of what collective security means.
>
> (UN, 2004: 11)

Among the half-dozen clusters of threats identified by the UN (2004: 12) panel were:

> Economic and social threats, including poverty, infectious diseases and environmental degradation. . . . Internal conflict, including civil war, genocide and other large-scale atrocities. . . . Transnational organized crime.

We conclude this section by noting the continuation of an intense debate around human security in the second decade of 21st century. Initially this revolved around narrower versus broader conceptualisation: freedom from fear versus freedom from want (MacLean et al., 2018). Initial formulations arising from the human development genre around the UNDP favoured the latter. By contrast, more cautious or conservative analysts oriented towards traditional international relations (IR) favoured the former. Reflective of such revisionist inclinations, MacFarlane and Khong (2006: 228) lamented the 'conceptual overstretch' around 'freedom from want' preferring to limit such security threats to those 'against their physical integrity' or 'organised violence'. This standoff was previewed by 21 analysts in a special section of *Security Dialogue* in the mid-decade (Burgess and Owen, 2004). As indicated in the next part, it entailed more than conceptual disagreement; it profoundly affected data and policy. But at the start of the third decade the field is being subjected to conceptual development around 'citizen security' (UNDP, 2012) as well as a radical critique of the humanitarian enterprise as being intended to advance and insulate global inequalities (Soderbaum and Sorensen, 2012).

Threats to human security in theory and policy

This second part juxtaposes a set of overlapping discourses which increasingly impact both the definition and implementation of human security at the start of the first second decade of the new century. In addition to freedom from fear/want, these include: uneven globalisation, 'African' international relations (IR) and redefinitions of both development and security. Taken together, this trio of debates has profoundly complicated the conceptual and empirical relationships between development and security.

First, 'globalisation' has clearly become more uneven than ever leading towards a new trio of 'worlds': i) the OECD states, ii) the 'emerging economies' of BRICS and N-11 and iii) the approximately 50 fragile states concentrated in Africa (Brock et al., 2012). Such inequalities are both intra-and inter-state, encouraging transnational alliances amongst the rich and the poor to either defend or challenge the status quo. Hence the privatisation of security on the one hand and the tendency towards political and/or religious radicalisation on the other.

Second, a further overview of 'African' IR as part of a global trend towards the 'discovery' of perspectives from the Global South (Cornelissen et al., 2012; Warner and Shaw, 2018) has raised the issue of whether such relations can be limited to inter-rather than intra-state relations of cooperation and conflict. This 'academic' discussion poses profound implications for security policy; for example, do only deaths from classical inter-state wars count? Reflective of conservative, traditional inter-state definitions, the first *Human Security Report 2005* from University of British Columbia (UBC; 2005) can claim an optimistic picture of a decline in conflicts and related deaths at century's turn. By contrast, data on 'internal' or intra-state wars and body-counts would be much less sanguine. Thus, Cornelissen et al. (2012: 12) note that a wide range of cross-border relations on the continent are unrecorded despite the several land-locked states: from economic and ecological, ethnic and criminal to disasters and viruses.

And third, as regional conflicts have continued in, say, Central Asia, the Great Lakes, Horn (from Darfur to Somalia), West Africa etc. (Boas and Dunn, 2007; Muggah, 2006), so two divergent responses can be identified. First, an extended version of the optimistic perspective on 'new multilateralisms' around 'new' security issues like landmines, blood diamonds, child soldiers, small arms (GIIS, 2011) etc.: further Ottawa and Kimberley Processes augmented by more robust peace-building as in, say, Sierra Leone along with longer-term developmental innovations like the Diamond Development Initiative. But second, a more critical or sceptical response based on the recognition of 'greed' rather than 'grievance': the 'political economy of conflict'. This cautions that peace-keeping per se cannot be efficacious as conflict is over resources and revenues rather than principles, notwithstanding efforts around corporate codes of conduct, Extractive Industries Transparency Initiative (EITI) etc. (Boege et al., 2006): the 'new' extended peace-keeping partnerships.

Growing scepticism about the feasibility of a 'liberal peace' is reinforced by the trend towards the securitisation of development and/or militarisation of security (Soderbaum and Sorensen, 2012; Duffield, 2001; Duffield, 2006). INGO policy analysts are increasingly apprehensive about this direction as it erodes the boundary between development and security, enabling national and private security organisations to claim to qualify for development funds under the DAC of the OECD.

Futures for peace-building and human security

Given the aforementioned analytic and empirical trends and debates, optimism about the future of peace-building partnerships and human security is a scarce commodity. Whilst the world of peace-building continues to expand given demand and supply, post-9/11 and 7/7 conflicts have eroded and compromised its niche. Continued enlightened new multilateralism and informed public diplomacy around the human development/security nexus are to be encouraged (MacLean et al., 2018). But the context is now complicated by improved prospects for the securitisation of development and the militarisation of security (Muggah, 2006) exacerbated by interventions in Iraq, Afghanistan and Syria. Moreover, the advocacy of citizen security in response to fear of crime and violence, especially around transnational organised crime (TOC) in the Caribbean and Central

America (UNDP, 2012), may widen policy options but complicate decision-making (Hanson et al., 2012).

Guide to further reading

Kabandula, A. & Shaw, T. M., 2018. Rising Powers and the Horn of Africa: Conflicting Regionalisms. *Third World Quarterly*, 39(12), pp. 2315–2333. An overview of how rising regional states are increasingly impacting NTS in the Horn, leading to migrations and diasporic involvements.

Kabandula, A. & Shaw, T. M., 2020. Rising Powers and the Horn of Africa: Conflicting Regionalisms. In: E. Parlar Dal, ed. *Rising Powers in International Conflict Management: Converging and Contesting Approaches*. 1st ed. London: Routledge. Regional transnational relations in the Gulf and Turkey are impacting human security in the Horn at the start of the third decade of the 21st century.

MacFarlane, S. N. & Khong, Y. F., 2006. *Human Security and the UN: A Critical History*. Bloomington: Indiana University Press. A comprehensive, informed but sceptical overview of the evolution of human security which advocates narrower freedom from fear rather than broader freedom from want.

MacLean, S. M., Black, D. R. & Shaw, T. M., 2018. *A Decade of Human Security: Prospects for Global Governance and New Multilateralisms*. Abingdon: Routledge. A reflective, somewhat revisionist history of the first decade of human security informed by Canadian perspectives.

References

Axworthy, L., 2001. Human Security and Global Governance: Putting People First. *Global Governance*, 7, pp. 19–23.

Boas, M. & Dunn, K. C., 2007. *African Guerrillas: Raging Against the Machine*. Boulder: Lynne Rienner.

Boege, V., Fitzpatrick, C. & Jaspers, W., 2006. *Who's Minding the Store? The Business of Private, Public and Civil Actors in Zones of Conflict*. Bonn: BICC.

Brock, L., Holm, H.-H., Sorenson, G. & Stohl, M., 2012. *Fragile States*. Cambridge: Polity.

Burgess, J. P. & Owen, T., 2004. Special Section: What is Human Security?. *Security Dialogue*, 35(3), pp. 345–387.

CIDCA, 2018. *China International Development Cooperation Agency* [Online]. Available at: http://en.cidca.gov.cn/2018-08/01/c_259525.htm [Accessed 26 January 2020].

Cornelissen, S., Cheru, F. & Shaw, T. M., 2012. *Africa and International Relations in the 21st Century*. London: Palgrave Macmillan.

Duffield, M., 2001. *Global Governance and the New Wars: The Merging of Development and Security*. London: Zed Books.

Duffield, M., 2006. Human Security: Linking Development and Security in an Age of Terror. In: *New Interfaces between Security and Development: Changing Concepts and Approaches*. Bonn: German Development Institute (DIE), pp. 11–38.

Economist, 2012. UN Troops in Africa: Blue Berets in the Red. *Economist*, 9 June, 403(8788), pp. 15–16.

Gates Foundation, 2020. *Bill & Melinda Gates Foundation* [Online]. Available at: https://www.gates-foundation.org/What-We-Do/Global-Health/Integrated-Development [Accessed 30 January 2020].

GIIS, 2011. *Small Arms Survey 2011: States of Security*. Oxford: OUP.

Hanson, K. T., Kararach, G. & Shaw, T. M., 2012. *Rethinking Development Challenges for Public Policy: Insights from Contemporary Africa*. London: Palgrave Macmillan.

ICISS, 2001. *The Responsibility to Protect: Report of the International Commission on Intervention and State Sovereignty*. Ottawa: International Development Research Centre.

Krahmann, E., 2003. Conceptualizing security governance. *Cooperation and conflict*, 38(1), pp. 5–26.

MacFarlane, S. N. & Khong, Y. F., 2006. *Human security and the UN: A Critical History*. Bloomington: Indiana University Press.

MacLean, S. M., Black, D. R. & Shaw, T. M., 2018. *A Decade of Human Security: Prospects for Global Governance and New Multilateralisms*. Abingdon: Routledge.

McRae, R., 2001. Human Security in a Globalized World. In: R. McRae & D. Hubert, eds. *Human Security and the New Diplomacy: Protecting People, Promoting Peace*. London: McGill-Queen's University Press, pp. 14–27.

Muggah, R., 2006. *No Refuge: The Crisis of Refugee Militarization in Africa*. London: Zed Books.

Peoples C. and Vaughan-Williams, N. (2015). *Critical Security Studies: An introduction*. p. 29– 46. London. Taylor Francis Group.

Shaw, T. M., 2008. Peace-Building Partnerships and Human Security. In: V. Desai & R. B. Potter, eds. *The Companion to Development Studies*. New York: Routledge, pp. 468–471.

SIDA, 2016. *Swidish Development Cooperation Agency* [Online]. Available at: https://www.sida.se/English/how-we-work/our-fields-of-work/conflict-resolution-peace-and-security/ [Accessed 20 January 2019].

Soderbaum, F. & Sorensen, J. S., 2012. The End of the Development-Security Nexus? The Rise of Global Disaster Management. In: F. Soderbaum & J. S. Sorensen, eds. *Development Dialogue*. Uppsala: Dag Hammarskjöld Foundation, pp. 7–20.

Stiglitz, J. E. & Kaldor, M., 2013. Protection from Violence. In: J. E. Stiglitz & M. Kalder, eds. *The Quest for Security: Protection Without Protectionism and the Challenge of Global Governance*. New York. Columbia Press, pp. 91–93.

UN, 2003. *Human Security Now: Commission on Human Security*. New York: United Nations.

UN, 2004. *A More Secure World: Our Shared Responsibility. Report of the High-level Panel on Threats, Challenges and Change*. New York: United Nations.

UNDP, 1994. *Human Development Report*. New York: Oxford Press.

UNDP, 2012. *Caribbean Human Development Report: Human Development and the Shift to Better Citizen Security*. New York: UNDP.

UNECA, 2011. *Economic Report on Africa 2011: Governing Development in Africa – The Role of the State in Economic Transformation*. Addis Ababa: United Nations Economic Commission for Africa.

Warner, J. & Shaw, T. M., 2018. *African Foreign Policies in International Institutions*. New York: Springer Nature.

World Bank Group, 2018. *Open Learnning Campus* [Online]. Available at: https://olc.worldbank.org/content/security-development-nexus [Accessed 15 January 2019].

Part 8

The changing landscape of development. Editorial introduction

This section aims to advance understandings of relationships among space, place, people, and the environment. Situated within the wider context of this edition of the Companion, this section unpacks questions around conceptualising enduring tensions in development such as around land, commodification, urbanisation, and displacement.

The chapters presented in this section examine the changing relationships among people, place, and space in the context of development processes and outcomes. We have used the term 'landscape' to describe the complex entanglements between physical geographies and social interactions, recognising that these landscapes continuously shape and are shaped by development trajectories. Landscape here refers not to mere passive scenery but rather to dynamic social spatial relationships, moulded by the interplay of environmental processes and human endeavours. From this perspective, landscapes reflect and are imbued with social, political, economic, and cultural relationships.

Power dynamics are etched into the very contours of the landscape. Infrastructure provides one example, as investments linking peripheral regions to cities may spur growth yet also serve privileged interests, transforming spatial relationships in the process. Thus, a unifying theme across these chapters is that landscapes mirror broader development patterns while crystallising shifting power relations and human values. Analysing landscape reveals insight into how space and place take shape through socio-political and cultural forces.

The phenomena examined in this section of the Companion are diverse, ranging from urbanisation to the commodification of natural resources. Nevertheless, several cross-cutting themes unite these chapters. As in other sections of the Companion, power remains firmly in focus. Here the emphasis is on how relations of power play out and become embedded in landscapes, shaping development outcomes.

A second theme concerns shifting rural-urban dynamics. In its nascence as an academic field, development studies often tended towards the rural, emphasising agrarian reforms and rural poverty alleviation. This partiality reflected the Eurocentric tendencies of many development thinkers, who equated progress with industrialisation and urbanisation. Rural areas were often dismissed as bastions of backwardness requiring modernisation or upliftment. However, the rapid urbanisation of recent decades and the challenges associated with it have challenged these visions of development.

Mass migration from rural to urban areas has blurred previously rigid boundaries. Cities rely on rural resources, from food to labour, even while remittances flow back to villages. Urban expansion displaces agriculture yet also provides markets for rural

DOI: 10.4324/9780429282348-87

products. Analysing these flows of people, resources, and capital spotlights the interconnections binding rural and urban areas, contrary to notions of discrete realms.

Fractures within both rural and urban landscapes have also become more salient. Spatial inequality manifests clearly in cities, where marginalised groups often reside in substandard housing lacking infrastructure access.

Rapid urbanisation has intensified the spotlight on cities. Often associated with economic dynamism and innovation urban areas are also marked by inequality, illbeing and displacement. Strategic planning is needed to ensure affordable housing and efficient public transport that connects marginalised residents to opportunities, and yet these often end up serving those with political power rather than those most in need of them.

Infrastructure is another key theme. As the often declared backbone of development, infrastructure both bridges and exacerbates divides. Infrastructure investments can facilitate economic integration and growth. For instance, electrification enables industrial production and illuminates homes. However, infrastructure also manifests political agendas, serving entrenched interests even as they are contested. Its innately political nature means it manifests priorities of the powerful, even as it remains contested. Road construction may better integrate peripheral areas into networks of trade or displace communities without adequate compensation or recourse. Dams supply electricity yet can displace people through flooding. While infrastructure is the backbone for growth, it can also spur grassroots resistance. Such movements highlight how infrastructure anchors broader relationships among states, markets, and civil society. Analysing infrastructure provides insight into the exercise of power and its impacts on landscape development. Infrastructure analysis thus provides vital insight into how power shapes landscape development, as investments erect physical networks that anchor broader state-market-society relationships.

Another recurring theme is the transformation of environmental features like forests, water, and land into commodities. Development often catalyses shifts from subsistence to commercial uses, such as clearing mangroves for aquaculture. These dynamics highlight evolving patterns of ownership, access, and use that often exclude marginalised groups. For instance, the enclosure of common grazing lands in Britain dispossessed peasant farmers. Similarly, mineral extraction on Indigenous territories may generate revenue but limit communities' ability to sustainably use areas they have traditionally stewarded. Changing resource dynamics reveal how landscapes become imprinted with power relations. As resources are commodified, they become imbued with social meanings and values that reflect dominant interests. Analysing these human environment relationships – and others – helps to reveal how landscapes become imprinted with power relations. When brought together, the chapters in this section of the Companion underscore the complex interplay among space, place, and power.

Chapter 80

Urban bias

Gareth A. Jones and Stuart Corbridge

Students of development and development practitioners have long concerned themselves with urban-rural relationships. A familiar assumption has been that ambitious people will move to urban areas to improve their lot. In the 1950s and 1960s an apparent bias in favour of urban-industrial models of development was justified by three key ideas. W. Arthur Lewis (1954) proposed that there is disguised unemployment or underemployment in rural areas of poorer countries, where the marginal productivity of labour is often very low. Men and women move to the city to find more productive work and to pull their families out of poverty. Hans Singer (1950) and Raoul Prebisch (1950) further argued there is a long-run tendency for the terms of trade, the ratio of export commodity prices to import commodity prices, to move against primary commodities like foodstuffs and raw materials. Import-substitution industrialisation was commended partly on this basis. And, national planners argued that goods and services would most efficiently be diffused from major cities to smaller cities and rural areas. Cities benefited from – and generated – economies of scale. This, after all, had been the experience of most Western countries.

All of these arguments have been disputed. There is no reason why manufacturing or service-sector jobs must be based exclusively or even mainly in urban areas. China and Taiwan have each generated large numbers of industrial jobs in rural areas over the past 40 years. In addition, countries like Australia, Canada and Norway became rich largely as exporters of food and raw materials. In practice, however, for both economic reasons (including the benefits of industrial clustering) and political reasons (including a disposition to think of the urban as modern), most non-agricultural jobs *have* been based in urban areas of developing countries. The question is whether this matters – and if so why and for whom and in what terms?

One answer to this question is that concentrations of urban-industrial power damage the 'authenticity' of a country. Gandhi (1997 [1908]) believed that the soul of India was to be found in its villages. For him, large-scale industrialisation was a form of social evil. Some deep ecologists also think in these terms, views that have informed some recent debates on climate change and proponents of alternatives to development. In the 1960s Michael Lipton began to develop a more positive (or testable) account of urban bias in the process of world development. His urban bias thesis (UBT) was formally presented in his book *Why Poor People Stay Poor: A Study of Urban Bias in World Development* (1977). The UBT proposes that urban classes in poorer countries use their social power to bias (distort) a range of public policies against members of the rural classes. Lipton argued that urban bias involves i) an allocation, to persons or organisations located in

DOI: 10.4324/9780429282348-88

towns, of shares of resources so large as to be inefficient and inequitable or ii) a disposition among the powerful [urban classes] to allocate resources in this way.

The urban bias thesis and its critics

In its first iteration, the UBT made five main claims: i) rural areas of developing countries suffer from too little spending on education and healthcare (relative to population); ii) these inequalities, combined with excessively 'urban' forms of teaching and curriculum development, pull bright young people to the cities; iii) people in rural areas are forced to pay a higher share of national taxes than is fair; iv) a series of government-imposed price twists causes inputs into rural areas to be overpriced when compared to a market norm and which causes outputs from rural areas to be correspondingly under-priced; and v) this combination of *distributional urban bias* (i, ii and iii) and price-twisting is not only unfair but inefficient: at the margin, Lipton maintained, a given sum of government money will generate higher marginal returns in the countryside – specifically the small 'family farm' agriculture sector – than it will in cities or large-scale urban-based industries.

Two principal critiques were levelled at this first iteration of UBT. A first line of criticism was that the UBT lacked empirical validity and was overly generalised. The UBT failed to prove that the inter-sectoral terms of trade moved everywhere against rural areas. Critics also said that Lipton was inattentive to the issue of urban poverty; that most of the rural poor were sellers of labour who benefited from cheap food; and that the UBT neglected the power of rural elites (Byres, 1979). A second line of critique was theoretical: it challenged Lipton's accounts of class, power and policy formation. Keith Griffin (1977) took exception to Lipton's attempt to account for intra-sectoral differences in wealth and power by counting members of the rural elite as members of the urban class and members of the urban poor as part of the rural class. This seemed like sophistry to Griffin.

Early support for the UBT came from the political scientist Robert Bates and from Elliott Berg at the World Bank. Bates (1981) argued that food production problems in sub-Saharan Africa (SSA) were mainly the result of non-democratic governments using urban-biased policies to discriminate against smallholding agriculturalists. Governments used marketing boards and over-valued exchange rates to procure food cheaply from the countryside or overseas. They also made it difficult for farmers to sell food and other crops to private merchants and spent large sums on industrial protection. Bates suggested that many regimes in SSA imposed urban-biased policies to keep the lid on unrest in towns and cities. The urban working class was bought off with cheap food, while city-based bureaucrats extracted large rents from systems of licenses and quotas. In the medium-term, however, rational farmers responded by producing less food for sale. This caused precisely those surges in food prices that triggered the urban food riots or even coups that Africa's ruling coalitions were keen to avoid.

Bates argued that international donors should exploit moments of political or economic crisis in SSA to push more farmer-friendly policies. This view found support in the World Bank's 1981 report on *Accelerated Development in Sub-Saharan Africa* – the so-called Berg report – and later became standard World Bank policy when the debt crisis opened the way for structural adjustment loans that required recipient governments to reduce subsidies on consumption and remove restrictive institutions such as compulsory purchasing programmes. Michael Lipton, for his part, accepts that many structural adjustment programmes (SAPs) have improved the inter-sectoral terms of trade in SSA and elsewhere: price twists are no longer so damaging to the countryside, although they remain significant at the global scale, given EU, Japanese and US farm-support policies. Lipton believes, however, that the World Bank's focus on price twists alone caused it to ignore the more deeply rooted inequalities in social and spatial power that promote urban bias. More should have been done, he argued, to enhance education and health-care provision and promote investment to small farmers in order to raise productivity in the rural economy.

The revised urban bias thesis and new criticisms

Lipton's later iterations of the UBT argued that distributional urban bias has increased precisely at the same time as a series of successful neoliberal assaults on the urban-rural terms of trade. In a paper with Robert Eastwood, he claimed that: i) overall within-country inequality increased significantly after 1980–1985, following adjustment policies; ii) that these increases have not been offset by declining rural-urban inequality; iii) that this absence of offset, save for in a few countries in Latin America, must be accounted for by a rise in distributional urban bias at a time of reduced price twists against the countryside (Eastwood and Lipton, 2000). In short, urban bias remained pervasive and damaging to rural livelihoods. There is considerable support for the view that urban/rural welfare ratios are not yet falling towards unity and may even be diverging in some countries that are enjoying rapid economic growth and even declines in aggregate poverty. China is the most notable example (Piketty et al., 2019). Nevertheless, we can identify at least four challenges to the UBT in its original and modified versions.

First, the UBT underestimates the scale and recent relative growth of urban poverty. Current estimates suggest about 550 million people in the cities of the Global South live in absolute poverty with nutritional and health conditions close to rural areas, and just under one billion live in 'slums' with limited access to the resource allocations supposedly biased in favour of cities. As Lucci et al. (2018) have argued, conventional and new multidimensional measures misidentify the poverty line in towns and cities, under recording non-food costs and over estimating access to services. Second, the UBT is beset by definitional and measurement problems. Definitions of urban and rural are not consistent. A large village in Bangladesh might count as a town in Peru. Moreover, in many parts of the world it is difficult-to-impossible to identify where the urban ends and the rural starts, while concepts such as peri-urban only complicate the landscape further. Such definitional inconsistency notwithstanding, there is simply no reliable way to measure the distribution of transfers, subsidies and in-kind benefits across the urban-rural divide. Many subsidies are hidden, for example when tax regimes fail to recoup the costs of public infrastructure provision from property developers, which might lead to an undercount of UBT.

Third, the UBT overestimates the possibilities for in situ improvement in agricultural livelihoods. In comparison to the arguments demonstrating the productivity of small farmers, research has identified constraints to farmer productivity beyond price twists or the concentration of services in large towns. Farmers are faced with fragmentation of land holdings and the threat from land grabbing, limited access to finance, imperfect market knowledge – especially as the global food economy becomes ever-more complex – and the multiple effects of climate change (Bryceson and Jamal, 2019). The thesis also underestimates (even ignores) the benefits of migration and ceaseless circulation and the construction of livelihoods based on the ability to move across the urban and rural divide (Flahaux and De Haas, 2016).

Fourth, there is a new challenge from the arguments of 'new economic geography'. This research suggests that many non-primate cities have grown not because of rent-seeking and distorted patterns of political access (urban bias in Lipton's terms) than because of the returns to scale and spill-over effects that are associated with the clustering of innovative economic activities (Duranton, 2015). The decision to concentrate public goods in cities, the argument goes, is a rational allocation of scarce resources. Cities bring together talented people, networks of knowledge and private investment, hence the allocation of resources to education, health, communications infrastructure, financial services, aims to further enhance innovation, enterprise and productivity (Duranton, 2015). The proliferation of 'smart cities' and incubator hubs is the clearest evidence that this argument has caught the attention of policy makers and investors (Karvonen et al., 2018). The implication is that resources are focused to where productivity gains and innovation frontiers are most likely. This is a bias of sorts, but it is different from the wilful distortion (of markets by states).

'Urban bias' and public policy

Where does all this leave us? It makes little sense to choose between the UBT and what might be called its opposite. There is much we still do not know, as Lipton readily concedes. Non-economic forms of urban bias are under-researched. Many rural people are stereotyped as backward and might experience 'urban bias' in terms other than those set out in Lipton's UBT, but which are nonetheless consistent with his view that key development actors are disposed to equate the urban with the modern ('dispositional urban bias'). It is thus prudent to retain the provocation set out by Lipton's thesis (location matters for welfare), while avoiding reference to urban bias either as a social fact or as a pathology that always needs correction. It would be too formulaic to suggest that if 70 per cent of the world's poorest people reside in rural areas, then it is always and everywhere rational that at least 70 per cent of development spending should be spent in rural areas. Nevertheless, whether some of development organisations and governments have moved to an implicit or explicit anti-urban bias in the last 30 years is not clear or how this matters in a more 'urban age'. There are good reasons for thinking that many cities in developing countries are growing on the back of a strong commitment to economic innovation and the production of dynamic growth clusters and not on the backs of rural people or as a result of rent-seeking by politicians. Not all 'bias' is bad, and many people in the countryside will rightly want to make their way to the city. In any case, many of the policies that Lipton and others want to see enacted – including better provision of primary education and healthcare in the countryside – can be argued for without resort to a generalised model of the exploitation of the countryside by the city. In this specific respect the UBT can be unhelpful.

Suggestions for Further Reading

Jones, G.A. and Corbridge, S.E., 2010, 'The Continuing Debate About Urban Bias: The Thesis, Its Critics, Its Influence, and Its Implications for Poverty Reduction Strategies', *Progress in Development Studies* 10(1): 1–18. A broad critical review of the Urban Bias Thesis from inception to contemporary relevance.

Lipton, M., 1977, *Why Poor People Stay Poor: A Study of Urban Bias in World Development*, London: Temple Smith. The classic text, argued as both an eloquent narrative and through economic theory.

Mitlin, D. and Satterthwaite, D., 2013, *Urban poverty in the Global South: Scale and Nature*, London: Routledge. A comprehensive review of urban poverty, including analysis of measures and alleviation.

References

Bates, R.H., 1981, *Markets and States in Tropical Africa*, Berkeley: University of California Press.

Bryceson, D.F. and Jamal, V. (eds.), 2019, *Farewell to Farms: De-Agrarianisation and Employment in Africa*, London: Routledge.

Byres, T.J., 1979, 'Of Neopulist Pipe Dreams', *Journal of Peasant Studies* 6: 210–244.

Duranton, G., 2015, 'Growing Through Cities in Developing Countries', *World Bank Research Observer* 30(1): 39–73.

Eastwood, R. and Lipton, M., 2000, 'Pro-Poor Growth and Pro-Growth Poverty Reduction: Meaning, Evidence, and Policy Implications', *Asian Development Review* 18(2): 22–58.

Flahaux, M.L and De Haas, H., 2016, 'African Migration: Trends, Patterns, Drivers', *Comparative Migration Studies* 4(1).

Gandhi, M.K., 1997 [1908], *Hind Swaraj, (Published with an Editorial Introduction by Anthony Parel)*, Cambridge: Cambridge University Press.

Griffin, K., 1977, 'Review of "Why Poor People Stay Poor"', *Journal of Development Studies* 14: 108–109.

Karvonen, A., Cugurullo, F. and Caprotti, F. (eds.), 2018, *Inside Smart Cities: Place, Politics and Urban Innovation*, London: Routledge.

Lewis, W.A., 1954, 'Economic Development with Unlimited Supplies of Labour', *Manchester School of Economics and Social Studies* 22: 139–191.

Lipton, M., 1977, *Why Poor People Stay Poor: A Study of Urban Bias in World Development*, London: Temple Smith.

Lucci, P., Bhatkal, T. and Khan, A., 2018, 'Are We Underestimating Urban Poverty?', *World Development* 103: 297–310.

Piketty, T., Yang, L. and Zucman, G., 2019, 'Capital Accumulation, Private Property, and Rising Inequality in China, 1978–2015', *American Economic Review* 109: 2469–2496.

Prebisch, R., 1950, *The Economic Development of Latin America and Its Principal Problems*, New York: UN-ECLA.

Singer, H., 1950, 'The Distribution of Gains between Investing and Borrowing Countries', *American Economic Review* 40: 473–485.

World Bank, 1981, *Accelerated Development in Sub-Saharan Africa*, Washington, DC: World Bank.

Chapter 81

Studies in comparative urbanism

Colin McFarlane

Cities have always been understood comparatively. While comparison may appear as a prosaic set of methodological questions around case studies, in practice it is a critical part of how understanding, theory and research about cities are produced and contested. Recent years have witnessed not just a resurgence of comparison but a new experimentalism with comparative thinking and methodologies. This is in part a response to the globalisation of urban policy, planning, economies, cultures and ecologies, but it is also an attempt to internationalise urban geography and development by thinking across intellectual and imaginative divides that have traditionally separated out the cities of the Global North from those of the Global South. Our inherited conceptions of the city are often premised on the experiences and theoretical work based upon cities in Western Europe and North America, accompanied by the often implicit slippage between claims about certain cities (e.g., New York, Los Angeles, Barcelona, Berlin, Paris, or London) and claims about 'the city' as an abstract, generalised category. Part of the revival of comparison has been to widen the range of urbanisms that constitute urban theory.

This upsurge in comparative research has been multi-faceted and includes – and this is by no means exhaustive: efforts to compare one city with several, (Nijman, 2007a, on Miami); comparing two cities (Huchzermeyer, 2007, on the production of informal settlements in Sao Paulo and Cape Town); exploring how specific processes or features recur or diverge in different cities (Gulger, 2004, on world cities in the South, or Roy, 2005, on planning and citizenship across North and South); research outlining frameworks for comparative urban research (Brenner, 2001); and work developing an explicitly postcolonial (Robinson, 2006, 2011; McFarlane, 2010) or relational (Ward, 2010) revisioning of comparative urbanism. In short, comparison is firmly on the agenda of urban studies, whether as a way of experimenting with the diversity of cities across and beyond inheritances of Global North/South or global city/megacity, as a means for thinking through the relations between case studies and wider processes, or – increasingly – as a resource for locating difference rather than similarity (McFarlane and Robinson, forthcoming).

The emphasis in much of these debates has been on comparison as a set of practical, methodological and typological questions, including Abu-Lughod's (1999) variation-finding comparison that seeks to explain differences across New York, Chicago, and Los Angeles as global cities; Nijman's (2007b) 'multiple-oriented comparative approach', which aims to better understand a particular case through individual comparisons with other (multiple) selected cases by seeking out both idiosyncrasies and analogies; and Ward's (2010) positioning of comparison as embedded in urban networks and flows

DOI: 10.4324/9780429282348-89

rather than discrete or self-enclosed cities. These debates have also included important efforts to consider the assumptions through which objects of urban comparison are arrived at and pursued. Drawing on postcolonial and development scholarship, three closely inter-related sets of emerging concerns are crucial here: *theory culture*, *learning*, and *ethico-politics*. Theory culture, following Mufti (2005: 475), is 'the *habitus* that regulates 'theory' as a discrete set of practices' within and sometimes between specialisms and regions (and see Connell, 2007, on *Southern Theory*). Comparative research *across* theory cultures prompts reflection not just on contrasting spaces or processes but on the ontological and epistemological framings that inform how the world is being debated, how knowledge is being produced and questioned, and about the purpose of knowledge, research, and theory. This means also considering the role of a whole range of institutional actors, from journals to forms and patterns of citation to modes of writing and dissemination that co-constitute theory cultures. By exploring how different theory cultures debate, for example, the city, politics, infrastructure, modernity, or globalisation, there is potential to develop more pluralised understandings through comparison.

In relation to learning, a key tension in comparative thinking, as Mbembe (2004: 375) has argued in relation to Johannesburg, is the 'temptation of mimicry' – the desire to copy, to learn directly from another urban experience. In contrast, thinking of urban comparison as learning-through-difference positions comparison as uncertain and provisional, because in widening the discursive field of cities the occurrence of unlikely translations involves the increased traversing of unfamiliar and unpredictable terrain. There is an important challenge here in trying to contextualise and understand knowledge from places unfamiliar to the researcher – texts that emerge from theory cultures that may require, for example, greater effort for an outsider to grasp (Connell, 2007).

Finally, efforts to compare between theory cultures raise ethical and political considerations (Jazeel and McFarlane, 2010). There is an ongoing challenge to engage, on as close a level playing field as possible, with the work of thinkers in different places: 'If a cosmopolitan urban theory is to emerge' writes Robinson (2002: 549–550), 'scholars in privileged western environments will need to find responsible and ethical ways to engage with, learn from and promote the ideas of intellectuals in less privileged places'. This requires a critical epistemic interrogation and reworking, such as that found in Appadurai's (2000) formulation of 'strong internationalisation'. He writes:

['Strong internationalisation'] is to imagine and invite a conversation about research in which . . . the very elements of this ethic could be the subjects of debate. Scholars from other societies and traditions of inquiry could bring to this debate their own ideas about what counts as new knowledge and what communities of judgement and accountability they might judge to be central in the pursuit of such knowledge.
(Appadurai, 2000: 14)

The demand here is an ethical commitment to learning and unlearning comparatively through different theory cultures. Drawing on Spivak (1993), McEwan (2003: 384) argues that 'unlearning' involves working hard to gain knowledge of others who occupy those spaces most closed to our privileged view through open-ended conversations. Part of this unlearning involves articulating the Western intellectual's participation in the formation of categories like 'Third world city'. This requires an ongoing sensitivity to the relationship between power, authority, positionality, and knowledge, but it is a set of problems that cannot simply be acknowledged away; the positions, privileges, and ways of seeing that help shape the comparisons we make cannot be stepped around.

The notion of strong internationalisation involves a particular and reflexive engagement with, for example, different regimes of academic knowledge production formulated through distinct patterns of collection, citation, or judgement. There is a challenge here for academics to connect more closely and more frequently with the worlds, languages, and vocabularies of disparate scholars (Desbiends and Ruddick, 2006). These ethico-political questions point to a range of practical challenges in working across different theory cultures, including negotiating new forms of collaboration for comparative research through journals, refereeing, and editorships or in supporting scholarship and writing from different contexts. Other examples of strong internationalisation might include developing personal contacts and resources, such as a fund for translations to deal with language barriers, universities and departments investing in graduate language skills, creating funds for exchanges, and encouraging more collaborative postgraduate programmes.

Then is, then, an emerging debate around thinking of comparison not just as a research method – as crucial as that is – but as a mode of thought and as a *strategy* for international urban studies. If we are interested in a more international or postcolonial conception of the city – a conception that attempts to grapple with the multiplicity of different cities and ways of knowing the city across the Global North-South divide – then it is inevitable that we examine what our implicit objects of reference are when we write urbanisms and that we consider how we might bring other urban experiences, knowledges, and theories into a more horizontal comparative field. What this opens is an expansive reading of comparison where a key question at stake is: what might be the implications for urban theory when we take comparison not just as a method but as a mode of thought and set of institutional practices that inform how urban theory is constituted? Comparative thinking can be a strategy for, first, revealing the assumptions, limits and distinctiveness of particular theoretical or empirical claims and second for formulating new lines of inquiry and more situated accounts.

The three overlapping areas of theory culture, learning, and ethico-politics matter if comparison is to assist in producing research that reflects a more global understanding of urbanism. This outline of comparison as a strategy can be part of a wider effort to foster the collaborative formation of research projects, where the effort is to avoid – under ongoing constraints of history, positionality, and unequal power relations – a privileging of one context over another and one that seeks not to assume that one theory culture represents a norm or standard of knowledge over others. It seeks to offer a route to alternative forms of comparative thinking and research that expands the field of inquiry to contribute to new ways of understanding contemporary urbanism.

References

Abu-Lughod, J.L. (1999) New York, Chicago, Los Angeles: America's Global Cities. Minneapolis. University of Minnesota Press.

Appadurai, A. (2000) 'Grassroots Globalization and the Research Imagination'. *Public Culture*, 12, 1, pp. 1–9.

Brenner, N. (2001) 'World City Theory, Globalization, and the Comparative-Historical Method: Reflections on Janet Abu-Lughod's Interpretation of Contemporary Urban Restructuring'. *Urban Affairs Review*, September, pp. 124–147.

Connell, R. (2007) *Southern Theory: The Global Dynamics of Knowledge in Social Science*. Sydney: Allen and Unwin.

Desbiends, C. & Ruddick, S. (2006) 'Guest Editorial: Speaking of Geography: Language, Power, and the Spaces of Anglo-Saxon Hegemony'. *Environment and Planning D: Society and Space*, 24, pp. 1–8.

Gulger, J. (2004) *World Cities Beyond the West: Globalization, Development and Inequality*. Cambridge: Cambridge University Press.

Huchzermeyer, M. (2007) 'Tenement City: The Emergence of Multi-Storey Districts through Large Scale Private Landlordism in Nairobi'. *International Journal of Urban and Regional Research*, 31, 4, pp. 714–732.

Jazeel, T. & McFarlane, C. (2010) 'The Limits of Responsibility: A Postcolonial Politics of Academic Knowledge Production'. *Transactions of the Institute of British Geographers*, 35, pp. 109–124.

Mbembe, A. (2004) 'Aesthetics of Superfluity'. *Public Culture*, 16, 3, pp. 73–405.

McEwan, C. (2003) 'Material Geographies and Postcolonialism'. *Singapore Journal of Tropical Geography*, 24, 3, pp. 340–355.

McFarlane, C. (2010) 'The Comparative City: Knowledge, Learning, Urbanism'. *International Journal of Urban and Regional Research*, 34, 4, pp. 725–742.

McFarlane, C. & Robinson, J. (forthcoming) 'Rethinking Comparative Urbanism'. *Urban Geography*.

Mufti, A. (2005) 'Global Comparativism'. *Critical Inquiry*, 31, pp. 427–489.

Nijman, J. (2007a) 'Place-Particularity and "Deep Analogies": A Comparative Essay on Miami's Rise as a World City'. *Urban Geography*, 28, pp. 92–107.

Nijman, J. (2007b) 'Introduction: Comparative Urbanism'. *Urban Geography*, 28, pp. 1–6.

Robinson, J. (2002) 'Global and World Cities: A View from off the Map'. *International Journal of Urban and Regional Change*, 26, 3, pp. 513–554.

Robinson, J. (2006). *Ordinary Cities: Between Modernity and Development*. London: Routledge.

Robinson, J. (2011) 'Cities in a World of Cities: The Comparative Gesture'. *International Journal of Urban and Regional Research*, 35, 1, pp. 1–23.

Roy, A. (2005) 'Urban Informality: Towards an Epistemology of Planning'. *Journal of the American Planning Association*, 71, 2, pp. 147–158.

Spivak, G.C. (1993) *Outside in the Teaching Machine*. London: Routledge.

Ward, K. (2010) 'Towards a Relational Comparative Approach to the Study of Cities'. *Progress in Human Geography*, 34, 4, pp. 471–487.

Further reading

McFarlane, C. (2010) 'The Comparative City: Knowledge, Learning, Urbanism'. *International Journal of Urban and Regional Research*, 34, 4, pp. 725–742.

Nijman, J. (2007) 'Place-Particularity and "Deep Analogies": A Comparative Essay on Miami's Rise as a World City'. *Urban Geography*, 28, pp. 92–107.

Robinson, J. (2011) 'Cities in a World of Cities: The Comparative Gesture'. *International Journal of Urban and Regional Research*, 35, 1, pp. 1–23.

Roy, A. (2005) 'Urban Informality: Towards an Epistemology of Planning'. *Journal of the American Planning Association*, 71, 2, pp. 147–158.

Ward, K. (2010) 'Towards a Relational Comparative Approach to the Study of Cities'. *Progress in Human Geography*, 34, 4, pp. 471–487.

Chapter 82

Understanding land as fictitious capital in financial capitalism

Sarah E. Sharma and Susanne Soederberg

Introduction

In this chapter, we evaluate the treatment of land as an object of speculation under financial capitalism and the social, economic, and environmental outcomes that arise from this reality.

In mainstream economics, land is generally seen as a resource that encompasses the natural resources used in production, especially resource extraction (mining, agriculture). Land can be private property or part of the public commons owned, entrusted, and managed by states (national and city parks, social housing complexes). Unlike economic understandings of land, we adopt a Marxist perspective that views land as a commodity under financial capitalism. Financial capitalism refers to a dominant way of accumulating wealth by selling money to make more money. As a commodity in financial capitalism, land is not just a thing but also a historical social relation of class-based power (landlord and tenant). Since the 1980s, there has been more emphasis placed on buying land – both private and public – in the hopes that its price will increase. The transformation of land into a financial asset is particularly striking in urban geographies, where an increasing number of people require offices in which to work, houses in which to live, and stores in which to shop. These commercial and personal built environments require land on which they may be constructed. Land speculators make their money gambling on which location will yield the greatest increase in price.

We argue that land – and the relations of power that permeate its ownership, distribution, and access – must feature heavily in understandings of present-day development, particularly at the urban scale. We highlight urban land in order to examine and critique how it has become a key area of interest for finance capital. Further, in advancing this argument, we emphasise that in the scholarly discussions regarding land as a financial asset, urban centres in the Global South have been overlooked, stressing the need for further inquiry in these spaces of financial capitalism. Finally, we highlight that land as fictitious capital – by which we mean the land is valued on its future earning potential, see later – leads to uneven forms of displacement of vulnerable populations such as refugees, migrants, and low-income groups, as housing and the land underneath it is seen primarily as an investment vehicle rather than a location for communities.

Understanding land

The historical analysis of land in development studies, primarily in the Global South, has principally examined rural land. Pivotal understandings of land have been developed by

DOI: 10.4324/9780429282348-90

post-colonial scholars such as Tania Li (2014). Li (2014) centres her understanding of land in the post-2008 financial crisis 'global land rush', where foreign direct investment in agricultural land escalated dramatically in search of investor profits from the transformation of arable land into agricultural enterprises. Defining land as a social and material object, Li (2014: 589) writes that as a resource it 'is a provisional assemblage of heterogeneous elements including material substances, technologies, discourses and practices'.

Li grapples with land as a central economic and social element in human life and highlights the tensions inherent to its increasing commercialisation for industrial, residential, and tourist uses. This analysis highlights global dynamics of power, bringing us closer to understanding land as an international asset being valued for its ability to accumulate capital for its owners, whether through productive (agricultural) or financial (future rents) channels.

More recently in urban geography there has been a resurgence of examining land as a financial asset. Much of this work draws on Marxian debates from the late 1970s and 1980s, considering the role of land and its connection to rent (Harvey, 1982, 1989). Notably, urban scholars such as Harvey (1982) have argued that urban space is where the contradictions of capital accumulation seek to resolve themselves through spatio-temporal fixes. Further central to the present-day understanding of land as a financial asset is Anne Haila's (1988) foundational work on the relationship between finance and urban real estate markets. Land is understood as a form of fictitious capital, which does not indicate that as an asset it is not real but rather that the land does not entail intrinsic value, yet it can assume a price. Land is fictitious because, in itself, it Is not capital (profit derived through the production process involving the exploitation of labour power) but merely debt claims (Marx, [1984] 1981). The value of land is realised in its future profit potential (promised titles on future ground rent).

Drawing on this insight, Kaika and Ruggiero (2013) define land as a financial asset, adding that it is also a lived process. They root this analysis within a historical and empirical analysis of the original factory grounds of the tire corporation Pirelli in Milan, Italy. By tracing transformations of this land from a site of production for goods and services (alongside benefits for workers engaged in labour such as housing) to a financial good from which landowners can accumulate capital on global markets starting in the 1980s, Kaika and Ruggiero argue that urban land has been enrolled into global financialised circuits of capital by corporate actors seeking profit in the context of declining revenue. Through this process, labour is increasingly displaced.

The displacement of surplus populations due to the status of land as a fictitious commodity is similarly discussed by Soederberg (2018) in her analysis of refugee housing in Berlin. Soederberg explains that as land becomes an asset for investment and speculation (Haila, 2016) and states further entrench austerity measures, housing becomes increasingly commoditised and out of reach for vulnerable and displaced groups. Underpinning this reality is the status of land as fictitious capital under market-based governance mechanisms, which has led to an inability for the market to meet housing demands at the urban level, particularly for refugees, migrants, and low-income groups.

Christophers (2018) helps us understand the role of the state in the commodification and privatisation of land. Through examining the privatisation of what was once state-owned land across the United Kingdom, he draws on Harvey (1982) to explain how, by selling land to private actors that treat land as a financial asset, the state is promoting and producing the policies and property rights that enable land to be envisioned primarily for capital accumulation purposes.

Defining land in the contemporary global political economy

Drawing on the previous authors and in contrast to mainstream economic views, we define land as a socially constructed object, a social relation of power between classes, and a lived experience. It is socially constructed as an object of finance in the manner in which future imagined rents become inherent to its present-day value (Harvey, 1982) and it is a social relation of power as the structuring, governance of, access to, and relationships with land are dictated by relations of power. Finally, it is a lived experience as there are social relationships, dialogue, and contestation surrounding its transformation into a financial asset between governments, corporations, civil society, and communities throughout history (Kaika and Ruggiero, 2013). The state is vital to examine when examining land at both the urban and rural level, as it mediates the policies and practices that lead to the ability for land to be envisioned primarily as an object of accumulation over social needs such as housing, green space (for recreational activities and carbon off-setting), or food.

In understanding land in the previous manner, we now turn to why land as a fictitious capital is imperative in development studies – particularly in the understudied Global South – in order to draw out critical theoretical and empirical implications of land as fictitious capital.

Urban land in the Global South

The previous literature on urban land as fictitious capital primarily considers spaces in the Global North, while the literature concerning rural land largely focuses on agricultural environments in the Global South. Following our argument, we advocate for further focus on how land in urban centres in the Global South are becoming increasingly commodified under the aegis of development – by states, international organisations, and private actors. Within this context, it is vital to investigate the relationships of power in the global political economy that work to transform land into a financial asset within the Global South. To do so, first, transnational narratives justifying land as a form of fictitious capital across the Global North and South must be critically analysed. Second, examining the treatment of land as a financial asset and the social implications therein highlight tensions in the previous rhetoric. Investigating how land as fictitious capital is socially produced through class-based relations of power and produces uneven forms of displacement and marginalisation across social, economic, and environmental lines ultimately sheds light on why and how urban land markets are a central feature of capitalist development.

The processes associated with transformation of land into fictitious capital address the ways in which land and real estate markets are intertwined through securitisation of mortgages and rents. Briefly, securitisation describes a financial practice in which various types of contractual debt (fictitious capital, such as mortgages, rents, credit cards) are pooled together and sold on secondary markets. Haila (2016) argues that at the urban level, when land is treated first and foremost as a financial asset, the built environment becomes submissive to market forces and financial speculation, and thus urban development becomes dependent on external forces such as debt crises, sovereign debt, and lending, and the prices of real estate and housing become difficult to control. Finally, in this phase of land regimes, states and municipalities become motivated to sell land they own as a source for fiscal revenue. Although some states, such as Singapore, Haila's case study, segregate land for housing and commercial purposes, this is uncommon, particularly in the Global South as land markets are becoming more valuable and financial markets grow around them.

Relations of power at the urban level determine *what* spaces and *who* (people who live in these spaces of capitalism) are understood to be lucrative investments. On the other end of the spectrum, displacement occurs when land is purchased for redevelopment and less-powerful groups are evicted from this space. Urban growth, in this manner, causes enhanced investment and purchasing of land and can result in the displacement of the less powerful for the benefit of upper classes, land developers, real estate developers, and property owners. In cities in the Global South with environmental threats, evicted groups can often become more vulnerable to facing the effects of natural hazards. The role of the state facilitates how land is able to be bought and sold and what groups benefit from redevelopment, indicating its important role in mediating relations of power between capital and labour.

Thus, in order to understand urban growth and development, the question of land must be investigated, particularly the increasing nature of land as fictitious capital. This includes the relations of power underpinning who gets access to land and the social, political, and environmental implications of the transformation of urban land into fictitious capital. We have argued that urban transformations in the Global South and the related questions of housing and environmental safety rely heavily on powerful understandings of land as a vehicle for capital accumulation and advocate for further research that outlines powerful discourses that facilitate land as a fictitious capital and the resulting implications for vulnerable groups.

Bibliography

Christophers, B. 2018. *The New Enclosure: The Appropriation of Public Land in Neoliberal Britain*, New York, Verso Books.

Fairbairn, M. 2014. "Like gold with yield": Evolving intersections between farmland and finance. *Journal of Peasant Studies*, 41, 777–795.

Haila, A. 1988. Land as a financial asset: The theory of urban rent as a mirror of economic transformation. *Antipode*, 20, 79–101.

Haila, A. 2016. *Urban Land Rent: Singapore as a Property State*, Oxford, UK, John Wiley & Sons Ltd.

Harvey, D. 1982. *The Limits to Capital*, Oxford, Basil Blackwell Publisher Limited.

Harvey, D. 1989. *The Urban Experience*, Baltimore, MD, The John Hopkins University Press.

Kaika, M. & Ruggiero, L. 2013. Land financialization as a "lived" process: The transformation of Milan's Bicocca by Pirelli. *European Urban and Regional Studies*, 23, 3–22.

Li, T. 2014. What is land? Assembling a resource for global investment. *Transactions of the British Institute of Geographers*, 39, 589–602.

Marx, K. [1984] 1981. *Capital: Volume III*, London, Penguin Books.

Soederberg, S. 2018. Governing global displacement in austerity urbanism: The case of Berlin's refugee housing crisis. *Development and Change*, 1–25.

Additional resources

Atkinson, R. & Bridge, G. 2004. *Gentrification in a Global Context*, London, Routledge.
Brenner, N. & Schmid, C. 2015. Towards a new epistemology of the urban? *City*, 19, 151–182.
Desai, V. & Loftus, A. 2013. Speculating on slums: Infrastructural fixes in informal housing in the Global South. *Antipode*, 45, 789–808.
Polanyi, K. [1944] 2002. *The Great Transformation*, Boston, MA, Beacon Press.
Ward, C. & Aalbers, M. B. 2016. Virtual special issue editorial essay: "The shitty rent business": What's the point of land rent theory? *Urban Studies*, 53, 1760–1783.

Chapter 83

Land grabs

Pádraig Carmody and Adwoa Ofori

Some of the most important structural features of the global political economy driving land grabbing are the continued and generalised nature of global economic growth, consequent resource demand, and the finite nature of terrestrially based natural resources. Sometimes this is described as the 'ecological contradiction', where the global capitalist economy is dependent on continued economic growth but runs up against resource constraints. This contradiction is partly spatially displaced through power relations, as those with less power may experience dispossession in order to ensure continued resource access for more powerful social actors and forces. There are also other structural drivers of resource and land demand, such as global population growth – which is largely driven by patterns of poverty and inequality produced through the current global economic system – and ecological 'scarcity', further exacerbated by climate disruption or departure and attempts to offset it through the creation of carbon plantations or forests mostly in the developing world in order to sequester it (Hunsberger et al., 2017), with often scant regard for social impacts. These processes then represent forms of spatial fixing where the impacts of systematic contradictions are offshored or out-sourced to the poor in the Global South. As such there is a form of conflict between what economists call use value (how useful something is for someone, such as land to produce food for family subsistence) and exchange value (how profitable a resource may be for a company or individual). Land as a commodity may be traded to realise profit (or exchange value) or grow plantation crops for export, for example, which can be considered a form of production value. Conflicts over land and its grabbing centre over the different types of value and the relative social power of people invested in the different types of value.

The recent spike in land grabbing around the world has also been driven by more conjunctural factors or second order drivers, such as the North Atlantic Financial Crisis of 2008/2009 which pushed investors to diversify their investment holdings into land, food, and (bio)fuel for example, which are tangible assets which retain value, unlike paper-based ones which may become worthless, such as stock market shares. The details of many of the deals prompted by this conjuncture are provided in Pearce (2013), for example.

Whereas large, often Western-based, corporate land deals have received much media attention (Kaag and Zoomers, 2014), there have also been many instances of 'local' elites grabbing land in response to price appreciation or for speculative purposes based on potential mineral deposits, for example (Hall, 2011). In contrast to many media accounts, states also often play an important role in the land grabbing phenomenon by allowing investors territorial access, clearing land of previous occupants or in some cases taking land back from 'squatters', in forest reserves for example. Land grabbing is then

DOI: 10.4324/9780429282348-91

often an outcome of particular networks and assemblages of power rather than solely driven or constructed by isolated actors and may be facilitated by both national and international institutions favouring 'liberal' (for some) market policies.

The globalisation of land

Land is what the famous economic historian Karl Polanyi (1944) called a 'fictitious commodity', with distinctive characteristics. Its fixity and scarcity mean that investors must go where the land is rather than being able import it, as they do for many other commodities. As argued earlier, investors are interested in land for purposes such as agriculture and speculation on its appreciating value or the value of other commodities it may enable access to, such as minerals.

Land has different meaning and purposes for different social groups. For small farmers and pastoralists land is often not only a source of livelihood but also assumes additional social importance as it is imbued with cultural meaning and consequently becomes territory which results from the conjunction of space, identity, and authority (Sassen, 2013). Consequently land grabbing, including under colonialism, results in changes in structures of authority and governance, with often long-lived and wide scale effects.

While much of the recent literature has focussed on the direct impacts on people who have been displaced by land grabbing there are also a wide variety of other effects, including on indirect land use change, as clearing of forests to make way for biofuel plantations, for example, may release substantial amounts of carbon dioxide, despite the purportedly 'green' or eco-friendly nature of these products (Van der Laan et al., 2017). Fargione et al. (2008) put the number of years it takes to become carbon neutral from the land use change of Brazilian Cerrado to sugarcane ethanol at 17 years. This coupled with the fact that the majority of ethanol produced, is destined for Europe adds further carbon emissions to the product, thus limiting its carbon reducing benefits an increasing the time it takes to pay back carbon released and may also deprive locals of access to biomass which they need for their own sustenance or fuelwood (Sheridan, 2010: 41).

The geography and impacts of land grabbing

The proliferation of land grabs, as well as its history, has been well documented in academic literature. These acquisitions have been taking place worldwide, though notably the majority occur on the African continent. This is for a number of reasons. For example, concern by African governments to raise revenue from these transactions, given the paucity of other sources, to pay off debts and allow for continued expenditure, has spurred interest in land deals with both domestic and international investors. Furthermore, the continent is deemed as having unexploited and underutilised land and water resources which need investment in order to drive development. Proponents of these grabbing processes often argue that large-scale agriculture is more productive and modern than that of small-holders, although this is belied by the fact that for most crops small farmers tend to produce more output per hectare and the fact that they also still account for the majority of the world's food production. Academics and agencies more favourable to these processes tend to refer to 'large-scale land acquisitions', rather than 'grabbing', although the two are not incommensurable as in some cases land is bought from willing sellers. Nonetheless the majority of academic research has unquestionably established that there are generally substantial negative livelihood implications to local communities as a result of land grabs. This is because numerous populations depend directly on the land for their livelihoods and sustenance and are adversely affected by the loss of access, decreased land sizes for farming, reduced harvests and incomes, reductions in food security, and the inability to access forest products, such as fuel or forage. Acquisitions therefore have 'become one of the most hotly debated current development issues' (Schoneveld, 2014 cited in Kleeman and Thiele, 2015: 269).

While the scale of global land grabbing has sometimes been sensationalised and over-stated, as noted earlier, it has global environmental and social impacts. It often represents what is sometimes referred to as 'accumulation by dispossession' and, by transferring resources and assets from the poor to the rich, increases national and global inequality. The accompanying commodification of land also produces dispossessed or 'free' labour, thereby potentially deepening capitalist social relations more generally. However, if formal labour demand is low this results in deepening informalisation with attendant problems of lack of regulation, insecure incomes, and the absence of a social contract between the governed and the government, whereby consent to be governed and tax revenues are implicitly exchanged for public services and social order. Thus land grabbing also has impacts on the social order worthy of further research.

Additional Resources

Farmland Grab: https://www.farmlandgrab.org/
Focus on Land in Africa: http://www.focusonland.com/
GRAIN: https://www.grain.org/
Land Matrix: https://landmatrix.org/
Via Campesina: http://viacampesina.org

References

Fargione J., Hill, J., Tilman, D., Polasky, S., & Hawthorne, P. (2008). Land clearing and the biofuel carbon debt. *Science, 319*(5867), 1235–1238.
Hall, R. (2011). Land grabbing in Southern Africa: the many faces of the investor rush. *Review of African Political Economy, 38*(129), 510–510.
Hunsberger, C., Corbera, E., Borras, S. M., Franco, J. C., Woods, K., Work, C., & Vaddhanaphuti, C. (2017). Climate change mitigation, land grabbing and conflict: towards a landscape-based and collaborative action research agenda. *Canadian Journal of Development Studies-Revue Canadienne D Etudes Du Developpement, 38*(3), 305–324.
Kaag, M., & Zoomers, A. (2014). *The Global Land Grab: Beyond the Hype.* London: Zed Books.
Kleeman, L., & Thiele, R. (2015). Rural welfare implications of large-scale land acquisitions in Africa: a theoretical framework. *Economic Modelling, 51,* 269–279.
Pearce, F. (2013). *The Land Grabbers: The New Fight over Who Owns the Earth.* Boston: Beacon Press.
Polanyi, K. (1944). *The Great Transformation: The Political and Economic Origins of Our Time.* Boston: Beacon Books.
Sassen, S. (2013). Land grabs today: feeding the disassembling of national territory. *Globalizations, 10*(1), 25–46.
Sheridan, K. (2010). *The Environmental, Social and Economic Effects of Biofuel Production in Tanzania: Case Studies in Kagera and Bagamoyo. (Masters in Environment and Development).* Dublin: Trinity College Dublin.
Van der Laan, C., Wicke, B., Verweij, P. A., & Faaij, A. P. C. (2017). Mitigation of unwanted direct and indirect land-use change – an integrated approach illustrated for palm oil, pulpwood, rubber and rice production in North and East Kalimantan, Indonesia. *Global Change Biology Bioenergy, 9*(2), 429–444.

Chapter 84

Gentrification

Ernesto López-Morales

Gentrification: a generic definition

Gentrification means transforming a working-class or vacant area in a city's inner or peripheral area for upper-income, residential or commercial use. In 1964, the sociologist Ruth Glass incepted the term to depict an emerging new 'urban gentry' class taking over declining inner neighborhoods in times of post-industrial economic renaissance:

> Shabby, modest mews and cottages two rooms up and two down have been taken over, when their leases have expired, and have become elegant, expensive residences.
> (Glass, 1964: xviii)

Ever since, gentrification has been a bone of contention for its alleged positive and negative outcomes, notably spatial displacement. Gentrification has also captured the attention of a broad spectrum of geographers, sociologists, anthropologists, urban economists and political scientists (Lees et al., 2008) all over the world, resulting in a substantial and diverse literature encompassing increasingly variegated cases at a planetary level that make the concept considerably mutate from its original definition.

Although generically gentrification represents the power imposition of a particular social class taking over and physically transforming urban space, disfavoring and excluding disfranchised existing social classes from that space, the process should not be reducible to one social dimension. The role of class, race, gender, sexuality and physical or mental ability – intersected or not – is crucial for the interplay between revalorization and devaluation and, on the other hand, between the powerful and the powerless (Slater, 2021), thus representing a challenge for gentrification studies.

The changing nature of gentrification

Ruth Glass only gave this process a name. Nevertheless, a common misconception is that gentrification began in the 1960s, as the phenomenon can be traced back centuries ago. Massive-scale gentrification was the 'Haussmannization' of Paris in the 19th century. As a Napoleon III court member, Baron Haussmann commanded an ambitious and technologically sophisticated plan to demolish unsound city quarters and clear space for brand new urban redevelopment. He incepted the Paris' boulevards network and polynuclear urban structure as we know them today. However, Friedrich Engels claimed this technological modernization depended on the destruction of working-class areas:

DOI: 10.4324/9780429282348-92

By "Haussmann" I mean the practice which has now become general of making breaches in the working class quarters of our big towns, and particularly in those which are centrally situated . . . the result is everywhere the same: the scandalous alleys and lanes disappear to the accompaniment of lavish self praise from the bourgeoisie on account of this tremendous success, but they appear again immediately somewhere else and often in the immediate neighborhood.

(Engels, 1872)

Gentrification was being interrogated in even earlier historical periods. Archeologist Andrew Dufton (2019) analyses archaeological evidence of different housing layers in Roman cities in North Africa, seeing elite residences built on top of the earlier modest ones, reflecting the desirability of wealthier inbound residents transforming not just single properties but broader neighborhoods with the significant displacement of less-affluent populations.

Gentrification is a dynamic urban phenomenon linked to economic and societal changes. In 2001, Jason Hackworth and Neil Smith presented a three-wave gentrification model (for New York City) to glimpse the changing role of the state in gentrification. The first wave existed under the justification of ameliorating urban decline (in NYC, this was visible in the form of landlord abandonment and arson). Gentrification was then sporadic and highly localised but already significantly funded by the state. The second wave started in the late 1970s and implied federal programs scaled back as gentrification took a more laissez-faire form and expanded geographically, covering a more significant part of the city, also expanding internationally by neoliberal urban policies. Either implicitly or explicitly, arts and culture functioned as a 'soft factor' attracting new capital flow into the targeted neighborhoods. In the early 1990s, the third wave started as gentrification expanded within inner-city and remote neighborhoods beyond the immediate core. Globalization in the real estate industry had set a larger context for more prominent developers to orchestrate expensive redevelopment. Effective resistance to gentrification declined as the working class had been continually displaced from the inner city. The state became more involved in the process than in the second wave. Waves changed due to significant economic crises, and developers and state policies also did. The wave theory reasserts the close connection between gentrification and the economy.

Two notorious examples of third-wave gentrification are the mass-scale renovation of the Down-Under-Manhattan-Bridge-Overpass (DUMBO) in Brooklyn, NYC and the uplifting of the De Wallen neighborhood in Amsterdam turned into the Red-Light District. In both, the city mobilized public-private coalitions to make these areas safe for investment without scaling down existing social problems but simply moving them to other, less visible locations.

The seminal book *Gentrification* by Lees, Slater and Wily presents a fourth gentrification wave related to the dot-com crash (1999) that led to a massive switching of capital into real estate, which eventually cumulated in the global financial crisis that started in the United States in 2007 ('Subprime crisis'), spread internationally and mutated into several other crises (Aalbers, 2019). This fourth wave combined 'an intensified financialization of housing . . . with the consolidation of pro-gentrification politics and polarized urban policies' (Lees et al., 2008: 179). According to Manuel Aalbers (2019), after the 2007 global crisis, current fifth-wave gentrification comprises corporate landlords and real estate as an asset class, investment by transnational wealth elites and middle classes, platform capitalism, global touristification of cities, global mortgage debt and a generalized lack of housing affordability amidst global economic financialization.

Shenjing He (2019) claims China has gone through a three-wave gentrification path, starting in the mid-1990s when the state attempted bounded areas of redevelopment in Beijing, Shanghai, and Guangzhou to a current, more encompassing third wave of national schemes under state-led financialization, nationwide resettlement housing provision and Shantytown Redevelopment Schemes (SRS; see Desai and Loftus, 2013).

Debates over causes and effects

By the 1980s, a service-producer economy and the decline of the manufacturing industry created the perfect storm for gentrification in the industrialized world. David Ley (1994) saw post-industrial societies and cities reasserting the role of individuality and a growing sensuous and aesthetic philosophy transforming residential environments and creating habitus. Meanwhile, Chris Hamnett proposed that an ascended 'professional class' was consuming (in) the city. Gentrification was far from predictable, and its implications were not related to class contradictions alone but cultural motivations and individual desires, as we see in the two examples next.

First, in the early 1980s, Damaris Rose saw sexual dissidence reclaiming territories in cities like San Francisco in the United States. She claimed 'marginal gentrifiers' found safer living environments among tolerant gentrifying neighborhoods than in the more traditional working-class suburbs. Rose's ideas resonate with current anti-neoliberal housing movements in some Latin American cities, intermingling marginal gentrifiers, sexual dissidents, indigenous and subaltern strata (López-Morales et al., 2021). Second, in an entrepreneurial way, international consultant Richard Florida recommended attracting a 'creative class' (gays, young people, bohemians, artists etc.) into declining cities to turn them more friendly to technologized urban economic reinvigoration. Florida's ideas closely connect with 'placemaking.'

By 1979, Neil Smith had severe doubts about gentrification being motivated by the desires and consumptions patterns of these new, post-industrial social classes. Thus, he presented a more material causal explanation: gentrification was mainly a 'back to the city' movement of capital, not people. Over the years, countless scholars have followed this perspective, which can be summarized in the following four key points:

1 Gentrification is a profit-driven transformation of places that have historically suffered from often deliberate disinvestment and abandonment.
2 It tends to occur in areas where commercial and residential land is cheap or 'informal' – where there is vast potential to make profits through upgrading existing structures and/or building new ones (i.e., Neil Smith's 'rent gaps').
3 It is caused by private developers, landlords, businesses and corporations, supported or led by governments (via rezoning, public subsidies, regeneration schemes, etc.).
4 Gentrification results in the displacement/exclusion of low-income and other disfranchised people.

So, there is little natural about gentrification, but often it comes from long-term policies of devaluation (e.g., redlining, meaning certain buildings or areas prevented of receiving maintenance or 'territorial stigma', meaning negative symbolic power over people because of the place where they live; Slater, 2021) and redevelopment. For the appropriation of the rent gap, the state and private owners and investors have specific roles to play. The former creates the economic, legal, and administrative framework, usually of a neoliberal guise; the latter respond to their private interests over land rent accumulation. The rent gap is a material example of 'creative destruction', an idea the Austrian economist Joseph Schumpeter brought in the 1930s to explain the pace of technological advancement under capitalism based on the deliberate destruction of existing fixed capital and its replacement by state-of-the-art technologies. Recent rent gap narratives have identified housing sharing economies like Airbnb creating gentrification as short-term tourist renters displace deprived, long-term tenants.

Critics like Asher Ghertner claim that linking gentrification to the rent gap obscures other related social factors, particularly informal land tenure regimes in broad areas of the Global South. From India, Sapana Doshi claims that in slum redevelopment, the threshold between gentrifiers and gentrified ones is thin, as ethnic and gender power dominance arises on top of economic interests. In Israel, policies of ethnic dominance, housing immigration on urban peripheries and the war conflict shape gentrification beyond its economic motivations.

Spatial displacement has been at the forefront of gentrification debates. By 1985, Peter Marcuse classified different displacement types. Direct displacement encompasses landlords cutting the heat in a building, increasing rents or people's eviction by the police. 'Exclusionary displacement' means residents who cannot access housing or neighborhoods anymore as it has become too expensive, while 'displacement pressure' is suffered by people during the neighborhood's transformation. Symbolic displacement represents the forced erasure of symbolic aspects inherent to the social composition of places and the loss of sense of place for original users (Mindy Fullilove calls it 'root shock'). However, other authors disagree: Lance Freeman and Chris Hamnett have said the connection between gentrification and displacement is weak and stress that gentrification represents community recovery in deprived ethnic enclaves or ghettoes and overall social class mobility in a city. Still, observing gentrification-led displacement is challenging as the displaced subjects from a specific area are often almost impossible to trace or at least extremely hard to follow (see Adey et al., 2020).

Conclusions: ideas for a future agenda

Gentrification has become planetary, as housing inaccessibility has soared at the hand of multinational equity firms, shell corporations, vulture investors, global real estate companies, highly specialized intermediary financers, REIT and the like. Besides, state policies do not regulate but often aim at gentrification-like policies under pro-growth macroeconomic agendas and neoliberal planning instruments. Changes due to COVID-19 and climate change may reinvigorate gentrification in unexpected places, like perceived healthy or safe urban or peri-urban areas.

In this vein, future gentrification scholars might require decolonizing some of its definitions. Beyond the usual suspects of Europe and North America, new reference points are East Asia, the Middle East and Latin America. Further, recent evidence of gentrification in ancient urban periods suggests the need to reconsider the confinement of gentrification within a specific time and place, the wave-gentrification approach and even capitalism as a necessary condition for gentrification. Crucial gentrification aspects like race, gender, ethnicity and religion invite questioning definitions imposed by Global North scholars. Racialized dispossession has often been examined in the Western context, but it is little known and looks different in South America, the Middle East and East Asia, where informality and urban marginalisation compensate the historical absence of a welfare state or it responds to nationalistic, ethnic dominance.

Bibliography

Aalbers, M. (2019). Revisiting 'The Changing State of Gentrification' – Introduction to the Forum: From Third- to Fifth-Wave Gentrification. *Tijdschrift voor Economische en Sociale Geografie*, *110*(1), 1–11. https://doi.org/10.1111/tesg.12332

Adey, P., Bowstead, J., Brickell, K., Desai, V., Dolton, M., Pinkerton, A. & Siddiqi, A. (Eds.) (2020) *The Handbook of Displacement*. London: Palgrave Macmillan, p. 825.

Desai, V. & Loftus, A. (2013). Speculating on Slums: Infrastructural Fixes in Informal Housing in the Global South. *Antipode*, *45*(4), 789–808.

Dufton, A. (2019). The Architectural and Social Dynamics of Gentrification in Roman North Africa. *American Journal of Archaeology*, *123*(2), 263–290. https://doi.org/10.3764/aja.123.2.0263

Engels, F. (1872). The Housing Question (Chapter 2). https://www.marxists.org/archive/marx/works/1872/housing-question/ch02.htm

Glass, R. (Ed.) (1964). *London: Aspects of Change*. MacKibbon and Kee, London.

Hackworth, J. & Smith, N. (2001). The Changing State of Gentrification. *Tijdschrift poor Economische en Sociale Geografie*, *92*(4), 464–477.

He, S. (2019). Three Waves of State-Led Gentrification in China. *Tijdschrift Voor Economische En Sociale Geografie*, *110*(1), 26–34. https://doi.org/10.1111/tesg.12334

Lees, L., Slater, T. & Wyly, E. (2008). *Gentrification*. Routledge, New York.

Ley, D. (1994). Gentrification and the Politics of the New Middle Class. *Environment and Planning D: Society and Space*, *12*, 53–74.

López-Morales, E., Ruiz-Tagle, J., Santos Junior, O., Blanco, J. & Salinas Arreortúa, L. (2021). State-Led Gentrification in Three Latin American Cities. *Journal of Urban Affairs*, *45*(8), 1397–1417. https://doi.org/10.1080/07352166.2021.1939040

Slater, T. (2021). *Shaking Up the City: Ignorance, Inequality, and the Urban Question*. University of California Press, Oakland.

Weber, R. (2002). Extracting Value from the City: Neoliberalism and Urban Redevelopment. *Antipode*, *34*(3), 519–540.

Additional resources

"#Gentrifícamelo!", Valparaíso Inamible, Valparaíso, Chile, August 2016. https://www.youtube.com/watch?v=VGY8HrYT3l4

"Gentrification and What Can be Done to Stop It", Loretta Lees, TEDxBrixton, December 2014. https://www.youtube.com/watch?v=gMz1x5_yF2Q

"Interview with Prof Chris Hamnett", Centre for Liveable Cities Singapore, April 2017. https://www.youtube.com/watch?v=d__RebAupKA

"The Roots of Gentrification in New York City", The Laura Flanders Show, March 2016. https://www.youtube.com/watch?v=TDmD4atDUVI

"Rose Street and Revolution: A Tribute to Neil Smith (1954–2012)", Tom Slater. https://blogs.ed.ac.uk/tomslater/neil-smith-tribute/

Chapter 85

'Slums and modernity'

Syed Haider

Publications on the phenomenon of slums abound. Mike Davis' *Planet of Slums* (2006) may be best known but it was certainly not the first. Kalpana Sharma's *Rediscovering Dharavi: Stories from Asia's Largest Slum* predates both Davis' volume and Danny Boyle and Loveleen Tandan's film, *Slumdog Millionaire* (2008). Indeed, the interest in slums has a long lineage from Dominique Lapierre's *City of Joy* (1985) to the 1970 edition of *Problems of American Society* (with its focus on slums); from Seebohm Rowntree's investigations in York published in 1901 to Jacob Riis' 1890 publication, *How the other half lives*. As such, slums persist as a problem and fascination in the global imaginary, and this is what this chapter focuses on. Ultimately, the argument presented here is that the persistence of slums in the global imaginary owes much to the fact that the slum acts as a nexus bringing together concerns and considerations around urbanism, capitalism, modernity, and more recently, environmental degradation.

This chapter is divided into two parts. The first part interrogates the term *slum* and explores the concern raised by some (Mayne, 2017; Roy, 2011; Slater, 2021) that the word itself is complicit in delimiting the way we think about those who live in such places. Concerns around language, representation, and the discourse built around slums runs throughout this chapter drawing attention to slums and their place in the mental geographies of a global imaginary. Building on this, the second part focuses on India and Dharavi, India's most (in)famous slum. Here the discussion unpicks connections between slums and modernity by treating Dharavi and economic development in India as a loose case study. The chapter concludes with the suggestion that slums *haunt* global imaginaries and that it is this *haunting* which accounts for the persistence of the slum in cultural, political, and academic discourse.

The word 'slum'

Whilst the term slum is a feature of both popular and academic accounts of urban dwelling, the term is contested. Alan Mayne (2017) challenges the uses of the word slum and argues that it is a term used by urban planners to dismantle working class communities. Slum-clearances for Mayne are more about accessing the land on which informal dwellings are built than supporting people who live there. Tom Slater (2021) makes a similar point claiming the term is often used to justify top-down interventions in informal settlements, without regard for the needs and perspectives of the people who live there. What these authors draw attention to is the political nature of language and, specifically, the act of labelling and classifying as a tool of power and coercion. Mayne criticises the

DOI: 10.4324/9780429282348-93

language behind initiatives such as the United Nations' Millennium Declaration (2000) and its adopted slogan, 'Cities without slums'. For these writers, such initiatives mis-characterise urban social disadvantage because the relationship between cities and slums is misshapen. This is what Mayne means when he claims the term 'slum' is a deceitful construct:

> 'Slum' is an especially unhelpful word. It misrepresents the complex realities of urban social inequality, whether in New Delhi today, in nineteenth-century London where the word was first coined, or in countless other places across time and space. It marginalizes poor people and low-income areas as supposedly deficient and dysfunctional by-products of urban development.
>
> (10)

In contrast, Gulliver (2018) strikes a sceptical tone questioning whether the emphasis Mayne and others place on language overstates the case. Is it not likely, she wonders, that 'whatever well-intentioned euphemism is adopted . . . [it] will soon acquire the same pejorative connotations, and be freighted with the same assumptions?' Surely the problem for those who live in slums has more to do with 'the lack of clean water and adequate shelter . . . [than] the fact someone describes their residence as a "slum" in a UN report'.

Gulliver's scepticism centres on her view that slums are not evidence of growing poverty but evidence of the growth of cities.

It is city wealth – and employment – that breeds slum development. Residents of slums – historically, as now – are often newcomers, either from the countryside or even from another country. In many cases, they have chosen to come. This is not to say that their living standards should not be improved, but rather to assert that slum dwelling is not necessarily some random fate befalling the hapless.

Whilst Gulliver and Mayne present competing points of view, what is clear is that language has political implications and plays a part in how economic inequalities are imagined. An approach that is insensitive to this fact is in danger of overlooking the structural complicity of certain kinds of urbanism that produce and sustain slums. Gulliver implies that slums are merely places of arrival and points of departure, a space for people on their journey from an informal city to the formal city. But what if informal settlements are structurally concomitant with the formal city? Approached this way, slums signify a mismatch between aspirations for an urbanism of equality and increasingly inequitable economic structures, both local and global. As such, slums act indexically to reveal the mental geography of a global imaginary.

The term 'global imaginary' refers to the shared images, symbols, and narratives that shape how people imagine the world beyond their immediate experience. These images and narratives can be influenced by factors such as media representation, politics, and cultural exchange. In this context, mental geography refers to the ways in which people mentally map and navigate the global imaginary. For example, people may have mental maps that prioritise certain regions or countries over others based on their cultural, political, or economic influence. They may also have mental maps that reflect their personal experiences and interactions with people from different parts of the world. Studying mental geographies of the global imaginary can provide insights into how people understand and interact with the world beyond their immediate environment. It can also reveal how global images and narratives are constructed and disseminated and how they shape perceptions and beliefs about the world. Demarcated in this way, the word slum warrants scrutiny. We should pay attention to practices of language (and labelling), representation, and discourse which coalesce around the term because they offer us important insights regarding our interconnected presents and potential futures. This is what the next part of this chapter will focus on, and it will look at the cultural representation of Dharavi and its history.

Dharavi, Mumbai

Slumdog Millionaire begins with a chase sequence of police officers pursuing a group of children in Dharavi. The low angle shot captures mountains of rubbish across which the children run, followed by an unsteady camera and rapid editing that captures the energy and speed of the chase. The happy faces of the children as they run, some jumping off roofs while others run round corners of ramshackle huts in windy streets, imbues Dharavi with a carnivalesque character; the children are in their element, running expertly around the slum, outwitting the plump middle-aged police officers, one of whom is showered with plastic rubbish that clearly abounds there. The slum is a mix of energy, hubbub, and friendly conviviality even as it is overly congested and mired in poverty, dirt, and drudgery.

The controversy that grew around Boyle and Tandan's film and the postcolonial anxiety it fuelled certainly owed to the fact that it was a project led by a Western filmmaker framing India. The international attention was welcome and the global exposure even courted, but the ambivalence arose from the fact that what drew the world to the image of India was not 'India shining' as many metropolitan Indians would have desired but the apparent underbelly of a fast-developing nation.

The roots of this ambivalence and anxiety, however, may be interpreted as running deeper if we see the place of the slum in an Indian imaginary (and beyond) as actually representing the uncanny self – not an aberration of modernity but its constitutive and excluded alter-ego. As Ashish Nandy has said, 'Is the shock many "enlightened" Indians claim to feel when facing the aesthetics of the slum, the shock of seeing one's own face in a convex mirror?' (1998: 3) But this shock is not merely confined to the Indian imagination. It extends to the shock of all moderns – East and West – for the slum is the shadow cast by enlightenment and the modernity associated with it.

In his spatial analysis of slum-living, Jan Nijman outlines briefly the history of Dharavi. He begins his essay by describing the colonial spatiality of Bombay in which the southern peninsula was divided between European residential quarters and native residential quarters, the latter having a considerably higher population density than the former and much less in the way of infrastructure. Census reports in the colonial era speak of the inadequacy of housing and sanitation, but colonial administrations, Nijman says, did little in terms of policy until an outbreak of the bubonic plague in 1869. 'It was decided', writes Nijman, 'that the gradual expansion of polluting industries in Native Town had to be halted and moved to a distant tannery town in the north (Dossal 1991, p. 202). That town was Nijman' (2010: 8).

The act of excluding, excising, and expelling can be seen as the constituting gesture for the birth and continuation of slums like Dharvi. Born as a result of the expulsion of noxious activities from the city to the peripheries, it mushroomed further following independence, when rapid modernisation, urban development, and industrialisation intensified rural-urban migration. Dharavi became the 'favourite dumping ground for companies operating across the city' (8) and a space to absorb surplus labour and human traffic. Like all slums, Dharavi is not only a product of our historically received and fashioned modernity (which is to say, merely an unfortunate by-product); it is in fact a sign of modernity.

This is where the ambivalence and anxiety over slums like Dharvi emerge from – an ambivalence and anxiety over the forces of capitalism and modern urban topography that shape and require slums. Citing architect and social activist Jai Sen, Ashish Nandy makes a similar point about the slum as the '"unintended city" – the city that was never a part of the formal "master plan" but was always implicit in it'. 'The official city cannot survive without its unintended self' writes Nandy, 'but it cannot own up to that self either' (1998: 2).

This dialectic of being constitutive but excluded (or better still, of constituted through exclusion) presents the slum as a complicated space, concept, and reality, best captured by the fact that the vastness of various slums is juxtaposed by their impermanent status (nobody who lives in Dharavi has any substantial legal right to be there). This dialectical relationship that slums share with their surroundings is what the cultural forays into slums as a topic of documentaries, films, novels, and other cultural products highlights. Reading such products in this way helps implicate the idea of progress with the existence and continuation of slums themselves. It exposes an aporia at the heart of modernity: progress has two faces; one, the material affluence of consumption and the success of economies grown on the logic of consumption and two, the degradation consumption wreaks on people and environments.

In Channel 4's 2010 documentary, *Slumming it*, we see presenter Kevin McCloud torn between the obvious squalor of the slum and the ingenuity of its inhabitants. These two themes run throughout the two episodes but they only account in part for the persistence of slums as problem and fascination in the global imaginary. The other reason has to do with that convex mirror Ashish Nandy spoke of and this is illustrated when McCloud ventures to a dumping ground from where waste is transported to Dharavi for recycling.

McCloud's visceral reaction, his confusion and inability to 'think straight' reflects the ambivalence and anxiety born of the reality of slums. If we recognise the ingenuity of the people who live there, the entrepreneurial spirit and raw capitalist energy – trappings of a capitalist ideology for which India has signed up – then we have to also recognise slum living and scenes such as the one which authorities did not want McCloud to film as two interconnected sides of the same picture. The waste that confronts the Channel 4 team may be in India, but McCloud cannot help thinking of Britain. The plastic bags used and discarded in Mumbai are not dissimilar to those used and discarded in Britain – or elsewhere for that matter. A global modernity premised on consumption – and a global economy based on consumerism – is something in which we all participate and so the slums that exist across the globe are intimately connected with us.

It is for this reason that the global imaginary is drawn to slums; they speak to us because we recognise our modernity in them. Yet it is only by excluding this truth or ignoring it or displacing the anxiety it gives rise to that we can produce and re-produce a capitalist world. This is particularly true in India where the official narrative is of India as an emerging superpower. In this story of India's rise, it is malls and multiplex cinemas that are signs of 'India shining'; to have to admit that slums and other inconvenient realities like horrendous air pollution are also signs is uncomfortable to say the least.

Ambivalence and anxiety around slums index the paradoxes of modernity as we have inherited and fashioned it. If slums like Dharavi were born of colonial and postcolonial modernity, and if India pursues modernisation as it is, then slums are the future that haunts it. That being the case, the ubiquity of slums in the global imagination may be part of the problem of stereotyping the Global South; such representations may not be political enough and may even slip into poverty porn, but the reason they persist may be because the global imaginary itself is similarly haunted.

Bibliography

Dossal, M. 1991. *Imperial designs and Indian realities: the planning of Bombay City*, 1845–1875. Oxford. Oxford University Press.

Gulliver, Katrina. 2018. *Sydney Review of Books*. https://sydneyreviewofbooks.com/review/slums-the-history-of-global-injustice-alan-mayne. (Accessed: 5 September 2022)

Mayne, Alan. 2017. *Slums: The History of a Global Injustice*. London: Reaktion Books.

Nandy, Ashish. 1998. *The Secret Politics of Our Desire*. London: Zed Books.

Nijman, Jan. 2010. "A Study of Space in Mumbai's Slums", *Journal of Economic and Human Geography*, 101: 1, pp. 4–17.

Roy, Ananya. 2011. "Slumdog Cities: Rethinking Subaltern Urbanism", *International Journal of Urban and Regional Research*, 35: 2, pp. 223–238.

Slater, Tom. 2021. *Shaking up the City: Ignorance, Inequality and the Urban Question*. Oakland, CA: University of California Press.

Suggested Further Reading

Davis, Mike. 2006. *Planet of Slums*. London: Verso.

Linke, Uli. 2012. "Mobile Imaginaries, Portable Signs: Global Consumption and Representations of Slum Life", *Tourism Geographies*, 14: 2, pp. 294–319.

Simone, AbdouMaliq. 2018. *Improvised Lives: Rhythms of Endurance in an Urban South*. New York: Polity.

Audio visual material

Channel 4. 2014. *Slumming it*. [Online Video]. https://www.youtube.com/watch?v=Im0tHRs9Bng (part 1); https://www.youtube.com/watch?v=l-yjpvzGKZQ (part 2). (Accessed: 10 September 2022)

Slumdog Millionaire. 2008. [Film]. Danny Boyle and Loveleen Tandan. dirs. UK: Film4.

Chapter 86

Urban health

Sustainable development and the healthy city

Jennifer Cole

Introduction

Throughout the 20th century, global health traditionally focussed on morbidity and mortality that could be reduced by a combination of improved access to clean water and sanitation (WaSH) and to trained healthcare professionals and healthcare systems. Throughout life and across a number of conditions, health tends to be poorer in lower-income individuals and communities, disproportionately affecting rural regions of low-income countries, often exacerbated by poor hygiene and lack of access to healthcare. The rapid urbanisation seen over recent decades has, however, seen an increase in the numbers of people living in absolute poverty within urban settings and within middle- as well as low-income countries. In addition, health is dependent on more than just access to doctors and hospitals: the quality as well as quantity of food available, the quality of the air we breathe and the water we drink, and opportunities for exercise and social interaction all have a profound impact on health.

Low-income urban areas can face health risks from poor sanitation, overcrowded conditions that enable disease to spread rapidly, poor housing, and lack of access to fresh and nutritious food. In addition, if public healthcare provision does exist or does not expand at the same rate as the city, residents may be left dependent on private healthcare providers whose costs can be out of reach for the urban poor. A lag in public provision of clean drinking water and waste management can further add to the health burden, but in rapidly developing countries, provision of infrastructure and municipal services does not always keep up with the pace of change.

The urban growth predicted to take place in developing countries between today and 2050 is predicted to increase the number of people living in peri-urban, slum, and informal settlements from 881 million in 2015 to 2 billion in 2030 [1]. Such settlements present many challenges to accessing urban healthcare systems. Residents who live on the periphery of a city experience greater travel times to healthcare facilities when seeking treatment than those living in more central areas, for example, and higher generic diagnoses of "fever" compared with identification of the specific cause of disease, presenting barriers to surveillance and prevention strategies.

An increasing focus on the (poor) health of cities in recent years, highlighted in reports and initiatives such as the 2010 Global Forum on Urbanization and Health, UN Habitat, the WHO Healthy Cities Programme, and the WHO/UNICEF report *Preventing Disease Through Unhealthy Environments* has led to considerable concern over the health of urban citizens in developing regions due to the environmental risk factors they face.

DOI: 10.4324/9780429282348-94

Health is strongly influenced by socioeconomic determinants and environmental factors as well as access to healthcare.

Health and sustainability in the SDGs

In response to this, a more integrated approach to health and its socioeconomic and environmental determinants has been taken in the Sustainable Development Goals (though the WHO Healthy Cities programme dates back to 1986 and the Ottawa Charter for Health Promotion). Many of the targets of SDG 3, Good Health and Well-Being, focus on improving urban health through upgrading slums; improving transport systems, air quality, and municipal waste management; and the need to consider the provision of urban space. Cities need to be planned with health in mind, remaining mindful of the requirements for a secure and nutritious food supply (SGD 2) and adequate sanitation (SDG 6) as well as simply providing healthcare systems and infrastructure.

Health in the city

Cities provide better economic opportunities, better access to healthcare facilities and services – particularly for mental health – and higher living standards. Worldwide, epidemiological and demographic data suggests that health outcomes, health services, and survival rates from morbidity at all stages of life are better in cities than in rural areas. But this does not mean that cities are inherently healthier places. Long-term poor health in later adulthood, from chronic conditions such as cancers, high blood pressure, heart disease, and diabetes, all tend to be worse in urban regions. The urban poor in densely populated cities may have increased life expectancy over their rural counterparts but it is not necessarily a healthier life.

The reasons for this are many, intersectional, and complex. Urban populations are more concentrated than those in rural communities, and this can create challenges with air pollution, the provision of clean water and waste disposal (globally, 2 billion people do not have access to waste collection services, while the amount of garbage produced annually is predicted to reach 4 billion tonnes by 2050). Only 12 per cent of cities worldwide reach pollution control targets. In poorly planned and managed cities, diseases can easily proliferate, including infectious diseases such as COVID-19, measles, tuberculosis, HIV and influenza; vector-borne diseases such as Zika and Nipah viruses; and those caused by water-borne pathogens that cause diarrhoea and worm and parasite infections.

City dwellers also face heightened risk exposure to the drivers of non-communicable diseases (NCDs) due to many interconnected factors including low-quality, processed diets; a greater likelihood of proximity to sources of pollution such as busy roads; lack of exercise due to poor availability of quality green space and the leisure time to use it; and higher likelihood of exposure to risk factors for stress and depression, such as high levels of noise and social isolation.

Violence and injuries, including road traffic accidents – one of the top ten conditions in the Global Burden of Disease Study – are also higher in cities. In LMICs roads are not always well maintained, while speed limits, wearing seat belts, and motorcycle helmets may not be enforced.

Urban areas also tend to exhibit more marked socio-economic inequality than rural areas. Residential Welfare Associations (RWAs, also known as Neighbourhood Associations) can have a strong influence on infrastructure and urban services, but they tend to be dominated by the urban middle-classes who press for a cleaner, greener, safer, and healthier environment for their own neighbourhoods. Poorer residents often find themselves having to do so through intermediaries such as NGOs, councillors, or employers.

Metrics and monitoring urban health

Understanding how socioeconomic and ecological factors intersect with health and can be used to promote and maintain good health throughout life will be essential to ensuring the sustainable development of cities into the future, but many of the current indicators we use for measuring health in the Global Burden of Disease Study (GBD) such as Disability Adjusted Life Years (DALYS) and Years of Life Lost (YLL) due to preventable premature mortality measure ill-health rather than good health. They record the problems but do not necessarily provide solutions. Consideration for how urban health is defined in the future led the WHO, in 2013, to establish a Reference Group on Statistics, which amongst other activity produces a Global Reference List of 100 Core Health Indicators [2]. This includes information on health-related SDGs and covers issues such as universal health coverage, non-communicable diseases, environmental risk factors, and socioeconomic risk factors. Health indicators are grouped into four key domains: health status, risk factors, service coverage, and health systems, providing a starting point for how urban environments can be made more conducive to health. The WHO Healthy Cities programme provides further guidelines and best practice on how to integrate health into urban planning, investments, and policy decisions through its New Urban Agenda.

Urban food systems and adequate nutrition

Poor quality diet is the main risk factor for many non-communicable conditions in adulthood including obesity, diabetes, high blood pressure, heart disease, and many cancers. Urban dwellers are more likely to receive a secure supply of food than their rural counterparts but not necessarily a better quality one. The food in their diets, particularly those of the urban poor, tends to be highly processed and contains too much sugar, salt, and fat. Displaced people and refugees moving into low-quality urban housing without adequate cooking facilities face a particular nutritional challenge.

Without the space to grow fruit and vegetables or to keep livestock, urban citizens' nutritional status over time can be poorer than that of their rural neighbours. The prevalence of childhood stunting (being smaller than the recommended average height for age) is higher in rural settings than urban ones, but poor urban children are more likely to be obese, which can be just as damaging to long-term health, particularly in those who are simultaneously stunted and overweight for their height. Cultural preferences for a more vegetarian-based diet in developing countries such as India may provide some protection against a shift towards fast-food and processed diets prevalent in North America and Europe, but food systems need to be designed to ensure accessible and affordable fresh food to urban centres.

Sanitation and waste

Second only to an adequate supply of nutritious food in ensuring human health is access to clean drinking water and well-managed sanitation. In low- and middle-income countries,

78 per cent of urban dwellers had access to at least basic sanitation services in 2015, 31 per cent more than rural dwellers [3]. Sanitation must be considered as an end-to-end service, which takes into account how human waste is collected and managed, as well as just whether or not the household has a private toilet, however. Sanitation that drains into open sewers and rivers, ponds, and other watercourses used for washing clothes, bathing, or fishing can be hotspots of disease transmission.

Access to safe water and sanitation in peri-urban areas can be particularly poor, especially where municipal infrastructure lags behind urban expansion. In temporary and informal settlements, support for faecal sludge management systems (FSM), in which human waste from managed septic tanks is collected and taken for treatment, can provide interim solutions until better infrastructure can be provided.

Approached systematically, sanitation can support other development opportunities, such as provision of clean energy, through biogas digestors that capture methane and reuse it as clean household and vehicle fuel. Developing regions such as Karnataka in India have combined municipal sewerage treatment with clean energy generation to push forward both agendas.

Air pollution and respiratory health

Creation of clean energy from waste can help to alleviate a third key health challenge in developing cities: air pollution and its impact on respiratory health. In 2017, 97 per cent of cities in low- and middle-income countries with more than 100,000 inhabitants did not meet WHO air quality guidelines [4]. Improvements in socioeconomic conditions can be cancelled out by the impact of poor air quality, particularly as it has a disproportionate impact on the very young, the poor, and the very old (the world's largest 750 cities will have an extra 150 million residents over the age of 65 by 2030). In developing cities, air pollution from household fuels such as wood and coal mix with fumes from traffic, waste incineration, and industrial manufacturing, all of which may be subject to less rigorous environmental regulation or enforcement than in high-income countries.

Planning urban transport well can improve air pollution from transport emissions considerably whilst also improving social inclusion and physical exercise, particularly through cycle lanes and well-serviced pedestrianised areas. The percentage of urban dwellers with convenient access to public transport (defined as being within 500m of a bus stop) is more than 70 per cent in most developed countries, for example Australia, but just 18 per cent in Sub-Saharan Africa.

Poor air quality can also be mitigated to some extent by designing parks and open spaces that act as "lungs" of a city and provide a calming environment that helps mitigate stress from traffic noise.

Urban greening and well-being

Low levels of physical activity and stress can lead to poor mental health. As weight gain is linked to the use of antidepressants, this can create a vicious cycle more likely to be prevalent in cities. "Urban greening" – making space for more natural environments in the midst of urban areas – brings long-term health benefits, including greater longevity, better-developed immune systems, and reduced risk of cardiovascular disease and can improve access to fresh food through urban gardens and allotments. Green space provides opportunities for active recreation, relaxation, contemplation, and neighbourhood social cohesion and can be particularly beneficial to the physical and mental health of urban residents with lower socio-economic status, who are less likely to have opportunities for countryside holidays, to have their own gardens, or to able to keep pets at home.

A reduction In the prevalence of several NCDs and their risk factors can be linked to the quantity, proximity, and usability of natural spaces in urban environments, but many developing cities have less than the 9m² per person of green space suggested by WHO.

Conclusions

Over the coming decades, the focus on urban environments will increase as rural populations continue to migrate to growing cities: urban health and urban sustainability cannot be disentangled.

Ensuring the health of developing urban populations depends on seeing the components of the urban city and its inhabitants as part of the same socioecological system, from which people cannot and must not be separated from nature and in which the air, water, and soil quality needs to be maintained within the urban environment. Managed well, developing cities have the opportunity to lock in excellent sustainability initiatives. By linking sanitation provision with clean energy generation, providing sustainable and efficient housing developments, and ensuring sufficient green space for leisure, physical exercise, community activities, urban gardens, and stress relief, it is perfectly possible to build healthier environments for future generations.

Further reading

Cole, J., 2019. *Planetary Health: Human Health in an Era of Global Environmental Change.* CABI.

Prüss-Üstün, A., Wolf, J., Corvalán, C., Bos, R. and Neira, M., 2016. *Preventing Disease through Healthy Environments: A Global Assessment of the Burden of Disease from Environmental Risks.* World Health Organization.

U.N. Habitat, 2016. *World Cities Report: Urbanization and Development: Emerging Futures.* United Nations Habitat.

World Health Organization, 2016. *Health as the Pulse of the New Urban Agenda: United Nations Conference on Housing and Sustainable Urban Development.* Quito, October 2016.

References

1. Ezeh, A., Oyebode, O., Satterthwaite, D., Chen, Y.F., Ndugwa, R., Sartori, J., Mberu, B., Melendez-Torres, G.J., Haregu, T., Watson, S.I. and Caiaffa, W., 2017. The history, geography, and sociology of slums and the health problems of people who live in slums. *The Lancet, 389*(10068), pp. 547–558.
2. World Health Organization, 2018. *2018 Global Reference List of 100 Core Health Indicators (Plus Health-Related SDGs) (No. WHO/HIS/IER/GPM/2018.1).* World Health Organization.
3. World Health Organization and UNICEF, 2017. *Progress on Drinking Water, Sanitation and Hygiene: 2017 Update and SDG Baselines.* Geneva: World Health organisation.
4. World Health Organization, 2016. *Ambient Air Pollution: A Global Assessment of Exposure and Burden of Disease.* Geneva: World Health organisation.

Chapter 87

Infrastructures for development

Stephanie Ketterer Hobbis

The promise of infrastructures

In December 2021, the European Commission announced a €300 billion fund for investments in infrastructure projects worldwide. Earlier in 2021, the US president and the G7 announced another new global infrastructure initiative, 'Build Back Better World' or 'B3W', in collaboration with multilateral institutions and private sector companies. Both efforts present themselves as alternatives to China's 'Belt and Road Initiative' or 'BRI' which, since its launch in 2013, has already spent over US$1 trillion on infrastructure projects in over 70 countries. All three reflect a broader global commitment to infrastructure development. Infrastructures also have a dedicated Sustainable Development Goal – SDG9: Infrastructure and Industrialisation – and they are central to a wide variety of other funding schemes from multilateral and national development agencies alike.

As material connections, between and among individuals and especially between states and their citizens and between economies, producers and consumers worldwide, infrastructures are deemed central for addressing a wide range of vital global challenges from urban-rural divides to public health and sustainability. Comprehensive telecommunication networks, for instance, enable banks to reach remote areas through mobile money services. Moreover, improved solid waste management can substantively reduce the prevalence of infectious diseases such as cholera or typhoid. The construction of new hydropower promises to cut greenhouse gas emissions as much as dependencies on fossil fuel. In other words, not only does the development potential of infrastructure projects seem endless, but development without adequate infrastructures even seems impossible.

Euro-American, Chinese and other commitments to infrastructure development then also entail a fundamentally apolitical dimension. Different actors regularly disagree about technical details linked to infrastructures, how infrastructures should be financed or where particular infrastructures should be built. Yet, at a most basic level, these actors concur that better, more reliable and comprehensive infrastructures are generally desired and needed by everyone who aims to realize development goals. Even international military interventions/peacebuilding initiatives such as the US counter-insurgency doctrine of 'clear, *build*, hold' (my emphasis) in Afghanistan increasingly prioritize infrastructure projects. Infrastructures promise immediate, tangible improvements to people's lives on the ground and, thus, offer a literally 'concrete' way to winning 'hearts and minds' in conflict-affected environments (Bachmann and Schouten, 2018).

DOI: 10.4324/9780429282348-95

The politics of infrastructures

Scholars have similarly acknowledged their centrality for achieving both global and local development goals, echoing the prioritization of infrastructures in development practice. However, they have also cautioned against a technical, apolitical perspective on infrastructures. Instead, research has paid close attention to the politics of infrastructures and how infrastructural investments often strengthen, 'cement' or shift existing power structures.

Euro-American policy makers regularly deny that their 2021 initiatives aim to curtail China's increasing influence in BRI participating countries from low income countries such as South Sudan or Nepal to high income countries such as Singapore or Italy. Still, there are clear political and economic implications linked to large-scale infrastructure investments. By creating new material networks, such as new interlinked roads and port systems, BRI projects actively foster new regional integrations around new economic connections and political collaborations between China and BRI participating countries (Flint and Zhu, 2019). These new economic connections may not necessarily compete with Euro-American interests. Nonetheless, fueled by the depoliticized narratives surrounding infrastructures, they create the material foundations for a potential global hegemonic shift towards Chinese political and economic interests. An analytical focus on the political economy of large-scale infrastructure investments can, thus, help better explain why infrastructure development has become so critical for government initiatives and, in this particular case, why the newest Euro-American infrastructure initiatives appear so urgent.

At the same time, critical development scholars have emphasised how a macro-level focus on infrastructure development can run the risk of masking more complex, messy politics in particular places, at specific times and beyond the larger economic and political goals associated with individual projects. For example, in a detailed study of the construction of the BRI-funded trans-Laos railway, Chen (2020) uncovered how delays in government payments had a multitude of contradictory ripple-effects that eventually marginalised, above all, Chinese actors: first, the Chinese enterprises executing the project had to take out personal loans and, second, the Chinese workers that replaced the Lao labourers who protested against delayed wage payments and exploitative working conditions. Fairhead's (1992) examination of state-sponsored road expansion in eastern Zaire offers another example of the potentially contradictory outcomes infrastructure development. While the road was meant to strengthen market integration of rural areas and central state authority, Fairhead found that it further alienated rural populations from both by intensifying extortion by state agents and land-grabbing by elites. In other words, the day-to-day realization of infrastructure projects on the ground can solidify or bring about new experiences of marginalisation, irrespective of the broader developmental goals these initiatives aim to realize.

Infrastructures are also regularly found to be designed to materially distance, rather than connect, particular groups and peoples. In apartheid South Africa infrastructures were central to the enactment of racially separated mobilities. Exemplified in segregated public transportation, apartheid infrastructures aimed to contain and divide South Africa's populations and, thus, to realize the 'separate development' vision and containment of black opposition by the state (von Schnitzler, 2018). Similar processes of infrastructural racial division can be found across the world, including the United States or

Canada, where indigenous settlements continue to have substandard access to, among others, water and sanitation infrastructures.

Moreover, when official narratives emphasise connectivity, infrastructure projects can keep populations purposefully apart. For example, Pedersen and Bunkenborg (2012) observed how new roads between Mongolia and China accelerate the movement of goods and peoples between China and Chinese resource extraction sites in Mongolia, as intended. Simultaneously, they effectively minimize Sino-Mongolian interactions, also as intended. The roads essentially allow for Chinese labourers to travel between China and their Mongolia-based worksites without stopping and, specifically, without having to engage with Mongolians living along the way. According to Pedersen and Bunkenborg (2012), this lack of encounter reflects both parties' desires who prefer not to encounter each other in their day-to-day lives. In other words, infrastructures may be designed to connect some people. Still, often they keep the groups they connect apart from others, as a form of active marginalisation, mutually preferred dissociation or anything in between.

At least partial Mongolian indifference towards better connections with China and their economic activities points to another key political dimension of infrastructure development. Dominant development policies and practices assume an overwhelming desire for better infrastructural integration with state systems, national and global trade networks. However, this is not necessarily the case. Tarekegn and Overton (2011), for instance, show how an Ethiopian village rejects road infrastructures in favour of an isolation-centric development strategy. Infrastructural disconnection here aspires to preserve local values and fend of the modern influences that roads are feared to bring. Similarly, in Papua New Guinea a telecommunication tower was torn down, at least in part, because mobile phones are seen as promoting immoral values (Hobbis, 2018). Instances of rejected infrastructures challenge the perceived apolitical nature of global commitments to infrastructure developments. They uncover their essentially ideological underpinnings, industrialist, often capitalist definitions of development, including an underlying conviction that infrastructures also bring 'civilization' to 'backwards' peoples (Flower, 2004).

The temporalities of infrastructure

How is it possible to make sense of the unwavering belief in the apolitical promise of infrastructure development despite overwhelming evidence for the messy political entanglements of infrastructures? In response to this seeming contradiction, development scholars have increasingly turned to research on the temporalities of infrastructures: the past, present and futures associated with infrastructure development.

The promise of infrastructure development, including its 'civilizing' aspirations, reflects the intense future orientation of infrastructure initiatives. Infrastructures promise to transform what currently is into something better – why else would they be built? These promises often enchant project funders and often also the proposed beneficiaries. Harms (2012), for instance, found that residents of Ho Chi Minh City supported extensive urban redevelopment plans because of the promised future infrastructural beautification it entailed even though it meant their eviction with inadequate resettlement plans. Hetherington (2016) even goes so far as to argue that the promise of better future infrastructures is so enticing that it overshadows previous failed infrastructural promises. The future potential of infrastructures is said to be kept alive through new somewhat modified promises, new technologies, techniques, administrative systems or legal frameworks.

Despite this overwhelming focus on the future, development researchers have highlighted the importance of examining on-the-ground discourses surrounding infrastructural pasts. These discourses can reveal otherwise often hidden 'truths' about the broader political economic contexts in which infrastructural promises are made, received and negotiated. In particular, infrastructures that have stopped working or are in urgent need of maintenance are material reminders of 'what could have been'. In various states of disrepair these infrastructures offer insights into how especially the proposed beneficiaries of these initiatives experience and negotiate in their lives the state of more elusive entities such as the nation or even just a neighbourhood (Trovalla and Trovalla, 2015). In other words, the interlinked temporalities of past infrastructural promises, present infrastructural realities and future hopes for infrastructures can reveal the sense of belonging individuals or particular groups feel towards those who made, remake or ruin the infrastructures that shape their lives.

Finally, potentially unexpected infrastructural adaptations may be uncovered by considering infrastructures in their multiple temporalities. Infrastructures may fail to achieve a particular, hoped for goal, but they may quite easily realize different visions for the present as much as the future. For example, some rural Solomon Islands have adopted solar home systems not to better their participation in the formal economy, as envisioned in official development narratives. Instead, rural Solomon Islanders often use the limited electricity they generate to strengthen social networks in their present lives, lighting up communal spaces underneath houses to encourage everyday socialities with familiar visitors and strangers alike (Hobbis, 2021). In this case, the envisioned future is one of persistent centrality of reciprocal social reproduction, rather than the promised future of capital- and individual-centric development. Hence, this alternative perspective on the future highlights the significance of paying close attention to competing temporal perspectives on infrastructure development and the competing, often unequal power relations informing infrastructure initiatives.

Concluding reflections

By nature of their design infrastructures are integral to contemporary development. Infrastructures often disappear into the backgrounds of day-to-day life. Politicians may organise lavish ribbon-cutting ceremonies when opening a new stretch of road or a new hydropower dam. Yet, immediately after these hyper-visible events, infrastructures are meant to disappear into the background of their users' lives. Roads users should no longer actively think about the road and the condition it is in. Instead, they should rely on it subconsciously, focusing instead on what the road enables, such as access to far-away markets.

Their desired and ideally realized invisibility is why infrastructures are so central to development initiatives – without them, it seems impossible to facilitate development materially. Simultaneously, because of their intended invisibility, they are politically advantageous. They allow material establishment but also hide political agendas and, at times, exclusionary goals. This, in turn, makes infrastructures so significant to better understand the hidden dynamics of international development: invisible, yet omnipresent and deeply embedded in political economic structures and ideologies, infrastructures materialize and bring into being the many challenges that characterise contemporary development goals and initiatives. As the term itself suggests, *infrastructures* do, indeed, underly it all.

Guide to Further Reading

Anand, N., Gutpa, A. and Appel, H. (eds) (2018). *The Promise of Infrastructure*. Durham: Duke University Press.

Harvey, P. and Knox, H. (2015). *Roads: An Anthropology of Infrastructure and Expertise*. Ithaca: Cornell University Press.

Larkin, B. (2008). *Signal and Noise: Media, Infrastructure, and Urban Culture in Nigeria*. Durham: Duke University Press.

Von Schnitzler, A. (2016). *Democracy's Infrastructure: Techno-Politics and Protest after Apartheid*. Princeton: Princeton University Press.

Winther, T. (2008). *The Impact of Electricity: Development, Desires and Dilemmas*. Oxford: Berghahn.

References

Bachman, J. and Schouten, P. (2018). 'Concrete approaches to peace: Infrastructure as peacebuilding', *International Affairs*, 94(2): 381–398.

Chen, W.K. (2020). 'Sovereign debt in the making: Financial entanglements and labor politics along the Belt and Road in Laos', *Economic Geography*, 96(4): 295–314.

Fairhead, J. (1992). 'Paths of authority: Roads, the state and the market in Eastern Zaire', *The European Journal of Development Research*, 4(2): 17–35.

Flint, C. and Zhu, C. (2019). 'The geopolitics of connectivity, cooperation, and hegemonic competition: The Belt and Road Initiative', *Geoforum*, 99: 95–101.

Flower, J.M. (2004). 'A road is made: Roads, temples, and historical memory in Ya'an Country, Sichuan', *The Journal of Asian Studies*, 63(3): 649–685.

Harms, E. (2012). 'Beauty as control in the new Saigon: Eviction, new urban zones, and atomized dissent in a Southeast Asian city', *American Ethnologist*, 39(4): 735–750.

Hetherington, K. (2016). 'Surveying the future perfect: Anthropology, development and the promise of infrastructure', in P. Harvey, C.B. Jensen and A. Morita (eds) *Infrastructures and Social Complexity: A Companion*. London: Routledge, 40–50.

Hobbis, S. (2018). 'Mobile phones, gender-based violence, and distrust in state services: Case studies from Solomon Islands and Papua New Guinea', *Asia Pacific Viewpoint*, 59(1): 60–73.

Hobbis, S. (2021). 'Beyond electrification for development: Solar home systems and social reproduction in rural Solomon Islands', *Asia Pacific Viewpoint*, 62(2): 151–163.

Pedersen, M.A. and Bunkenborg, M. (2012). 'Roads that separate: Sino-Mongolian relations in the Inner Asian Desert', *Mobilities*, 7(4): 555–569.

Tarekegn, T. and Overton, J. (2011). 'Isolation as a development strategy: Perspectives from an Ethiopian village', *African Geographical View*, 30(2): 35–52.

Trovalla, E. and Trovalla, U. (2015). 'Infrastructure as a divination tool: Whispers from the grids in a Nigerian city', *City*, 19(2–3): 332–343.

Von Schnitzler, A. (2018). 'Infrastructure, Apartheid technopolitics and temporalities of "transition"', in N. Anand, A. Gutpa and H. Appel (eds) *The Promise of Infrastructure*. Durham: Duke University Press, 133–154.

Part 9

Approaches to policy and practice. Editorial introduction

The first section of this book focused on theories of development, engaging with foundational questions about what development might mean and how it might take place. This was primarily a theoretical exploration. This section engages with a similar set of questions but takes as its entry point the concrete realities of policy and practice. This is not to say that theory and practice exist in isolation of one another. Far from it. Indeed, a focus on the dynamic interplay among theory, policy, and practice is perhaps what marks development studies out as distinctive. Another hallmark of development studies is its emphasis on the normative. Development studies operates beyond just academic discussions or strict disciplinary boundaries, being fundamentally preoccupied with making judgements about what kind of change is desirable and how such change might be achieved.

That said, development studies is a diverse field. Within this diversity there is variety in terms of how normative concerns come to be positioned relative to the analytical. Development studies is perhaps characterised by three broad traditions. The first is an instrumentalist approach, action-oriented and centred on analysing policies and interventions to evaluate their effectiveness, often with a view to designing more effective interventions. Next, discourse-based, critical approaches take a more reflective stance. They aim to identify and interrogate the ideas, concepts, and assumptions that underpin development as process and practice, questioning dominant narratives, and providing alternative framings. The third set of approaches is both retrospective and forward-looking, aiming to situate present and future challenges, and the policies and practices that seek to address them in broader contexts of social and historical change. These approaches tend to view development as processes of long-term social change, encompassing both historical and contemporary processes. They seek to understand our past and present, as well as possible futures. Here the focus is less on analysing intervention per se but on the structures, processes, relationships, and historical moments that frame them. While these categories are a useful heuristic device, there are many points of overlap and interconnection, such that approaches to understanding policy and practice seldom rigidly conform to them. Nevertheless, each of these traditions can be found within the chapters that constitute this section.

Beyond these traditions, several overarching themes emerge in this section, which include the legacies of colonialism that loom large over contemporary approaches to policy and practice. Crucially, this historical context is not just a backdrop but a critical lens through which to analyse present-day approaches. Several chapters in this section

DOI: 10.4324/9780429282348-96

emphasise the need to address structural inequalities rooted in colonial encounters. Connected to this, social justice is another powerful thread that cuts across this section. The chapters highlight the inequalities that shape policy processes and the effects they might have, not only on the intended 'beneficiaries' but societies more broadly. Power has long been a major theme in development studies, and this is reflected in the chapters brought together here. The chapters highlight the relations of power that shape processes of policymaking and intervention and the outcomes they have, not just in material terms but in defining and ordering the world and those who inhabit it.

In conclusion, this section not only bridges theoretical discussions with practical implications but also underscores the transformative potential (*for better or worse*) of development studies in the real world. The chapters in this section are firmly focused on the real-world approaches, processes, objectives and outcomes of development policies, critically evaluating the underlying assumptions and frameworks that guide them and the challenges and opportunities they present for change.

Chapter 88

How to manage for effective aid? The recent emergence of three management approaches

Brendan S. Whitty

How should we manage for effective aid? In this contribution to the Companion, I trace how the UK development policy has generated a variety of answers to this question since – if not necessarily because of – the Global Financial Crash. I suggest three competing management approaches emerged within the discursive space to displace existing models of aid effectiveness. Each was a coherent approach to the challenge of managing aid; each had its own underlying rationalities, tools and understandings of the aid business and its challenges. This contribution traces how these three approaches – broadly, results-based management, evidence-based project designs and adaptive management – emerged and how they were absorbed into the UK government's development apparatus. I start, however, with the prior aid effectiveness regime.

The rise and fall of the 'aid effectiveness' paradigm

Before the New Labour landslide election victory of 1997, the Overseas Development Administration was tucked away in the Foreign and Commonwealth Office. Its budget was relatively low; its delivery was often smudged by British commercial and diplomatic interests, 'tied' to delivery through British companies and afflicted by serious scandals. That was to change in 1997. Clare Short insisted that she would take the Secretary of State for International Development role only if it came with a Cabinet post. The political heft this afforded meant that a new aid management paradigm could emerge within the new ministry, the Department for International Development (DFID). It had several key pillars: DFID would have a solely anti-poverty focus, freed from commitments to British interests; spending on aid would have a budgetary input target of 0.7 per cent GNI; it would seek a set of output targets in the form of the Millennium Development Goals and lastly, spending would be defined by an aid approach that committed to country ownership, the use of developing country systems and partnerships.

This last was to be enshrined in the Paris Declaration on Aid Effectiveness under the auspices of the OECD DAC. It meant that DFID would (nominally at any rate) follow the developing country state priorities and would support the developing country to achieve these goals. The aid budget, its purposes and the mechanisms by which it would be held to account would travel through country systems – independent projects were discouraged, although a significant portion of British aid continued to flow through these mechanisms. For the new sector and budget support modalities, identifying attributable contributions to the UK aid pound through sector level support or budget support modalities was impossible. The data did not permit sifting out British contribution from

DOI: 10.4324/9780429282348-97

the state's own revenue streams other donors – and the data on outcomes was often lacking. The government's own corporate targets did not capture aggregated and attributable results targets but intermediate assessments of project success which were based on attainment of Logframe outcomes.

The consensus underpinning this model of aid came under considerable pressure when the financial crisis set in. Public support for aid in the UK began to decline (Henson et al., 2010). The contribution and partnership model was subject to increasingly sceptical reviews from government scrutiny such the Treasury and NAO, as well as from wider opinion across the political spectrum. Internal advocates – particularly from DFID's influential economics cadre – were arguing for increased rigour. As the dominant model faltered, other management models took their place.

Results-based management

In 2007, Shriti Vadera – one of the new DFID ministerial team who was a former investment banker and who had worked with Gordon Brown in the Treasury – challenged the department. She asked why she was getting submissions asking to approve spending but without an indication of its value for money (interview, DFID economist). The challenge resulted in an internal report called 'Using Numbers' which reviewed critically DFID's production and use of quantitative data. One official involved noted that in many of the Logframes, the page number was the only number (interview, DFID official). Shortly thereafter, work started to introduce a business change programme which would improve corporate decision-making and public communication (National Audit Office, 2009). Three key reforms were introduced: first, much more stringent guidance was given to include quantified targets and annual milestones in the Logframe (DFID's key planning document), articulated ex ante and against which a project was to be held to account; second, standardised indicators were produced and aggregated at the corporate level; third, the economic appraisal process was greatly strengthened through new guidance and new business case format.

The initial moves towards results-based management made under the labour administration were super-charged by the new Conservative Secretary of State Andrew Mitchell following the election of 2010. Mitchell saw the results agenda as crucial to "a centre-right development policy" and to justifying the commitment to the aid volume commitment of 0.7 per cent per cent GNI. Results were about what was achieved with the British pound, not airy references to contributions and practices: "for every pound of taxpayers' money we spend, we will demonstrate 100 pence of value" (Mitchell, 2010). Internal approval, planning and reporting mechanisms were tightened further. Value for money and results targets were integrated throughout corporate management and discourses – and, in their more adventurous forms, were harnessed to new funding modalities, financing models and contracting practices.

Evidence-based approaches to aid

As the 'aid effectiveness' agenda was reaching its high point in 2005, an epistemic community composed chiefly of public health specialists and economists were asking why development was so poor at learning from its failures. They were asking and concluding that there was an *evaluation gap* at the centre of development management (Savedoff et al., 2005). Drawing on methods with roots in evidence-based medicine (Claridge and Fabian, 2005), they proposed to close that gap with rigorous evidence. Its significance was to be acknowledged by the Nobel Prize for Economics in 2019 to three of the most influential advocates (Esther Duflo, Abhijit Banerjee and Michael Kremer).

In the UK, the epistemic community resonated with efforts to depoliticise and to move to a post-ideological technocratic 'third way', based on 'what works'. It fit with the New Labour way of doing things. A number of key influential individuals – some in very senior positions within the department – were in a position to champion the evidence-based public management expenditure (interview, IACDI member). They sought to institute a public management principle which insisted that expenditure could be justified only by rigorous evidence – professional judgement and unsubstantiated theory would simply not be enough.

This rationality took root across the department and DFID's 13 independent expert cadres by: a significant organisational consolidation of learning and evidence functions into a new Division; a significant increase in resources for rigorous evaluations, carefully targeted; the mainstreaming of new tools across the department, including intervention designs, pilot studies, impact evaluations, experimental and quasi-experimental methods, evaluability assessments and systematic reviews and significant efforts to insist on and support evidence use in projects, including through controls. Yet at the heart was a change in philosophical, discursive and practical orientation towards the question of what the organisation needs to know in advance about an intervention's likely workability before it commits to public spending.

Complexity and adaptive management

In 2008, Ben Ramalingam and Harry Jones of the Overseas Development Institute published a paper that identified a series of ten ideas drawn from 'complexity science'. The paper challenged the idea that development processes were linear or simple in their causality or – in many cases – could be predicted in advance. Drawing on ideas from complexity science, it suggested instead that the aid interventions achieved their goals or not through their engagement with complex, multi-dimensional, social and physical systems and networks whose attributes were often unknown and whose reactions to new forces might be unpredictable and contingent. Their target was not the old aid effectiveness regime but traditional project interventions – and crude results-based management approaches – whose understanding of social change tended to be linear and simplistic. Linear intervention logics were under attack from as broad an array of disciplinary starting points as institutional reform (Andrews, 2013; Pritchett and Woolcock, 2004), evaluation (Rogers, 2009; Woolcock, 2009), public health systems (De Savigny and Adam, 2009) and organisational learning (Mitleton-Kelly and Ramalingam, 2011) – amongst others. Each offered a challenge to management logics that required detailed ex ante planning, in favour of more exploratory methods.

Advocates for complexity models came to together to convene coalitions and platforms such as Doing Development Differently (convened in 2014) which brought together strands of thinking and proposing – amongst other reforms – for an approach termed 'adaptive management'. As both results-based management and evidence-based approaches tend to emphasise ex ante controls and planning, so the adaptive management advocates sought to re-empower the implementer within a structure. Professional groupings seeking to foster institutional change – whether in markets or the states, whether the Private Sector Development or the Governance cadre – found intellectual sustenance from the complex and adaptive systems understanding of the world.

Concluding remarks

In the aftermath of the GFC, three broad management approaches emerged in reaction – at least in part – to the loss of credibility of the previous aid effectiveness paradigm and the increasing public and political scrutiny of aid. Each had its own rationalities, tools, lineages and sub-divisions. Each is to some extent incompatible with the others. Thus identifying a project's theory and context as complex challenges the assumption that results and targets are predictable and therefore can be meaningfully stated; the evidence on results-based management modalities is shaky and evidence-based management approaches' methodological conception of rigour would rule out expenditure of many exploratory interventions or ones reliant on adaptive techniques.

Yet all three emerged more or less contemporaneously. Two points occur. First, the emergence of the three management approaches reveals the *regularity* of the aid sector: regardless of how the sector manages for effective aid, it must depoliticise. Even adaptive management – an approach specifically tailored to interventions driving institutional, behavioural and political change – works by stripping out recognition of public contestation over political values, boiling down profoundly political processes to a mechanical procedure of experimentation and adaptation.

Second, aid is a disciplinary diverse landscape: how people propose to manage for effective aid will be driven by their position within that landscape. DFID (now the Foreign, Commonwealth and Development Office) is a global organisation, working across diverse and sometimes tumultuous environments. How far is control of their operations by distant London possible or desirable? Evidence- and results-based approaches tend towards more control: to replicable and predictable projects with established underlying intervention theories which can be subject to centralised audit. Complexity and adaptive management tend towards more flexible approaches and towards decentralised management. Further, evidence-based and results-based approaches tend towards the fields of development intervention which permit replicable, 'large-n', multiple-intervention projects – health and education, for example – rather than more exploratory, singular processes seeking for example to foster or drive institutional change – governance and private sector development, for example. How we manage is contested in part because development expertise is diverse.

References

Andrews, M. (2013). *The Limits of Institutional Reform in Development.* Cambridge: Cambridge University Press.

Claridge, J. A., & Fabian, T. C. (2005). History and Development of Evidence-based Medicine. *World Journal of Surgery, 29*(5), 547–553. doi:10.1007/s00268-005-7910-1

De Savigny, D., & Adam, T. (2009). *Systems Thinking for Health Systems Strengthening.* Geneva: World Health Organization.

Henson, S., Lindstrom, J., Haddad, L., & Mulmi, R. (2010). Public Perceptions of International Development and Support for Aid in the UK: Results of a Qualitative Enquiry. *IDS Working Papers, 2010*(353), 1–67. doi:10.1111/j.2040-0209.2010.00353_2.x

Mitchell, A. (2010). *Full transparency and New Independent Watchdog will give UK Taxpayers Value for Money in Aid [speech]*. https://www.gov.uk/government/speeches/full-transparency-and-new-independent-watchdog-will-give-uk-taxpayers-value-for-money-in-aid

Mitleton-Kelly, E., & Ramalingam, B. (2011). Organisational Learning and Complexity Science: Exploring the Joint Potential. In *The SAGE Handbook of Complexity and Management*, 349–365. London: Sage.

National Audit Office (2009). Performance of the Department for International Development 2008-09. House of Commons International Development Committee, November 2009. London: National Audit Office.

Pritchett, L., & Woolcock, M. (2004). Solutions when the Solution is the Problem: Arraying the Disarray in Development. *World Development*, 32(2), 191–212.

Rogers, P. J. (2009). Matching Impact Evaluation Design to the Nature of the Intervention and the Purpose of the Evaluation. *Journal of Development Effectiveness*, 1(3), 217–226. doi:10.1080/19439340903114636

Savedoff, W. D., Levine, R., & Birdsall, N. (2005). *When Will We Ever Learn? Recommendations to Improve Social Development through Enhanced Impact Evaluation*. Washington, DC: Center for Global Development.

Woolcock, M. (2009). Toward a Plurality of Methods in Project Evaluation: A Contextualised Approach to Understanding Impact Trajectories and Efficacy. *Journal of Development Effectiveness*, 1(1), 1–14.

Chapter 89

Participatory development

Giles Mohan

Introduction

Over the past 30 years a wide range of organisations have started involving local people in their own development, so much so that it has become a new 'orthodoxy' (Cornwall, 2002a). This chapter begins by looking at different definitions of participatory development and examines through what sorts of organisation it is achieved. As there are many possible approaches I have included case studies which demonstrate different facets of participation. This brings us on to a critique and an overview of where things seem to be heading, particularly linking participation to citizenship.

Participatory development in theory

The emergence of participatory development (PD) is tied into critiques of both theory and practice.

The emergence of participation

According to the strongest advocates of PD, 'normal' development is characterised by Eurocentrism, positivism, and top-downism which constitute 'epistemological disenfranchisement' (Connell, 2007: 107). The tendency is to equate development with the modernity achieved by 'Western' societies with the flipside that 'non-expert', local people were sidelined.

As it became apparent that programmes had yielded limited benefits, the volume of criticism grew. In the 1970s, radicals such as Paulo Freire (1970) advocated Participatory Action Research which created new learning environments for people to express their needs and achieve development. Green (2010) shows how in the case of Tanzania, participation in development became a deep-rooted political culture in the post-independence period allied to experiments in African socialism. Even mainstream organisations like the World Bank pushed for Basic Needs which targeted marginalised groups. Added to this were academics, most notably Robert Chambers, who argued that 'putting the last first' was necessary for rural development.

Contested definitions

Participation is generally deemed 'a good thing', but it has multiple meanings, which makes it amenable to different interpretations and uses while appearing to be speaking

DOI: 10.4324/9780429282348-98

about the same thing. Thus, Green (2010) terms it a 'boundary object', which creates 'the possibilities for groups with divergent perspectives and interests to enter into temporary collaborations around shared objects of management' (p. 1242). Therefore, defining these divergent meanings is important for assessing its possible (ab)uses and impacts. In terms of development a key question is, if people participate, what are they aiming to gain by participating? One view is about *efficiency and effectiveness* of 'formal' development programmes (Cornwall, 2002a). The goals of development are valid although the institutions are malfunctioning but can be improved by involving the beneficiaries. Another view concerns *mutual learning*, in which participation entails understanding where others are coming from and, ideally, learning from one another to achieve a better outcome (Chambers, 1997). Others take this further in seeing participation as more *transformative* (Hickey and Mohan, 2005). That is, 'development' is flawed and only by valorising other voices can meaningful social change occur. It is in this sense that the recent emphasis on participation as citizenship, which I discuss later, is aimed.

Despite these differences, there has been a growing acceptance regarding the importance of local involvement from both neo-liberals and radicals, what Dagnino (2008) terms a 'perverse confluence'. Underlying this 'consensus' is the belief in not relying solely on the state. So, it is not accidental that PD gained popularity around the same time as the neo-liberal counter-revolution of the 1980s with its discourse of self-help and individualism and has remained popular under the slightly more state-friendly 'inclusive liberalism' of the new millennium (Golooba-Mutebi and Hickey, 2010).

Powerful processes

It needs emphasising that whichever approach to participation we adopt, PD is fundamentally about power (Nelson and Wright, 1995). Cornwall (2002b) usefully distinguishes between 'invited' and 'claimed' spaces of participation. Invited spaces are the more formal events where development agents create forums for stakeholders to contribute and, ideally, reach a consensus. By contrast, claimed spaces are more organic and involve the poor taking control of political processes, without necessarily being invited in. In practice, political struggle usually has elements of both invitation and claiming with subaltern agents resisting and subverting these political processes in creative ways (Shakya and Rankin, 2008).

Participatory development in practice

In this section I discuss the institutional arrangements involved in PD and the processes through which it attempts to change power relations.

Grassroots civil society

In rejecting the statism and top-downism of 'normal' development, the focus for PD has become the grassroots level which permits a plurality of developmental goals to be realised as well as giving communities the self-determination they need. Hence, PD has become associated with civil society. If state structures are bureaucratic and unaccountable, then civil society organisations are believed to be more accountable and hands-on. Although civil society has multiple meanings, it has largely been interpreted as the realm of non-governmental organisations (NGOs), with many Southern-based ones relying on funding and institutional support from Northern partners, and increasingly states use NGOs as vehicles for certain forms of social development (Dagnino, 2008).

New knowledges

The first step in reversing the biases marginalising the poor concerns rethinking knowledge generation. The expert systems of modernity relied upon scientific approaches so that the recipients of development were treated as passive. PD reverses this. The research methods for accessing local knowledges were inspired by Paulo Freire and have grown into a veritable industry (Chambers, 1997), but all centre upon trying to see the world from the point of view of those directly affected by the developmental intervention.

The most widely used methodology is Participatory Rural Appraisal (PRA). As Chambers (1997: 103) explains:

> The essence of PRA is change and reversals – of role, behaviour, relationship and learning. Outsiders do not dominate and lecture; they facilitate, sit down, listen and learn. Outsiders do not transfer technology; they share methods which local people can use for their own appraisal, analysis, planning, action, monitoring and evaluation. Outsiders do not impose their reality; they encourage and enable local people to express their own.

PRA relies on many visual and oral techniques for generating knowledge because it is felt that the medium of written language is prejudicial to free expression. So, PD seeks out diversity rather than treating everybody as uniform objects of development.

Participation in action

So far I have outlined the theory of PD, but what happens when it is practised in the 'real' world? These brief case studies demonstrate different facets of PD. The Aga Khan Rural Support Programme (India) has used participation to enhance the effectiveness of pre-determined projects. The participatory approach aimed at 'consensus-building' and 'to find a meeting ground to negotiate terms of collaboration' (Shah, 1997: 75). In a dam scheme the farmers were not given an option regarding water payments, but the participatory exercise helped reach mutually agreeable solutions. As Shah (1997: 77) concludes, this was 'Certainly not true empowerment where villagers decide and prioritise development proposals with minimal external support and facilitation'. Shah suggests that while transformatory participation might be desirable it is rarely viable where external agents are time-bound and accountable to funders. But that is not to say they are dictatorial and that the lack of true empowerment detracts from real benefits. As Corbridge (2007: 201) argues, also based on Indian examples, 'good practical (indeed political) arguments can be made in favour of particular development policies that might seem reformist or hopelessly pragmatic'. Corbridge makes a plea for not basing our analysis and prescriptions on a normative 'high ground' of politics divorced from the messy realities of actually existing struggle; a theme I return to later.

A similar issue is raised where participatory approaches have been 'scaled up'. In the mid-1990s the major donors initiated Poverty Reduction Strategies (PRSs), which responded to the criticism of Structural Adjustment Programmes for being imposed on countries. Instead, formulation of PRSs is to be 'owned' by the countries concerned, which means scaling up the invited spaces in which citizens and their representative organisations have a voice (Lazarus, 2008). However, these mass participation exercises are often piecemeal, late in the policy process, and involve only 'safe' civil society organisations who will not question the neo-liberal logic of PRSs.

By contrast, Esteva and Prakash (1998) see the Mexican Zapatistas as a political force pushing for a different understanding of development through novel forms of participation. However, more low key and less combative approaches focus on civic engagement in urban service delivery through such things as school boards whereby participation in one institution has knock on effects that transform the process of local governance (Fung, 2004). Taken together, it becomes clear that these different uses of participation are not exclusive and means that in any given situation we need to be realistic and specific about the nature of participation that is either envisaged and/or possible.

The problems of participatory development

Having looked at these case studies it is worth drawing together some of the inter-related problems that have emerged with PD.

The first is tokenism. As PD has become popular, some agencies use the rhetoric of participation with limited empowerment. In many cases PRA has become so routinised that many agencies treat it as a 'rubber stamp' to prove their participatory credentials. In the PRS process, which champions participation and ownership, most representatives of civil society are in fact middle-class activists who are hand-picked to ensure agreement (Golooba-Mutebi and Hickey, 2010). Allied to this is that reliance on a toolkit approach to knowledge creation has tended to produce endless local studies, which should reflect local contexts but end up looking very similar (Lange, 2010).

Second, much PD has treated communities as socially homogeneous (Robins et al., 2008). While community empowerment might be an improvement on unresponsive bureaucracies, there have been cases where support for 'the community' has meant that resources have passed to elites.

Third, the emphasis on civil society can create competition and overlap between local organisations. With aid being channelled through such organisations it is the better organised or more acceptable which capture resources, often those run by and for middle classes. The result is that weaker organisations or those more genuinely championing the poor are further undermined. Allied to this is that many 'partnerships' between Northern and Southern NGOs are heavily loaded in favour of the former. Such problems are repeated when the relationships are not NGO-to-NGO but state-to-NGO, for as Lange (2010) shows, participatory schemes in Tanzania established parallel political structures which by-passed local government actors and institutions who are, despite all the weaknesses of electoral democracy, accountable to local people.

Fourth, is whether participation is an end in itself or also a means to an end. From a democratic perspective simply being able to participate is a major achievement, but for the poor their lack of resources means that any participatory process must yield tangible benefits. Furthermore, as Brett (2003) warns, simply participating is meaningless unless there is some institutionalised accountability. He argues we should focus on "the nature of the institutional constraints that determine how much leverage users can exercise over agencies, whether these operate in the state, market or voluntary sector" (Brett, 2003: 18).

The final problem is broader and relates to the causes of underdevelopment. PD seeks to give local people control, but many processes affecting their (or our) lives are often not readily tackled at the local level. For example, it is very hard for a small co-operative in Africa to change the rules governing international trade when the World Trade Organization is dominated by the developed economies. The emphasis on grassroots society can leave important structures untouched and do nothing to strengthen states and make them more accountable to their citizens (Green, 2010).

Taken together these operational critiques of participation add up to a 'depoliticisation' of the idea. It promises some quite radical approaches to development but ends up disconnecting genuine needs and political struggles from circumscribed interventions of development agents. It is, as Murray Li argues (cited in Green, 2010: 1251), about 'rendering technical' the conflictual realities of poverty and its amelioration.

Citizenship and the future of participatory development

It becomes clear that while PD has brought benefits to some communities it has been abused and does little to address extra-local processes. This recognition that development will involve broader questions of citizenship and sovereignty has been part of the 'inclusive liberalism' which sees agencies building the capacity of the state rather than by-passing it and empowering civil society (Golooba-Mutebi and Hickey, 2010). This involves bolstering citizenship.

This reframing of participation as citizenship gained ground from the turn of the millennium and situates PD in a broader range of socio-political practices – or expressions of agency (Gaventa, 2002) – through which people extend their status and rights as members of particular political communities, thereby increasing their control over socio-economic resources. This unites a 'liberal' theory of citizenship, stressing formal rights and political channels, with 'civic republican' approaches that emphasise the collective engagement of citizens in the determination of their community affairs. The focus here is on substantive rather than procedural forms of citizenship, a participatory notion that offers the prospect that citizenship can be claimed 'from below' through the efforts of the marginalised.

While the citizenship turn promised to break from the voluntarism of PD as discussed earlier, it too suffers from depoliticisation and neo-liberal co-option. In many cases citizenship action was reduced to being a consumer within a market and so undermined communal notions of rights and moralities (Dagnino, 2008). Like PD it tended to imagine a 'pure' world of citizenship where the poor could relatively straightforwardly secure access to state services (Robins et al., 2008).

The reality for the poor in the Global South is one of 'indeterminacy' and struggle which means that singular and theoretically pure forms of political practice are impossible. As Robins et al. (2008: 1079) argue 'In the scramble for livelihoods and security, poor people tend to adopt plural strategies; they occupy multiple spaces and draw on multiple political identities, discourses and social relationships, often simultaneously'. It is in these multiple practices that the poor can leverage gains from more formal participatory schemes as Corbridge (2007) and Golooba-Mutebi and Hickey (2010) note. In turn this means we should not necessarily ditch PD based on a universal condemnation of its depoliticising effects but rather as Lazarus (2008) notes start with the political realities of poor people and not a normative ideal, however well-meaning.

The following text references provide the basis for further reading

Brett, E.A. (2003) 'Participation and Accountability in Development Management', *Journal of Development Studies*, 40, 2, 1–29.

Chambers, R. (1997) *Whose Reality Counts? Putting the First Last*, London: Intermediate Technology Publications.

Connell, R. (2007) *Southern Theory: The Global Dynamics of Knowledge in Social Science*, London: Polity Press.

Corbridge, S. (2007) 'The (Im)possibility of Development Studies', *Economy and Society*, 36, 2, 179–211.

Cornwall, A. (2002a) *Beneficiary, Consumer, Citizen: Perspectives on Participation for Poverty Reduction, SIDA Studies no. 2*, Stockholm: Swedish International Development Cooperation Agency.

Cornwall, A. (2002b) *Making Spaces, Changing Places: Situating Participation in Development, IDS Working Paper 170*, Brighton: Institute of Development Studies.

Dagnino, E. (2008) 'Challenges to Participation, Citizenship and Democracy: Perverse Confluence and Displacement of Meanings', in Bebbington, A.J., Hickey, S. and Mitlin, D. (eds) *Can NGOs Make a Difference? The Challenge of Development Alternative*, London: Zed Books, 55–70.

Esteva, G. and Prakash, M. (1998) *Grassroots Post-Modernism: Remaking the Soil of Cultures*, London: Zed Books.

Freire, P. (1970) *The Pedagogy of the Oppressed*, New York: The Seabury Press.

Fung, A. (2004) *Empowered Participation: Reinventing Urban Democracy*, Princeton: Princeton University Press.

Gaventa, J. (2002) 'Exploring Citizenship, Participation and Accountability', *IDS Bulletin*, 33, 2, 1–11.

Golooba-Mutebi, F. and Hickey, S. (2010) 'Governing Chronic Poverty under Inclusive Liberalism: The Case of the Northern Uganda Social Action Fund', *Journal of Development Studies*, 46, 7, 1216–1239.

Green, M. (2010) 'Making Development Agents: Participation as Boundary Object in International Development', *Journal of Development Studies*, 46, 7, 1240–1263.

Hickey, S. and Mohan, G. (2005) 'Relocating Participation within a Radical Politics of Development', *Development and Change*, 36, 2, 237–262.

Lange, S. (2010) 'The Depoliticisation of Development and the Democratisation of Politics in Tanzania: Parallel Structures as Obstacles to Delivering Services to the Poor', *Journal of Development Studies*, 44, 8, 1122–1144.

Lazarus, J. (2008) 'Participation in Poverty Reduction Strategy Papers: Reviewing the Past, Assessing the Present and Predicting the Future', *Third World Quarterly*, 29, 6, 1205–1221.

Nelson, N. and Wright, S. (1995) 'Participation and Power', in Nelson, N. and Wright, S. (eds) *Power and Participatory Development: Theory and Practice*, London: Intermediate Technology Publications, 1–18.

Robins, S., Cornwall, A. and von Lieres, B. (2008) 'Rethinking "Citizenship" in the Postcolony', *Third World Quarterly*, 29, 6, 1069–1086.

Shah, A. (1997) 'Developing Participation', *PLA Notes*, 30, 75–78.

Shakya, Y. and Rankin, K. (2008) 'The Politics of Subversion in Development Practice: An Exploration of Microfinance in Nepal and Vietnam', *Journal of Development Studies*, 44, 8, 1214–1235.

Chapter 90

Cash transfers and HIV prevention in Africa

Kevin Deane

Introduction

Cash Transfers (CTs) have become increasingly central to social protection programmes in the Global South. Originating in Latin America in the 1990s, they are now being implemented in most regions of the world, especially in Africa where there has been a significant expansion in the number of CT programmes in recent years. CTs involve regular payments to programme beneficiaries. They can be conditional, in which beneficiaries receive the payment in return for meeting a specific behavioural condition (such as attending school or a clinic) or unconditional in which there are no conditions attached to the payment and thus the beneficiaries are free to use the cash as they wish. In most cases the beneficiaries of CT programmes are women, with programmes aiming to provide targeted household budget support to alleviate extreme poverty. However, increasingly CTs are also being used to target a wide range of development-related issues that involve changing behaviours, from addressing infant malnutrition through changes in infant feeding practices to reducing HIV incidence through sexual behaviour change.

The application of CTs to the HIV epidemic in sub-Saharan Africa is motivated by several factors. First, the epidemic in Africa is highly gendered, especially with respect to adolescent girls and young women (AGYW) aged 15–24. The most recent evidence shows that whilst AGYW comprise around 10 per cent of the population in sub-Saharan Africa, they account for 25 per cent of all new infections (UNAIDS, 2020). They are viewed by UNAIDS as a 'key population' for HIV-related programming. Second, whilst there are biological reasons that make women more susceptible to heterosexual transmission of HIV than men, the gendered nature of the epidemic in Africa is also rooted in prevailing gender inequalities, both social and economic. CT programmes seek to address what have become known as the structural drivers of HIV.

CTs and HIV vulnerabilities

CT programmes that aim to reduce HIV are primarily targeted at poorer AGYW and are typically comprised of the CT plus a range of additional training and information activities. For example, a recent CT project called *Sauti* implemented in Northwest Tanzania included a transfer of approximately $31 every 3 months (across an 18-month period) alongside financial literacy education, entrepreneurial training, an individual/group savings and loan scheme, and 10 compulsory hours of behaviour change communication sessions that focused on HIV risk gender-based violence and reproductive health (Pettifor

DOI: 10.4324/9780429282348-99

et al., 2019). CT programmes aim to empower AGYW to reduce pressures for them to engage in sexual behaviours that are related to enhanced HIV risk, which include transactional sex, age-disparate sex, and unprotected sex more generally. In theory, these 'risky' behaviours are reduced through a range of interrelated mechanisms. For example, the cash can increase the economic independence of AGYW, enabling them to make financial contributions to their household and to afford basic needs and consumer goods that they previously had to rely on male partners (both marital and extra/non-marital) for access to. This reduces the need for them to engage in transactional sex and rebalances power in existing relationships, making it easier for AGYW to negotiate and enforce condom use. The cash can also support investment in informal sector business activities which can underpin longer term economic independence following the end of the trial (Pettifor et al., 2019). The cash can also enable AGYW to remain in school in settings where a lack of resources needed to buy books and stationery can contribute to AGYW dropping out of school earlier than their male peers, which limits access to education and future life chances. The related training activities can also improve knowledge and understanding about HIV and reproductive health leading to improved health-related decisions, reflections on prevalent gender roles, enhanced self-esteem, and the creation of important peer-support networks.

Evidence

There is a growing body of evidence that seeks to quantify the impact of CTs on HIV incidence and sexual behaviours amongst AGYW (Stoner et al., 2021). Randomised controlled trials (RCT), widely seen as the 'gold standard' methodological approach for evaluating biomedical and behavioural interventions, are used to evaluate the impact of CT programmes by comparing HIV and behavioural outcomes between a treatment group (AGYW who receive the CT) and a control group (AGYW who do not) over a specific time period (often two to three years). In some cases, the RCT also compares different configurations of the intervention, for example comparing outcomes for three groups: AGYW who receive the CT and Gender Training (CT+), AGYW who receive just the CT, and the control group. These RCTs involve conducting baseline surveys with all trial participants that collect data on demographic details, reported sexual behaviours, and include a HIV test. The baseline survey is conducted at the beginning of the trial. The survey is then repeated at specific points in the trial (for example after 12 months, 18 months, and 24 months) which enables changes in the outcomes of interest to be quantified and compared across the different intervention groups. One key aspect of any RCT is that the treatment group(s) and control group are similar in terms of demographic characteristics so that any differences in outcomes can be attributed to the intervention rather than differences in the characteristics of the participants in each group.

Whilst CTs are being promoted by UNAIDS and the World Bank as an important intervention for HIV prevention, the empirical evidence from RCTs is more mixed. A recent systematic review (Stoner et al., 2021) found that only 3 out of 8 RCTs that quantified the impact of CTs on HIV incidence reported a statistically significant reduction in HIV. The systematic review found stronger evidence that CTs can have a positive impact on sexual behaviours, with 10 out of 18 studies reporting a statistically significant impact on delaying sexual debut but weaker evidence for other behavioural outcomes. This limited evidence base raises questions as to how and why CTs receive so much attention in the literature and suggests an overly optimistic reading of existing evidence. On their own terms, it has been suggested that the lack of statistically significant results may be explained by low overall incidence rates in trial populations, the relatively short length of most RCTs (typically between 24–36 months and sometimes shorter), and the relatively small value of the cash that is being distributed in the trial.

Critical remarks

Whilst the empirical evidence for the effectiveness of CTs to prevent HIV is mixed, they remain popular amongst policy makers because they are easy to implement, requiring few significant design changes for different contexts, and they lend themselves to evaluation through RCTs. This illustrates how methods of evaluation can a priori influence the intervention agenda. However, there are a number of limitations that can help explain why CTs are no silver bullet for HIV prevention amongst AGYW. First, it is unrealistic to expect CTs to address deep-rooted historical social and economic inequalities that include unequal access to land and other productive assets, inheritance regimes that mean that women may not inherit their partners' land and assets, social norms regarding employment and informal sector activities that women engage in, alongside the gendered division of labour that continues to ascribe the burden of most household work and caring responsibilities to women. The limited additional economic power that CTs may enable are unlikely to significantly rebalance unequal power relations, especially with respect to sexual interactions with older and wealthier men that can involve both social and economic coercion. Related to this, the theory of change (i.e. how the intervention is supposed to have an impact on the outcome of interest) that underpins CTs and HIV prevention fails to confront male power in a meaningful way, which also increases the attractiveness to (male) policymakers and gatekeepers in the African context. All too often prevention activities such as CTs that focus on AGYW do not include engaging with men. The result is that CTs are located in an ahistorical and apolitical vacuum. Whilst some small marginal gains may be made for some, it is unlikely that CTs will form the basis for longer lasting social change.

Another important aspect of the debate is the extent to which CTs are a response to an incorrect or incomplete diagnosis of the 'problem'. For example, in relation to transactional sex, recent work makes it clear that transactional sex is engaged in for a wide range of reasons by AGYW of all social classes and not just those in poverty. Whilst the CT may initially enable AGYW access to basic goods and a degree of financial independence, in a more dynamic sense without significant changes in economic opportunities, access to other consumer goods and luxury goods will remain unattainable without reliance on male partners. Whilst this is not to suggest that consumption needs are insatiable, consumption often leads to and prompts further consumption – for example a mobile phone requires network credit and creates other possibilities for consumption. It is unrealistic to expect AGYW to be satisfied with their consumption levels once basic needs have been accounted for, especially if AGYW have children to provide for. Consumption practices also conform to social pressures and are linked to aspirations of social mobility and social status. In the longer term, the need to engage in transactional sex therefore may not significantly reduce. Further, transactional sex is a complex social practice. The CT theory of change views transactional sex through an economic lens, ignoring social norms and expectations around transactional sex, the role of transactional sex in love and ongoing relationships, and the fact that transactional sex is not a practice that is necessarily stigmatised or to be avoided.

CT are also conceptually flawed as a response to the structural drivers of HIV. Whilst on the surface they aim to engage with structural factors such as gender inequality and poverty, the focus of CTs are individuals. The case for CTs draws on social epidemiological work that maps pathways from broader structural factors to individual behaviours, and so whilst interventions initially aim to engage with structures, they often do not do so. This conceptual conundrum has been termed the prison of the proximate (O'Laughlin, 2015). Indeed, the conceptual model of behaviour underpinning cash transfers draws more on behavioural economics and nudge theory than more systemic theoretical perspectives that may offer alternative views on how to address questions of power. Related to this is the charge that CTS embed neoliberal values by placing consumption and increased market power at the centre of processes of empowerment and change.

More broadly, CT have come to dominate the social policy agenda and represent a reductionist, neoliberal view of development in which objectives such as structural transformation and institutional change have been replaced with targeted poverty-reduction programming. The focus on the poor also eschews other forms of social policy and collective provisioning that have formed the bedrock of social policy in the Global North (where means-tested social policy is only one aspect of the social policy agenda; Adesina, 2020). In relation to HIV, CTs play a similar role as a key intervention that addresses the structural drivers of HIV in lieu of other interventions or policies that may present more of a challenge to male power and gender norms and that may not be as easily subjected to RCTs to quantify their impact given that social change is rarely a smooth, linear process. The policy space is, therefore, limited by the dominance of approaches like CTs (and microfinance). Future directions in HIV prevention for AGYW require a deeper engagement with social theory as well as a more universalistic approach that addresses gender relations for all AGYW in the African context.

Suggested Further Reading

Bastagli, F., J. Hagen-Zanker, L. Harman, V. Barca, G. Sturge, T. Schmidt and L. Pellerano (2016). *Cash Transfers: What Does the Evidence Say? A Rigorous Review of Programme Impact and of the Role of Design and Implementation Features.* London: ODI, OPM and DFID.

Harman, S. (2011). "Governing Health Risk by Buying Behaviour." *Political Studies* 59(4): 867–883.

Johnston, D. (2015). "Paying the Price of HIV in Africa: Cash Transfers and the Depoliticisation of HIV Risk." *Review of African Political Economy* 42(145): 394–413.

Stoebenau, K., L. Heise, J. Wamoyi and N. Bobrova (2016). "Revisiting the Understanding of 'Transactional Sex' in Sub-Saharan Africa: A Review and Synthesis of the Literature." *Social Science & Medicine* 168: 186–197.

Bibliography

Adesina, J. O. (2020). "Policy Merchandising and Social Assistance in Africa: Don't Call Dog Monkey for Me." *Development and Change* 51(2): 561–582.

O'Laughlin, B. (2015). "Trapped in the Prison of the Proximate: Structural HIV/AIDS Prevention in Southern Africa." *Review of African Political Economy* 42(145): 342–361.

Pettifor, A., J. Wamoyi, P. Balvanz, M. W. Gichane and S. Maman (2019). "Cash Plus: Exploring the Mechanisms through which a Cash Transfer Plus Financial Education Programme in

Tanzania Reduced HIV Risk for Adolescent Girls and Young Women." *Journal of the International AIDS Society* 22(S4): e25316.

Stoner, M. C. D., K. Kilburn, P. Godfrey-Faussett, P. Ghys and A. E. Pettifor (2021). "Cash Transfers for HIV Prevention: A Systematic Review." *PLoS Medicine* 18(11): e1003866.

UNAIDS (2020). *Seizing the Moment: Tackling Entrenched Inequalities to End Epidemics. Global AIDS Update*. Geneva: UNAIDS.

Chapter 91

Social protection in development context

Sarah Cook and Katja Hujo

What is social protection?

Social protection (SP) is generally understood to be concerned with preventing and managing situations that adversely affect people's well-being. SP programmes assist individuals and households in maintaining basic consumption and living standards when confronted by contingencies such as unemployment, illness, maternity, disability or old age, as well as economic crisis or natural disaster. They may include contributory insurance schemes, tax- or aid-financed social assistance programmes, and labour market interventions. Broader definitions may include the provision of essential public or social services (water, housing, education, healthcare, transport), as well as interventions such as tax credits, micro-finance and micro-insurance, subsidies for smallholder farmers, and provision by non-state actors.

The idea of SP is not new: informal or community-based mechanisms existed before European welfare states adopted formal instruments to deal with foreseeable income shortfalls and transitory experiences of poverty within relatively stable life trajectories. The right to social security was enshrined in the 1948 Universal Declaration of Human Rights and elaborated in several conventions of the International Labour Organization (ILO), in particular the Minimum Standards Convention No. 102 of 1952, which established norms for medical care, unemployment, old-age, family, and other benefits.

In the early 21st century, propelled by a global commitment to poverty reduction represented by the Millennium Development Goals, SP emerged on the international development agenda as the dominant response to addressing poverty and vulnerability in low income contexts. The period up to the global financial crisis of 2008 witnessed a dramatic expansion in social assistance coverage, dominated by cash transfers albeit with significant variations in programmes across countries. While the global SP agenda has been further strengthened through the SP Floor initiative (ILO Recommendation No. 202) and inclusion of a SP target in the Sustainable Development Goals (SDG 1.3), it increasingly appeals to national governments as an effective instrument for poverty reduction and inclusion.

Social protection in development contexts: the rise of cash transfers

Responding to the failures of development policy of the 'Washington Consensus' era and a renewed global concern with poverty, innovative mechanisms for SP emerged initially

DOI: 10.4324/9780429282348-100

from the Global South. In Latin America, conditional cash transfers (CCTs) – designed to support household consumption and improve welfare in contexts of widespread poverty and informal employment relations and to overcome obstacles to development through facilitating investments in health and education – became the instrument of choice for several governments in the region. These programmes require recipient households to comply with specific conditions such as regular school attendance and health check-ups for children.

Promoted by international organisations and donors and with conditionalities often proving acceptable in national political debates, CCTs rapidly expanded across the developing world.

Other instruments found greater traction elsewhere, including employment and public works programmes (notably NREGA in India), unconditional transfers such as social pensions and child grants, or 'productive safety nets' linking transfers (cash or in-kind) to agricultural productivity (Table 91.1). Nonetheless, the popularity of 'just giving cash to the poor' among the donor community facilitated the spread of cash transfers including to Asia and more recently sub-Saharan Africa (SSA).

Some middle-income countries have achieved universal coverage of cash transfer programmes and increased access to social services within comprehensive systems. Argentina, Brazil, and Chile have reached near universal coverage of children and elderly persons with cash transfers (ILO, 2017). In SSA, cash transfer programmes are largely unconditional – in part due to supply-side constraints on health and education services and limited administrative capacity to ensure conditions can be implemented. Initiated as small-scale pilots led by international organisations and often targeted at particular groups (such as Aids affected populations), an increasing number of countries are now adopting national programmes (Table 91.1).

Table 91.1 Social assistance in developing countries: selected programmes, objectives, and impact

Instruments	Programmes (programme start date)	Objectives	Impacts
Unconditional cash transfers			
Income transfers targeted to poorest	Kalomo pilot social transfer scheme (2004)/National Social Cash Transfer, Zambia (2010); Mchinji pilot social transfer, Malawi (2006)/National SCTP Mtukula Pakhomo (2012), China *dibao* programme (1999)	Reduce poverty and vulnerability among poorest households without economic capacity and with children	Zambia: increased school enrolment, nutrition, reduction of household debt, increased asset ownership (Arruda and Dubois, 2018) Malawi: decreased poverty, increased school enrolment, health, food security, reduced child labour (Arruda, 2018)

(Continued)

Table 91.1 (Continued)

Categorical income transfers: social pensions and child grants	Social pensions in Bangladesh, Bolivia, Botswana, Brazil, Chile, India, Lesotho, Mauritius, Namibia, Nepal, South Africa Child Support Grant (CSG), South Africa (2001), *Asignación Universal por Hijo*, Argentina (2009)	Reduce poverty and vulnerability among older people and their households Reduce poverty and facilitate investment in schooling, break poverty across generations	CSG: improved height-for-age scores for children and positive educational outcomes; decreased likelihood of child labour or risky health behaviours (DSD et al., 2012); women's empowerment: increased decision-making in household but employment limited by care role (Patel et al., 2013)
Income transfer conditional on work			
Public works, cash-for-work, employment guarantee	Employment Guarantee Scheme (NREGA), India (2006); *Jefas y Jefes*, Argentina (2002) Productive Safety Net Programme (PSNP), Ethiopia (2006)	In rural areas, to smooth seasonal income fluctuations. In urban areas, to reduce poverty caused by unemployment and underemployment	NREGA: significant job creation; improved rural infrastructure and agricultural productivity (UNDP, 2010) PSNP: positive nutritional impact, asset accumulation (Gilligan et al., 2009; Devereux and Nzabamwita, 2018)
Income transfers conditional on human capital investment			
Human development–targeted conditional transfers	*Bolsa Familia* (2001/2005), Brazil; *Oportunidades/Prospera* (1997/2004), Mexico, *Familias en Acción*, Colombia (2001), *Bono de Desarrollo Humano*, Ecuador (2003); *Keluarga Harapan* Programme, Indonesia (2007); *Pantawid Pamilyang Philipino* Programme, Philippines (2008); Ghana LEAP programme (2008); Tanzania PSSN (2013)	Improve consumption for poorest households; facilitate investment in nutrition, health, and schooling; reduce intergenerational poverty	*Oportunidades*: higher birth weight among participating women (Barber and Gertler, 2008). *Bolsa Familia*: positive impact on school attendance and drop-out rates, positive impact on equality (Veras Soares et al., 2007.)
Integrated poverty reduction/ eradication programmes targeting the extreme poor	Targeting the Ultra Poor, Bangladesh (2002) Chile Solidario, Chile (2004)	Stabilise consumption of poorest households; improve human and productive assets; achieve minimum thresholds for: income, employment, housing, health, education, etc.	Bangladesh: Positively impacted occupational choices, increasing self-employment and sustainable entrepreneurship activities (Bandiera et al., 2013).

Table 91.2 Coverage of social protection programmes

Type of programme	Selected countries	Coverage	Unit of coverage and year
Conditional cash transfers	Bolivia	61.5	Percentage of national popu-
Selected CCT programmes	Ecuador	13.0	lation in 2015
Bolsa Familia	Brazil	27.9	Millions of households in 2016
Progresa, Opportunidades	Colombia	21.7	Millions of households in 2017
and Prospera (component	Mexico	24.1	
with conditionality)	Latin America	20.9	
	Brazil	13.6	
	Mexico	6.1	
Social pensions	Brazil (Rural Pension)	5.8	Million beneficiaries in 2018
	Mexico	5.1	
	Thailand	8	
	South Africa	3.1	
	Bangladesh	3.1	
Child support grant	South Africa	12	Million beneficiaries in May 2017
	Argentina	11.4	Million beneficiaries in 2017.
Employment guarantee schemes	India	332	Millions of households 2019

Source: ECLAC, 2017; HelpAge, 2018; Mpedi, 2017; MRD, 2019; UNICEF, 2018

Social protection: key issues and debates

Coverage and adequacy: the ILO estimates that only 29 per cent of the global population has access to comprehensive social security, while 55 per cent remain unprotected (ILO, 2017). While social assistance coverage has expanded significantly over recent years, in Latin America this is now declining – from 22.7 per cent of households in 2010 to 20.2 per cent in 2016 (ECLAC, 2017). Whether the benefit is sufficient to satisfy basic needs – and the regularity and predictability of payments – are critical for poverty reduction and livelihood promotion. Non-contributory programme benefits vary considerably, with low benefit levels often exacerbated by problems of access to and quality of essential social services.

Sustainability: coverage and adequacy are closely linked to the availability and allocation of financial resources, with fiscal space being negatively affected by economic crises and austerity policies (ILO, 2017; UNRISD, 2016). Quantity and composition of SP expenditures differ widely: developed and transition countries spend on average more than 15 per cent of GDP on SP (excluding health); this figure is below 5 per cent for large parts of Asia and Sub-Saharan Africa, albeit with significant variation among countries at similar income levels (ILO, 2017, see Figure 91.1). In low-income countries financial constraints may be eased by international aid. In many cases there is space to shift the financing mix within existing public expenditures – from subsidised schemes for formal workers towards more progressive public expenditures; as well from less efficient individual solutions (such as out-of-pocket payments), towards collective schemes with greater scope for risk pooling, economies of scale, and redistribution. Ultimately, long term sustainability will require improvements in domestic resource mobilisation, including taxation, but this can encounter strong domestic opposition particularly where programmes are targeted to the poor. Sustainability is further secured through political commitment,

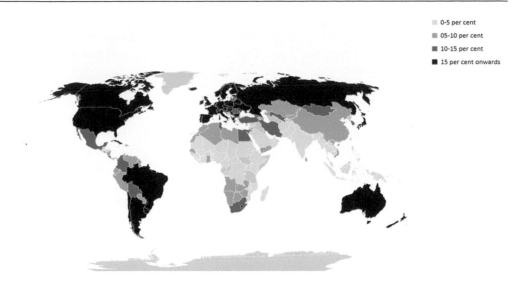

0-5 per cent
05-10 per cent
10-15 per cent
15 per cent onwards

Figure 91.1 Social protection expenditure (excluding healthcare) as a percentage of GDP, latest available year

Source: authors' elaboration based on data from ILO (2017) and ILO (2014)

institutionalisation of programmes based on national law, and participation of the non-poor in universal schemes.

Principles of eligibility: extending coverage also depends on the basis on which people access benefits – whether through contributions, on the basis of need, vulnerability, or dependency status, or by right as citizens or residents. Eligibility according to need or status implies *targeting,* whether categorical (by population group) or by income (means-testing). The latter is often favoured on the grounds of containing expenditures and minimising adverse incentive effects, but it entails high administrative costs, usually significant errors of inclusion or exclusion, while potentially stigmatising beneficiaries. By delinking rights to social protection from citizenship, targeted schemes may enhance the discretionary power of authorities to assign benefits, increasing the risk of clientelism and corruption; they may also foster segmentation in social service access and quality by class or income. Categorical targeting (social pensions, child benefits) is sometimes more politically acceptable and less associated with adverse labour market incentives; it can help overcome structural inequalities or vulnerability based on individual or collective characteristics (gender, disability, ethnicity). Such affirmative targeting of excluded groups may be essential to ensure universal coverage (Mkandawire, 2005).

Beyond safety nets: are social protection programmes meeting their development promise?

Over the past decade growing evidence has been generated regarding the individual and household level impacts of SP programmes (such as poverty, nutrition, and schooling, see Table 91.1). Factors leading to better outcomes include programme design, the size, regularity, and predictability of payments and complementary supply-side investments (such as health, education, and transport). Evidence is more limited for broader poverty and inequality impacts, especially when programmes are targeted to the poorest households, are small in terms of coverage and fiscal investment, and complementary supply-side interventions are insufficient.

Future Challenges: while the expansion of social assistance coverage is widely recognised as contributing to a reduction in poverty and vulnerability, related issues increasingly demand attention. These include weak links to labour markets; the fiscal and political sustainability of programmes; insecurity linked to financialisation (such as the use of regular payments as collateral for debt); the privatisation or commodification of social services; and the gendered nature of unpaid care work which is mainly delivered by women and girls. Other challenges stand out for the future: how to address the protection and care needs of aging populations; insecurity associated with climate change and natural disasters; and political resistance arising from middle classes towards funding programmes for the poor. New approaches are emerging in response, including strengthening investment in human capabilities through 'social investment' and advocacy for Universal Basic Incomes.

To ensure that the new generation of SP programmes moves beyond minimal safety nets to achieve the development promise, several conditions must be met. Guarantees of protection grounded in values of human rights are needed to ensure inclusion of vulnerable social groups; effective pathways are needed from social assistance into employment-based or contributory social protection through strategies that expand economic opportunities for the poor, including for women; and innovative approaches to social protection will need to respond to environmental and climate risks, new technologies, migration, and aging populations.

Acknowledgment

Excellent research assistance from Tejal Ambardekar is gratefully acknowledged.

References

Arruda, Pedro. 2018. *A Brief History of Malawi's Social Cash Transfer Programme (SCTP). IPC-IG Working Paper No. 172.* Brasilia: UNDP International Policy Centre for Inclusive Growth.

Arruda, Pedro and Laura Dubois. 2018. *A Brief History of Zambia's Social Cash Transfer Programme. IPC-IG Research Brief 62.* Brasilia: UNDP International Policy Centre for Inclusive Growth, Brasilia.

Bandiera, O., R. Burgess, N. Das, S. Gulescid, I. Rasule and M. Sulaiman. 2013. *Can Basic Entrepreneurship Transform the Economic Lives of the Poor? Working Paper April 2013.* London: International Growth Centre, London School of Economics and Political Science.

Barber, Sarah L. and Paul J. Gertler. 2008. "The impact of Mexico's conditional cash transfer programme, Oportunidades, on birthweight." *Tropical Medicine and International Health*, 13: 11, 1405–1414.

Devereux, S. and J. Nzabamwita. 2018. *Social Protection, Food Security and Nutrition in Six African Countries. IDS Working Paper 518.* Brighton: Institute of Development Studies.

DSD, SASSA and UNICEF. 2012. *The South African Child Support Grant Impact Assessment: Evidence from a Survey of Children, Adolescents and Their Households.* Pretoria: UNICEF South Africa.

ECLAC (Economic Commission for Latin America and the Caribbean). 2017. *Conditional Cash Transfer Programmes in Latin America and the Caribbean: Coverage and Investment Trends. Report. Series in Social Policy No. 224.* Santiago: United Nations Publication.

Gilligan, Daniel O., John Hoddinott and Alemayehu Seyoum Taffesse. 2009. "The impact of ethiopia's productive safety net programme and its linkages." *Journal of Development Studies*, 45: 10, 1684–1706.

HelpAge. 2018. "Social pensions database." March. http://www.pension-watch.net/social-pensions-database/social-pensions-database--/. Accessed 30 July 2019.

ILO (International Labour Organization). 2014. *World Social Protection Report 2014–2015: Building Economic Recovery, Inclusive Development and Social Justice*. Geneva: International Labour Office.

ILO (International Labour Organization). 2017. *World Social Protection Report 2017–19: Universal Social Protection to Achieve the Sustainable Development Goals*. Geneva: International Labour Office.

Mkandawire, T. 2005. *Targeting and Universalism in Poverty Reduction. Social Policy and Development Paper No. 23*. Geneva: UNRISD (United Nations Research Institute for Social Development).

Mpedi, L.G. 2017. *The Child Support Grant in the Republic of South Africa: Does It Comply with The ILO Social Protection Floors Recommendation 202 of 2012*. Inaugural Address as Executive Dean of Faculty of Law, University of Johannesburg, South Africa. https://www.uj.ac.za/newandevents/Documents/2017/Executive%20Dean%20Inauguration%20Lecture%20-%20Letlhokwa%20Mpedi%20(20%20November%202017).pdf. Accessed 30 July 2019.

MRD (Ministry of Rural Development), Government of India. 2019. "Mahatma Gandhi national rural employment guarantee act: Households benefitted." July. https://www.nrega.nic.in/netnrega/mgnrega_new/Nrega_StateReport.aspx?typeN=5. Accessed 30 July 2019.

Patel, Leila, Tessa Hochfeld and Jacqueline Moodley. 2013. "Gender and child sensitive social protection in South Africa." *Development Southern Africa*, 30: 1, 69–83.

UNDP (United Nations Development Programme). 2010. *What Will It Take to Achieve the Millennium Development Goals?* New York: UNDP.

UNICEF (United Nations Children's Fund) Argentina. 2018. *Universal Child Allowance (AUH): A Contribution to the Discussion on the Role of Conditionalities. Report*. Buenos Aires: UNICEF Publication.

UNRISD (United Nations Research Institute for Social Development). 2016. *Policy Innovations for Transformative Change. Implementing the 2030 Agenda for Sustainable Development*. Geneva: UNRISD.

Veras Soares, F., R. Perez Ribas and R. Guerreiro Osório. 2007. *Evaluating the Impact of Brazil's Bolsa Família: Cash Transfer Programmes in Comparative Perspective. Evaluating Note No. 1*. Brasilia: UNDP International Poverty Centre for Inclusive Growth.

Additional Resources

https://socialprotection.org/: An online member-based knowledge sharing and capacity building platform, open to social protection practitioners, policy makers and experts, as well as academics and students. The platform directs up-to-date content on social protection, with a focus on low- and middle-income countries.

https://socialprotection-humanrights.org/: Provides a space to convene leading thinkers around the complex challenges of implementing social protection from a rights perspective, as well as practical guidance for policy makers and practitioners.

http://www.social-assistance.manchester.ac.uk/: Consists of a harmonised panel dataset of social assistance programmes in low- and middle-income countries from 2000–2015. Detailed information about social assistance programme design, objectives, implementation, annual programme budgets, institutionalisation, and coverage is provided in the dataset.

https://www.ids.ac.uk/programme-and-centre/centre-for-social-protection/: Supports a global network of partners working to mainstream social protection in development policy and encourage social protection systems and instruments that are comprehensive, long-term, sustainable, and pro-poor.

https://transfer.cpc.unc.edu/: Evidence from impact evaluations of cash transfer programmes in sub-Saharan Africa.

Audio-visual materials

https://www.unsdglearn.org/microlearning/3-minutes-3-messages-social-policy-as-a-tool-for-trans-formation/

https://www.unsdglearn.org/microlearning/3-minutes-3-messages-financing-sustainable-develop-ment/

https://transfer.cpc.unc.edu/multimedia/

http://www.fao.org/3/a-i6460e.pdf

Chapter 92

Universal basic income

Elizaveta Fouksman

Introduction

Universal basic income (UBI) is a proposal to give individuals a regular and repeated cash payment, large enough to ensure a basic livelihood. This income is *universal* in that the payment would be made to individuals without any behavioral conditions or targeting and would thus go to all residents or citizens of a country or region. And it is *basic* by providing a basic but adequate livelihood.

UBI is one of a number of different schemes that could provide a *universal minimum income guarantee* or a *social protection floor*. Other names for the proposal include basic income grant (BIG), citizen's income, social dividend, commons dividend, basic capital and demogrant. A closely linked proposal is a *negative income tax*. Rather than disbursing the same income to all, a negative income tax tops up individuals' income so that no one earns below a set minimum.

A closely connected policy, which has become popular as a development intervention, is cash transfers. Like UBI, cash transfers give cash directly, often (though not always) to individuals. Unlike UBI, cash transfers are often targeted (frequently at the poor, sometimes at an age group – such as the elderly – or a gender). They vary in size and frequency; some are one-off grants, some are disbursed monthly for a fixed period of time or are ongoing. Cash transfers sometimes come with behavioral conditions, such as sending children to school or vaccinations.

The political, economic and social implications of UBI are highly dependent on its size, its justification and its financing, along with the social policies that would accompany it. These details vary widely; in this sense UBI is more 'a field of debate' (Purdy, 1994) rather than one single policy. Many politically progressive UBI proponents add two additional characteristics to UBI: that it is large enough in size to provide a real alternative to waged work and that it does not replace – but rather complements – other public goods and services.

Justifications and benefits

Arguments justifying UBI can be divided into two broad camps: those that focus on economic justice and rights and those with a focus on poverty eradication and distributional efficiency (Fouksman and Klein, 2019). Justice-based arguments make the case that UBI would be radically emancipatory, guaranteeing 'real freedom' (van Parijs, 1995) for all to develop their capabilities. This tradition builds a case for UBI on the basis of economic

DOI: 10.4324/9780429282348-101

justice, arguing that UBI should be seen as a right to communal assets. Such communal assets can be understood as land and natural resources or more broadly as the common heritage of the collective wealth and productive capacity of society as a whole. In this tradition, UBI is a way of disbursing one's 'rightful share' (Ferguson, 2015) of shared and collectively and historically generated wealth.

An alternative tradition focuses on UBI as an efficient way of achieving poverty reduction and providing social security compared to current welfare models. In this view, 'just giving money to the poor' (Hanlon et al., 2010) is far more effective than other development or social policy interventions. Because there is no means-test or behavioral conditionalities, the administrative and bureaucratic burden of UBI is far less than standard conditional welfare programs. Uptake rates would also be higher than with current programs, and welfare cliffs and poverty traps would be avoided.

Besides the justice- and freedom- enhancing and the poverty-reducing benefits listed earlier, the benefits of UBI also include lowering inequality (provided that the policy is funded through a redistributory tax regime), enabling better working conditions through supporting individual and collective worker bargaining (if large enough, a UBI could function as a kind of permanent strike fund), economically supporting unpaid carework and other undervalued occupations such as the arts, and mitigating the risks involved in entrepreneurship. Extensive experiments and pilots with UBI-like policies have also shown that a guaranteed minimum income can lead to better physical and mental health outcomes, greater autonomy and empowerment for underprivileged populations (in particular women), lower crime rates and higher school attendance rates (Widerquist, 2018).

Common concerns and critiques

Universality and cost

A common objection to universal basic income focuses on its universality: why give the income to the rich if they do not need it? Linked to this is the cost objection: surely a UBI is too expensive?

The response to both of these objections is that while a UBI is *disbursed* universally, it is returned to the state by those that pay taxes. Thus while an entire population receives a UBI, only some fraction of the population are actually *net beneficiaries* of the policy (meaning their income increases after taxes and transfers), and the better-off are *net contributors* (meaning their income decreases after taxes and transfers). Though universal, UBI is simultaneously redistributory, once its funding is taken into account.

This also deals with the cost objection – many cost estimates focus on the *gross cost* of UBI, simply multiplying the size of the population by the size of the UBI. This figure is inaccurate – it does not demonstrate the actual amount of money that has to be distributed away from one set of people to another, as the taxes paid by the rich would be offset by the UBI received by the rich. The root of this frequent conceptual confusion is that with most nationally funded UBI proposals, some people would both be receiving money and giving money to the policy at the same time. The true cost of UBI is thus its *net cost*: the amount of money actually transferred from one population to another.

This can be illustrated with a simple example: imagine two people decide to implement a basic income of $10 per person amongst themselves. They need to raise $20 to do so – this is the gross cost. They decide to tax the richer individual of the two $20. They each then get their $10 UBI. The richer person thus gave $20 and received a $10 UBI, leaving them $10 poorer than before the taxes and transfers. The poorer person simply received $10, and is $10 better off. The true, net cost of the policy is only $10 – the amount ultimately transferred from the richer person to the poorer. This is why, arithmetically and distributionally, a UBI can be identical to a negative income tax. The policy described here is the same as a $10 negative income tax which goes to the poorer person, funded by a $10 positive tax on the richer person.

An accurate cost estimate will always include a discussion of net contributors and net beneficiaries, and the rate at which beneficiaries become contributors as their earnings go up (sometimes called the claw-back rate). Depending on the exact scheme, the net cost is typically a small fraction of the gross cost. For instance, in Karl Widerquist's (2017) 'back-of-the-envelope' calculation, a $1,000 per month UBI in the US has the annual gross cost of $3.415 trillion. However, combined with a claw-back rate of 50 per cent (meaning for every dollar earned a recipient will give $0.50 of the UBI back in tax, until they reach the point at which they have returned the full UBI in tax, after which the claw-back no longer applies), the actual, net cost is $539 billion per year, or 15.7 per cent of the gross cost.

Inflation

Another common concern is that of inflation: will the cost of goods go up with a UBI, thus negating any increases to standards of living? Broadly speaking, the answer is no: in classical economic theory, inflation is caused by more cash chasing the same amount of goods (for instance when a government prints money). With a redistributory UBI funded by taxation, the same amount of money within an economic system is following the same amount of goods – it has simply been redistributed. This is supported by evidence from UBI-like interventions, such as the Alaska Permanent Fund and other large-scale cash transfer programmes, which do not seem to have long-term or country-wide inflationary effects.

However, some empirical studies indicate that a redistributory UBI could potentially cause price inflation of some poor-specific goods. In particular, studies with cash-like interventions targeted at the poor show that the price of low-cost housing can increase, with landlords and property owners capturing much of the transfer. This is an argument in favor of additional programs, such as state-mediated low-income housing programs, that run alongside a UBI.

Work disincentives

A third key concern around UBI is its effect on labour markets. If a UBI is large enough for people to survive without working, won't many people choose to stop working and free-load off those who continue?

There are multiple ways of addressing this concern. The first is through empirical evidence. There have been large numbers of UBI-like experiments and pilot programs from the 1970s to the present, in both the Global North and South. Nearly all of these trials look at labour market effects and most have found either statistically negligible or indeed positive impacts of UBI or cash transfers on economic participation. Many trials have in fact found that recipients are more likely to become economically active by using the cash to get further training, migrate to areas with jobs or invest in the capital needed to start enterprises. And in the few cases where a small negative effect on working hours *has* been found, this is largely due either to recipients choosing to stay in education and training, primary caregivers (such as mothers with small children) spending more time on unpaid carework, or people between jobs choosing to wait longer to start working again (presumably because they can be more choosy and wait for a better job offer).

There is both an economic and cultural logic behind these findings: most people would prefer to have more than just a *basic* income. And anthropologists and sociologists of work have shown us that people engage in economic activities not simply for monetary gain, but also for status, meaning, hierarchy, community and sociality and thus will continue to do so even when basic needs are met. This is reinforced by the observation that most well-off people continue to work, as do most lottery winners (though they do often change professions).

There is also a philosophical, justice-based response to this concern, which argues that regardless of people's choices, it is unethical to force people into work with the threat of destitution, starvation or homelessness. A human-rights based perspective also echoes this logic, making the case that having enough resources to prevent hunger and homelessness is a basic right of being human, regardless of behavior. As Michael Howard has argued, we provide those we punish with imprisonment with unconditional shelter and food, regardless of their behavior – surely we should provide the same basic minimum to every member of society.

Advocacy and interest in the Global South

Versions of the UBI proposal have been around for at least two centuries. Some of the best-known older debates around the policy emerged in Europe and North America. These include Thomas Paine's proposal in *Agrarian Justice* in 1797 for a universal inheritance, James Mead's and G. D. H. Cole's proposals in the first part of the 20th century and Milton Friedman and James Tobin's negative income tax in 1960s USA.

Yet the most recent surge in interest and attention to the policy has been global in scope, with key nodes in both the Global North and South. In the Global South, this interest includes prominent political movements for a UBI in the late 1990s and early 2000s in South Africa, Namibia and Brazil. Brazil passed a law in 2004 that promised to gradually introduce a UBI, while both South Africa and Namibia rejected a basic income as state policy. Yet the policy outcome in all three countries has largely been the same: all three have instituted state-run and nationally funded policies of targeted cash transfers to the poor. These are particularly extensive in South Africa, where a third of the population receives some form of social grant from the state (most of these are child support grants, which are disbursed to the carers of needy children) and in Brazil through the Bolsa Familia program. More recently, India has seriously politically debated implementing a UBI, with the prominent Congress Party pledging to implement a partial UBI if elected in 2018 (which it was not).

Besides such political deliberations, which have occurred largely in middle income countries, the Global South has been the site of extensive trials and experiments with UBI and cash transfers. Sites of UBI-related pilots and experiments include Namibia, India and Kenya. The last has a large and ongoing randomized control trial of UBI, run by the US-based and donor-funded NGO GiveDirectly. Alongside these have been many experiments with various cash-transfers, some approaching a UBI in terms of scope or size.

The Global South does pose some specific implementation challenges for UBI. In particular, while middle income countries could generally implement a modest UBI on the basis of their own tax base, lower-income countries might struggle to have the fiscal resources to do so. This is why cash transfers in lower-income countries are largely aid-funded, small-scale and temporary. One potential solution to such problems is to implement a *global UBI*, which would redistribute wealth between richer countries and poorer ones. A global UBI could fit into both of the justifications for basic income discussed earlier: it could be understood as an effective aid or poverty eradication policy or alternatively it could be framed as a way of more justly distributing wealth or even as reparations for the injustices of colonialism and the slave trade.

Basic income and wage labour

A particularly lively debate around UBI has been in conjunction to the past, present and future of wage work. Often such debates center around the potential future impacts of automation on wage labour, particularly with the rise of machine learning and artificial intelligence. Yet such speculative debates are in many ways unnecessary, as regardless of the future, in the present wage labour is often failing to ensure livelihood and well-being. The anthropologists James Ferguson and Tanya Li (Ferguson and Li, 2018; Li, 2013) have both argued high levels of precarity, underemployment and low wages around the world should lead us to question the central role of wage work in both our economic, distributory and social imaginary. UBI can be one way to imagine livelihood and distribution outside of wage labour.

Indeed, Ferguson (2015) argues that there is an emergent *new politics of distribution* in the Global South (particularly South Africa), with what he sees as increasing popular demand and expectations for forms of distribution outside of wages. Such interest in the future of work and post-work – and the role of UBI in decentering and decommodifying wage-labour – is rooted in a longer tradition of both feminist thinking, which highlights the importance of unpaid care and ecological economics, which underscores the necessity of no-growth or degrowth futures. Both of these challenge the centrality of a productivist capitalist work ethic and suggest that UBI and shorter working hours could de-emphasise the capitalist drive towards ever-increasing production and consumption.

References

Ferguson, James. 2015. *Give a Man a Fish: Reflections on the New Politics of Distribution*. Durham and London: Duke University Press.

Ferguson, James and Tania Murray Li. 2018. *Beyond the "Proper Job": Political-Economic Analysis after the Century of Labouring Man*. Cape Town: Institute for Poverty, Land and Agrarian Studies (PLAAS).

Fouksman, E. and E. Klein. 2019. "Radical Transformation or Technological Intervention? Two Paths for Universal Basic Income." *World Development* 122.

Hanlon, Joseph, Armando Barrientos and David Hulme. 2010. *Just Give Money to the Poor: The Development Revolution from the Global South*. Sterling, VA: Kumarian Press.

Li, Tania Murray. 2013. "Jobless Growth and Relative Surplus Populations." *Anthropology Today* 29(3): 1–2.

Purdy, David. 1994. "Citizenship, Basic Income and the State." *New Left Review* 208(November/ December): 30–48.

van Parijs, Philippe. 1995. *Real Freedom for All: What (If Anything) Can Justify Capitalism?* Oxford: Oxford University Press.

Widerquist, Karl. 2017. "The Cost of Basic Income: Back-of-the-Envelope Calculations." *Basic Income Studies* 12(2).

Widerquist, Karl. 2018. *A Critical Analysis of Basic Income Experiments for Researchers, Policy-makers, and Citizens.* Cham, Switzerland: Palgrave Macmillan.

Additional Resources

Basic Income Earth Network, basicincome.org

Bastagli, Francesca, Jessica Hagen-Zanker, Luke Harman, Valentina Barca, Georgina Sturge, Tanja Schmidt and Luca Pellerano. 2016. *Cash Transfers: What Does the Evidence Say?* London: Overseas Development Institute.

Standing, Guy. 2017. *Basic Income: A Guide for the Open Minded.* New Haven: Yale University Press.

Chapter 93

Making social work visible in social development in Nigeria

Challenges and interconnections

Uzoma Okoye and Susan Levy

Introduction

Social work is a global, social justice profession committed to the UN call to *Leave No One Behind* (LNOB), 'the central, transformative promise of the 2030 Agenda for Sustainable Development' (United Nations Systems, 2017). This chapter contributes to strengthening understanding of social work and its role in social development. We introduce the reader to social work within the context of social development, explore the interconnections between the two and some of the challenges of situating social work more centrally within social development work. Recommendations are made for integrating social work into social development narratives to achieve change in the lives of many. A clear message threaded throughout the chapter is that social work, as a profession directly involved with tackling poverty; inequalities; oppression; and working to address the UN Sustainable Development Goals (SDGs) must be more visible in development work.

Relevance and influence of social work and social development in Africa

Social work is a global profession focused on social justice and social change, achieved through enhancing and promoting well-being, diversity, and empowerment as set out in the Global Definition of Social Work:

> Social work is a practice-based profession and an academic discipline that promotes social change and development, social cohesion and the empowerment and liberation of people. Principles of social justice, human rights, collective responsibility, and respect for diversities are central to social work. Underpinned by theories of social work, social sciences, humanities and indigenous knowledge, social work engages people and structures to address life challenges and enhance wellbeing.
>
> (IAASW and IFSW, 2014)

Social workers work with the social dimensions of a range of different issues, including gender inequality and gender-based violence; refugees and migrants; disability and mental health; and child and family welfare. Poverty and marginalisation underpin and are contributing factors to much social work, both of which have structural antecedents, and as Ugiagbe (2017) states, are best understood through a social development lens.

DOI: 10.4324/9780429282348-102

Within the African context, these structural factors have roots in colonial histories and the continued prevailing dominance of Western knowledge and practices. Such practices, orientated at an individual level, largely based on casework and therapeutic approaches (Gray, 2017) and designed for use in countries with an established welfare system, create tensions and challenges when situated in a different socio-cultural and political context. These tensions and challenges are centred on 'relevance' and 'influence'.

'Relevance' refers to the applicability of 'imported theories and models' (Ugiagbe, 2017: 271) to address African social problems. Levy et al. (2022: 10) report that African social work students find that curricula lack cultural relevance with a call to 'indigenise social work to suit my practice' and better prepare for working with cultural diversity. African everyday life happens at a communal rather than an individual level, which creates a dissonance between imported theories and models, and local interventions that value and speak to the unique strengths and culture of the African people. Progress, albeit slow, is beginning to situate indigenous knowledge into African practice and decolonise curricula. However, to achieve meaningful change and a voice for social work in social development, alongside decolonising social work, structural factors need addressing. This leads onto questions of 'influence' and the impact social workers can achieve in influencing change at a policy level and with the key players in the field of development (Midgley and Conley, 2010): international financial institutions World Bank, International Monetary Fund (IMF), and African Development Bank, as well as international, national, and local non-governmental organisations (NGOs).

> To make a real impact on social development, social work has to strengthen its visibility within multidisciplinary networks, develop new partnerships and alliances, and make a stronger contribution to policy development and implementation.
>
> (Gray, 2017: 8)

The previous discussion highlights that the boundary between social work and social development in Africa is fluid. The values and essence of social work along with areas of practice align with those of social development. Both are working towards ameliorating social and economic problems and achieving impact through addressing the UN Sustainable Development Goals (SDGs). However, despite these interconnections, social work in Africa remains peripheral within social development discourses. Social work is the active but 'unseen' profession at the forefront of working to achieve a paradigm shift to a more equal society. In the following section, we expand and contextualise the issues raised in this section within social work and social development in Nigeria.

Profession of social work and social development in Nigeria

Nigeria is a country of extremes; it is the most populous and wealthiest country in Africa. It is also a country with high levels of income inequality and poverty, and social welfare is largely undeveloped, poorly funded, and neglected. The available services are grossly inadequate compared to the number of people in need and the significant social problems which centred around poverty cover internal displacement; migration; environmental disasters; gender-based violence; and inequality. Prevailing structural barriers keep people in poverty in Nigeria; curtail the efficacy of government investment and international aid directed at eradicating poverty and other social problems; and importantly, limit the effectiveness of social workers. As Ugiagbe (2017) argues, no matter how concerned social workers are with the social well-being of the vulnerable, marginalised, and oppressed, change can be limited. The role of social workers in social development has not received adequate attention since the inception of social work practice in Nigeria and other parts of sub-Saharan Africa. Whilst there is a clear role for social work in achieving inclusive and sustainable development, there are challenges and opportunities in making this a reality on the ground. Three of these challenges/opportunities are explored in the following paragraphs.

First, social work across Africa is an emerging profession. Social workers are trained through undergraduate and postgraduate social work degrees that prepare them for social work practice. However, social work in Nigeria, as in much of Africa, is not recognised as a profession; it is unregulated (Okafor et al., 2017), and lacks the legal frameworks to support practice. Regulation of social work by law, as happens in most of the Global North, places a statutory obligation and commitment on developing the profession. The unregulated nature of social work in Nigeria – and the lack of professional legitimacy – is interconnected with the paucity of visibility of social work in social development discourses. To position the profession as a role player in social development, social work needs scaffolding in the form of a political mandate supported by social welfare policy targeted at addressing the increasing social problems and to be embedded in a development approach.

A second, interconnected challenge stems from a development orientation in social work not being formalised or recognised at government level. Ugiagbe (2017: 277–278) notes that 'social workers have been largely excluded from policy formulation and implementation because they are caught in remedial services fashioned after the British colonial model'. The role and impact of social work in social development sit outside of the frame of vision of policymakers and are thus not seen as vital in social development programmes and projects (Onalu and Okoye, 2021). Whilst social workers are actively working in government and non-governmental organisations (NGOs), they remain peripheral and largely unacknowledged. Having been assigned a very narrow role in the development agenda, the role of social workers and the work they do is too easily overlooked or carried out by non-social workers (Kurevakwesu et al., 2022).

A third challenge/opportunity lies in social work education and research being a driver for change through creating culturally relevant curricula and generating indigenous knowledge (Patel and Hochfeld, 2013). Social work education, including continuous professional development (CPD), is yet to align with the development agenda. More needs to be done to embed a development discourse and assume a development approach that aligns with the salience of social work achieving inclusive and sustainable development and addressing present and emerging issues (Dhemba and Nhapi, 2020). Raising awareness and enhancing understanding of the interconnections between social work and social development will support re-orienting social work practitioners towards social development (Anucha, 2008; Kakowa, 2016).

Ways forward

This chapter has introduced social work as a global, social justice profession working with issues centred around poverty and inequalities with people who can be marginalised, vulnerable, and oppressed. As with social development, social workers are directly involved in achieving impact through addressing the UN Sustainable Development Goals (SDGs). However, despite these interconnections, social work in Africa remains peripheral, lacking visibility within social development discourses. The following are three key recommendations for strengthening the role of social work in social development narratives:

1 Indigenous Knowledge: to be embedded within culturally relevant curricula and continuous professional development, ensuring 'relevance' for practice.
2 Voice at Policy Level: ensure social work is positioned to 'influence' development agendas.
3 Professional Legitimacy: for the social work profession.

Moving forward, social work in Africa cannot and should not be detached from the wider dynamics of development initiatives; social work must engage with and be a key partner in the development agenda.

References

Anucha, U. (2008) 'Exploring a New Direction for Social Work Education and Training in Nigeria', *Social Work Education*, 27(3): 229–242.

Dhemba, J. and Nhapi, T. G. (2020) 'Social Work and Poverty Reduction in Southern Africa: The Case of Eswatini, Lesotho, and Zimbabwe', *Social Work & Society*, 18(2): 1–13.

Gray, M. (2017) *The Handbook of Social Work and Social Development in Africa*, London: Routledge International Handbooks.

IFSW and IAASW (2014) *Global Definition of Social Work*, International Federation of Social Workers and International Association of Schools of Social Work, Global Definition of Social Work – International Federation of Social Workers. ifsw.org

Kakowa, F. (2016) 'Nuturing professional social work in Malawi', *Journal of Social Work*, 6(2): 1–5.

Kurevakwesu, W., Chikwaiwa, B. and Mandau, M. (2022) 'The struggle for social work professional identity in contemporary Zimbabwe: A study on abuse of the social work title', *Qualitative Social Work*, 22(2): 200–216. https://doi.org/10.1177/14733250211061827

Levy, S., Okoye, U. O. and Ingram, R. (2022) 'Making the "local" visible in social work education: Insights from Nigeria and Scotland on (re)balancing and contextualising indigenous and international knowledge', *British Journal of Social Work*, 52(7): 4299–4317. https://doi.org/10.1093/bjsw/bcac028

Midgley, J. and Conley, A. (eds) (2010) *Social Work and Social Development: Theories and Skills for Developmental Social Work*, New York: Oxford University Press.

Okafor, A., Onalu, C., Ene, J. and Okoye, U. (2017) 'Social work in health care setting', in Okoye, U. O., Chukwu, N. and Agwu, P. (eds) *Social Work in Nigeria: Book of Readings*, Enugu: University of Nigeria Press, pp. 1–15.

Onalu, C. E. and Okoye, U. O. (2021) 'Social justice and social work curriculum at the University of Nigeria Nsukka', *Research on Social Work Practice*, 31(6): 576–583. https://doi.org/10.1177/10497315211001532

Patel, L. and Hochfeld, T. (2013) 'Developmental social work in South Africa: Translating policy into practice', *International Social Work*, 56(5): 690–704.

Ugiagbe, E. O. (2017) 'Decolonising social work practice in Nigeria: Moving beyond indigenisation to development', in Gray, M. (ed) *The Handbook of Social Work and Social Development in Africa*, London: Routledge International Handbooks, pp. 271–280.

United Nations Systems (2017) *Leaving No One Behind: Equality and Non-Discrimination at the Heart of Sustainable Development*, CEB Equality. framework-A4-web-rev3.pdf; unsceb.org

Further Reading

Gray, M. (2017) *The Handbook of Social Work and Social Development in Africa*, London: Routledge International Handbooks.

Midgley, J. and Conley, A. (eds) (2010) *Social Work and Social Development: Theories and Skills for Developmental Social Work*. New York: Oxford University Press.

Chapter 94

Technological innovation and development

Theo Papaioannou

Introduction

The concept of technological innovation refers to the continuous efforts of human beings to provide novel technological solutions to practical problems and facilitate change, transforming their technical and socio-economic environment (Papaioannou, 2018). In this sense, the concept of technological innovation is rather broad. It is not only concerned with basic and applied research which lead to development of 'physical technologies' but also with social practices and institutional structures which constitute 'social technologies' (Chataway et al., 2010). Both types of technologies can be transformative of peoples' lives and capabilities, promoting international development.

Since the 1780s, when the first industrial revolution began in Britain and from there spread around the world, successive technological revolutions have introduced innovations that have had a tremendous impact on the material living conditions of the world population, their economies, and societies. These innovations range from industrial textiles, railways, and electricity to medicines, food processing, automobiles, information, and communication technologies (ICTs) and, more recently, to digital technologies, 3D printing, artificial intelligence (AI), nanotechnology, genomics, and biotechnology (Juma, 2016; Papaioannou, 2011; Freeman and Soete, 1997). They have induced transformations extending from the 18th right to through to the 19th, 20th and 21st centuries during which most parts of the world entered industrialisation, capitalism, and globalisation.

Examples of transformation through technological innovation

There several examples of transformation induced by technological innovation. Take the case of the so-called Green Revolution. The international effort to increase agricultural production and especially supply of stable crops such as rice and wheat (through the development and dissemination of new seed varieties) in the 1960s and 1970s led to massive increases in yields of staple crops across South Asia. Another example is the impact of vaccine technologies on diseases such as Polio in developing countries and more recently Ebola. In the case of Polio, the generation of innovative and effective vaccines in the 1950s and in the 1960s, and more importantly their diffusion to developing countries in the 1970s (through national immunisation programmes) led to controlling the disease worldwide. Such vaccines are exemplary of pro-poor innovations because of their affordability and accessibility by low-income countries. In the case of Ebola, the licensing of effective vaccines in 2019 and in 2020 – and their subsequent approval for use in affected

DOI: 10.4324/9780429282348-103

countries such as Democratic Republic of Congo and Central African Republic – have had positive impacts on public health and the economy of these developing countries.

Apart from physical technologies such as vaccines, the development and diffusion of social technologies such as managerial techniques, standards, and processes of reducing risks in innovations in different sectors, e.g., food and health, have had induced transformations, including increase of safety in many contexts. All these examples indicate complex interrelations between technological innovation and development and hence they are highlighted in the United Nations (UN) Millennium Development Goals (MDGs) and the subsequent Sustainable Development Goals (SDGs) which consistently emphasise the importance of equitable access to technological innovation.

Technological innovation and social exclusion

Although there is agreement among innovation and development experts that technological progress has radically changed the lives of most people across the world, enabling developing countries of the Global South to modernise or catch up with advanced countries of the Global North, in general it has tended to exclude the very poor (both in the Global North and Global South), thereby increasing inequalities (Soares and Cassiolato, 2013). In addition, ecological problems and other negative consequences of technological change have put in question the relationship between innovation and development. For example, innovative production and industrialisation in various sectors have led to environmental degradation and climate crisis. In agriculture, systematic cultivation, mechanised irrigation, and use of various agricultural chemicals for insecticides or fertilisers have had tremendous impact on soil and water quality as well as on plants and animals. According to Hochstetler (2021), expanding agricultural production through innovation often means bringing more marginal lands under cultivation, increasing negative environmental impacts. In manufacturing, the dependency on fossil fuels and the exploitation of natural resources, especially in developing countries, had negative effects, ranging from air pollution to toxic wastes. For example, the introduction of smartphones on the one hand increased demand for rare earths and nickel (i.e., key materials for smartphones chips) and on the other pushed for constant innovation of new generation of devices and models which increase toxic waste that ends up processed in poor developing countries.

Since the World Summit for Social Development in Copenhagen in 1995, the relationship between innovation and development has been increasingly challenged. It has become apparent that social progress is not possible without fairer distribution of benefits and risks of new technologies, global market regulation, and ecological sustainability. The neo-liberal orthodoxy according to which unequal generation and diffusion of technological innovation through the global market is necessary to foster growth and sustainability proved to be a myth. Data shows that although global poverty (i.e., the number of people living below $1.99 per day) has fallen the last two decades due to innovation-led growth in regions such as China and India, several fast-growing countries have either experienced little poverty reduction or increasing poverty (Fosu, 2011). One reason for this is the contradictory nature of technological revolutions or the 'creative destruction' they cause by leaving many people behind (Perez, 2002). For example, the so-called 4th industrial revolution of data-driven technologies and automation leaves behind many people who lack advanced digital skills and/or access to fast Internet. This is particularly so in developing countries' youth populations. Although the adoption of smartphone technologies has reduced this digital divide between the Global North and the Global South, there is growing evidence that suggests the main beneficiaries of accelerated innovation in the 21st century are a minority of highly skilled workers located in the Global North. The impact of digital and automation technologies on the Global South is negative given that there are more manual jobs to be lost through labour-substituting innovation. In 2019, the World Bank estimated that around 1.8 billion jobs could be threatened by digital and automation technologies in the Global South.

Another reason for the slow progress in poverty reduction in the 4th industrial revolution is the increasing inequality in the generation and distribution of innovation-led growth. Even in China, the fact that the economy enjoys relatively high percentage of growth per capita GDP says nothing about how this growth is distributed within the country. In India, the fact that the economy enjoys high annual growth has no overall impact on poverty reduction (Kaplinsky, 2011). In his recent book *Global Inequality*, Banko Milanovic (2016) confirms this trend, arguing that for the first time since the first Industrial Revolution, global inequality is being driven by rising gaps within countries and not between countries. In fact, with the increases of incomes in countries such as China and India, the gaps between countries have been growing. Therefore, no matter how crucial innovation-led economic growth is for reducing extreme poverty, the shift from this indicator to innovation-led equality is necessary for achieving development.

The shift from innovation-led growth to innovation-led equality

The adoption of MDGs and the follow-up agreement on SDGs has confirmed the normative and practical necessity of shifting focus from innovation-led growth to innovation-led equality, redefining the mission of technology and innovation as the reduction of poverty and inequality and the elimination of extreme deprivation. Achieving SDGs by 2030 requires approaches which put technological innovation at the centre of development policy and practice. Some governments in developing countries have already developed such approaches. For example, governments in South Africa and Kenya have facilitated increase of Internet use and mobile broadband submissions by their populations. The same holds for governments of Asian countries such as India and Bangladesh. However, the expenditure of these governments in R&D and related policies remains disappointing. In contrast, China has massively increased their R&D expenditure and policies for technological innovation. Hence, it is on track for achieving several SDGs, including SDG 2 (Zero Hunger) and SDG 13 (Climate Action).

Developing countries need technology but more importantly they need to develop the skills and knowledge to use technology for development. Considerable high proportion of knowledge and technology remain in the ownership of advanced developed nations in the Global North. Therefore, they should be made available to developing countries in the Global South. This argument of global redistribution of knowledge and technology implies the need for global change of knowledge and technology governance structures such as intellectual property rights (IPRs). It is these global structures which enable private appropriation of new scientific breakthroughs in the Global North and exclusion of the Global South. Take for example the recent unequal generation and diffusion of COVID-19 vaccines which have prevented more than 85 poor countries from securing access to these technologies before 2023. Most of these poor countries are in Africa and Asia, relying on COVAX (an initiative by the WHO, aiming at securing 6 billion doses of

COVID-19 vaccines) for immunising their populations. This 'vaccine apartheid' in combating COVID-19 was caused by IPR claims of big pharmaceutical companies located in the Global North. The so-called Trade Related Aspects of Intellectual Property Rights (TRIPS) agreement needs to change to enable manufactures around the world to use innovative technologies to produce affordable vaccines for all. This argument also seems to echo Harry Truman's famous Inaugural Address about development:

> We must embark on a bold new program for making the benefits of our scientific advances and industrial process available for the improvement and growth of underdeveloped areas. More than half of the people of the world are living in conditions approaching misery. . . . For the first time in history, humanity possesses the knowledge and skill to relieve the suffering of these people.
>
> (Public Papers of the Presidents of the United States, 1964: 114–115)

In Truman's narrative, radical technological innovation goes hand in hand with economic development and therefore is able to drive people out poverty, cure diseases, and promote peace and prosperity. However, this overlooks the fact that incremental technological change and innovation which occur during the actual process of production are also crucial for development. Kaplinsky (2011) reminds us of the historical fact that one response to the northern-focused radical innovation trajectory was the so-called Sussex Manifesto (SM; Singer et al., 1970) and the Appropriate Technology Movement (ATM; Schumacher, 1973).

The SM argued that the focus of research and development (R&D) agendas ought to be on the needs of low-income countries. The ATM promoted the development of incremental and appropriate technologies (AT) to address these needs and improve poor infrastructures in low-income countries. For example, in agriculture, the focus of ATM was on small scale farming, promoting simple agricultural technologies and irrigation techniques to increase supply of food in local communities in India, Kenya, Thailand, etc. In energy, the ATM recently supported technologies such as photovoltaic (PV) solar panels and biofuels to reduce dependency on fossil fuels and achieve environmental benefits at local level.

Of course, even though it is true that the ATM offered the prospect of more equitable and environmentally friendly growth, it is also true that it faced several problems. According to Kaplinsky (2011) the first problem was that most ATs were 'economically inefficient' in the sense that input of capital and labour exceeded the output of products. The second problem was the contextual nature of 'appropriateness' and the disappointment of AT expectations. Finally, the social context of innovation was different from that of developed countries and therefore ATs could not scale up.

The emergence of inclusive models of innovation and development

Kaplinsky (ibid) observes that, although the ATM of the 1970s and early 1980s failed to keep the promise of more equitable and balanced growth, there are new factors emerging which can potentially disrupt the dominance of global system of technological innovation. These include: the dynamism of emerging economies such as China and India; the emergence of new radical technologies such as ICT; new green technologies;

biotechnology and nanotechnology; the diffusion of innovation capabilities; and the emergence of disruptive entrepreneurs. However, the question that arises is whether potential disruption of the dominance of global innovation can automatically transform international development.

First, in Asian economies such as China and India there is a large market of poor consumers. Therefore, the hypothesis that low-income driving consumption will induce what we have called elsewhere (Kaplinsky et al., 2009) below the radar innovation (BRI) seems to be correct. However, the context of pro-poor innovation is not the same in all developing regions. In Latin American countries, for instance, there is more scarcity of infrastructures, skills, and socio-economic institutions than in Asian countries. Thus, in the former it is more likely to have what Srinivas and Sutz (2008) call scarcity-induced innovations (SII) than in the latter.

Second, the emergence and growth of new radical technologies such as ICT, AI, green technologies, blockchain, biotechnology, and nanotechnology can have substantial impact on meeting the needs of the poor and promoting more equitable growth. However, the rapid diffusion of these technologies through the global market is neither equal nor fair. This is because the global market diffuses innovation according to effective demand (i.e., the intention to purchase novel technologies by people who have money to pay for them) and not according to human need (i.e., the necessity to use novel technologies to satisfy basic human needs such as health, food, energy, etc.). Let's take for example biotechnology. Many developing regions lack access to important innovations such as molecular diagnostics, antibody-coated dipstick tests, recombinant vaccines, and improved methods of drug delivery. This has become even more apparent during the COVID-19 crisis in Africa and in Latin America. Such innovations are potentially crucial for these developing regions' populations to meet their basic needs.

Third, there has been an increase in capabilities and learning-by-doing but this has happened in some low-income economies which increased their share of global manufacturing, namely China and India. Other countries, for example in the Latin American region, face huge scarcity in terms of innovative capabilities (Srinivas and Sutz, 2008). This is not only because they lack cognitive, institutional, and socio-economic capabilities but also because their firms are not scaled up to the point that they can participate in global value chains. Scaling up to this point requires abundance of inputs, and unless firms are able to find innovative ways to proceed (e.g., experimentation routines at firm level), they cannot get share of export markets.

In fact, there are emerging models of innovation which are directly linked to development. Perhaps the most important ones are: 'frugal' innovations, i.e., simplified versions of existing technological products; and 'grassroots or below the radar' innovations, i.e., low- and middle-income group generated innovations drawing on traditional knowledge and available technologies. These models of innovation have been crucial for combating the COVID-19 pandemic. Several firms in the Global South have modified their products to supply hospitals with personal protective equipment such as masks and sanitisers and even ventilators. In India and Pakistan railways organisations allowed the modification of train cars into mobile isolation wards. In China, a 1,000-bed hospital was constructed in just 10 days. Medical researchers produced 'new drugs' from 'old drugs'. All these innovations follow the principle of doing more with less for the many.

In fact, 'frugal' and 'grassroots or below the radar' innovations can be regarded as novel models of innovation and development because they embody a number of social and technical changes. Heeks et al. (2014: 2) summarise these changes as follows:

1 Significant involvement of the private sector and global value chains in innovation for the poor.
2 The development of poor consumers as an accessible mass market.
3 Growth and technological capabilities within developing countries.
4 The involvement of new technologies, especially information and communication technologies such as mobile phones.

Emerging models of innovation respond to the dominant paradigms of innovation. The most crucial characteristic of these models is that they engage poor people as both consumers and producers who actively participate in driving innovation and growth. According to OECD (2012), emerging models of innovation for or by low- and middle-income groups are essential for inclusive growth. In this sense such models enable countries to achieve key SDGs, including SDG1 (No Poverty) and SDG2 (Zero Hunger).

References

Chataway, J., Hanlin, R., Mugwagwa, J. and Muraguri, L. (2010) Global health social technologies: Reflections on evolving theories and landscapes. *Research Policy* 39: 1277–1288.

Fosu, A. K. (2011) Growth, Inequality, and Poverty Reduction in Developing Countries: Recent Global Evidence. *BWPI Working Paper 147*. http://www.un.org/esa/socdev/egms/docs/2011/Growth-Inequality.pdf [last accessed 15 February 2023]

Freeman, C. and Soete, L. (1997) *The Economics of Industrial Innovation*, London: Pinter.

Heeks, R., Foster, C. and Nugrohoa, Y. (2014) New models of inclusive innovation for development. *Innovation and Development* 4(2): 175–185.

Hochstetler, K. (2021) Environmental degradation and sustainability. In T. Allen and A. Thomas (eds) *Poverty and Development in the 21st Century*, Oxford: Oxford University Press.

Juma, C. (2016) *Innovation and Its Enemies: Why People Resist New Technologies*, Oxford: Oxford University Press.

Kaplinsky, R. (2011) Schumacher meets Schumpeter: Appropriate technology below the radar. *Research Policy* 40(2): 193–203.

Kaplinsky, R., Chataway, J., Hanlin, R., Clark, N., Kale, D., Muraguri, L., Papaioannou, T., Robbins, P. and Wamae, W. (2009) Below the radar: What does innovation in emerging economies have to offer other low-income economies? *International Journal of Technology Management and Sustainable Development* 8(3): 177–197.

Milanovic, B. (2016) *Global Inequality: A New Approach for the Age of Globalisation*, Cambridge, Massachusetts and London, England: The Belknap Press of Harvard University Press.

OECD (2012) *Innovation and Inclusive Development, Conference Discussion Report*, Cape Town: Organisation of Economic Cooperation and Development. https://web-archive.oecd.org/2013-02-27/216206-oecd-inclusive-innovation.pdf

Papaioannou, T. (2011) Technological innovation, global justice and politics of development. *Progress in Development Studies* 11(4): 321–338.

Papaioannou, T. (2018) *Inclusive Innovation for Development: Meeting the Demands of Justice through Public Action*, London: Routledge.

Perez, C. (2002) *Technological Revolutions and Financial Capital: The Dynamics of Bubbles and Golden Ages*, Cheltenham and Northampton, MA: Edward Elgar.

Public Papers of the Presidents of the United States (1964) *Harry S Truman, Year 1949*, Vol. 5, Washington, DC: United States Government Printing Office.

Schumacher, E. F. (1973) *Small is Beautiful: A Study of Economics as if People Mattered*, London: Blond & Briggs.

Singer, H., Cooper, C., Desai, R. C., Freeman, C., Gish, O., Hill, S. and Oldham, G. (1970) *The Sussex Manifesto: Science and Technology for Developing Countries during the Second Development Decade*, Brighton: IDS.

Soares, M.C.C. and Cassiolato, J.E. (2013, July) Innovation systems and inclusive development: Some evidence based on empirical work. In dsa *Information, Technology and Development Study Group Meeting: "New Models of Innovation for Development"* (pp. 4-5).

Srinivas, S. and Sutz, J. (2008) Developing countries and innovation: Searching for a new analytical approach. *Technology in Society* 30: 129–140.

Additional Resources

Cozzens, S. and Thakur, D. (2014) *Innovation and Inequality: Emerging Technologies in an Unequal World*, London: Elgar.

Heeks, R., Foster, C. and Nugrohoa, Y. (2014) New models of inclusive innovation for development. *Innovation and Development* 4(2): 175–185.

Kaplinsky, R. (2011) Schumacher meets Schumpeter: Appropriate technology below the radar. *Research Policy* 40(2): 193–203.

Kaplinsky, R., Chataway, J., Hanlin, R., Clark, N., Kale, D., Muraguri, L., Papaioannou, T., Robbins, P. and Wamae, W. (2009) Below the radar: What does innovation in emerging economies have to offer other low-income economies? *International Journal of Technology Management and Sustainable Development* 8(3): 177–197.

Papaioannou, T. (2014) How inclusive can innovation and development be in the twenty-first century? *Innovation and Development* 4(1): 187–202.

Smith, A., Fressoli, M. and Thomas, H. (2014) Grassroots innovation movements: Challenges and contributions. *Journal of Cleaner Production* 63: 114–124.

Audio-visual materials

Inclusive Innovation: Challenges and Paths Forward: https://www.youtube.com/watch?v=HPCCdVLJhOU

Rethinking Society for the 21st Century: Developing a Science and Technology Studies perspective: https://www.youtube.com/watch?v=EPAYvmbENoM

SCOTONOMICS Ep 38: Innovation in a modern economy: https://www.youtube.com/watch?v=q6pmFV4R5bE&t=65s

Strategies for Innovation and Development, William Lazonick, SOAS University of London: https://www.youtube.com/watch?v=Cxat5t-FTSk

Chapter 95

Information and communication technology for development (ICT4D)

Azadeh Akbari

In the last decades, the speed of technological developments, especially information and communication technologies (ICTs), has fundamentally changed all aspects of our lives. We can process complicated calculations in a matter of milliseconds with computers. The internet has transformed access to knowledge and informational resources. Social media have revolutionised interpersonal communication and societal interactions. Smartphones have become an extension of our bodies and have remodelled our being in this world. If everything could be transformed, improved, and accelerated, why not development efforts?

The idea of making development happen faster, more efficiently, and justly through ICTs reflected the zeitgeist of the early 90s. The fascination with technological advances and solutions was dominant throughout this decade and is also evident in the first manifestations of the so-called Information and Communication Technology for Development (ICT4D). United Nations (UN) Commission on Science and Technology for Development established its Working Group on ICT4D in 1997. Their first report immediately called on all developing countries to "establish a national ICT strategy" and required each UN agency to "review the financing, production and use of ICTs for social and economic development in their area of responsibility" (UN Commission on Science and Technology for Development, 1997: 4). Despite the immediacy of the call to action targeted to the national governments and UN agencies, the UN Millennium Development Goals only mention ICTs under goal eight, indicating, "in cooperation with the private sector, make available benefits of new technologies, especially information and communications."

This inconspicuous approach to the ICTs and development intrinsically changes in the formulation of UN's Sustainable Development Goals in 2015. Target 9.c of the SDGs directly sets its objective to "significantly increase access to information and communications technology and strive to provide universal and affordable access to the Internet in least developed countries by 2020." Target 4.b encourages higher education scholarships in the area of the ICTs, and two other targets mention ICTs as "enabling technologies" for promoting the empowerment of women (target 5.b) and for technology and innovation capacity-building (target 17.8). Although these targets seem to set necessary objectives, a glance at the evaluation indicators shows how these measures are inadequately defined. For example, women empowerment through ICTs is assessed based on the "proportion of individuals who own a mobile telephone, by sex," and ICT capacity building with the "proportion of individuals using the Internet" (United Nations, n.d.). Researchers and practitioners are increasingly criticising such quantified techno-centric understanding of the role of ICTs in development in the field. The following section summarises the

DOI: 10.4324/9780429282348-104

academic trajectory of the ICT4D as a research field moving parallel and entangled with the governmental and international discourse on ICT4D.

Development of ICT4D as a research field

The first instances of ICT4D research appeared between the mid-1980s to mid-1990s and were primarily influenced by the Information Systems (IS) approaches and conceptual frameworks (Walsham, 2017: 3). The 1988-conference of "social implications of IS in developing countries" in India was the birthplace of the International Federation of Information Processing's (IFIP) working group 9.4, focusing on the implications of information and digital technologies for development. IFIP 9.4 is still an active player in shaping the ICT4D scholarship and hosts an annual conference in developing countries. The early phase of ICT4D research engages thematically with "participative and cooperative design . . . the need for indigenous development" (Walsham, 2017: 4), and the importance of considering contextual properties and social systems in designing information systems. The second phase from the mid-1990s to mid-2000s is concurrent with a global increase in access and usage of ICTs. As a result, the scope and range of ICT4D research grow, and the research approaches and projects become more interdisciplinary (ibid.). Within these approaches, major identifiable themes are 'local adaptation and cultivation of ICTs, standardization versus localization of technology, and in-depth studies of particular technologies such as GIS' (Walsham, 2017: 5).

The first criticisms also emerge in this era scrutinising the development discourse and the designated role of ICTs in achieving certain conceptualisations of development. The criticism also sharply pointed at the absence of gender and gender relations from the ICT4D literature and research (Gillard et al., 2008). The third phase – or the so-called proliferation period – from the mid-2000s to the present is majorly influenced by the 'mobile phone revolution' (Walsham, 2017: 5). The technological advances and the ease of access to the ICTs are mirrored in interdisciplinary and multidisciplinary research approaches to ICT4D across journals, conferences, and scholarly debates. This phase of ICT4D research champions fresh approaches contrary to traditional understandings of development, such as the capabilities approach – focusing more on individual aspects of what development might mean – and women's empowerment through technological means.

A fourth phase must be added to this chronological categorisation that focuses on the technologies and the socio-technological phenomena such as 'datafication' that starkly revolutionise social, economic, and political aspects of life. Datafication and the invasive gathering of data by private and public entities means that any phenomenon is transformed to 'a quantified format so it can be tabulated and analyzed' (Mayer-Schoenberger and Cukier, 2013: 78). Datafication does not only target human life, behaviour and sentiments but fundamentally datafies 'everything' (Mayer-Schoenberger and Cukier, 2013). The datafication debate has opened up new perspectives in revising every component of ICT4D by offering concepts such as data justice, where ideas of justice and fairness are highlighted 'in the way people are made visible, represented and treated as a result of their production of digital data' (Taylor, 2017). Other emerging research areas underline the datafication of work by arguing for the 'platformisation of labour' as platforms increasingly replace 'preexisting modes of economic coordination' and business models (Casilli and Posada, 2019: 294), primarily through precarious working conditions for the so-called gig workers in developing countries.

In addition to the technological transformations and their effects on the ICT4D research, global trends and events such as the COVID-19 pandemic have also significantly affected the field. The pandemic not only paved the way for a boom in using tracking and surveillance technologies to control the disease spread but it also accelerated digitalisation and datafication of the health sector, education, public services, and alike in an unprecedented way. These rapid changes and the responses to them in developing countries were addressed in the IFIP 9.4's 2021 virtual conference on 'Resilient ICT4D,' scrutinising the paradigmatic shift in understanding the role of ICTs in development strategies, developing countries, and their crisis responses.

Critique

Although the majority of ICT4D research has been focusing on the ICT part of the phrase, the criticism against the conceptualisation and theorisation of the D has increased sharply. Similar to the development critique mentioned in the book at hand, the ICT4D researchers and practitioners also criticised the 'top-down, urban-based, centralised model of development' (Unwin, 2009: 11) in their field. Additionally, people-centric approaches to social change increasingly question the utopian accounts of the digital society and the empty promise of overcoming structural inequalities through technocentric methods (Toyama, 2015). Consequently, ICT4D research has been criticised for the lack of 'academic rigor and quality contributions that could help practitioners', 'lack of critical perspectives and critical theories in research (e.g., post-colonial theory)', and insufficient 'impact assessment of ICT4D projects and long-term studies' (Schelenz and Pawelec, 2021: 6). Inadequate innovative research, practical approaches, and inter-, trans-, and multidisciplinary exchange (Schelenz and Pawelec, 2021: 5) are also reflected in the politics of ICT4D research, where researchers are insufficiently involved in policy formulation, indigenous and local voices are not integrated into research community and policymaking, and impact-focused research for the improvement of marginalised people's situation is scarce (Schelenz and Pawelec, 2021: 6).

The ICT4D practices are also under criticism for being short-term, narrow, not sustainable, or generally for the issues of 'scalability, accountability', and ethics (Schelenz and Pawelec, 2021: 8). Other issues such as 'power inequalities, corporatization of ICT4D, and top-down stakeholderism' in ICT4D practices (Schelenz and Pawelec, 2021: 8) might contribute to reinforcing existing structural inequalities. The role and place of the users in ICT4D practices is an important point of criticism since user participation is not sufficiently integrated into the project design, implementation, and evaluation. Although ICT4D could be considered an inherently interconnected area among academia, policymaking, and civic engagement, the gap between theory and practice is still a pressing issue in the ICT4D community (Unwin, 2009). The new direction of ICT4D tries to distance itself from its Eurocentric and modernity bias (Schelenz and Pawelec, 2021: 10) and its technosolutionism – prioritising technological solutions instead of tackling fundamental social, political, and economic challenges.

ICT4D future

ICT4D aims to integrate ICTs in development strategies, planning, implementation, and evaluation as well as to minimise the digital divide – the distinction between those who have 'access to ICTs, as well as the related skills that are needed to take part in the information society' (Eurostat, 2019). This section has offered a chronological account of ICT4D research along with a thorough categorisation of the critique. However, the future direction of ICT4D is significantly dependant on the rapid technological progress in the areas of platformaisation and datafication. These techno-social changes pave the way for a paradigm shift from ICT4D to a more all-encompassing concept of 'digital-for-development' (Heeks, 2020). These recent accounts are inspired by critical development theories and new approaches to society and technology studies. The insufficiency of setting global development plans, especially in the dynamic field of ICTs, international crises such as the COVID-19 pandemic and the climate change, and the increasing use of surveillance technologies for authoritarian purposes in developing countries have instigated the need for a more inclusive understanding of ICTs in the context of our globalised societies. For an ethical, inclusive future, ICT4D research needs to reflect critically on the entanglements between the sociopolitical changes and technological advances, not only in the context of individual countries but also amid technological streams and movements between the Global South and North.

References

Casilli, A. A. & Posada, J., 2019. The Platformization of Labor and Society. In: M. Graham & W. H. Dutton, eds. *Society and the Internet: How Networks of Information and Communication are Changing Our Lives*. Oxford: Oxford University Press, pp. 293–306.

Eurostat, 2019. *Glossary: Digital Divide* [Online]. Available at: https://ec.europa.eu/eurostat/statistics-explained/index.php?title=Glossary:Digital_divide [Accessed 28 December 2021].

Gillard, H., Howcroft, D., Mitev, N. & Richardson, H., 2008. "Missing Women": Gender, ICTs and the Shaping of the Global Economy. *Information Technology for Development*, 4, 262–279.

Heeks, R., 2020. ICT4D 3.0? Part 1 – The Components of an Emerging "Digital-for-Development" Paradigm. *The Electronic Journal of Information Systems in Developing Countries (EJISDC)*, 86(3).

Mayer-Schoenberger, V. & Cukier, K., 2013. *Big Data: A Revolution that Will Transform How We Live, Work, and Think*. London: John Murray Publishers.

Schelenz, L. & Pawelec, M., 2021. Information and Communication Technologies for Development (ICT4D) Critique. *Information Technology for Development*, 28(1), 165–188.

Taylor, L., 2017. What Is Data Justice? The Case for Connecting Digital Rights and Freedoms Globally. *Big Data & Society*, 4(2).

Toyama, K., 2015. *Geek Heresy*. New York: Public Affairs.

UN Commission on Science and Technology for Development, 1997. *Report of the Working Group on Information and Communication Technologies for Development* [Online]. Available at: https://digitallibrary.un.org/record/233314?ln=en#record-files-collapse-header [Accessed 28 December 2021].

United Nations, n.d. *Department of Economic and Social Affairs: Sustainable Development* [Online]. Available at: www.sdgs.un.org [Accessed 28 December 2021].

Unwin, T., 2009. Development Agendas and the Place of ICTs. In: T. Unwin, ed. *ICT4D: Information and Communication Technology for Development*. Cambridge: Cambridge University Press, pp. 7–38.

Walsham, G., 2017. ICT4D Research: Reflections on History and Future Agenda. *Information Technology for Development*, 23(1), 18–41.

Suggested Further Reading

International Federation of Information Processing's (IFIP) working group on the implications of information and digital technologies for development (IFIP 9.4), 2013–present. *Conference Proceedings* [Online]. Available at: http://ifiptc9.org/wg94/ifip-9-4-conferences/

The Principles for Digital Development Forum, 2015 (ongoing). *Principles for Digital Development* [Online]. Available at: https://digitalprinciples.org/

Dearden, A. & Kleine, D., 2018. *Minimum Ethical Standards for ICTD/ICT4D Research* [Online]. Available at: https://ictdethics.wordpress.com/

Three journals specifically publish papers on the ICT4D research and practice.

- Journal of Information Technology for Development since 1986
- Electronic Journal of Information Systems for Developing Countries since 2000 [open access]
- Journal of Information Technologies and International Development since 2003 [open access]

Chapter 96

Decolonising global health

Julia Ngozi Chukwuma

Introduction

People across the world enjoy different degrees of health. To some extent, this may be the consequence of 'biological' attributes, such as women rather than men experiencing period pain or the elderly being more likely to suffer from a non-communicable disease. Yet, what is problematic is the immense disparity in health outcomes of different people experiencing the same condition. For example, maternal deaths of Black mothers in the US are 2.5 times higher compared to White mothers (MacDorman et al., 2021). There is no plausible biological explanation why this would be the case. Rather this is an example of disparities in health outcomes between different groups of people, which are not only unfair but the product of systemic inequalities and injustices that privilege some groups of people over others. Such unjust and avoidable inequalities in health and access to health-care also exist between countries with, e.g., the probability of a woman living in a low-middle income country surviving pregnancy and delivering her child being much lower in comparison to a woman living in a high-income country (Yaya and Ghose, 2019).

Addressing health inequalities is, in theory, at the heart of the field of global health, concerned with the health needs of all people across the world as opposed to the health needs of people within a particular nation only (Brown et al., 2007). However, the COVID-19 crisis highlighted afresh existing biases within global health which disproportionally accords attention to the health concerns of people living within countries of the Global North (Yegros-Yegros et al., 2020). At the same time, the pandemic has shown the underestimation of global health responses and practices of countries of the Global South (Fofana, 2021). These biases make it imperative to engage with the history of global health and its precursors (colonial medicine, tropical medicine, and international health) as the basis for engaging in health policymaking, which is genuinely committed to health equity and finding just solutions to global health challenges.

Defining global health

Although a commonly used term today, a consensus on how best to define global health remains elusive. According to some scholars, the field of global health can be divided into two regimes (see, notably, Lakoff, 2010). On the one hand, the global health security regime is concerned with cross-border threats to health (mainly interested in keeping wealthy countries safe from infectious diseases feared to emanate from the Global South). On the other hand, the humanitarian biomedicine regime focuses on alleviating

DOI: 10.4324/9780429282348-105

individuals' suffering regardless of where they are based or to which social group they belong, yet focusing particularly on health problems in countries of the Global South (Lakoff, 2010).

Breaking away from the division into two regime types, proponents and activists committed to adopting a social justice standpoint conceptualise global health yet again differently, putting health equity at the centre of global health. From this perspective, the field of global health regards health as a human right and is concerned with evoking policies that seek to put into place universal and equitable health systems, which enable *all people*, within both lower and higher-income countries, to access medical solutions and healthcare (Abimbola, 2018; Birn et al., 2017; Taylor, 2018). Perhaps most importantly, this perspective pays attention to the uneven distribution of power and resources and intersecting social inequalities (Holst, 2020). From this position, global health differs decisively from international health (the term more commonly used until the 2000s), with the former being characterised by its commitment to health equity and social justice rather than merely focusing on addressing health issues in lower-income countries on a case-by-case basis (Farmer, 2004). While the focus of international health throughout the 20th century was the transfer of solutions to health problems from higher to lower-income countries, global health recognises the existence of health inequalities everywhere in the world, including in higher-income countries (Abimbola, 2018). In order to appreciate existing challenges and diverging views within the field of global health of today, it is crucial to understand its colonial roots and to re-historicise global health.

The colonial roots of global health and its after-effects

With the COVID-19 pandemic and the resurgence of the Black Lives Matter (BLM) movement following the brutal murder of George Floyd in 2020, endeavours to decolonise global health have returned to centre stage (Hirsch, 2021). Global health evolved from colonial and tropical medicine – which, in the 19th and 20th centuries, sought to keep White European settlers living in colonies healthy and local Brown and Black populations productive (Abimbola, 2018; Hirsch, 2021; Holst, 2020; Khan et al., 2021). At the same time, colonial and tropical medicine served as a platform to discuss and conduct research on clinical and sanitation issues pertinent to Europeans living in African and Asian colonies (Manton, 2011).

At the beginning of the 20th century, colonial and tropical medicine metamorphosed into international health, notably with the establishment of the International Health Commission (IHC) by the Rockefeller Foundation in 1913. At the time, the imperative was to pursue disease control interventions – such as the hookworm or yellow fever campaigns (Packard, 2016) – and to put in place a worldwide system of infectious disease control in order to keep rich nations secure from diseases seen to more commonly originate in poorer regions of the world (Holst, 2020). Since then, international health was replaced in common discourse by global health. It maintained its commitment to global health security, which often happens to the detriment of its mandate of humanitarian biomedicine and, crucially, the wider need to address structural inequalities and injustices determining global health inequities (O'Laughlin, 2016).

More generally, there has been a shift in how health policy, as an area of social policy, is conceptualised. In the 1970s, many African countries that had newly gained their independence from colonial powers perceived the provision of social services as a way of supporting their nation-building efforts (Adesina, 2015; Kpessa et al., 2011). Notably, the focus was on assuring the public provision of comprehensive primary healthcare services to every person. However, attaining health for all was deemed only possible, if, simultaneously, global efforts were aimed at fighting neo-colonialism and imperialism and correcting international inequities and injustice. As a result, the *Declaration on the Establishment of a New International Economic Order* was adopted in 1974 with the explicit ambition to fight neo-colonialism and imperialism (UNGA, 1974). A few years later, in 1978, the *Declaration of Alma-Ata* formally stipulated the right to health for all, echoing the political climate at the time.

Still, the consensus achieved at the conference during which the Alma-Ata declaration was signed did not stand strong for long. Critical voices emerged soon, worrying that the ideal of universal, comprehensive primary healthcare for all by 2000 was not affordable (Stuckler et al., 2010). Subsequently, many governments turned away from the Alma-Ata principles, notably with the rise to prominence of neo-liberal ideology and the ensuing push for budgetary discipline throughout the 1980s. Since then and up until today, the privatisation of healthcare service delivery has been pushed forward by proponents, who believe that market mechanisms can more economically and efficiently provide public goods and services. For instance, a report published in 2019 by a consortium of progressive NGOs picked up on this reality, highlighting that discussions around strengthening public-private-partnerships are being held in 50 developing countries; and, in spite of absent proof of their advantageousness, 'there has been a large push by the multilateral banks and the IMF . . . to leverage private sector investments for development purposes' (Ortiz and Cummins, 2019: 43).

This trend towards the continuous privatisation of social service delivery, however, continues to be met with caution. Notably, in April 2019, the African Union member states came together in Egypt, adopting a resolution voicing concern regarding the reality that

> the growth of private actors' involvement in health and education services delivery often happens without the consideration of human rights resulting in growing discrimination in access to these services, a decrease in transparency and accountability, which negatively impact the enjoyment of the rights to health and education.[1]

Crucially, the resolution highlights African national governments' discontent about the mounting pressure of 'bilateral donors and international institutions . . . to privatize or facilitate access to private actors in . . . health and education sectors'.

Global health in a globalised world

Global health during times of neo-liberal capitalism oftentimes fails to prioritise strategies striving to guarantee the right to health and to promote health equity and social justice (Taylor, 2018). More commonly, today's (so-called) global health efforts are frequently spearheaded by private foundations or international organisations. These are targeted at 'fixing technical problems', which, supposedly, requires biomedical solutions and the transfer of knowledge, technology, and (alleged) best practices from the Global North to the Global South (Birn et al., 2017; Packard, 2016). By doing so, contemporary leading global health institutions perpetuate colonial attitudes, concentrating expertise and resources in institutions of the Global North to 'deliver solutions' to poorer regions while reinforcing existing power imbalances simultaneously (Khan et al., 2021).

Moreover, the shortcoming of this technical (as opposed to a more system-focused) approach has been made visible by the COVID-19 health emergency. Industrialised countries are believed to have passed through an epidemiological transition, where ill-health is more likely to occur as a result of old-age and lifestyle than because of exposure to an infectious pathogen (O'Laughlin, 2016). COVID-19, however, has highlighted how easily a virus can be transmitted across the world, illustrating the danger to everyone of failing to address global health inequalities. At the same time, the pandemic has highlighted (and exacerbated) existing inequalities that appear to be the outcome of the current capitalist economic model and contemporary power relations (Stevano et al., 2021). While African countries may not have had similar headcounts of people dying from COVID-19 per se, the economic impact of the crisis is hitting lower-income countries hardest and, within these countries, the most disadvantaged people. Capital outflow, production and supply chain disruptions, mounting debt, and imposed austerity measures are likely to negatively impact people's health status, livelihood, and well-being (World Bank, 2022). Moreover, the reality that, as an example, the pharmaceutical company *Pfizer* sold 80 per cent of its vaccine doses, which it was able to produce until the end of 2021, to mainly rich governments within days of announcing the existence of an effective vaccine – despite calls by civil societies across the globe to ensure a fair and just distribution of a potential COVID-19 vaccine – is a blatant demonstration of the unwillingness of rich nations to make global health truly 'global'.[2] While rich nations in the West continued to stockpile vaccines and expand their vaccination plans to younger segments of the population, the most vulnerable people in the Global South remained without access to the vaccine.

Conclusion

The legacy of colonialism remains palpable in global public health today. During colonialism, European colonial powers, rooted in racist beliefs, made use of Africa and Africans for their medical experimentation (aside from causing ill-health amongst Africans due to colonial exploitation, including, for instance, through forced work in mines; Tilley, 2016). Recent remarks by French medical researchers – who suggested testing the extent to which the Bacillus-Calmette-Guérin/tuberculosis vaccine could protect people from COVID-19 in Africa 'where there are no masks, no treatment, no intensive care'[3] – reminded many of a such an attitude (Affun-Adegbulu and Adegbulu, 2020). Moreover, the relatively widespread astonishment about lower rates of COVID-19-related casualties on the African continent and the disbelief that this, to some extent, may have been the result of appropriate and timely policy response by (some) African governments furthermore demonstrates the persistence of attitudes of 'Western' superiority. Against this backdrop, it is crucial to ensure that health equity and social justice concerns remain at the centre of global health practice and rhetoric. Furthermore, it is critical to continue fighting skewed power dynamics relating to leadership and knowledge production as well as prejudiced views, while acknowledging the legacy and impact of colonialism, racism, sexism, and exploitative capitalism on health equity – all of which is at the heart of the *decolonising global health* agenda (Büyüm et al., 2020; Khan et al., 2021).

Notes

1 The resolution is available here: https://www.achpr.org/sessions/resolutions?id=444 (Accessed: 31/05/2022).
2 See, for instance, Ghosh's analysis of "vaccine apartheid": https://www.project-syndicate. org/commentary/pfizer-vaccine-doses-claimed-by-rich-countries-weakens-covax-by-jayati-ghosh-2020-11 (Accessed: 31/05/2022).
3 Verbatim, Dr Jean-Paul Mira, exchanging with Professor Camille Locht, was reported to have said: "If I can be provocative, shouldn't this study be done in Africa, where there are no masks, no treatment, no intensive care, a bit like it is done in some studies on AIDS or among prostitutes. We try things because we know they [sex workers] are highly exposed and they don't protect themselves. What do you think about that?" (See: https://hbr.org/2020/09/white-coats-black-scientists [Accessed: 31/05/2022]).

References and further readings

Abimbola, S., 2018. On the meaning of global health and the role of global health journals. *International Health* 10, 63–65. https://doi.org/10.1093/inthealth/ihy010

Adesina, J.O., 2015. Return to a wider vision of social policy: Re-reading theory and history. *South African Review of Sociology* 46, 99–119. https://doi.org/10.1080/21528586.2015.1077588

Affun-Adegbulu, C., Adegbulu, O., 2020. Decolonising global (public) health: From Western universalism to global pluriversalities. *BMJ Global Health* 5, e002947. https://doi.org/10.1136/bmjgh-2020-002947

Birn, A.-E., Pillay, Y., Holtz, T.H., 2017. *Textbook of Global Health, Textbook of Global Health*. Oxford: Oxford University Press.

Brown, T.M., Cueto, M., Fee, E., 2007. The World Health Organization and the transition from 'international' to 'global' health. In: Bashford, A. (Ed.), *Medicine at the Border: Disease, Globalization and Security, 1850 to the Present*. London: Palgrave Macmillan UK, pp. 76–94. https://doi.org/10.1057/9780230288904_5

Büyüm, A.M., Kenney, C., Koris, A., Mkumba, L., Raveendran, Y., 2020. Decolonising global health: If not now, when? *BMJ Global Health* 5, e003394. https://doi.org/10.1136/bmjgh-2020-003394

Farmer, P., 2004. Global health equity. *AMA Journal of Ethics* 6, 191–193.

Fofana, M.O., 2021. Decolonising global health in the time of COVID-19. *Global Public Health* 16, 1155–1166. https://doi.org/10.1080/17441692.2020.1864754

Hirsch, L.A., 2021. Is it possible to decolonise global health institutions? *Lancet London and England* 397, 189–190. https://doi.org/10.1016/S0140-6736(20)32763-X

Holst, J., 2020. Global health – emergence, hegemonic trends and biomedical reductionism. *Global Health* 16. https://doi.org/10.1186/s12992-020-00573-4

Khan, M., Abimbola, S., Aloudat, T., Capobianco, E., Hawkes, S., Rahman-Shepherd, A., 2021. Decolonising global health in 2021: A roadmap to move from rhetoric to reform. *BMJ Global Health* 6, e005604. https://doi.org/10.1136/bmjgh-2021-005604

Kpessa, M., Béland, D., Lecours, A., 2011. Nationalism, development, and social policy: The politics of nation-building in sub-Saharan Africa. *Ethnic and Racial Studies* 34, 2115–2133. https://doi.org/10.1080/01419870.2011.574717

Lakoff, A., 2010. Two regimes of global health. *Humanity: An International Journal of Human Rights, Humanitarianism, and Development* 1(1). https://doi.org/10.1353/hum.2010.0001

MacDorman, M.F., Thoma, M., Declcerq, E., Howell, E.A., 2021. Racial and ethnic disparities in maternal mortality in the United States using enhanced vital records, 2016–2017. *American Journal of Public Health* 111, 1673–1681. https://doi.org/10.2105/AJPH.2021.306375

Manton, J., 2011. Tropical medicine. In: Berridge, V., Gorsky, M., Mold, A. (Eds.), *Public Health in History*. Maidenhead, Berkshire: Open University Press.

O'Laughlin, B., 2016. Pragmatism, structural reform and the politics of inequality in global public health. *Development and Change* 47, 686–711. https://doi.org/10.1111/dech.12251

Ortiz, I., Cummins, M., 2019. *Austerity the New Normal: A Renewed Washington Consensus 2010–24*.Working Paper October 2019. IPD, ITUC, PSI, EURODAD and BWP.

Packard, R.M., 2016. *A History of Global Health: Interventions into the Lives of Other Peoples*, Illustrated edition. Baltimore: Johns Hopkins University Press.

Rondinelli, D.A., Mccullough, J.S., Johnson, R.W., 1989. Analysing decentralization policies in developing countries: A political economy framework. *Development and Change*, 57–87.

Stevano, S., Franz, T., Dafermos, Y., Van Waeyenberge, E., 2021. COVID-19 and crises of capitalism: Intensifying inequalities and global responses. *Canadian Journal of Development Studies/ Revue canadienne d'études du développement* 42, 1–17. https://doi.org/10.1080/02255189.20 21.1892606

Stuckler, D., Feigl, A.B., Basu, S., McKee, M., 2010. The political economy of universal health coverage. Presented at the First global symposium on health systems research, Montreux.

Taylor, S., 2018. 'Global health': Meaning what? *BMJ Global Health* 3, e000843. https://doi.org/10.1136/bmjgh-2018-000843

Tilley, H., 2016. Medicine, empires, and ethics in Colonial Africa. *AMA Journal of Ethics* 18, 743–753. https://doi.org/10.1001/journalofethics.2016.18.7.mhst1-1607

UNGA, 1974. *Declaration on the Establishment of a New International Economic Order*. New York: United Nations General Assembly.

World Bank, 2022. *World Development Report 2022: Finance for an Equitable Recovery*. Washington, DC: The World Bank.

Yaya, S., Ghose, B., 2019. Global inequality in maternal health care service utilization: Implications for sustainable development goals. *Health Equity* 3, 145–154.

Yegros-Yegros, A., van de Klippe, W., Abad-Garcia, M.F., Rafols, I., 2020. Exploring why global health needs are unmet by research efforts: The potential influences of geography, industry and publication incentives. *Health Research Policy and Systems* 18, 47. https://doi.org/10.1186/s12961-020-00560-6

Chapter 97

Navigating the institutional gaps, mismatch, and neglect

Exploring the landscape of non-communicable diseases in the developing countries

Pallavi Joshi and Dinar Kale

1.1 Introduction

This chapter discusses how institutional gaps act as barriers in addressing access to health technologies for noncommunicable diseases in low-and middle-income countries (LMICs) by using the cases of cancer care in India and African countries.

In the last few decades, non-communicable diseases have emerged as the leading cause of mortality at the global level, reflecting epidemiological and demographic transitions and underlying exposure changes in risk factors. To put this into perspective, total NCDs mortality is estimated to be 41 million which is 74.37 per cent of global mortality (56.5 million) in 2019 (GBD, 2019). The burden of NCDs is disproportionately higher in LMICs, mainly as 85 per cent of premature mortality occur in these countries, i.e., deaths between ages 30 and 70 years from cardiovascular diseases, cancer, diabetes, and chronic respiratory diseases.[1] In terms of cancer, the case in point, the rising burden of premature mortality has even more serious concerns for LMICs. It is observed that LMICs record a lower incidence of cancer compared to advanced countries, however, the mortality rate is much higher (GBD, 2019; IARC, 2020). This 'cancer paradox' is attributed to resource constraints present in LMICs that hinder the early detection of cancers resulting in around 70 per cent of cancer patients being diagnosed at advanced stages when treatment is either too expensive to continue or no longer effective (IARC, 2020).

The literature suggests that timely availability of and access to affordable early detection methods lies at the heart of creating an effective treatment for cancer (World Health Organization, 2017). However, institutional gaps in the form of mismatch in health, industrial and regulatory policies pose a significant barrier to accessing key health technologies for cancer (Ganju et al., 2020; Mackintosh et al., 2018; Srinivas, 2012). Scholars linking health and industrial policies for different countries suggest that government interventions in the health sector can promote industrial development in the pharmaceutical sector and make health policies more effective (Banda et al., 2021). The demand patterns and investment incentives caused by healthcare and health policies provide an 'implicit' industrial policy for manufacturers of pharmaceuticals and medical products (Mackintosh et al., 2016). However, in case of LMICs, the institutional gaps have either weakened the effectiveness of such linkages or propelled their siloed evolution.

In the next two sections, this chapter discusses some contrasting scenarios demonstrating how the impact of institutional gaps in access to appropriate health technologies different in India and Africa and some emerging approaches to mitigate the institutional gaps in access to cancer care. This chapter concludes by highlighting efforts such as

DOI: 10.4324/9780429282348-106

'Choosing Wisely' as one of the key perspectives for developing value-based accessible solutions that are relevant for treating NCDs in LMICs

1.2 Institutional gaps in accessing effective care for cancer in LMICs

The area of institutional gaps affecting effective cancer care in LMICs can be highlighted using Srinivas' (2012) qualitative heuristic of a 'triad' of institutional domains, production, consumption, and delivery (Figure 97.1).

This heuristic helps to define the linkages between health and industry by pointing out the strong of role of these domains in development of health sector performance and industrial growth. However, the cumulative interactions among these three domains remain neglected in health system research and industrial policy analysis. Using this framework, our research exploring issues of effective cancer care in Africa point outs that gaps and failures in the relationship between the three domains has significant impact on facilitating the access to potentially life-saving technologies and essential care for local populations. The majority of African countries rely on costly imports to satisfy local needs of drugs, vaccines, diagnostics, and devices used for the treatment of cancer and other NCDs creating significant issues of affordability. For instance, Kenya is entirely import dependent for cancer medicines, medical devices, and supplies, with large gaps in need, demand, and access to cancer care. This lack of local manufacturing and heavy reliance on imports creates an additional constraint on the access to diagnosis and treatment due to supply shortages of essential medicines and medical devices in African countries (Banda et al., 2021). Further, the lack of broader industrial base and local presence of multinational companies makes it challenging to maintain and repair high-tech diagnostic equipment in the local primary healthcare centres. These gaps have been more prominent during the supply chain disruptions and challenges accessing healthcare during the COVID-19 pandemic. Therefore, substantial disintegration at all three domains persists with respect to cancer care. There is need to broaden the health industries' industrial base in Africa, including in medical devices and diagnostics for effective cancer care (ibid).

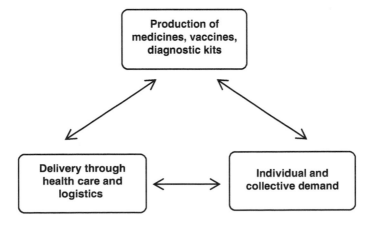

Figure 97.1 The triad

A recent study suggests that enhanced health-industry linkages can play a significant role in tapping the underutilized Kenyan industrial and innovation potential to improve cancer care (Mandaku et al., 2022). Towards this end, one of the key steps would be consolidating demand by promoting local procurement of essentials health technologies that are competitively produced in Kenya (ibid). While some signs of local manufacturing are emerging in African countries, strong institutional changes are required to combat heavy reliance on imports and cases of donated medical device and diagnostics that are incompatible with local conditions, health needs, and healthcare delivery systems.

The Indian scenario presents a contrasting challenge compared to the African countries. In the case of India, it is often argued that despite being a leading manufacturer of medicines and receiving international recognition as the pharmacy of the Global South, access to medicines remain a significant challenge in India (Chaudhuri, 2007). Srinivas (2012) points out the lack of effective integration between interventions for industrial production with both demand and delivery as a root cause of access issues. Even in terms of medical device and diagnosis, the highly stringent procurement policies in India created challenges in accepting frugal devices in the healthcare system and discouraged local innovators. Furthermore, for a long period of time there have been inconsistent policies and the lack of an appropriate regulatory environment, which has had a negative impact in creating positive synergies between local technological innovation and better access to diagnosis and treatment (Srinivas and Kale, 2022). The failures in interlinking technology development, markets, and procurement policies affect affordability, reliability, and ultimately access to cancer care. However, in recent years, some convergence in the science, technology and innovation, health, and industrial policies in cancer care, along with some key strides in regulatory processes, have channelled all three domains of the triad. This change is mainly observed with the emerging ecosystem for medical technologies for cancer care that focus on addressing unmet clinical needs and increasing affordable cancer diagnosis and management in India (Joshi, 2022).

1.3 A novel approach to tackle institutional bottlenecks – a way forward

There is a need for innovative solutions to address these challenges to enhance access to locally relevant health technologies for cancer care in a sustainable manner. One such pathway is a stronger collaborative arrangement among different stakeholders and emphasising the need for bridging the institutional gaps. It is possible to broaden and create access to cancer care through a strategy that links health sector innovation to local industrial investment, innovation, and technological upgrading, supported by active collaboration among LMICs. Sullivan et al. (2017: 328) point out,

> Cancer 'moonshots' may improve individual outcomes in high-income countries with strong governance, but they will not solve the rising economic and social burden of cancer globally. What we need are 'earthshots' that focus on building infrastructure and delivering affordable, equitable and effective care.

This understanding has spurred the establishment of the 'Choosing Wisely' perspective and work of National Cancer Grid in India. National Cancer Grid is an interesting evidence-based interdisciplinary collaboration of major cancer centers, research institutes, patient groups, and charitable institutions that have facilitated shared decision-making to identify patient-centric practices and overuse of unnecessary and potentially harmful investigations and treatments to patients (Pramesh et al., 2019). The Choosing Wisely India project of the National Cancer Grid involved Indian representatives from surgical, medical, and radiation oncology from public and private sectors and representatives from patient advocacy groups. Using a multidisciplinary framework-driven consensus process they identified low-value or potentially harmful practices relevant to the Indian cancer healthcare system to facilitate high-quality, value-based cancer care in India (ibid). Such collaborative actions provide a robust framework and facilitate uniform standards of patient care for prevention, diagnosis, and treatment of cancer, specialized training and education in oncology, and collaborative basic, translational, and clinical research in cancer. Hence, these 'earthshots' could provide a resolution to the perineal conundrum of lack of access to appropriate health technologies faced by policymakers in resource constraints by channelling the industrial capabilities to address local health needs

1.4 Conclusion

This chapter focused on the institutional challenges associated with access to health technologies for non-communicable diseases, specifically cancer care in LMICs. Using the case of India and African countries, this chapter highlights the lack of linkages among the three domains of production, delivery, and demand as the main hindrance in creating access to essential care for local populations. In addressing some of these institutional challenges, we highlight the need for stronger collaborative arrangements among different stakeholders and the importance of innovative solutions rooted in the local industrial capacities. Stressing the significance of 'earthshots', we conclude that efforts like 'Choosing Wisely' are some of the key perspectives for developing sustainable and value-based accessible solutions that are relevant for treating NCDs in LMICs.

Note

1 https://www.who.int/data/gho/indicator-metadata-registry/imr-details/3411 (last accessed on 10 March 2022).

References

Banda, G., Mugwagwa, J., Wanjala, C., Mackintosh, M., & Kale, D. (2021). Local manufacturing, local supply chains and health security in Africa: Lessons from COVID-19. *BMJ Global Health*, 6(6), e006362.

Chaudhuri, S. (2007). The gap between successful innovation and access to its benefits: Indian pharmaceuticals. *The European Journal of Development Research*, 19(1), 49–65.

Ganju, A., Goulart, A. C., Ray, A., Majumdar, A., Jeffers, B. W., Llamosa, G., . . . Wang, Y. P. (2020). Systemic solutions for addressing non-communicable diseases in low-and middle-income countries. *Journal of Multidisciplinary Healthcare*, 13, 693.

Global Burden of Disease Study. (2019). *Global Burden of Disease Study 2019 (GBD 2019) Results*. Seattle, United States: Institute for Health Metrics and Evaluation (IHME).

IARC. (2020). World cancer report – cancer research for cancer prevention. In C. P. Wild, E. Weiderpass, & B. W. Stewart (Eds.), *International Agency for Research on Cancer*. Lyon.

Joshi, P. (2022). Studying development and adoption of inclusive innovations for the early detection of cancer in India: An evolutionary perspective. Conference paper, Globelics Academy 2022, Tampere, Finland.

Mackintosh, M., Mugwagwa, J., Banda, G., Tibandebage, P., Tunguhole, J., Wangwe, S., & Karimi Njeru, M. (2018). Health-industry linkages for local health: Reframing policies for African health system strengthening. *Health Policy and Planning*, 33(4), 602–610.

Mackintosh, M., Tibandebage, P., Kungu, J. K., Njeru, M. K., & Israel, C. (2016). *Health Systems as Industrial Policy: Building Collaborative Capabilities in the Tanzanian and Kenyan Health Sectors and Their Local Suppliers. In Making Medicines in Africa* (pp. 147–165). London: Palgrave Macmillan.

Manduku, V., Wanjala, C., Njeru, M. K, Mokua, S., Nyandieka, L., & Mackintosh, M. (2022). Strengthening Kenyan industrial development to improve cancer care. *ICCA Policy Brief*.

Pramesh, C. S., Chaturvedi, H., Reddy, V. A., Saikia, T., Ghoshal, S., Pandit, M., Babu, K. G., Ganpathy, K. V., Savant, D., Mitera, G., Sullivan, R., & Booth, C. M. (2019). Choosing Wisely India: Ten low-value or harmful practices that should be avoided in cancer care. *The Lancet Oncology*, 20(4), e218–e223.

Srinivas, S. (2012). *Market Menagerie: Health and Development in Late Industrial States*. Stanford: Stanford University Press.

Srinivas, S., & Kale, D. (2022). New approaches to learning and regulation in medical devices and diagnostics: Insights from Indian cancer care. *Innovation and Development*, 1–24.

Sullivan, R., Pramesh, C. S., & Booth, C. M. (2017). Look beyond technology in cancer care. *Nature*, 549(7672), 325–328.

World Health Organization. (2017). *Guide to Cancer – Early Diagnosis*. Geneva: World Health Organization.

Chapter 98

What vocational education and training for what development?

Simon McGrath

Throughout human history people have learnt to do things, such as hunt, gather, farm and weave, and in the process they have become more integrated and integral members of their communities. Over time, these processes became more strongly formalised in many parts of the world through structures such as apprenticeships and guilds. With the maturing of industrialisation, community learning structures were joined by state-organised systems of formal vocational education and training (VET). Emerging in the first industrial nations, these formal VET systems were then exported to colonies and ex-colonies through the second half of the 20th century as part of industrialisation drives.

Whilst all national VET systems in reality are palimpsests of multiple VET traditions, the commonly understood 'national VET systems' of the present day are predominantly designed to serve a theory of development that is centred on state-led industrialisation, overlaid with neoliberal governance reforms from the 1990s on. Yet, neither big industrial development nor the Washington Consensus approach are viable development models for the current moment, let alone for the future.

What then might vocational education and training look like if it is to be responsive to more viable development theories? Space precludes any systematic and detailed analysis, so I will briefly offer four development lenses through which VET might be viewed differently.

A revised notion of VET responsiveness

Public VET, especially in the South, is roundly criticised for being unresponsive to labour market realities. For the largely neoliberal proponents of this view, this means an unresponsiveness to formal employment opportunities, in spite of the very limited nature of such opportunities in cases where the age cohort of school leavers is as big as the formal economy and many times the size of the demand for new workers.

Noting that massified VET systems face a major problem of numbers is not to argue that there is no place for high-quality public VET that does respond to formal labour market opportunities. However, a better notion of responsiveness can be found by drawing on the evolutionary economics tradition (e.g., Nelson and Winter, 1982), which has been revisited by a number of development economists (e.g., Chang and Grabel, 2014). Evolutionary economics, and, specifically, the national innovation systems approach strongly associated with it, offers distinct benefits in conceptualising VET's developmental role due to its stress on the importance of education, skills, work, innovation and production for economic development. Importantly, the focus on organisations and networks

DOI: 10.4324/9780429282348-107

offers a new scalar level and methodological purchase on the VET-economic development relationship. Together, these theoretical and methodological lenses offer very different policy implications and possibilities when compared to more conventional neoclassical approaches (Kruss et al., 2015).

A focus on competences and capabilities (in the innovation literature sense) enriches an analysis of how responsive provider institutions are by giving a set of new analytical categories. Moreover, the attention of the innovation account on organisational learning helps move from the deficit-laden snapshot of a responsiveness survey towards a richer sense of institutions that are dynamic, even if that dynamism is fallible.

Furthermore, the insight of the innovation systems approach that organisations should be understood in a dynamic and relational manner crucially moves our thinking away from the atomised understanding of the responsiveness orthodoxy. It draws our attention to how VET institutions interact with other actors and reminds us that they exist within wider networks, both of national (and provincial) policies and structures and of firms, universities, private training providers and, crucially, sectoral intermediary organisations. It directs us to an analysis of how these constellations have evolved over time; what place they give to public VET provision; and what potential there is for strengthening that place. The focus on intermediary organisations reinforces the importance of looking at ways of building capacity and communication at institutional, place-based and sectoral levels.

This all has practical implications for thinking about how to improve responsiveness at institutional and VET system levels. It reiterates the need to look at how public providers are responsive but offers important new insights into what that responsiveness might consist of that is more about capacity development, learning and interaction within wider sectoral networks. This has potentially critical implications for how colleges are funded, staffed and organised and about their degrees of autonomy. It also highlights the need to think about sectoral intermediary organisations as crucial to college capacity development rather than a tradition in which they are largely marginal if not invisible in the calculations of VET policymakers. This emphasises the dual logics of sector and region in thinking about skills and innovation and this also has implications for policy processes that seem to only be conceived in terms of the centralisation-decentralisation axis.

Majority labour markets

Whilst the previous section argues that there is a place for innovative thinking about high quality public VET provision for the formal sector in Southern contexts, the majority of young people are likely to find work in the informal or subsistence economies in many countries.

After a flowering of literature on VET and the informal sector in the 1980s, there has been almost no attention to either rural subsistence or urban informal sectors and VET in the last quarter-century.

Nonetheless, some literature does exist. Work by authors such as Zeelen et al. (2010) stresses the need to locate VET debates in rural settings in the wider contexts of capitalist development and its effects on rural underdevelopment and experiences of educational exclusion as a result of the failings of the education for all movement. By focusing on how young people experience both labour markets and education systems, it is argued it is possible to think about how alternative systems can be developed that build from existing knowledges and viable opportunities for craftsmanship and entrepreneurship.

There is also some very recent work that is beginning to revisit skills in informal economies. Importantly, this literature both places greater emphasis than the earlier tradition on structural dynamics – such as gender and poverty – and the persistent effects of colonialism. DeJaeghere (2019) also usefully insists on the need to distinguish between types of entrepreneurship. She argues that much of what is provided in the South as skills for entrepreneurship actually misreads the nature of entrepreneurship being experienced. Rather than a form of Northern 'opportunity entrepreneurship' where educated, middle class youth are empowered by the state and other structures to follow their entrepreneurial dreams; the reality for much of the South is one of 'necessity entrepreneurship'. Whilst the young people engaged in this may also aspire to much of the entrepreneurial dream, they face poverty, marginalisation and a disenabling state and are unlikely to be able to move beyond subsistence. More important than a notion of becoming a successful individual businessperson is gaining the skills and income to be recognised in their communities as adults. What is required, therefore, is an approach to skills for entrepreneurship that is grounded both in structural constraints and broader agentic aspirations towards human flourishing.

Human development

This concern with aspirations towards human flourishing takes us to a new theoretical approach to VET and development that draws on the human development and capabilities approach (Sen, 1999). Echoing DeJaeghere, a key figure in this literature, it addresses both inequality in skills development and how we move away from a narrow focus on immediate employability and production towards a wider view of VET as supporting human flourishing (Powell and McGrath, 2019).

The approach insists on foregrounding poverty in order to better understand many young people's challenging lived experiences. In keeping with the wider capabilities tradition, it sees poverty as being multidimensional. Given that most who enter VET are from poor backgrounds, this insistence on a careful analysis of how they experience poverty seems essential. Furthermore, the approach draws on feminist literatures to stress how women experience intersectional disadvantages that shape the decisions that they make about education and work throughout their lives and the outcomes they achieve. Whilst it is concerned to understand structural obstacles to human flourishing, this approach emphasises the centrality of human agency. This leads to a strong focus on what individuals aspire to, why they attend VET and what they value in their VET experience. Unlike the orthodoxy, this literature finds that many, both youth and adult returners, actively choose VET and that they find value in it that includes but transcends its immediate labour market value.

Related to this, the approach insists on a broad conception of work. It argues that work is not only about income/production but should also be about self-identity and self-worth (Sen, 1975). This leads to a stress on how work's potential to fulfil wider human needs can be maximised. Thus, the notion of 'decent work' needs to be protected and expanded. However, the approach notes that access to and status of VET and many forms of work are unequal and structured profoundly by class, gender and race. Given the intended tight relationship between VET and work, these perspectives are crucial for vision of transformed and transformative VET.

All of this implies the need for a reassessment of what counts as success for vocational providers. Rather than emphasise pass, throughput or employment rates, important though these are, it calls for evaluation to focus primarily on the extent and ways in which institutions and the system support the flourishing of learners.

Just transitions

There are signs in the latter two literatures of a need to address environmental concerns while also addressing societal and economic concerns. This section will explore the complexity of sustainable development and how it must be addressed in the VET arena.

Historically, mainstream VET has been disconnected from a concern with the environment.

Colonial land expropriation underpinned a bifurcation of skills formation systems between an industrial-focused VET mainstream and a separate agricultural and natural resource management-focused skills system. Whilst there were some post-independence efforts in some countries to adopt a more participatory and sustainable approaches to rural skills, this had relatively little impact on the industrial VET model.

Since the Rio Earth Summit in 1992, a discourse of green skills has emerged at the international policy level. However, there have continued to be two major problems with this. First, there remains a fundamental tension between a drive from environmental ministries towards green skills and a VET policy reform approach characterised by new public management and employability. At the same time, the drive towards environmental skills has tended to neglect the technical level, ignoring the types of skills most usually found in VET.

More recently, there has been an attempt to bring just transitions thinking into the VET mainstream (e.g., McGrath et al., 2019). This provides a way to move beyond the jobs-versus-environment argument and intersects with the environmental justice and climate justice movements to provide a broad framing that supports an expanded scale of considerations across economic, social and environmental dimensions.

A system that is more proactively constituted for addressing complex and as yet unknown consequences of climate change and water scarcity is needed. Here some studies are emerging to better assess the skills demand that probe the unlocking of green skills that are latent and hidden, yet possible within current systems of operation, e.g., via adopting circular economy principles in local industrial or agricultural contexts and within the wider frame of just transitions (Rosenberg et al., 2015).

Overall, this approach highlights the ways in which colonial and post-colonial political economies have resulted in VET systems that largely ignore the needs of the majority of people and of the environment. It argues that VET needs urgent transformation if it is to be fit to support pressing processes of just transitions. It also points to a need to theorise VET in more complex ways that consider history, ecology and context as well as national policy imperatives, political economy, inclusivity, transformative curriculum and pedagogy and capabilities. This approach suggests that just transition principles, processes and practices are needed to support a fundamental rethinking of the systemic organisation of VET systems, the world of work and the ways in which they have excluded the majority.

Conclusion

'Modern' VET has been inextricably linked to faith in the inevitability of industry-led development, in spite of the waning of such faith elsewhere in development thinking. To better support development, VET needs to be far better aligned to both existing labour markets and to the possibilities of transforming these towards the challenge and opportunity of just transitions. In so doing, it is important also that VET's close relationship to thinking about work does not ignore the longstanding counter-tradition in VET that stresses VET's role in building people and communities and not just workers.

References

Chang, H.-J. and Grabel, I., 2014. *Reclaiming Development*. London: Bloomsbury.

DeJaeghere, J., 2019. A capability approach to entrepreneurship education. In McGrath, S., Mulder, M., Papier, J. and Suart, R. (eds.) *Handbook of Vocational Education and Training*. Basel: Springer.

Kruss, G., McGrath, S., Petersen, I.-H. and Gastrow, M., 2015. Higher education and economic development. *International Journal of Educational Development* 43, 22–31.

McGrath, S., Ramsarup, P., Zeelen, J., Wedekind, V., Allais, S., Lotz-Sisitka, H., Monk, D., Openjuru, G. and Russon, J.-A., 2019. Vocational education and training for African development. *Journal of Vocational Education and Training* (In press).

Nelson, R.R. and Winter, S.G., 1982. The Schumpeterian tradeoff revisited. *The American Economic Review* 72(1), 114–132.

Powell, L. and McGrath, S., 2019. *Skills for Human Development*. Abingdon: Routledge.

Rosenberg, E., Lotz-Sisitka, H., Ramsarup, P., Togo, M. and Mphinyane, A., 2015. *Green Skills for the Mining Sector*. Rhodes University and MQA. www.greenskills.co.za

Sen, A. 1975. *Employment, Technology and Development*. Oxford: Clarendon.

Sen, A. 1999. *Development as Freedom*. Oxford: Oxford University Press.

Zeelen, J., van der Linden, J., Nampota, D. and Ngabirano, M. (eds.), 2010. *The Burden of Educational Exclusion*. Rotterdam, Boston and Taipei: Sense.

Additional Resources

Bonvin, J.-M., 2019. Vocational education and training beyond human capital. In McGrath, S., Mulder, M., Papier, J. and Suart, R. (eds.) *Handbook of Vocational Education and Training*. Basel: Springer.

DeJaeghere, J., 2017. *Educating Entrepreneurial Citizens*. Abingdon: Routledge.

McGrath, S., 2012. Vocational education and training for development. *International Journal of Educational Development* 32(5), 623–631.

Glossary

Adaptive Management   A flexible management approach that allows for continuous learning and adaptation of strategies based on ongoing results and changing conditions.

Aid Effectiveness Paradigm   Refers to the set of principles and practices designed to ensure that foreign aid is used as efficiently and effectively as possible to meet development objectives.

BRICS   An acronym for Brazil, Russia, India, China, South Africa, representing emerging economies that have become significant international development aid donors and recipients.

Basic Needs Approach   A development approach focused on meeting the fundamental needs of all individuals, often including elements like healthcare, nutrition, education, and housing.

Belt and Road Initiative   China's global infrastructure initiative, often described as a 21st-century silk road, that has invested in infrastructure projects in nearly 70 countries since 2013.

Big D Development   Characterises the post-1945 international aid system, financial institutions, and NGOs. It represents institutional interventions in development that often support the expansion of capitalism.

Bio-psycho-social Model of Disability   the understanding of disability that underpins the UNCRPD. Disability results from the interaction between long-term physical, mental, intellectual or sensory impairments and various barriers that may hinder the full and effective participation in society of persons with disabilities on an equal basis with others.

Bounded Rationality   The concept that decision-making is limited due to cognitive constraints, such as processing capacity and time costs, affecting rationality.

Brain Drain   The phenomenon where educated, skilled, and dynamic individuals leave their country or region, leading to a loss of talent and potential economic development.

Bretton Woods   an area in New Hampshire, United States, where a conference of Allied nations took place in 1944 to discuss the post-war international monetary and financial order. One outcome was the establishment of the International Monetary Fund and the World Bank, together known as the Bretton Woods twins.

Capability Approach   A central aspect of Amartya Sen's theory, focusing on the capabilities or potential life choices individuals have, beyond economic or material well-being. It considers a person's freedom to pursue different life paths and achieve well-being.

Cash transfers: (conditional or unconditional)   Conditional cash transfers combine a regular income transfer to poor or vulnerable households with conditions with which beneficiaries must comply. Conditionalities (sometimes referred to as co-responsibilities) attached to the provision of cash benefits may require beneficiaries to perform certain verifiable actions (such as regular school attendance or health check-ups). Unconditional cash transfers are provided without such conditions.

Circular Economy   An economic system aimed at eliminating waste and the continual use of resources, is relevant.

Citizen Consumer   A concept redefining citizenship to include political consciousness, agency, and action through daily consumption practices. It moves away from traditional state-centric citizenship models, integrating ethical consumerism as a form of political participation and social change.

Civil Society   A broad category encompassing various forms of collective action and associational life outside the market, state, and family, where people organise, debate, and act. Civil society includes NGOs, media, human rights defenders, and social movements, among others.

Climate Justice   The application of justice principles, primarily distributional and recognition-based, to climate policy related to global production, distribution, consumption, and security, including energy policies.

Common Property Resources   Natural resources that are owned and managed collectively by a community or society, as opposed to individual ownership. This communal management often involves shared rules and responsibilities.

Common-pool Resources   Natural or artificial resources available to all members of a community or society, often managed collectively. They can be owned by governments, communal groups, or private entities, and when unowned, they are used as open access resources.

Community Policing   A police reform strategy aiming to make policing more democratic by reflecting community desires. The approach is characterised by its adaptability and has become a common component of police reform programs worldwide, though its exact definition and successful implementation can be ambiguous.

Community and Public Engagement (CPE)   The processes by which organisations, actively involve communities and the general public in their decision-making processes, project design, and implementation.

Composite Indices in Development   Aggregated measures that summarize multiple dimensions of human development, such as the Human Development Index (HDI). While informative, they are criticized for not fully capturing the complexities and trade-offs between different development aspects.

Conflict Resources   Resources that have been sourced from conflict zones and sold to sustain the conflict.

Conservation   The protection and management of natural resources, wildlife, or landscapes.

Corporate Social Responsibility (CSR)   The practice and philosophy where companies take responsibility for their impact on society. It often encompasses activities by Global North companies in the Global South, often aligned with Sustainable Development Goals and private-public partnerships.

Criminalisation   The process whereby the actions of particular individuals or groups – or who or what they signify – become viewed or understood as crimes or criminal. This can reflect a change in legal definition (such as declaring those living informally as illegal) or can refer more abstractly to the language or practices of those in power against particular groups, such as street children or gang members.

Data Justice   A concept focusing on fairness in how people are represented and treated in the digital world, particularly concerning the production and use of digital data.

Datafication   The process of transforming various aspects of human life into data, enabling quantitative analysis and decision-making.

Decolonisation   A movement aimed at challenging dominant theories, methods, and knowledges from the Global North that marginalise the Global South. It seeks to deconstruct racist, gendered, and capitalist norms in development and address structures that create marginalisation.

Demographic Transition   The shift in population dynamics where a country moves from high fertility and mortality rates to lower rates, leading to an increasingly older population. This transition, observed in many developing countries, has significant implications for social and economic policies, especially in terms of aging populations and poverty reduction strategies.

Dependency Theory   An approach to development theory that suggests that underdevelopment in certain regions is the result of exploitative economic relations, especially in a global context.

Developmentalism   A critique within World-systems Theory, developmentalism suggests that societies progress sequentially through feudalism, capitalism, socialism, to communism. It posits that societies can be individually analysed and transformed, separate from the global system.

Digital Divide   The gap between individuals, households, businesses, and geographic areas at different socio-economic levels with regard to access to, use of, or knowledge of information and communication technology.

Digital Division of Labour A phenomenon where the Internet, rather than democratising knowledge sharing, reinforces the visibility, voice, and power of the world's economic cores, leading to a digital division of labour and perpetuating inequalities.

Digital Inclusion Efforts to ensure people have access to and can effectively use information and communication technologies.

Disruptive Philanthropy A modern approach to philanthropy by new foundations aiming to challenge the status quo and drive radical change through innovative and often technology-focused interventions. It views social systems as malleable and subject to transformative strategies.

Economic Commission for Latin America (ECLA) A United Nations body that contributed to the initial theorisation of Dependency Theory with a structuralist perspective.

Energy Justice The application of distributional, recognition-based, and procedural justice principles to energy policy, production, systems, consumption, activism, security, and climate change politics.

Environmental Defenders Individuals or groups who actively protect ecosystems and communities from adverse effects such as land grabs, pollution, and industrial operations. They often play a crucial role in preserving biodiversity, ecologies, and human/more-than-human species and relations, recognised even by the United Nations for their significance.

Epidemiological Transition A process where a population's dominant health challenges shift from infectious and malnutrition-related diseases to predominantly non-communicable diseases, such as heart disease and cancer, typically accompanying socio-economic and technological development.

Ethical Consumerism The practice of using market power to influence positive changes in production and trade processes. It encompasses efforts to address social issues like worker's rights and environmental stewardship through conscious consumption choices.

Evidence-Based Project Design An approach to project design that relies on empirical evidence and rigorous analysis to make decisions, with a view to ensuring interventions are grounded in demonstrable and proven methods.

Exploitation Unfair treatment of individuals to derive benefit from their efforts or resources.

Extractivism A production form under neoliberal financialization where global commodity chains externalize social and environmental costs. It contrasts with production for localised needs and emphasises community ownership and control.

Faith-Based Organizations (FBOs) Religious entities that play a significant role in development, often involved in health, education, and community development. FBOs are categorised into congregations, national networks of congregations, and freestanding religious organisations. They have gained prominence in mainstream development and receive substantial governmental support.

Famine A widespread and prolonged disruption in food access resulting in acute malnutrition and mass mortality. Historically, famines have been linked to poor governance, natural disasters, colonial interventions, and economic changes.

Fictitious Capital In Marxian terms, accumulated claims on future earnings or assets. In the context of land, it refers to valuing land based on its potential future earnings.

Financial Capitalism a dominant way of accumulating wealth by selling money to make money.

Financialisation The increasing influence of financial markets, motives, actors, and institutions in the domestic and international economies, reflecting the growing weight of finance in the world economy.

Food Security The state where all people, at all times, have physical, social, and economic access to sufficient, safe, and nutritious food that meets their dietary needs and food preferences for an active and healthy life

For-Profit Development Contractors and Consultants Firms contracted by development agencies and foundations to provide development services. These range from management consultancies to specialist development-oriented firms, playing an increasingly significant role in the development sector.

Forced Labour A situation where a person is compelled to work or provide services against their will, under threat of penalty.

Fourth Industrial Revolution The current era of technological advancement characterised by a fusion of technologies blurring the lines between the physical, digital, and biological spheres, including advancements in AI, robotics, and the Internet of Things (IoT).

Fragile State A term describing states with weak structures and institutions, often linked to patterns of violent conflict, and seen as crucial in development studies for understanding state-society relationships.

Frugal Innovations Cost-effective, simplified versions of technologies designed to be accessible and useful in resource-limited settings, often in developing countries.

Gender-Based Violence (GBV) Violence where the motive is related to the victim's gender, encompassing violence against women, girls, men, boys, transgender people, and others with gender-nonconforming identities.

Gender Mainstreaming Making the concerns and experiences of women as well as men an integral part of the design, implementation, monitoring, and evaluation of policies and programs.

Gini Index for Income The Gini coefficient (or Gini index and Gini ratio), developed by Corrado Gini, is a measure of statistical dispersion that represents income inequality within a nation or a social group: the smaller the value, the lower the inequality.

Global Justice Movement A collective term for various movements opposing aspects of globalisation, including neoliberal policies.

Global Laboratory Approach in Philanthropy A method where philanthropical foundations experiment with technologies and social norms in developing countries. This approach treats the Global South as a testing ground for technological innovations and social progress.

Good Governance Agenda A World Bank initiative emphasising civil society's role in promoting democratic rights and pro-poor development. It focuses on participation, decentralization, and democratization to drive development alongside economic liberalisation.

Grassroots Innovations Innovations that originate from low- and middle-income groups, often leveraging traditional knowledge and locally available technologies, focusing on local needs and sustainability.

Greed vs Grievance Hypothesis Greed refers to the financial incentives associated with joining a rebellion; grievance, on the other hand, captures motives associated with group identity.

Green Revolution A period of significant increase in agricultural production due to the adoption of new technologies, high-yield crop varieties, and advanced agricultural practices.

Gross Domestic Product (GDP) A widely used indicator of economic development, representing the total value of goods and services produced in a country. While a standard measure for cross-country comparisons, it oversimplifies and can conceal inequalities and well-being aspects.

Historical-Empirical Approach A development theory approach that builds upon past events and experiences to formulate development strategies.

Human Development A broad approach to development that focuses on enhancing individual freedoms and capabilities, as well as overall human flourishing.

Human Development Index (HDI) A composite indicator measuring life expectancy, educational attainment, and GDP per capita. It provides a broad view of a country's development but does not account for internal distribution.

Human Security A concept prioritizing personal, economic, and social concerns over traditional national security issues, encompassing ecological, educational, food, habitat, and health priorities. It challenges the traditional focus on state security, emphasising individual and community well-being.

Human Trafficking The movement, recruitment, or harboring of individuals for the purpose of exploitation, often involving coercion or deception.

ICT Capacity Building Efforts to improve the skills and abilities of individuals and organisations in developing countries in using ICT effectively.

ICT4D Information and Communication Technology for Development (ICT4D) aims to integrate ICTs in development strategies, planning, implementation, and evaluation to close the digital divide, build ICT capacity for development, and empower individuals with the help of enabling technologies.

Immanent Development Refers to the natural progress and improvements in people's lives that arise independently, often associated with the inherent unfolding of capitalist growth and expansion.

Import Substitution Industrialization (ISI) A development strategy focused on domestic production under tariff protection to replace industrial imports, particularly prevalent in Latin America.

Social Inclusion   Efforts and policies designed to ensure equal access to resources and opportunities for all, particularly for marginalised or disadvantaged groups. It involves removing barriers to participation in society, whether they are economic, social, political, or cultural.

Inclusive Development   This term refers to development processes that ensure benefits are equitably shared across society and all groups of people, particularly the marginalised and disadvantaged.

Infrastructures   foundational material structures and facilities such as roads or electricity grids that connect individuals, and especially states and their citizens, and economies, producers and consumers worldwide.

Innovation   the introduction of a product or process that is new to the firm or new to the market or new to the world

Intentional Development   Differentiates from 'immanent' development by focusing on state-led reformist actions aimed at managing capitalism's undesirable consequences and bringing order to economic change.

International Monetary Fund (IMF)   established at the Bretton Woods conference, the IMF is mandated with the surveillance of the world economy. It also makes loans to countries undergoing economic crisis, usually in return for satisfying distinct policy conditions. The Fund is known for advocating liberal, market-oriented development policies.

Intersectionality   A concept that describes the complex interplay between different structures of difference, such as race, class, and gender. It recognises that these categories are co-constitutive and inseparable, shaping relationships of domination and subordination in specific historical contexts.

Intimate Partner Violence   Harmful behaviours within intimate relationships, including physical, sexual, and psychological harm.

Just Transitions   The concept of ensuring a fair and equitable shift towards sustainable practices, particularly in the context of labour markets and employment.

K-shaped Economic Recovery   A term describing how, particularly during crises like COVID-19, individuals at opposite ends of the income spectrum diverge further economically. Those with resources recover more easily, exacerbating existing inequalities.

Keynesianism   Economic theories advocating government intervention in the economy.

Kondratieff Cycles   Long-term economic cycles, about 50–60 years each, identified by Nikolai Kondratieff. They represent different phases of global capitalism, characterised by expansion (A-phase) and stagnation (B-phase) periods, influencing global business activities and economic trends.

Labour Market Policies   Policies aimed at ensuring basic standards and rights at work, e.g., collective bargaining and minimum wage policies.

Land Grabbing   The rapid acquisition of land, often resulting in the displacement of previous users or owners. This phenomenon is increasingly prevalent due to global socio-natural changes and is often accompanied by direct and indirect violence.

Livelihood   The assets (natural, physical, human, financial, social capital), activities, and access to these (mediated by institutions and social relations) that together enable individuals or households to earn a living.

Majority Labour Markets   This term refers to the predominant labour market sectors where most employment takes place, often including informal and subsistence sectors.

Microfinance   The provision of financial services to entrepreneurs and small businesses lacking access to banking and related services.

Millennium Development Goals (MDGs)   A set of eight targets established in 2000 by world leaders to reduce poverty by 2015. The MDGs emphasised the importance of meeting children's rights to education, health, and protection as essential components for achieving these goals.

Misinformation and Disinformation   Misinformation is false information spread without the intent to mislead, while disinformation is deliberately disseminated to deceive. Both have significant negative impacts on society, particularly in areas like climate change, where they can reduce climate literacy and polarize public opinion.

Modern Slavery   An umbrella term encompassing human trafficking, slavery, servitude, and forced labour, reflecting contemporary forms of exploitation.

Modernization Theory A theory in development studies that views development as a linear process of adopting more 'modern' practices, typically those of Western societies.

Monitoring Evaluation and Learning (MEAL) Systematic processes used in development projects and programmes to track progress, assess effectiveness, and glean insights for future improvement. Monitoring involves the continuous collection of data on project activities; Evaluation focuses on assessing the impact and outcomes; Learning incorporates the insights gained into future planning and decision-making.

Multidimensional Inequality Refers to the complexity of disadvantageous situations of individuals or groups within a given society. It highlights how unequal positions are shaped by the simultaneous disparities in income, laboured market participation, educational attainment, access to welfare, public policies, and multiple forms of segregation (by gender, race, ethnicity).

Multidimensional Poverty Index (MPI) An index that combines education, health, and standard of living indicators to measure poverty in multiple dimensions.

Negative Income Tax A tax system where people earning below a certain amount receive supplemental pay from the government instead of paying taxes.

Neo-Abolitionism A contemporary movement focused on eradicating modern slavery, drawing parallels to historical efforts against trans-Atlantic slavery.

Neo-developmentalism A contemporary approach to economic management in the Global South, retaining elements of neoliberal ideology.

Non-Aligned Movement (NAM) An alliance of newly decolonized states formed in 1956 to resist alignment with Cold War superpowers, advocating for sovereignty, non-aggression, and peaceful coexistence.

Non-Governmental Organisations (NGOs) Private, non-profit initiatives involved in development issues, ranging from large international charities to local self-help organisations. They play key roles in service provision, advocacy, and empowerment of the disadvantaged or marginalised.

Open Access Resources A condition where access to resources, such as fisheries, is unrestricted and available to all, without specific ownership or management controls.

Paris Declaration on Aid Effectiveness An international agreement that outlines principles for effective aid delivery, including ownership, alignment, harmonisation, results-based management, and mutual accountability.

Participatory Development (PD) An approach that involves stakeholders, especially local communities, in development projects' planning, implementation, and evaluation, valuing their knowledge and empowering them in decision-making.

Persons with Disabilities the term used in the UNCRPD to refer to people who have long-term physical, mental, intellectual or sensory impairments that, in interaction with various barriers, hinder their participation in society on an equal basis with others.

Philanthrocapitalism A term describing a new generation of philanthropists who employ capitalist skills to solve global problems. This approach emphasises efficiency, measurability, and business-oriented strategies, contrasting with traditional philanthropy's reactive practices.

Physical Technologies These are tangible, hardware-based technologies, such as machinery, equipment, and devices, used in various sectors like manufacturing, agriculture, and healthcare.

Plantation Economy Model A concept used to describe economies heavily reliant on a single agricultural commodity, characteristic of many Caribbean states during and after the colonial period.

Platformisation of Labour A phenomenon where digital platforms transform traditional labour and economic coordination, often leading to precarious working conditions,.

Positionality How one's identity, defined by factors like gender, race, class, and sexuality, and situated in a particular context and time, shapes understanding of the world.

Post-Development An approach that critically examines and challenges traditional Western-centric concepts of development. It argues that development should not be seen as a universal or linear process but rather as a diverse and context-specific phenomenon. This theory promotes alternative development approaches that are more culturally sensitive and locally driven, often questioning the dominance of Western models of economic growth and modernization.

Post-work Universalism The idea that, in a future with less emphasis on traditional wage labour due to automation and other factors, universal income support might become a key social feature.

Postindustrial Societies   Societies, where economies are driven more by services and knowledge industries rather than traditional manufacturing industries.

Public-Private Partnership (PPP)   These are collaborative arrangements between government agencies and private-sector companies used to finance, build, and operate projects like public transportation networks, parks, and schools.

Randomised Controlled Trials (RCTs)   A research method used to evaluate the effectiveness of interventions by randomly allocating participants into treatment and control groups.

Redistributory Tax Regime   A tax system designed to redistribute income or wealth within a society, often used to fund social welfare programs.

Relationality   The concept that identity constructions gain meaning in relation to other subjects. It implies that identities such as race, gender, and sexuality are often defined in opposition to an 'Other'.

Remittances   Financial transfers made by migrants to their family members or communities in their country of origin. These can significantly contribute to the economy of the receiving country.

Rent Gap   The economic potential that arises from the difference between the current use value of land and the potential value post-redevelopment. This concept is central to understanding the economic drivers of urban gentrification.

Resource Curse Theory   The paradox where countries with abundant natural resources tend to have less economic growth and worse development outcomes than countries with fewer natural resources.

Resource Wars   Conflicts largely driven by the control or pursuit of valuable natural resources, often associated with political and economic instability in resource-rich regions.

Results-Based Management   A management strategy focusing on performance and achieving specific outcomes. It involves setting goals, using evidence and feedback to guide decision-making, and evaluating outcomes against pre-set targets.

Rights-Based Approach   An approach to development that integrates human rights principles into development planning and processes. This approach emphasises that development should not only focus on economic growth but also on fulfilling and protecting individuals' human rights.

Security Sector Reform (SSR)   Refers to the reform and restructuring of security institutions, including police, as part of development initiatives.

Sexual and Reproductive Health and Rights (SRHR)   Health care and education services and rights relating to sexual activity and reproduction. It includes the right to healthy and respectful relationships, access to high-quality and confidential sexual health services, and the freedom to make informed and autonomous decisions about one's sexual and reproductive life.

Skills Formation Systems   Refers to the structures and processes involved in developing the skills of the workforce, including education and training systems.

Small d Development   The continuous process of capitalist growth and expansion resulting in developmental change but with uneven outcomes, including both winners and losers.

Social Assistance   Non-contributory transfers based on eligibility, such as low income, vulnerability, or rights as citizens or residents; also includes public employment or food for work programmes.

Social Capital   A concept linking social norms, rules, and reciprocal obligations to social and economic actions. It comprises various social structures that facilitate specific actions and is critical in understanding governance, democracy, and collective action. Social capital influences poverty, welfare, and livelihoods through access to networks, organisations, and relationships.

Social Investment   Policies and programs aimed at investing in human capital, including education, healthcare, and employment training, to improve long-term social and economic outcomes.

Social Movements   Coalitions of actors engaged in political struggle, working together to transform the status quo and effect social, political, cultural, or economic change. These movements use diverse tactics such as protests, civil disobedience, and policy critique to achieve their goals.

Social Protection Floors   Basic sets of social rights, services, and facilities that countries aim to guarantee to all citizens, including essential healthcare and income security.

Socio-Economic Rights   Rights that encompass both normative standards for society and institutional mechanisms for enforcement. These rights, beyond property rights, were recognised in the post-World War era, reflecting the evolving understanding of basic freedoms to include economic rights alongside traditional human rights like life, liberty, and conscience.

Stagflation   An economic condition where inflation and unemployment rise simultaneously, contrary to typical economic situations where they have an inverse relation. It challenges traditional economic theories and was particularly notable during the 1970s.

State Failure   The loss of a state's monopoly on violence and governance, often resulting from internal conflicts and seen as a critical factor in understanding violent conflicts and international security.

Structural Adjustment Loans   Loans from IMF and World Bank made in wake of the 1980s debt crisis to stabilise fiscal deficits, with conditional elements that obliged governments to adopt neo-orthodox pro-market policies, such as privatisation, deregulation and reduced subsidies.

Sustainable Development Goals (SDGs)   Established in 2015, replacing the Millennium Development Goals, the SDGs consist of 17 goals with 169 targets, aiming to address global, holistic, interdependent, and inclusive development challenges.

Technosolutionism   The belief that technology is the solution to complex social, political, and economic problems, often criticized for oversimplifying these issues.

Tenure Security   The secure and legal ownership or tenancy rights to land and property.

Third World   A Cold War-era term denoting newly decolonized states in Asia, Africa, and Latin America, characterised by political and economic vulnerabilities and a focus for international development policies.

Transnationalism   The operation and interaction of various social, economic, and political processes across national borders, reflecting a global perspective.

Twilight Institutions   Non-state actors engaged in everyday policing, often operating in a space between public authority and private action. These institutions, including vigilantes and private security providers, derive authority from various sources and play a significant role in maintaining order in their communities, sometimes outside formal legal frameworks.

United Convention on the Rights of Persons with Disabilities (UNCRPD   the United Convention on the Rights of Persons with Disabilities, a human rights convention that outlines the specific rights of persons with disabilities to participate in society on an equal basis with others. It was adopted in 2008 and rapidly ratified by many countries.

United Nations Convention on the Rights of the Child (UNCRC)   A significant international treaty adopted in 1989 that provides comprehensive rights for children. It was established due to the perceived inadequacy of previous non-binding declarations in protecting children's rights, reflecting an evolved understanding of children's unique needs and rights.

Universalism   As a normative principle, universalism is concerned with solidarity and the notion of social citizenship, which includes social rights alongside civil and political liberties and emphasises collective responsibility for individual well-being.

Urbanization   The process by which there is an increase in the proportion of a population living in urban areas compared to rural areas. This phenomenon is often associated with a variety of social, economic, and environmental changes, including shifts in employment from agriculture to industry and services, changes in living conditions, and challenges related to urban planning, infrastructure, and sustainability.

Urban Bias   The argument that governments and development organisations have given preference to cities over countrysides in the allocation of scarce resources and in the distribution of subsidies.

Urban Redevelopment   The process of renovating urban areas, often involving significant investment and re-zoning from lower to higher density development. It typically targets areas perceived as devalued or old-fashioned for transformation.

Violence Against Women and Girls (VAWG)   A widespread human and women's rights violation, public health concern, and impediment to development, involving physical and/or sexual violence against women and girls.

Vocational Education and Training (VET)   A system of education and training that provides individuals with skills and knowledge relevant to specific occupations or industries.

Waithood   A term describing the state of young people being delayed or held back from achieving adulthood milestones like securing a job, forming a family, or financial independence.

Washington Consensus   A set of economic policies emphasising trade liberalisation, fiscal discipline, and privatisation, particularly in developing countries. Developed in the neoliberal era, it became a standard reform package imposed by global institutions during economic crises.

Water Insecurity   Extends beyond access to safe drinking water, affecting income-generating activities, food security, health, hygiene, sanitation, and education. It emphasises water as a co-produced resource within physical and socio-political structures.

Water Security   A paradigm in the water sector since the 1990s, focusing on the physical availability of fresh surface and ground water, as well as social factors. It evolved from technical approaches to include multi-dimensional aspects of water management.

World Bank   established at the Bretton Woods conference, the World Bank lends funds to member governments towards development projects. The scope of its operations ranges from poverty reduction and infrastructure to administrative reforms. The Bank has offices around the world, employs thousands of researchers and often promotes liberal economic policies.

World Bank Income Classification   A system categorising countries based on their GDP per capita, classifying them as low, lower-middle, upper-middle, or high income. It is used for economic analysis and policy formulation.

World-systems Theory (WST)   A framework analysing global economic systems, emphasising how global processes, commodity chains, labour divisions, and geopolitical relationships shape individual countries' development prospects. It critiques the traditional focus on national development in isolation.

Index

Note: Page numbers in *italics* indicate a figure and page numbers in **bold** indicate a table on the corresponding page.